Engineering Economy:

Analysis of Capital Expenditures

Second Edition

Engineering Economy:

Analysis of Capital Expenditures

GERALD W. SMITH

The Iowa State University Press, Ames, Iowa, U.S.A.

GERALD W. SMITH is Professor of Industrial Engineering at Iowa State University, where he received his Ph.D. degree. Since 1964 he has served as an officer or director of the Engineering Economy Division of ASEE, and in 1971 he was elected its chairman. In 1968–1971 he was designated Iowa State's "Alcoa Professor"; in 1969 he was recipient of the E. L. Grant Award for best paper in *The Engineering Economist;* in 1970 he was named associate editor for book reviews of that same journal; and in 1970–1971 he served on the editorial board of *AIIE Transactions.* He has consulting and industrial experience and is a registered professional engineer.

Library of Congress Cataloging in Publication Data

Smith, Gerald W
   Engineering economy.

   Bibliography: p.
   1. Capital investments.   2. Capital budget.   3. Depreciation allowances —United States.   I. Title. HG4028.C4S6   1973   658.1'52   73-4581
ISBN 0-8138-0552-X

This book is dedicated to

the planner, who relinquishes quiet harbors of assurance and sets course for adventure over untraveled high seas of the future, sometimes with the guidance of stars and always with the perils of man and nature.

May his contemplations find a way through the darkness of doubt to the isles of partially realized goals. May his helmsman's wheel give the rolling sea of storm her due but no more; may the captain's log reveal his adaptiveness as well as imperfection. May his stars not always be masked by the fog and clouds of an unanswerable unknown. May his failures become buoys of future journeys, as well as monuments to fallibility; may incomplete success carry him to ports of new achievement. And finally when the call of quiet harbor comes, let him find his solace in having tamed a bit the thunderous sea.

Contents

Preface

Planning without action is futile,
action without planning is fatal.

THIS BOOK deals with the concepts, principles, techniques, and reasoning by which the planner can be guided in his decisions pertaining to long-term facilities. The planner may be a manager of capital equipment, an engineer responsible for its design, a citizen acting on behalf of his city, a businessman concerned about a project's impact, or a consultant acting on behalf of his client. The frequent role of the engineer and the objective of long-term economy are responsible for the widely used title "Engineering Economy." The book is intended as both a textbook and reference for students of engineering, management, business, and economics as well as practitioners who are already involved as decision-making planners.

Engineering economy relates or applies many concepts from economics, mathematics, accounting, statistics, and management; the text provides enough detail of these topics to be self-sufficient. For those who wish to pursue such topics more fully, a number of helpful references are noted in the Selected References.

The general plan of the book is to begin with very restrictive mathematical models of reality, then to relax these restrictions one at a time to focus attention on the primary topic at hand. To continue this process generally requires that succeeding chapters *not* repeat all the complicating factors of reality; thus the section on capital budgeting forsakes the more complex models of income tax, economic life, risk, and obsolescence so that the reader may concentrate upon the topic at hand. The problems of real life which *do* include simultaneous consideration of such factors are accompanied by monetary consequences which can justify a depth of study far beyond the practical limits of the classroom.

The audience to whom this book is addressed varies considerably in background, needs, and time available for its study. An attempt has been made to achieve some flexibility in meeting the varying needs of readers by the inclusion of material which a teacher may choose to include or exclude, according to the purposes and nature of the course involved. Sufficient material is offered for a two-term course. In an introductory one-term course the first eleven chapters may provide most of the coverage. Even in those eleven chapters the assignments might be limited as:

Chapter 4, through Table 4-2   Problems 1–8
Chapter 5, through Example 5-5   Problems 1–19
Chapter 10, through Table 10-3   No limitation on problems

Sample assignment sheets are available from the author.

In response to the enthusiastic comments of teachers and students, graphic presentations in this edition have been further utilized in the development of concepts. In addition to cash flow diagrams, break-even charts, decision trees, net present value (*PEX*) diagrams, sensitivity studies, utility functions, and probabilistic cash flows, readers will find many original graphic presentations:

- Corporate flows of cash.
- Network diagrams used to develop conditional decision responses under unspecified rate of return requirements.
- Relating multialternative net present value diagram to network diagram decisions.
- Probabilistic before-tax and after-tax cash flows.
- Summary of the Monte Carlo simulation of multivalued inputs.
- Determination of expected utility for probabilistic cash flows that are continuous functions.

- "Economic package" determinations, including decision tree analyses and the role of "expansion rate" and "economies of scale."
- "Marginal investment opportunity curve" for various decisions on economic package; deferred investment; optimal timing of acquisition, retirement, or replacement.
- Life cycle of a capital expenditure as a multistage series of sequential commitment decisions with time-variant "escape" costs. Time-variant avoidable (controllable) annual cost and the sunk cost concept.

Other unique features include:

- Early and continuing emphasis is placed on the integration of accounting information with the decision needs of the feasibility study, including the unifying perspective of the "Sources and Application of Funds Statement."
- The specially designed format of the compound interest tables, with a proven advance in symbol systems, provides both mnemonic and dimensional analysis aids in computation.
- The conversion between continuous cash flows and end-of-year cash flows is reduced to a simple convenient form not found in other texts.
- Depletion and depreciation are treated in a manner relevant to current business decisions.
- A newly developed approach to income tax computation simplifies and shortens analyses involving debt financing.
- Multioutcome (risk and uncertainty) analyses involving discrete or continuous probability functions are summarized in convenient form for expected outcome and expected utility criteria.
- The relevance of "marginal opportunity cost" as empirical evidence of the capital budgeting cutoff rate is stressed in the examples of many chapters, thus emphasizing the "system" impact of such decisions.
- Capital expenditure decisions are portrayed as imperfectly reversible, multistage, sequential decisions and permit the reader a depth of understanding and intuitive "feel." Probabilistic treatment of the risk of premature retirement is similarly treated.
- Typical contemporary treatment of optimal service life (economic life) is restricted to the case where acquisition age

is zero. That constraint as well as those relating to revenue level and duration are removed, thus permitting determination of optimal acquisition age and optimal retirement age.

- A noncomputer capital budgeting game permits students to gain insights through a brief encounter with capital budgeting decisions and cash flows in a realistic dynamic context.
- Optional suggested computer programming applications of various types of managerial analyses are included.

The objectives of the book and course should be not only to bring its reader to a point of competent analysis but also to help him recognize unsound reasoning, criteria, methods, or criticism and to respond to their defects.

I am indebted to my students whose enthusiasm for the subject has made its teaching a pleasure. I am grateful for the opportunity of association with the engineers and managers of public utility companies whose experience with capital expenditure planning has made obvious the increasing stake of the firm and of the nation. I am indebted to the students, business executives, and colleagues with whom I have been privileged to associate for their insights into new challenges and unsolved problems. I wish to thank Dr. Clair G. Maple for the derivation which led to his interesting discovery that capital recovery factors for group properties following the Iowa Type S_x survivor dispersions can be expressed as a Bessel function. I also wish to thank Professors Arthur Lesser, Jr., Loran E. Mohr, Jean C. Hempstead, and Edward J. Carney for reading and commenting upon sections of the book. I owe much to Professors J. K. Walkup and J. P. Mills for their guidance, both present and past. For bringing order from chaos, I thank Mrs. Nancy Bohlen; her editing efforts were most necessary and helpful.

To thank my wife for her typing is unduly restrictive; so finally, to my family, I acknowledge the value of the quiet, understanding, and encouragement they so patiently supplied.

GERALD W. SMITH

I

Enterprise Decision Making and Action in the Management of Capital Equipment

THE MONOGRAM of our national initials, which is the symbol of our monetary unit, the dollar, is almost as frequently conjoined to the figures of an engineer's calculations as are the symbols indicating feet, minutes, pounds, or gallons.

HENRY R. TOWNE in a paper before the American Society of Mechanical Engineers in 1886

I

Capital Expenditures and Analysis for Economy

Design for Function: Design for Economy

THE ENGINEER is a designer; what he designs—structures, processes, circuits, mechanisms, production environments, or systems of men and machines—is dependent upon his engineering specialization.

His design must accomplish some function and therefore meet the test, Does it work? However natural in logic, the test is nonetheless insufficient in scope. The majority of basic science, engineering science, and engineering analysis and design courses in the engineer's education are devoted to the study and understanding of *functional* aspects of design. We agree that function is a *primary* criterion of design but note that it is not a *unique* criterion. Even when an engineer is asked to design a system whose only specifications appear to be functional, it is most improbable that the cost of his system is inconsequential. With economy as one criterion of design, the engineer's task becomes one of finding the best way to achieve the functional goal. This means the engineer (and facility manager too) must *search for, recognize, and generate alternatives* and also *compare and evaluate* them. This book deals primarily with this latter need—the principles, concepts, techniques, factual data, and methods by which alternatives can be compared and evaluated. It should be self-evident that both search and comparison stages must be properly performed if meaningful results are to be produced.

Interestingly enough there is a tendency for the functional and economic criteria of a given design or facility to move from predominantly functional to predominantly economic with time. Even though efforts to travel through space are now primarily a search for *a way* to do so, the years ahead will see increasing effort to make space travel more *economic*. Analogous reductions in cost with time are apparent in the cost of television sets, air travel, or computers. Consider the trends with time in the cost of (1) television sets per square inch of viewing area, (2) air travel where speed (and consequently time saved) has increased much more rapidly than cost, or (3) computers where cost per unit of memory or per unit of internal speed continues to decline.

It seems unlikely that dependence upon an engineer's intuitive regard for cost is reasonable. We contend that engineering economy concepts are an essential element in the engineering science background upon which the engineer bases his designs. Criteria of design include function and economy as well as other criteria such as time, safety, and reliability.

Illustration of the Relationship of Function and Economy

One of our nation's problems today is that of assurance of an adequate supply of water which is both potable and palatable. In response to this need experimental efforts have included the design of facilities for conversion of sea water. Today there are plants which meet the functional criteria; these are largely experimental, however, because present methods of sea water conversion do not yet satisfy certain *economic* criteria.

One of the problems of the electric utility company is that the costly generating facilities must be sufficient to meet peak demand and then, during off-peak hours, a portion of high-cost facilities are idle. The natural suggestion here is the storage of electric energy. For many years we have been capable of storing electric energy as chemical energy via batteries; this, however, is not a satisfactory answer to the problem because, though functional, it is uneconomical. Engineers have, however, produced an alternative storage method; off-peak energy is used to pump water to an elevated reservoir and the process is reversed at peak load times, thus converting energy from electric to potential, then to electric again. This system has been found to meet the criteria both of function and of economy in a number of applications. Still other means include storage of energy as compressed air within the confines of suitable underground spaces.

Role of the Engineer

All decisions are decisions between alternatives. Even the single investment opportunity provides two alternative courses of action, acceptance or rejection of the opportunity. Many times, the alternatives from which we must choose can be described by an engineer only; his specialized training and knowledge permit him to describe alternatives and to make estimates with regard to them. The economic comparisons which are based on engineering alternatives are not exclusively an engineering activity.

Viewpoint

Decisions between alternatives can be made from the viewpoint of (1) an engineer in the midst of design or (2) a manager acting upon a number of investment opportunities and alternatives within each opportunity. The engineer has a role in this latter viewpoint also since his estimates, judgments, and alternatives are involved in those opportunities which have technical considerations. The organization of this book is based more on the latter viewpoint, although some examples will reflect the former.

Role of the Manager

Our nation's productive facilities change in response both to the affluence of people and to the technology of innovators. As productive facilities evolve from a dependence upon manual effort to a reliance upon mechanized, automated, computer-controlled systems, the responsibility of the manager changes from a man-oriented to a machine-oriented role. The manager must respond to his changing role; he must learn:

1. To interview, screen, and select equipment just as effectively as he interviews, screens, and selects people.
2. That, like people, equipment has both desirable and undesirable characteristics to be weighed.

3. To ask the right questions of his prospective equipment; once he decides to "hire" the equipment, he may find his acquisition is an irreversible process.
4. That some "team" concepts are equally applicable to equipment (selecting a team of nine excellent first basemen certainly must fail to optimize one's baseball prospects).
5. To assign carefully, according to ability, the task of production to the various items of equipment.
6. To devise systems of periodic checkups to verify that the equipment is still adequate and appropriate.
7. To be aware of new (equipment) talents which will be available in the foreseeable future.

We have in mind the manager of fixed assets, that is, plant and equipment; the decisions of managers of inventory are treated in other books (especially those of operations research); the decisions of managers of cash (comptroller and/or treasurer) are deferred to texts in the appropriate area of finance, control, or budgeting, except as such have relevance to the capital expenditure decision.

Objectives of Analysis

The ultimate objective of analysis is action. Action comes as a result of implementing a managerial decision. Frequently that decision is one of accepting or rejecting an investment opportunity and is based on comparisons which should always answer the question "Which is the best of the alternatives?" and sometimes answer the question "Is this 'best' alternative good enough?" Simply put, analysis permits identification of the best alternative and sometimes reveals economic feasibility; by this means, analysis facilitates managerial decision making and can ultimately lead to prescription of a course of action.

Consider next the question of how we might describe *best* or *good enough* in more meaningful ways. Approaching this question from the viewpoint of the enterprise, we can see that our response is dependent upon objectives, plans, and criteria of the enterprise as well as the political and economic structure of the nation.

Objectives, Plans, and Criteria of the Enterprise

If a comparison of alternatives is to yield results, the criteria by which the alternatives are to be evaluated must be specified.

Some possible criteria include:

1. Maximize profit.
2. Minimize cost.
3. Maximize the benefit-cost ratio.
4. Maximize profit rate.
5. Minimize risk of loss.
6. Maximize safety.
7. Maximize quality of service.
8. Maximize sales.
9. Minimize cyclic fluctuations of the firm.
10. Alleviate cyclic fluctuations of the economy.
11. Maximize growth rate of firm.
12. Maximize prestige of firm.
13. Serve community needs by providing both money and people.
14. Maximize economic, physical, and psychological security of the employees by recreation facilities, counselling services, challenging roles, generous wages, and so forth.
15. Create a certain type of public image.

Obviously the criteria are not entirely compatible; the reader is cautioned to note that generally we cannot simultaneously satisfy more than one of the objectives. This book will in general be based on the first objective, maximization of profit, even though we recognize the objectives of the firm may be more encompassing and altruistic. For the sake of simplicity and workability we will accept a trivial degree of suboptimality in our decisions. The more encompassing view is not entirely neglected, since utility of money concepts, consideration of non-monetized factors, and certain mechanisms in the capital budgeting process treated later will permit some relaxation of the apparent narrowness of our objectives.

The first two criteria frequently coincide and will be dominant in most of the problems and examples. The methods shown, however, also permit treatment of such projects as public parks where the first two criteria may not be appropriate. Likewise governmental agency actions may give prime consideration to Criterion 10; even here analysis can be helpful in providing some guidelines as to the ultimate cost of such programs.

Capital Expenditure or Investment Opportunity Alternatives

The object of the analyst's study is the capital expenditure opportunity; that is, a course of action which will require a present cash outlay in return for future benefits (generally two or more years distant). This description is broad enough to include certain expenditures for research, training, or advertising (excepting items which are trivial in cost and consequence).

Consider the many alternative equipment and facility possibilities in such problems or purposes as:

1. Alleviation of a shortage of potable water.
2. Fastening of two materials.
3. Systems of communication.
4. Production of a certain item.
5. Harnessing natural sources of energy.
6. Fabrication of a geometric shape.
7. Transportation of people from suburb to city to suburb.
8. Manufacture of a gear.
9. Reduction in the danger of flood.

Suboptimization and the Accept/Reject Decision

The majority of capital expenditure decisions of this book appear to be made on a one-by-one basis and as such suggest the possibility of nonoptimality from the viewpoint of the firm. In fact, however, the acceptance decision is a *preliminary* one (perhaps from a departmental viewpoint) with further screening and review to follow. Later, from a company viewpoint, the same projects will be reexamined (1) by a number of upper management persons each having familiarity with special considerations of the overall system; (2) simultaneously with many other projects so that a *balance* of interrelated projects can be attained, with approval tending to be based on company rather than department or project considerations; and (3) for consistency with long-range operating and financial plans of the company. Finally, we might note that it will be shown later that even the economic aspects of accept/reject decisions are not entirely separable from the overall capital budgeting decisions.

In theory the aggregate of decisions of individual firms, might also seem to create a possibility of nonoptimality of results for our nation as a whole; the alternative is, however, a very drastic departure from our present system and certainly has not shown itself to be free of difficulties. The issue of free-enterprise decisions versus centralized governmental planning is of course one of the characteristics by which our system of government differs from those of several nations where central governmental "five-year plans" are typical. This is not to say that our government lacks vigor or concern with regard to capital spending. Recent examples of direct government influences on capital spending include such legislation as the investment credit act, modified income tax rates, and bills which bear directly on specific industry groups. In response to the complex "balance of payments" problem, one governmental measure has been encouragement of voluntary rationing by industry of capital expenditures abroad. The suggestion that decisions made on a company-by-company basis cause nonoptimal allocation of resources on a national basis should be cautiously approached; nonoptimality in allocation may still be preferable to achieving a theoretical optimality of allocation through governmental actions which are nonoptimal in terms of national ideals and principles.

Capital Expenditures and the Firm

Time consumes productive facilities and necessitates their eventual replacement. Consumer demand for more goods and services necessitates an increase in productive facilities. Despite the inevitable consequences of time and growth, there is a surprising flexibility in the specific responses which the firm can offer to the needs of replacement, modernization, and expansion.

The amount of money which a firm or the aggregate of firms spends on new plant and equipment tends to be somewhat discretionary; in the short run, expenditures can be minimized by sacrificing some output capabilities and some cost-reduction opportunities. Indeed it appears that capital spending is very responsive to prospective productive demands and economic prospects and thus is subject to wide fluctuations as well as the accumulative effect of underspending or overspending in earlier periods. The typical capital expenditure decision involves the pressures of competition and limited time for making a decision about a course of action with an uncertain future about which only limited information is available and in which one must simultaneously satisfy corporate objectives, legal requirements,

TABLE 1-1. Selected Expenditures in the United States, 1970

Item	Expenditure
	billions
Trade, service, finance, and construction firms	$17
Manufacturers of durable goods	16
Manufacturers of nondurable goods	16
Public utilities	13
Communications	10
Transportation and mining	8
Total business expenditures for new plant and equipment	$80
Expenditures by professional, institutional, and real estate firms	17
Increased business inventories	4
Private nonfarm residential structures	29
Farm structures and equipment	6
Federal, state, and local expenditures for construction	28
Research and development expenditures by the federal government	16
Research and development expenditures by industry and universities	12
Corporate philanthropy	3

Note: See the most recent issues of *Economic Report of the President,* Washington, D.C., *Industrial Research,* Beverly Shores, Indiana, and *Federal Funds for Research, Development, and Other Scientific Activities,* National Science Foundation, Washington, D.C.

time limitations, and scheduling restrictions, all with a limited supply of capital, manpower, and other resources.[1]   The key word is *limited.*

Capital expenditures can be classified conveniently (not without overlap unfortunately) into three general groups: Those which are required to (1) maintain productive capacity, (2) mechanize or automate present facilities, and (3) expand product lines and productive capacity. Capital expenditures are important because the future well-being of the firm is dependent upon the ability of its planners to forecast accurately data of the future, and to evaluate properly the available courses of action.

Statistics of capital spending in the United States for the year 1970 showed the capital expenditures listed in Table 1-1.

These are the "dollars of decision" with which this book is

---

[1] For an excellent perspective of the changing role of the decision-maker analyst, see C. E. Bullinger, *Engineering Economy,* 3d ed. (New York, McGraw-Hill, 1958), pp. 33–43.

concerned. These dollars represent a very significant portion of our gross national product; our concern is that these dollars are efficiently and wisely spent. The concern of individual firms and agencies is evidenced by the books,[2] manuals, and pamphlets which they have authored and by the effort spent in education of personnel in engineering economy.[3]

Problems

1-1. a. Give an example of some current need for which no functionally adequate design has yet been demonstrated.

b. Give an example of some current need for which present designs meet functional criteria but not economic ones.

1-2. Find capital expenditures for each of the past five years for:

a. General Motors

b. DuPont

c. American Telephone and Telegraph

d. Exxon

e. International Business Machines

f. A company specified by your instructor

Possible sources of information are the annual report of the company and *Moody's Industrial Manual* (see also the utility, transportation, and other manuals when appropriate).

1-3. As the man who is predominantly a people manager becomes predominantly an equipment manager, his ability to control costs and activities of his department would seem to be reduced and his department's flexibility likewise diminished. He is also likely to encounter problems in reassigning people whose activities are assumed by automating equipment. Comment upon these disadvantages of automation.

1-4. Your decision to obtain a college education is somewhat analogous to a capital expenditure decision. Your investment of time and money during the years of your college education represent the present outlay; in return for this you might have some future benefits.

---

[2] For example, those by American Telephone and Telegraph, National Machine Tool Builders, Public Service Gas and Electric of New Jersey, Johns-Manville Co., Commonwealth Edison of Chicago, Long Island Lighting, International Business Machines Corporation, Wisconsin Power and Light, and Kaiser Engineers.

[3] Companies have conducted many classes, and university programs for company personnel are or have been offered by Stanford, Clemson, University of Colorado, Illinois Institute of Technology, Washington University (St. Louis), Rensselaer, Cornell, and Iowa State University.

a. Name several of the benefits you hope your education will provide.

b. Your present investment of time and money is quite *certain* while the prospective benefits are only *possible.* Is this not giving up a bird in the hand for two in the bush? What then is the incentive?

c. Would it be possible to list all the factors which should influence anyone in his decision to seek a college education? Why?

d. If it were possible to compose the list in Part c, would it be possible to produce a weighting scheme to weigh properly each of the factors for you? For anyone? Why?

e. If it were possible to list and weigh all factors influencing the decision by a person to seek a college education, could we apply a monetary value to each of these and produce a "prospective total benefit" to compare to "prospective total cost" for the purpose of making the decision? Why?

1-5. A college student has $60 with which he can buy either a suit or a typewriter. He feels he needs both but can afford only one.

a. Suggest some factors which he might consider in making his decision.

b. Assume that presently he does not have a typewriter and that he could make good use of one. How is his decision influenced by the number of suits he already owns?

1-6. Your automobile has just "thrown a rod," and you need to take some action. You have restricted your alternatives to those which involve either restoration of the present car or replacement of it. After considerable effort you have identified your alternatives:

A. Have the engine repaired for $200 (this figure is an estimate; the garage will not guarantee this price).

B. Buy a reconditioned engine having an installed cost of $300.

C. Sell the car in an "as is" condition for $100 to a firm which specializes in salvaging autos. Buy a used car costing $700.

D. Sell the car in an "as is" condition to a mechanic who is aware of the engine's condition and who has offered $200 for the car. Buy the same used car noted in C.

a. One of the alternatives above "dominates" (is preferable regardless of cash available, and so forth) one other alternative. Identify the dominating and dominated alternative.

    b. Are you always completely unrestricted on your selection decision regarding the remaining alternatives? Explain.

    c. Discuss the factors which affect the "pressure" to make a decision and to take action as in the instance above. (Do we aggressively seek situations about which to make decisions? Or do circumstances thrust upon us questions which we cannot decline? Are decisions made at times chosen by us at our discretion or forced upon us by circumstance?)

    d. What *one* additional bit of information regarding A and B would help you in making a decision?

    e. Discuss the factors which affect the "pressure" to take a certain course of action. Note for each whether the factor is one over which you can exercise some control.

1-7. You recently had constructed for you a duplex which you plan to use for rental purposes. The following costs were incurred:

Legal fees, city permits	$ 120
Purchase of lot	5,500
Survey of lot	50
Construction costs	26,000
Advertising for tenants	20
Interest costs prior to completion of duplex	140
Grading, seeding, sodding	40
Fire insurance for one year	60
Grass mowing and general preparation for showing	30
Architect's fee	1,300

What is your *investment* in the property?

1-8. In describing his work activity, a young engineer writes, "One of my recent problems has been the development of a means of communicating assembly techniques for prototype models to the technicians who produce the units. Because of the increasing emphasis on research and development activities the prototype construction is a problem of growing concern. The communication technique proving the most valuable was the use of recorded assembly instructions which the technician can control in speed, can reverse and repeat, or can stop, pending further steps in assembly. The speed of producing a prototype unit is frequently more important for 'contract opportunity' reasons than for reducing the cost of the prototype itself."

    Would this activity be one of those exceptional ones in which dollar considerations are of relatively little importance? Discuss.

2

Enterprise Action
and Prospective Costs

An Investment Opportunity, Course of Action, and Record of Action

IMAGINE that you have an investment opportunity to consider; it requires an estimated present investment of $1000 in equipment and an additional $500 cash (working capital) to produce net receipts of $500 per year for each of the next four years. At the end of the four years the working capital of $500 should remain, but the estimated value of the equipment is zero.

This book deals with the evaluation and comparison of such investment opportunities as the one you are asked to consider here. The concepts and techniques for evaluation and comparison appear in subsequent chapters, but the flow of decisions and actions as recorded in accounting can be of help in providing the framework of the business setting within which decisions are made.

There are at least two courses of action you can follow with regard to this investment opportunity; you can *accept* it or you can *reject* it. (We might also consider postponement or alteration of an alternative.) These courses of action are said to be *mutually exclusive* because choosing one rules out choice of the other.

Suppose that you do accept the investment opportunity and that this is your first business venture. Suppose further that your subsequent course of action is:

1. Organize a firm, supplying it with $1500 capital from your personal savings.
2. Purchase the equipment under consideration for a total cost (includes freight, installation, and so forth) of $1000.
3. Procure the years necessary labor and material, then manufacture. Note addition to inventory resulting from $630 of labor, $120 of material, and $200 worth of administrative services you have rendered.[1] Recognize these items as accounts due but not yet paid.
4. Recognize and record the partial consumption of capital (equipment) with the passing of time (one year).
5. Sell for cash all the goods produced this year, 1000 units @ $1.45 each.
6. Pay accounts payable.
7. Check income tax liability. Because you are a full-time university student, your other income is limited to an extent that no tax liability arises.

A simplified version of the accounting entries which would record exactly the same information in a different format is:

	Account	#1	#2	#3	#4	#5	#6	#7	Balance Sheet at End of Year One
					dollars				
Assets	Cash	+1500	-1000			+1450	-950	0	+1000
↑ These	Inventory			+950	+250	-1200		0	0
totals should	Equipment		+1000		-250			0	+750
always balance									
↓	Accounts payable			+950			-950	0	0
Liabilities and equity	Equity	+1500				+250		0	+1750

Accounting Records of Performance and Position

The preceding permits several observations which may be helpful in understanding (1) the methods employed by the ac-

---

[1] In a more detailed analysis, administrative expenses would bypass the inventory account and be handled through accounts not shown here.

countant, (2) the role of accounting as an important source of information, and (3) the impact of our decisions upon future accounting reports.

The accountant could employ a format identical to that shown except that the large number of *accounts* (where we show 5, the accountant may actually deal with 500) and the large number of *transactions* (where we show 7, perhaps 7000 is typical of a medium-sized firm) make the format impractical except for extremely limited operations. In response to this need accountants devised *journal entries,* a sequential listing of each transaction showing titles of the affected accounts and the numeric change in each, thus:

		Debit	Credit
#1	Cash	$1500	
	Equity		$1500
#2	Equipment	1000	
	Cash		1000
#3	Inventory	950	
	Accounts payable		950
#4	Inventory	250	
	Equipment		250
#5	Cash	1450	
	Inventory		1200
	Equity		250
#6	Accounts payable	950	
	Cash		950
#7	No action		

By use of *debits* and *credits* the need for plus and minus signs is satisfied. Each *journal entry* replaces one *vertical column* in the format of the simplified version. From the journal entries the accountant can then post results in appropriate *T accounts.*

Cash		Inventory		Equipment	
#1 1500	1000 #2	#3 950	1200 #5	#2 1000	250 #4
#5 1450	950 #6	#4 250			

Accounts payable		Equity	
#6 950	950 #3		1500 #1
			250 #5

Each *T account* replaces one *horizontal row* in the format of the simplified version.

A *balance sheet* is a summary of the various T account balances at some specified point in time; after entry #4 the following balance sheet can be constructed:

Assets		Liabilities and Equities	
Cash	500	Accounts payable	950
Inventory	1200	Equity	1500
Equipment	750		

The balance sheet in Figure 2-1 contains the added detail resulting from larger, more complex operations. Because annual sales per dollar of investment (the period over which investment is recouped) and the financial structure of the typical utility and industrial firms differ so drastically, the data for both are given in Figures 2-1, 2-2, and 2-3 with the hope of providing some helpful comparisons and insights. Only the utility data appears in Figure 2-4.

An *income statement* based on either the simplified format or the T accounts can also be constructed:

Operating revenues (sales)	+$1450
Operating costs (labor, materials, and administrative costs)	-950
Depreciation expense (capital cost)	-250
Interest on borrowed (debt) capital	0
Before-tax return on owner's (equity) capital = taxable income	+$ 250

It is important to recognize that net receipts (cash flows) of $1450 - $950 = $500 are not all profit. The accountant accomplishes this by allocating the $1000 cost of equipment over its four-year life, possibly in uniform annual amounts of $250 per year. For the investor, the $500 per year must provide for (1) repayment of the investment consumed, (2) return on the investment, and (3) payment of income taxes. Because repayment is annual, the remaining investment declines from year to year; the original investment of $1000 in equipment and $500 in working capital will diminish to an estimated $0 in equipment and $500 in working capital by the end of year four. By subtracting an allowance for investment consumed (depreciation expense) of $250 from the cash flow of $500, we are left with

Balance Sheet of the_____Company as of December 31, 19xx

(All figures are in millions of dollars)

Item	Utility	Industrial	Explanation
			The company owns:
CURRENT ASSETS			
Cash	+18	+ 38	Checking accounts and cash
U.S. government securities	+46	+ 98	Treasury bonds, notes, bills
Accounts and notes receivable	+40	+135	Amounts owed to company by customers and others
Inventories	+ 3	+268	Raw material, work in progress, and finished goods
FIXED ASSETS			
Property, plant, and equipment	+1055	+861	Land, buildings, machinery
Less accumulated depreciation	-226	-482	Reduction for accumulation of cost allocation to years of use
OTHER ASSETS	+64	+82	Patents, copyrights, goodwill, investments in subsidiaries, prepayments, and so forth
Total Assets	1000	1000	
			The company owes:
CURRENT LIABILITIES			
Accounts and notes payable			
Accrued liabilities			
Currently maturing long-term debt	+90	+187	Obligations which have been incurred but not yet paid
Income taxes payable			
Dividends payable			
Advance billing and customer deposits			
LONG-TERM DEBT	+ 290	+ 24	Borrowed money
EQUITY			
Common stock (at par)	+ 301	+ 79	Shares outstanding times par value
Capital surplus	+ 148	+ 75	Proceeds in excess of par value received from sale of stock
Earned surplus	+ 171	+ 635	Accumulation of prior years earnings less dividends
Total Liabilities and Equity	1000	1000	

Fig. 2-1. Balance sheet statement with data for utility and industrial firms.

a return of $250/$1000 = 25\%$ as provision for interest (return) on the average investment and income taxes.[2]

At the end of a year of operation, still another statement can be very helpful to us, the *sources and application of funds*

---

[2] By exact methods illustrated in Chapter 7, $i = 21.1\%$; the approximate nature of the result above is due to our neglecting to recognize that (1) the outstanding balance of investment declines more slowly in the early years of repayment and (2) the repayment amounts are end-of-year rather than continuous. Both factors slightly inflate the approximate rate.

Income Statement of the _____ Company for the year ended December 31, 19xx

(All figures dre in millions of dollars)

Item	Utility	Industrial	Explanation
Operating revenues (sales)	+333	+1670	Cash received from customers
Operating costs (cost of goods sold, selling and administrative expense)	-173	-1300	Cash paid for wages, material, fuel, property taxes, supplies, insurance
Depreciation expense	- 48	- 49	Noncash allocation of cost of plant and equipment
Interest on debt	- 11	- 1	Cash paid for use of borrowed funds
Taxable income	+101	+ 320	Basis for tax computation
Income taxes	- 47	- 151	Cash paid for income taxes
Operating return on equity (net income)	+ 54	+ 169	Earnings applicable to stock

Fig. 2-2. Income statement with data for utility and industrial firms.

Funds Flow Statement of the _____ Company for the year ended December 31, 19xx

(All figures are in millions of dollars)

	Utility	Industrial
SOURCE OF FUNDS		
Operating revenues—operating costs—income tax or After-tax cash flow or Net income + depreciation expense + interest on debt	113	219
Less: Interest on debt	- 11	- 1
Less: Dividends on equity	- 36	- 84
Disposal of plant and equipment	10	4
New debt and equity capital	74	2
Total	150	140
APPLICATION OF FUNDS		
Increase in working capital during period	13	31
Repayment of debt and equity capital	21	3
Capital expenditures for new plant and equipment	116	106
Total	150	140

(Net funds generated from operation — bracketing the After-tax cash flow, Less: Interest on debt, and Less: Dividends on equity rows)

Fig. 2-3. Funds flow statement with data for utility and industrial firms.

statement sometimes simply called the *funds flow statement.* A simple version of the funds flow statement for the hypothetical firm at the end of one year of operation is:

SOURCE OF FUNDS

$$\left.\begin{cases} \text{Operating revenues - operating costs - income tax} \\ \qquad\qquad\qquad\text{or} \\ \text{After-tax cash flow} \\ \qquad\qquad\qquad\text{or} \\ \text{Net income + depreciation expense +} \\ \quad\text{interest on debt} \end{cases}\right\} \quad \$500$$

Less: Interest on debt K_1
Less: Dividends on equity K_2
Disposal of plant and equipment K_3
New debt and equity capital K_4
   Total K_5

APPLICATION OF FUNDS
   Increase in working capital during period K_6
   Repayment of debt and equity capital K_7
   Capital expenditures for new plant and equipment K_8
     Total K_9

The statement above at least implies that you must make some decisions with regard to the funds of your firm; it also leads us into some of these possibilities. The funds flow statement reveals that your company's cash supply increased by $500 during the year; one question before you, then, is what to do with this extra cash. Some alternatives are:

A. Do nothing. (Interestingly enough this *is* a decision even if only the result of neglecting the question.) In this case your supply of working capital is now more than it was at the start of operations ($K_6 = \$500$).

B. Withdraw some of the excess funds from the business for your personal use, say $600. In this case $K_6 = -\$100$ and $K_7 = \$600$. This action causes your company's cash balance to drop $100 below your original plan of $500.

C. Expand operations with additional facilities costing $1000. Finance this by obtaining additional funds of $500 and by permitting your company's cash balance to drop $100 lower than your original plan of $500. Thus $K_4 = \$500$, and $K_8 = \$1000$.

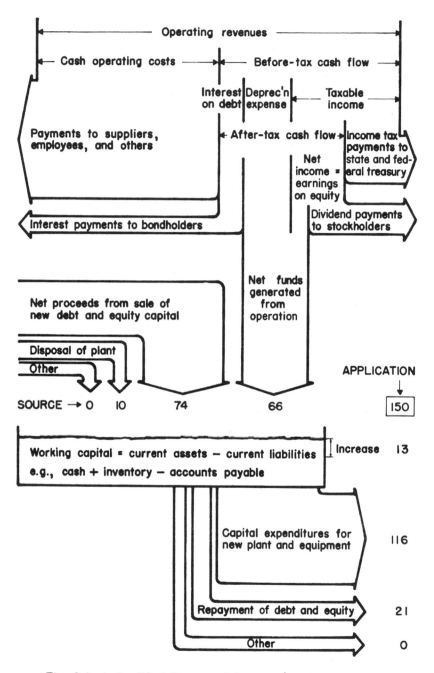

Fig. 2-4. A simplified diagram of the corporate flow of funds.

Working Capital

Working capital is an unavoidable form of investment for virtually all organizations. The prospective owner of a service station invests in equipment; he must also commit or "tie up" some of his funds in working capital. That working capital may include such current assets as (1) cash on hand and in the checking account; (2) accounts receivable (for merchandise he has sold and charged to customers whose cash payments will be made at some future date); and (3) an inventory of gasoline, oil, and other automotive products. Working capital will also reflect the offsetting effect of such current liabilities as "accounts payable to suppliers."

An inadequate supply of cash may cost the owner a missed opportunity for discount on goods purchased or a penalty cost for late payment. Inadequate inventory or too stringent a credit policy may result in reduced sales volume.

An oversupply of cash means that the owner is earning 0% on funds which might be profitably invested elsewhere. An oversupply of inventory may cost storage space and may tie up limited capital resources in slow turnover merchandise, thus denying the owner an opportunity for more profitable application.

Proliferation of sizes, types, models, and colors, imposes pressures for increased inventories. Managerial efforts to standardize merchandise offered or facilities used can deter such pressures. Tobacco companies typically have a very large inventory investment in a product that is being aged. Manufacturers of aircraft and autos typically have a very large inventory investment in a product that is costly.

Very significant changes in working capital are sometimes within managerial control. The manufacturer, observing a mounting inventory of finished goods to sell, may elect to curtail or shut down production, letting the continuing sales gradually consume the excess inventory.

Credit policies can vary from requiring prepayment of bills and customer deposits to very liberal collection policies that permit a 60- or 90-day delay in payment. Changing the collection policy or even the frequency of customer billing can have a considerable one-time effect. Consolidated Edison of New York once estimated that switching to meter reading and billing every month instead of every two months would provide the company a one-time cash surge of more than $13 million.[3]

---

[3] Letter to Public Service Commission of New York, October 15, 1970.

Perhaps our most dramatic and daring illustration comes from the life of Henry Ford.[4] During the last years of World War I, Henry announced he would cut dividends to the bone and put the bulk of the company's profits into reducing the price of the Model T and enormously expanding his plant. When stockholders John and Horace Dodge heard of Ford's plans they were disturbed and eventually brought suit for "reasonable" dividends. In February 1919 the Supreme Court ruled that nonpayment of dividends was, considering the remarkable profits, illegal and arbitrary. Ford was ordered to pay a delayed dividend of $19 million plus interest. Henry proceeded to remove the encumbrance of minority stockholders by purchasing their stock to gain total control. To accomplish the purchase Ford borrowed $75 million from two large banks, agreeing to repay the loan in April 1921. The depression of 1920 came and a crisis seemed imminent. The financial "tour de force" that finally pulled him out came when Ford produced 90,000 cars and shipped them to the dealers. Cash payment was demanded upon delivery of the unordered cars. The dealers turned to their local bankers, and the burden was transferred to hundreds of small banks who in effect were made to pay the two large banks, with the dealers as reluctant go-betweens.

Cost Concepts and Analysis

Based on your past experience with investments in capital goods, suppose you conclude that an investment must provide a before-tax rate of return of 20% or more to be attractive. To meet this threshold of acceptability, the investment in equipment and working capital discussed earlier must produce a before-tax cash flow of at least $486 (this sum is based on methods that will be introduced in Chapter 6). The $486 would be sufficient to provide for (1) repayment of the investment consumed ($250 per year for four years), (2) return on the investment, and (3) income taxes. We could then summarize the annual cost of producing 1000 units as follows:

Before-tax cash flow required		
(for repayment, return, and income taxes)		= $ 486
Operating costs: administrative	200 ⎫	
labor	630 ⎬	= $ 950
material	120 ⎭	
Annual equivalent cost		= $1436

[4] See Roger Burlingame, *Henry Ford*, (New York, New American Library of World Literature, 1954), pp. 67–74.

If your sales (and therefore production) volume were to increase from the present level of 1000 units per year, it is likely that labor and material costs would increase proportionately. Labor and material costs are generally referred to as *variable* costs because they tend to vary with the number of units produced. The administrative costs and the "before-tax cash flow required" tend to be unaffected by the number of units produced, hence they are *fixed* costs. Although Figure 2-5 illustrates the "perfect" case, it is more realistic to expect to find few if any costs which are either absolutely independent of or perfectly correlated to (dependent upon) the output level. The

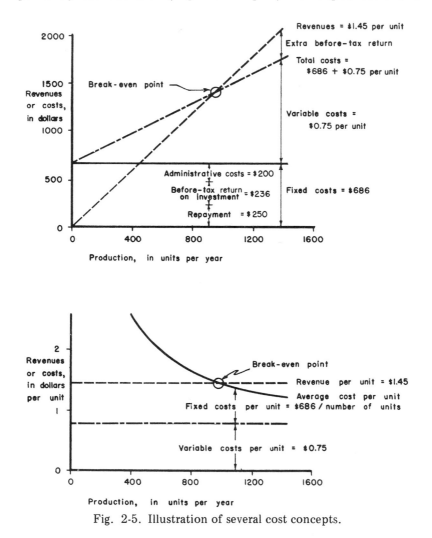

Fig. 2-5. Illustration of several cost concepts.

relative fixedness of a cost is dependent upon time; given enough time the manager can make virtually all costs respond to changing output needs. One common *time horizon* is the *short run* in which, by definition, management is not able to alter the equipment, commitments, and resources of the firm, but is able to vary the firm's output by changing such variable factors of production as labor or materials. The concepts of fixed costs, variable costs, and the short run have found considerable application in spite of definitions which at best are circular. A simple graphic relationship of variable and fixed costs reveals that expanded sales, if realizable without substantial price concessions, may provide additional profits; see Figure 2-5.

The same conclusion can be reached by comparing *per unit* costs and revenues. Here a number of additional concepts of cost are employed:

1. *Average cost* per unit is the average of the costs of each of the units produced. This average cost is large for low production rates because fixed costs must be averaged over a small number of units.
2. *Marginal cost* per unit (also called incremental, extra, or differential cost per unit) is the difference in cost of the Kth and the $(K + 1)$th unit of production.[5]
3. *Break-even point* is the production level at which revenues just equal costs (including the minimum attractive before-tax rate of return).

Charts like Figure 2-5 help to portray how such costs as those for labor and material should respond to changes in the production level. As a means of prediction, budgeting, and control of costs the concepts of fixed and variable costs can be helpful.

In the practical sense it may be necessary to restrict the use of our terms fixed and variable costs to a narrower range than the implied infinite range of Figure 2-5. At higher production levels we may require additional equipment, and at lower production levels we may have to pay premium prices or give up quantity discounts on our material purchases. Under such circumstances we can still speak of costs which tend to be fixed or variable within a certain specified *relevant range* of output.

---

[5] In the special case where variable costs per unit and fixed costs are exactly constant, as in Figure 2-5, variable costs per unit are also arithmetically equal to marginal, incremental, extra, or differential cost per unit.

Relationship of Accounting to Economy Studies

In the example cited, you have traced through the acquisition and one years operation of an investment alternative. This had an impact on three different accounting statements (income, balance sheet, and funds flow statement), all of which are very important to the understanding of the effects of an investment decision upon the firm. The example also illustrates that todays investment decision creates tomorrows need for still other decisions. Although the example is treated in isolation for a newly created firm, the details given are equally applicable for an established firm asking what effect this one decision would have; that is, what *extra* costs, savings, and future funds would arise if the investment opportunity were accepted?

While accounting provides much helpful information for economy studies, it is essential to recognize that the viewpoints of the accountant and the economy analyst are quite different. One such difference can be observed in the treatment of information for accounting and economy study purposes. While the accountant hopes to reveal for public and other perusal an exacting and objective picture of the firm's performance and position, the economy analyst is projecting estimates of a future performance and position which is generally for internal managerial purposes only; circumstances dictate that the less exact nature of his task and the somewhat different purpose of his role permit leniency in approximating the rules and conventions of accounting. More specifically, the economy analyst can often ignore the very real possibility of fluctuations in billings, payments, overdue accounts, or inventory levels (and therefore reported net income and even cash flow). This he does on the premise that year-by-year fluctuations are smoothed in the long-term view appropriate to capital expenditure analysis. He also notes that where an investment or action has an enduring effect upon the necessary inventory (or other working-capital item) level, he does recognize its effect by inclusion of "interest on investment in working capital" in his study.

The difference between accountant and economy analyst is also apparent in the acquisition of a facility. Prior to acquisition of a facility, *first cost* (the sum of the investments in equipment and related working capital, $1500 in our example) is a vital factor in the economy study. As soon as the property is acquired, the investments are past costs and the then relevant information is the cash alternative if ownership is relinquished. The concept that past costs are irrelevant to the economy study is called the *sunk cost* concept. Here the differing role of accountant and economy analyst causes differences in treatment; the accountant cannot dismiss past cost, the economy analyst must (except for income tax effects, if any). Accounting requires analysis and recording of actual data of the *past;* the economy study requires analysis of estimated data of the *future.* A former student expressed this difference in a succinct, if not ingratiating, manner, "Accountants are historians, economy analysts are fortune-tellers."

Problems

2-1. A corporation started the year with the following balance sheet:

	Account	Balance as of 1-1-xx
Assets	Cash	$10,000
	Inventory	6,000 (150 units costing $40 each)
	Net plant and equipment	40,000
Liabilities and equities	Accounts payable	3,000
	Owner equity	53,000

During the year the following transactions took place:

(1) Paid (cash) $3000 to creditors (accounts payable).
(2) Ordered, received, and used $12,000 worth of material; paid (cash) for $4000 of above and noted liability to creditor (accounts payable) for remaining $8000; paid (cash) for labor costs of $18,000 and acknowledged depreciation (reduced "net plant and equipment") of $10,000. Assume that labor, materials, and depreciation are the only costs incurred so that finished goods inventory is increased by 1000 units with an average cost of $40 each.
(3) Sold 1000 units for $54 each (cash). This transaction increases cash, reduces inventory, and increases equity by the pretax profit of $14 per unit.
(4) Paid (cash) $5000 for state and federal income tax. This transaction decreases cash and decreases equity.

a. Make an end-of-year *balance sheet.*

   b. How do you explain the fact that the firm had more
      cash at the end of the year than at its beginning?
   c. Make an *income statement* for the year just ended.
   d. Make a *funds flow statement* for the year just ended.
2-2. a. The "utility" and "industrial" data in Figure 2-1 is
      based on actual figures from two large U.S. firms and
      adjusted to percentages of total assets. By this device
      total assets of either are shown as 1000. Compare the
      data for the utility to that of the industrial to note
      those items for which the two types of firms differ
      quite radically.   Give a possible cause for the
      difference.
   b. Repeat the above for Figure 2-2.
   c. Repeat the above for Figure 2-3.
2-3. Certain annual costs of driving an automobile tend to be
     fixed, while other costs vary with the number of miles of
     operation.
   a. List all the costs and label them either fixed or
      variable.
   b. Since a Volkswagen can be operated for less money
      per mile than a Lincoln, Cadillac, or Imperial, how do
      you explain the choice by some people of the latter
      type of car?
2-4. Refer to Figure 2-5.
   a. Let q_1 = the production quantity at which costs, in-
      cluding the required rate of return, just equal revenues.
      Find q_1.
   b. Assume that the past sales and production level has
      been 1000 units per year and that an increased sales
      demand at the same selling price per unit is expected.
      Find the sales level q_2 which would double the before-
      tax return (taxable income).
2-5. Machine A rents for $100 per year and the material costs
     for its output are 8 cents per copy. Machine B rents for
     $300 per year and the material costs for its output are
     3 cents per copy. Labor costs of using either machine are
     unknown but identical.
   a. How many copies per year would justify a preference
      for Machine B?
   b. What other factors should be considered in your
      selection of a machine?
   c. Make a graph showing total cost per copy using the
      more economic machine at each level; consider the
      range 0–8000 copies per year in increments of 1000.
   d. Is it necessary to distinguish between *preference for*
      and *selection of* Machine B?
2-6. Blue, Inc., has made the following estimates for opera-
     tions for the coming year:

Sales	$34,000,000
Cash operating costs	25,000,000
Depreciation expense	1,300,000

The combined state and federal income tax rate is about 50%. Financial data includes:

1,000,000 shares of common stock
   100,000 shares of 6% preferred $100 par
   10,000 5% bonds @ $1000 each

a. Find the projected earnings per share.
b. If the company continues its past common stock dividend payout of about 60% and if the working capital is not substantially altered, how much additional cash from internal sources would be made available for capital expenditures by the one year of operation?

2-7. Green Corporation started the year with the following balances:

Account	Balance as of 1-1-xx
Cash	$100,000
Inventory (100,000 units)	100,000
New plant and equipment	400,000
Accounts payable	50,000
Owner equity	550,000

Transactions during the year were limited to:

(1) During the year, manufacture and add to inventory 300,000 units. This involves payment of $100,000 for labor and $150,000 for materials; note equipment depreciation of $50,000; observe that the units cost $1 each to manufacture.
(2) Sell 300,000 units for $2 each, cash.
(3) Purchase new equipment costing $70,000.
(4) Pay state and federal income tax of $160,000. Accounts payable at the end of the year were the same as at the beginning of the year.

a. Make an end-of-year *balance sheet.*
b. Make an *income statement* for the year just ended.
c. Make a *funds flow statement* for the year just ended.
d. Although things seem to be going well for Green

Corporation, it does have what may be a problem of growing concern. What is the problem? (**Hint:** The funds flow statement or balance sheet may be helpful in its identification.)

2-8. A retail store is reducing prices on selected items for a January sale of merchandise left over from the Christmas stock of goods. A quantity of Item X is left over; Item X cost the store $11 and has been priced at $19.

If the item is not sold at clearance prices, the store must hold the item in inventory (the distributor will not buy back the goods) until the next Christmas season. The holding action involves some costly charges against Item X for handling, storage, interest on money tied up, hazards of its being outmoded by some of next years models, insurance and property taxes, and the like; such costs are estimated at 50% of the item value (that is, an item worth $1 now and held until next Christmas season would have a total cost of $1.50 after adding in the costs of handling, and so forth).

The store personnel feel that by next year the item may have lost some of its timeliness and also may be undergoing stiffer competition. On this basis they have estimated next years costs as $9 and next years selling price as $15.

If an item is carried over, the number of these items sold probably will not increase but the store's new order for Item X would be smaller.

a. Suppose the store personnel decide to mark down the price of Item X in several stages. What is the final minimum clearance price that should be charged? Explain your reasoning.

b. What concept from this chapter is involved in this problem?

2-9. Answer each of the following by writing the appropriate equation.

a. What is the relationship between *before-tax cash flow* and *taxable income*?

b. What is the relationship between *after-tax cash flow* and *net funds generated from operation*?

2-10. The following information applies to the Strapped Corporation:

(1) Start the year with $100,000 in cash.

(2) During the year sell 400,000 units for $1 each.

(3) During the year spend 25 cents per unit for labor in the manufacture of 400,000 units.

(4) During the year spend 30 cents per unit for material in the manufacture of 400,000 units.

(5) Acknowledge depreciation of equipment of $50,000. (Assume material, labor, and depreciation make up the entire "cost of goods sold.")

(6) Pay state and federal income tax of $70,000. Accounts payable at the end of the year are approximately the same as at the beginning of the year. Financing is entirely by equity capital.

    a. Assume "cost of goods sold" consists only of material, labor, and depreciation expense. Find net income for the year.

    b. Find the end-of-year cash balance. How do you explain this?

2-11. Consult an accounting text or handbook in your library to answer the following questions.

    a. What is the difference between *earned surplus* and *capital surplus*? Why bother separating them?

    b. Give 3 examples of *intangible assets*.

    c. What is *dividend payout*?

    d. Distinguish between *depreciation* and *depletion*.

2-12. The following are balance sheet items. Show which of these accounts are included in working capital by labeling them: + if their presence adds to working capital, 0 if their presence has no effect upon working capital, and - if their presence reduces working capital.

Buildings	Reserve for uncollectible
Cash	accounts
Accrued taxes	Accounts payable
Common stock	Prepaid expenses
Depreciation reserve	Accrued interest receivable
Materials and supplies	Inventories
Customers' deposits	Land
Earned surplus	Capital surplus
Machinery	Accounts receivable
	Dividends declared and payable

2-13. a. Under what circumstances could labor be a fixed cost?

    b. Under what circumstances would fixed costs related to additions to capacity increase in stair-step rather than continuous fashion?

    c. Why might marginal cost per unit of output decrease with increases in the production rate, then increase as the production rate is still further increased?

2-14. Green Corporation has made the following estimates for the coming year. Capital expenditures will be $50,000; after-tax loss will be $30,000; no debt capital will be involved; dividends paid on equity capital will be $10,000; by increased sales efforts and a reduced production schedule, the company hopes to reduce its finished goods inventory by $60,000; depreciation expense of $40,000 will be charged; $20,000 will be realized from the disposal of old property and equipment; no new financing from debt, equity, or other such sources is planned; except for the "cash" account, no other changes in asset or liability accounts are anticipated.

If the preceding estimates prove to be correct, by how much will the "cash" account increase in the coming year?

2-15. Refer to the financial section of your library. See *Moody's* or *Standard and Poor's* or the annual report of a company selected by you or assigned by your instructor. Find operating revenue (sales) for the most recent year, then show the following data as *ratios per dollar of operating revenue.*

(1) Operating costs (do not include income taxes, depreciation expense, or interest on funded debt).
(2) Depreciation expense.
(3) Interest on debt.
(4) Income taxes.
(5) Operating return on equity (net income).
   Check: #1 + #2 + #3 + #4 + #5 = 100%.
(6) Dividends paid for the year.
(7) Internally generated funds (= #2 + #5 - #6).
(8) Externally derived funds (from new debt and equity less retired debt and equity).
(9) New plant and equipment (capital expenditures for the year).
(10) Total assets.

II

Time and Money

Yesterday is a cancelled check.
Tomorrow is a promissory note.
Today is cash.

AUTHOR UNKNOWN

3

Timing of Cash Flow

Repayment Alternatives

SUPPOSE that after one year of operation the business venture begun in Chapter 2 has been successful to the extent that you are considering expansion. You hope to finance the $1000 expansion (1) by reinvesting the cash flow of year one ($500) and (2) by borrowing $500 from your family. The family has agreed to lend you this amount at 6% interest and has left it to you to determine a plan of repayment. You hope to repay the loan as soon as possible, and yet you also hope to stretch repayments over a period long enough that payments will be easy to meet even if your cash flows fall below their predicted levels. This latter consideration is based on your wishing to avoid the embarrassment of having to revise the repayment plan later and also on your desire to have some excess cash with which to take advantage of fleeting opportunities. After such reasoning you decide to repay the loan over a five-year period; next you consider several repayment plans which might be employed.

Plan A. Repay the principal in uniform amounts at the end of each of the five years and also make simultaneous annual interest payments on the unpaid balance (principal).

Plan B. Repay the principal and interest in five uniform end-of-year payments.

Plan C. Repay the principal in a single lump sum at the end of five years and make annual end-of-year interest payments on the unpaid balance (principal).

Plan D. Repay the principal and interest in a single lump sum at the end of five years.

You next determine the size of each payment in the preceding set of alternatives and show these as Table 3-1.

TABLE 3-1. Several Plans for Repayment of a $500 Loan Having an Interest Rate of 6%

Alternative	Year	$\begin{pmatrix}\text{Outstanding} \\ \text{Balance at} \\ \text{Beginning of} \\ \text{Year}\end{pmatrix}$ +	$\begin{pmatrix}\text{Interest} \\ \text{Accrued} \\ \text{During} \\ \text{Year}\end{pmatrix}$ −	$\left(\begin{matrix}\text{End-of-Year} \\ \text{Payment}\end{matrix}\right)$ =	$\begin{pmatrix}\text{Outstanding} \\ \text{Balance at} \\ \text{End of Year}\end{pmatrix}$
Plan A	1	$500.00	$30.00	$130.00	$400.00
	2	400.00	24.00	124.00	300.00
	3	300.00	18.00	118.00	200.00
	4	200.00	12.00	112.00	100.00
	5	100.00	6.00	106.00	0.00
Plan B	1	$500.00	$30.00	$118.70	$411.30
	2	411.30	24.68	118.70	317.28
	3	317.28	19.04	118.70	217.62
	4	217.62	13.06	118.70	111.98
	5	111.98	6.72	118.70	0.00
Plan C	1	$500.00	$30.00	$ 30.00	$500.00
	2	500.00	30.00	30.00	500.00
	3	500.00	30.00	30.00	500.00
	4	500.00	30.00	30.00	500.00
	5	500.00	30.00	530.00	0.00
Plan D	1	$500.00	$30.00	$ 0.00	$530.00
	2	530.00	31.80	0.00	561.80
	3	561.80	33.71	0.00	595.51
	4	595.51	35.73	0.00	631.24
	5	631.24	37.87	669.11	0.00

How can you choose from among the repayment plans? You have four alternative plans and they are *mutually exclusive.* Your family has specified only that interest be computed at 6%; since all plans were determined with interest at 6%, the plans are *equivalent* to your family from their *point of view.* Notice that equivalence of the plans is contingent upon an interest rate of 6%, and the equivalence holds despite the fact that total repayment under each plan is different.

Plan	Total of Payments
A	$590.00
B	593.50
C	650.00
D	669.11

Why? Note that in Plan A, after the first interest period the outstanding balance (principal) is less than it is under the remaining plans; we simply have the use of a lesser amount of money. As a matter of fact, if the payments of Plan A were deposited to an account paying 6% interest, the balance at the end of the fifth payment would be $669.11, identical to the payment of Plan D. You could check as follows:

Year	Interest on Balance During Year	Deposit at End of Year	End-of-Year Account Balance
1	$ 0.00	$130.00	$130.00
2	7.80	124.00	261.80
3	15.71	118.00	395.51
4	23.73	112.00	531.24
5	31.87	106.00	669.11

The above may help to reinforce the concept of equivalence. When repayment plans are equivalent at some rate of interest, say 6%, we know that anyone whose item preference for money is that same rate will be *indifferent* as to the plans; he will simply have no preference. If your personal time preference for money is some rate other than the one for which the repayment plans were equivalent, you will not be indifferent in your preference among them. If your other investment opportunities offer only 5%, then the repayment plans are not equivalent, and you (the borrower) should prefer Plan A. Conversely, if many attractive opportunities with prospective earnings of 15% arise, it would be to your advantage to choose Plan D.

Describing Cash Flows

The verbal description of a unique set of cash flows can take many forms; the cash flows of Plan B from the lender's point of view are correctly described in all the following:

1. Find the uniform annual payment required to repay a present loan of $500 if payments are made at the end of years one, two, three, four, and five and if interest is computed at 6% (from the viewpoint of the lender).
2. Find the uniform end-of-year receipts over each of the next five years equivalent to a present expenditure of $500 when $i = 6\%$.
3. Find the uniform end-of-year cash savings over each of the next five years which would justify a present investment of $500. Assume interest is 6%.
4. If $500 is deposited today in an account paying 6% interest, what uniform withdrawal could be made at the end of each of the next five years to exhaust the fund?

All the preceding statements can be described by a single *cash flow diagram,* a graphic portrayal of the verbal statement of a problem:

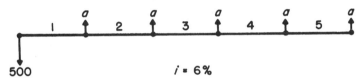

The cash flow diagram above is a complete description of the problem statements. Only after drawing the diagram is it obvious that the four problem statements referred to a single unique set of cash flows.

The cash flow diagram employs several devices and conventions:

1. The horizontal line is a *time scale* with the progression of time moving from left to right. The year labels are applied to *intervals* of time rather than points on the time scale. Note that the end of year two is coincident with the beginning of year three. Only if specific *dates* are employed should the points rather than intervals be labeled thus:

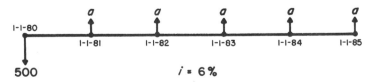

2. The arrows signify cash flows; downward arrows represent disbursements and upward arrows represent receipts. We will

treat upward arrows (receipts) as positive (+) cash flows and downward arrows (disbursements) as negative (-) cash flows.
3. The cash flow diagram is dependent upon *point of view*. The *lender's* diagram:

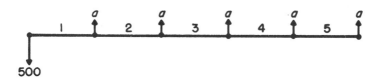

is just inverted from the *borrower's* diagram:

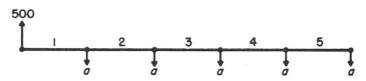

The repayment plans, A, B, C, and D, considered earlier could be portrayed by the cash flow diagram, Figure 3-1, and the cash flow table, Table 3-2. They are shown from your (the borrower's) viewpoint.

TABLE 3-2. Cash Flow Table for Several Plans for Repayment of a $500 Loan Having an Interest Rate of 6%

Cash Flow	End of Year					
	0	1	2	3	4	5
Plan A	+$500.00	-$130.00	-$124.00	-$118.00	-$112.00	-$106.00
Plan B	+500.00	-118.70	-118.70	-118.70	-118.70	-118.70
Plan C	+500.00	-30.00	-30.00	-30.00	-30.00	-530.00
Plan D	+500.00	0	0	0	0	-669.11

In addition to representing cash flows by verbal descriptions, cash flow tables, and cash flow diagrams, there is a fourth method which for Plan C simply would show:

(+$500.00, -$30.00, -$30.00, -$30.00, -$30.00, -$530.00)

The convention is one which simply starts with the end-of-year zero cash flow and separates each years cash flow by a comma. This latter method is probably the most concise; despite this real advantage we will generally employ the cash flow diagram on the grounds that it has (1) a pictorial advantage, (2) a me-

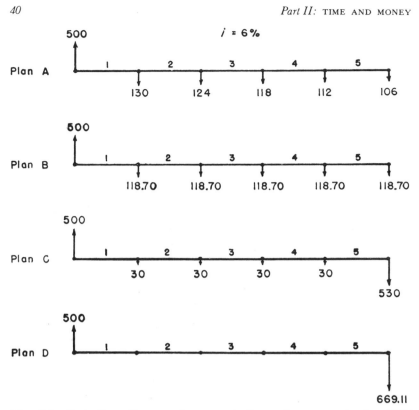

Fig. 3-1. Cash flow diagrams for several plans of repayment of a $500 loan having an interest rate of 6%.

chanical and readability advantage when n is large, and (3) a greater flexibility to describe cash flows which are more frequent than annual, continuous, or nonend-of-year.

In the four repayment plans considered there are at least two questions which might be raised:

1. Although the annual payment of Plan B ($118.70) does exactly repay the loan, how was this amount determined?
2. The method of Plan D permits us to observe that the $669.11 is a correct sum required for repayment, but the method would surely be cumbersome at best if the interest rate were fractional (say $5\frac{3}{4}\%$) and the number of periods were large (say 20). Is there some more convenient method of determining such a sum?

In the next chapter the mathematical relationships of various patterns of cash flow are developed; these relationships are presented in table form in Appendix I. Chapter 5 illustrates the application of the formulas and tables to problems involving cash flows.

Problems

3-1. You borrowed $10,000 and have agreed to repay this amount over the next five years at 10% interest. Make a table similar to Table 3-1 to show repayment accomplished by:

    a. Year-end payment of interest on the principal with the fifth payment to include repayment of the principal itself.

    b. Year-end payment of one-fifth of the principal ($2000) plus interest on the unpaid balance.

    c. Lump-sum repayment at the end of the fifth year.

    d. Year-end payments of equal size. (By methods developed and illustrated in the next two chapters, we can find the payment should be $2638 per year.)

    e. Draw a cash flow diagram for Part d.

3-2. a. Use a format similar to Table 3-1 to show repayment of a present loan of $10,000 with interest at 10%. Repayment is to be accomplished by payment of the entire principal plus accrued interest at the end of year two.

    b. Repeat Part a for repayment accomplished by payment of interest at the end of year one, and payment of principal plus interest at the end of year two.

    c. Repeat Part a for repayment accomplished by payment of interest plus half the principal at the end of year one, and payment of interest plus half the principal at the end of year two.

    d. Repeat Part a for repayment accomplished by uniform payments at the end of years one and two.

    e. Draw a cash flow diagram for Part a.

    f. How did you determine the payment in Part d?

    g. If the number of payments (and years) in Part d had been larger, it is quite unlikely that all payments could be uniform to a whole penny amount. What might a lending agency do in such circumstances?

3-3. Assume that you have three outstanding obligations and the interest cost on Obligation A is 3%, Obligation B is 7%, and Obligation C is 12%.

    a. If you were given a choice, which obligation would you try to repay first and why?

    b. If you expect to have a shortage of cash for the next several years, how would this affect your response to Part a?

c. If you expect to have a surplus of cash for the next several years and have some investment opportunities which will earn an estimated 8%, how will this affect your repayment plans?

3-4. Several different repayment plans were suggested in this chapter:

Plan A. Repay the principal with uniform annual install-ments and make annual interest payments on the unpaid balance.

Plan B. Pay a uniform annual sum so that interest and principal will be extinguished at the end of the specified period.

Plan C. Pay interest annually and repay principal at the end of the specified period.

Plan D. Repay entire principal and accrued interest at the end of the specified period.

Give an example of each plan, that is, name a type of security, investment, mortgage, and so forth, in which the repayment plan is employed.

4

Derivation of Interest Formulas

Time Value of Money

CAPITAL EXPENDITURE decisions generally involve a present investment of money based on the prospect of greater future receipts. Analysis of the economic advisability of such investment requires comparison of sums of money (cash flows) at various points in time. In turn this means we need methods which will enable us to take into account the time value of money (interest)[1] and the nonuniform purchasing power of money (inflation/deflation). In later chapters the effects of inflation/deflation will be considered; up to that time our problems will

---

[1] Interest concepts have not always enjoyed their present respectability. The concept of interest on money was rejected by Christians (prior to 1550) who equated interest and the usury prohibited in the Bible, by Aristotle who called money a barren thing incapable of reproducing itself, and by Marx who considered interest on money as a cause of inequality. In *The Merchant of Venice*, (Act I, Scene III) *c.* 1598, William Shakespeare wrote:

> *He lends out money gratis and brings down*
> *The rate of usance here with us in Venice,*
> *. . . he rails . . . on . . . well-won thrift,*
> *Which he calls interest.*

not deal with the impact of changing purchasing power levels except as a nonquantitative factor.

This chapter deals with the time value of money and the mathematical basis of interest formulas which permit conversion of dollars at a given point in time to an equivalent amount at some other point in time.

Interest Formula Symbols

Let i = interest rate per period

n = number of time periods

a = the periodic sum (usually annual) in a uniform series of sums discretely flowing at the end of each of n periods

f = a future sum discretely flowing at the end of the nth period

p = a present sum

g = gradient, the periodic increase in an arithmetically increasing series of sums discretely flowing at the end of each of n periods such that the periodic sums are $0g$, $1g$, $2g$, ..., $(n-1)g$

r = the rate of change in a geometrically increasing or decreasing series

c = a sum discretely flowing at the end of the first period of a series of sums of $c, c(1+r), c(1+r)^2, ..., c(1+r)^{n-1}$

Interest Formulas Relating Present, Future, and Periodic Sums

If p dollars are deposited now in an account earning i% per period, the account will grow to $p(1+i)$ by the end of one period, and by the end of two periods the account will be $p(1+i)(1+i)$; at the end of n periods the account will have grown to a future sum f, as given by

$$f = p(1+i)^n$$

so

$$f/p = (1+i)^n \qquad (4\text{-}1)$$

The cash flow diagram is

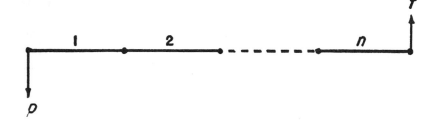

Where p, i, and n are known, f may be computed. The ratio of f to p is called *future worth of a present sum* and is denoted $(f/p)^i_n$. The reciprocal, *present worth of a future sum*, is denoted $(p/f)^i_n$.

$$p/f = 1/(1 + i)^n \qquad (4\text{-}2)$$

The cash flow diagram is the same as for Equation 4-1. Where f, i, and n are known, p may be computed.

If a dollars are deposited at the end of each period for n periods in an account earning $i\%$ per period, the future sum f accrued at the end of nth period is

$$f = a[1 + (1 + i) + (1 + i)^2 + \cdots + (1 + i)^{n-1}]$$

Multiply by $(1 + i)/a$:

$$(f/a)(1 + i) = [(1 + i) + (1 + i)^2 + \cdots + (1 + i)^n]$$

Subtract the equation below from the equation above:

$$(f/a) = [1 + (1 + i) + (1 + i)^2 + \cdots + (1 + i)^{n-1}]$$

and the result is

$$(f/a)i = (1 + i)^n - 1$$

or

$$f/a = [(1 + i)^n - 1]/i \qquad (4\text{-}3)$$

The cash flow diagram is

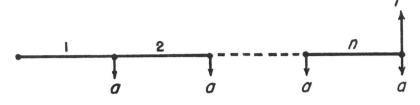

Where a, i, and n are known, f may be computed.

The ratio of f to a is called the *future worth of a uniform series* and is denoted $(f/a)_n^i$. The reciprocal, *uniform series worth of a future sum* or *sinking fund factor*, is denoted $(a/f)_n^i$.

$$a/f = i/[(1 + i)^n - 1] \tag{4-4}$$

The cash flow diagram is the same as for Equation 4-3; where f, i, and n are known, a may be computed.

When the present sum p rather than the future sum f is desired, Equations 4-2 and 4-3 can be combined to yield

$$p/a = (p/f)\,(f/a) = \left[\frac{1}{(1 + i)^n}\right]\left[\frac{(1 + i)^n - 1}{i}\right]$$

so

$$p/a = [(1 + i)^n - 1]/[i(1 + i)^n] \tag{4-5}$$

The cash flow diagram is

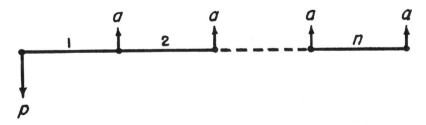

Where a, i, and n are known, p may be computed. The ratio of p to a is called the *present worth of a uniform series* and is denoted $(p/a)_n^i$. The reciprocal, *uniform series worth of a present sum* or *capital recovery factor*, is denoted $(a/p)_n^i$.

$$a/p = [i(1 + i)^n]/[(1 + i)^n - 1] \tag{4-6}$$

The cash flow diagram is the same as for Equation 4-5; where p, i, and n are known, a may be computed.

The six interest formulas just derived are compared in Table 4-1. The formulas have been solved for selected values of n and i and are given in Appendix I. Note that the ratio labels suggest their own application; thus when given a future sum and solving for its equivalent present sum,

$$p = f(p/f)$$

Alternatively, one can employ the ratio having the unknown amount as numerator and known amount as denominator.

TABLE 4-1. Interest Formulas Relating Present, Future, and Periodic Sums

To Find	Given	Equation Number and Equation	Formula Name	Use	Column # of Interest Tables in Appendix
f	p	(4-1) $f/p = (1 + i)^n$	Future worth of a present sum	Find future sum equivalent to present sum	7
p	f	(4-2) $p/f = \dfrac{1}{(1 + i)^n}$	Present worth of a future sum	Find present sum equivalent to future sum	4
f	a	(4-3) $f/a = \dfrac{(1 + i)^n - 1}{i}$	Future worth of a uniform series	Find future sum equivalent to uniform series of end-of-period sums	8
a	f	(4-4) $a/f = \dfrac{i}{(1 + i)^n - 1}$	Uniform series worth of a future sum or Sinking fund factor	Find uniform series end-of-period sum equivalent to a future sum	1
p	a	(4-5) $p/a = \dfrac{(1 + i)^n - 1}{i(1 + i)^n}$	Present worth of a uniform series	Find present sum equivalent to uniform series of end-of-period sums	5
a	p	(4-6) $a/p = \dfrac{i(1 + i)^n}{(1 + i)^n - 1}$	Uniform series worth of a present sum or Capital recovery factor	Find uniform series end-of-period sum equivalent to a present sum	2

Interest Formulas for Periodic Sums Increasing by an Amount g per Period

Some engineering economy problems involve receipts or disbursements that increase each period by a constant amount. Maintenance and repair expenses on specific equipment may increase by a relatively constant amount of change, g, each period. Even when moderate fluctuations from the exact model

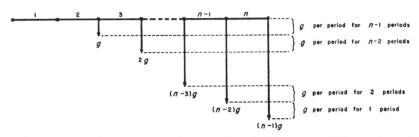

Fig. 4-1. Cash flow diagram for a series of disbursements increasing at the constant amount of change, g dollars per period.

of change are expected in practice, the model may be valuable as a convenient estimate of reality.

Figure 4-1 is a cash flow diagram of a series of end-of-period disbursements increasing at the constant amount of change, g dollars per period. For some amount of change, g, to exist there must be a minimum of two periods over which this amount of change is observed; hence the end-of-period-one amount is zero, while there are g dollars at the end of period two, and $(n - 1)g$ dollars at the end of period n.

This model is appropriate where there is a constant *amount* of change per period, and therefore the growth is *linear* and the progression is *arithmetic*.

The future sum (at the end of the nth period) equivalent to the gradient series shown in Figure 4-1 is

$$f = g\{(f/a)_{n-1}^i + (f/a)_{n-2}^i + \cdots + (f/a)_2^i + (f/a)_1^i\}$$

$$f = g\left\{\left[\frac{(1+i)^{n-1} - 1}{i}\right] + \left[\frac{(1+i)^{n-2} - 1}{i}\right]\right.$$

$$\left. + \cdots + \left[\frac{(1+i)^2 - 1}{i}\right] + \left[\frac{(1+i) - 1}{i}\right]\right\}$$

or

$$fi/g = (1+i)^{n-1} + (1+i)^{n-2} + \cdots + (1+i)^2 + (1+i) - n + 1$$

In deriving Equation 4-3 it was shown that

$$f/a = (1+i)^{n-1} + (1+i)^{n-2} + \cdots + (1+i)^2 + (1+i) + 1$$

so substituting

$$fi/g = (f/a) - n$$

or

$$f/g = 1/i \, (f/a - n) \qquad (4\text{-}7)$$

or

$$f/g = \frac{1}{i} \left[\frac{(1 + i)^n - 1}{i} - n \right] \qquad (4\text{-}7a)$$

The ratio of f to g is called the *future worth of a gradient series*, is denoted $(f/g)_n^i$, and is used to find the future sum equivalent to a gradient series of end-of-period sums. Multiplication of both sides of Equation 4-7 by a/f yields

$$(f/g) \, (a/f) = 1/i \, [(f/a) \, (a/f) - n \, (a/f)]$$

Simplifying,

$$a/g = 1/i \, [1 - n \, (a/f)] \qquad (4\text{-}8)$$

or

$$a/g = (1/i) - n/[(1 + i)^n - 1] \qquad (4\text{-}8a)$$

The ratio of a to g is the *uniform series worth of a gradient series*, is denoted $(a/g)_n^i$, and is used to find the uniform series of end-of-period sums equivalent to a gradient series of end-of-period sums. Multiplication of both sides of Equation 4-7 by p/f yields

$$(f/g) \, (p/f) = 1/i \, [(f/a) \, (p/f) - n \, (p/f)]$$

Simplifying,

$$p/g = 1/i \, [(p/a) - n \, (p/f)] \qquad (4\text{-}9)$$

or

$$p/g = \frac{1}{i} \left[\frac{(1 + i)^n - 1}{i(1 + i)^n} - \frac{n}{(1 + i)^n} \right] \qquad (4\text{-}9a)$$

The ratio of p to g is the *present worth of a gradient series*, is denoted $(p/g)_n^i$, and is used to find the present sum equivalent to a gradient series of end-of-period sums.

The three interest formulas involving a gradient just derived are compared in Table 4-2. The formulas have been solved for selected values of n and i and are given in Appendix I. Example 5-5 provides a sample application. Selected limits appear in Table 4-3.

TABLE 4-2. Interest Formulas Involving the Gradient Series

To Find	Given	Equation Number and Equation	Formula Name	Use	Column # of Interest Tables in Appendix
f	g	(4-7a) $f/g = \dfrac{1}{i}\left[\dfrac{(1+i)^n - 1}{i} - n\right]$	Future worth of a gradient series	Find future sum equivalent to gradient series of end-of-period sums	9
a	g	(4-8a) $a/g = \dfrac{1}{i} - \dfrac{n}{(1+i)^n - 1}$	Uniform series worth of a gradient series	Find uniform series end-of-period sum equivalent to a gradient series of end-of-period sums	3
p	g	(4-9a) $p/g = \dfrac{1}{i}\left[\dfrac{(1+i)^n - 1}{i(1+i)^n} - \dfrac{n}{(1+i)^n}\right]$	Present worth of a gradient series	Find present sum equivalent to a gradient series of end-of-period sums	6

TABLE 4-3. Limits of Interest Factors

Column # of Interest Tables in Appendix	Factor	Limit as $n \longrightarrow \infty$	Limit as $i \longrightarrow 0$	Limit as $i \longrightarrow \infty$
1	a/f	0	$1/n$	0
2	a/p	i	$1/n$	∞
3	a/g	$1/i$	$(n-1)/2$	0
4	p/f	0	1	0
5	p/a	$1/i$	n	0
6	p/g	$1/i^2$	$(n^2 - n)/2$	0
7	f/p	∞	1	∞
8	f/a	∞	n	∞
9	f/g	∞	$(n^2 - n)/2$	∞

Interest Computations for Periodic Sums Changing by a Rate, r, per Period

Some engineering economy problems involve receipts or disbursements that increase (decrease) each period by a constant rate of growth (decline). Revenues from sale of electric

energy by an electric utility company may increase by a relatively constant rate of change, $r\%$, each period as a result of
annual growth in number of customers served and power usage
per customer. Even when moderate fluctuations from the exact
model of change are expected in practice, the model may be
valuable as a convenient estimate of reality.

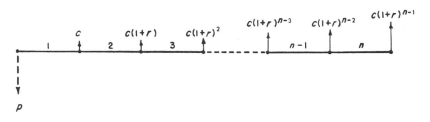

This model is appropriate where there is a constant *rate* of
change per period, and therefore the growth is *exponential* and
the progression is *geometric*. The gradient model treated earlier
is appropriate where there is a constant *amount* of change each
year, growth is *linear*, and the progression is *arithmetic*. The
distinction between the two models is essential to proper
application.

The present investment equivalent to the series of receipts
increasing at the constant rate, r, per period when valued at an
interest rate, i, is

$$p = c \left[\frac{1}{(1 + i)} + \frac{(1 + r)}{(1 + i)^2} + \cdots + \frac{(1 + r)^{n-1}}{(1 + i)^n} \right]$$

$$= \frac{c}{(1 + r)} \left[\frac{(1 + r)}{(1 + i)} + \frac{(1 + r)^2}{(1 + i)^2} + \cdots + \frac{(1 + r)^n}{(1 + i)^n} \right]$$

Let

$$1 + x = (1 + r)/(1 + i) \text{ when } r > i$$

then

$$p = [c/(1 + r)] [(1 + x) + (1 + x)^2 + \cdots + (1 + x)^n]$$
$$= [c/(1 + r)] (1 + x) [1 + (1 + x) + \cdots + (1 + x)^{n-1}]$$

In developing Equation 4-3, it was shown that

$$[1 + (1 + x) + (1 + x)^2 + \cdots + (1 + x)^{n-1}] = \frac{(1 + x)^n - 1}{x} = (f/a)_n^x$$

so

$$p = \frac{c}{(1 + i)} (f/a)_n^x \text{ when } r > i \text{ and } \frac{1+r}{1+i} = 1 + x$$

Let

$$1/(1 + x) = (1 + r)/(1 + i) \text{ when } r < i$$

then

$$p = \frac{c}{(1 + r)} \left[\frac{1}{(1 + x)} + \frac{1}{(1 + x)^2} + \cdots + \frac{1}{(1 + x)^n} \right]$$

It can be shown, as it was similarly with Equation 4-3, that

$$\left[\frac{1}{(1 + x)} + \frac{1}{(1 + x)^2} + \cdots + \frac{1}{(1 + x)^n} \right] = \frac{(1 + x)^n - 1}{x(1 + x)^n} = (p/a)_n^x$$

so

$$p = \frac{c}{1 + r} (p/a)_n^x \text{ when } r < i \text{ and } \frac{1+r}{1+i} = \frac{1}{1+x}$$

Summarizing:

*When $r > i$ and $\dfrac{1+r}{1+i} = 1 + x$, then $p/c = \dfrac{1}{(1 + i)} (f/a)_n^x$ (4-10)

When $r < i$ and $\dfrac{1+r}{1+i} = \dfrac{1}{1+x}$, then $p/c = \dfrac{1}{(1 + r)} (p/a)_n^x$ (4-11)

*When $r = i$, then $p/c = n/(1 + r) = n/(1 + i)$ (4-12)

---

*Where growth rate r equals or exceeds valuation rate i, the limit is finite only if n is finite.

The r and i for a given ratio such as $^r(p/c)_n^i$ will be given as a presuperscript and postsuperscript, respectively, and in a problem where $r = 5\%$ and $i = 10\%$ would be given as $^{5\%}(p/c)_n^{10\%}$. Note that the change rate r may be negative as well as positive. Note too that the first end-of-period amount is c and may be thought of as a "base period" upon which the rate of growth or decline is referenced.

Without Equations 4-10, 4-11, and 4-12 additional tables would be necessary since the tables would require the specific values of i and r rather than the simple generalized ratio, $(1 + r)/(1 + i)$. Example 5-10 illustrates application of these equations.

By using the tables of factors already formulated, the number of tables required is held to a minimum. When x as calculated for Equations 4-10 and 4-11 is not tabled as such, interpolation between available tables produces an approximation of sufficient accuracy for the problems dealt with here.

Compounding Frequency

In most economy studies interest is accounted for as if compounding occurs once per year. In practice the interest computation may take place more frequently, so it is important to note the effects of compounding frequency and to treat properly those problems where the assumption of annual compounding of interest is not appropriate. Table 4-4 presents some examples of compounding frequencies.

TABLE 4-4. Examples of Various Interest Compounding Frequencies

Frequency of Compounding	Example*
Annual	Certain long-term loans having a lump-sum repayment provision
Semiannual	U.S. savings bonds, series E, credit union deposits
Quarterly	Savings accounts
Monthly	Real estate mortgage loans, credit union deposits
Weekly or daily	Certain loans which are very large scale and/or short term
Continuous	Business cash flow and reinvestment could be treated as if all money invested would earn a return from the moment it first is in the hands of the business

*This is not to imply that compounding frequency is invariable for the example given.

To demonstrate the effects of compounding frequency:

Let f = payment due one year from now
i_a = effective rate of interest per year
i = rate of interest per compounding period
m = number of compounding periods per year
p = present sum borrowed

It follows that

im = nominal interest rate per year
n/m = number of years

Next, assume that a present sum, p, is borrowed for one year with a stated (nominal) annual interest rate of $im\%$ per year. At the end of the year the outstanding principal plus interest is

$$f = p(1 + i)^m$$

Thus

$$\text{Effective annual rate of interest} = (f - p)/p = i_a$$
$$= (1 + i)^m - 1 \quad \text{(4-13a)}$$

so

$$\text{Nominal annual rate of interest} = im = m[(1 + i_a)^{1/m} - 1]$$
$$\text{(4-13b)}$$

and

$$\text{Nominal periodic rate of interest} = i = (1 + i_a)^{1/m} - 1 \quad \text{(4-13c)}$$

For any given nominal interest rate ($im > 0$) per year, the effective interest rate is dependent upon the compounding frequency (which may or may not coincide with the cash flow frequency discussed next). Given a nominal interest rate of 6% per year, various compounding frequencies may be substituted in Equation 4-13a to find the equivalent annual rate of interest shown in Table 4-5 for $m < \infty$. When m increases without limit, $1/i$ increases without limit, the compounding is continuous, and the effective annual rate of interest is

$$i_a = (1 + i)^m - 1 = (1 + i)^{im/i} - 1$$

The base of natural logarithms ($e = 2.718\ldots$) can be expressed

$$e = (1 + i)^{1/i} \text{ as } 1/i \text{ increases without limit}$$

So when compounding is continuous ($i \rightarrow 0$, $m \rightarrow \infty$, but im is constant),

$$\left(\begin{array}{l}\text{Effective annual}\\\text{rate of interest}\end{array}\right) = i_a = e^{im} - 1 \qquad \text{(4-14a)}$$

and

$$\left(\begin{array}{l}\text{Nominal annual}\\\text{rate of interest}\end{array}\right) = im = \ln(1 + i_a) \qquad \text{(4-14b)}$$

TABLE 4-5. Compounding Frequency and Its Impact upon Effective Annual Interest Rate (Nominal Annual Rate = 6% = *im*)

Frequency of Compounding	*m*	*i*	Effective Annual Rate of Interest, i_a
			percent
Annual	1	6	6.00
Semiannual	2	3	6.09
Quarterly	4	1.5	6.14
Monthly	12	0.5	6.17
Continuous	$\longrightarrow \infty$	$\longrightarrow 0$	6.18

Cash Flow Frequency and Patterns

The actual flows of cash in a venture tend to have a wide range of frequency, uniformity, and regularity. Cash receipts from sales may tend to be continuously received during each moment of operation (thus frequency is continuous), yet the receipts may be nonuniform over the year (consider the sales of the retail store where pre-Christmas sales virtually match sales for the remainder of the year) even though they occur regularly. Disbursements such as those for repairs may be infrequent, nonuniform, and quite irregular. Some typical cash flows and frequencies appear in Table 4-6.

The task of building a mathematical model which would exactly account for each of the appropriate cash flows generally requires an uneconomic amount of analytic effort. Generally all cash flows are treated as if they occur as a lump-sum end-of-year amount. The lump-sum end-of-year approximation is frequently defended on the grounds that the error introduced is generally negligible, and the cost of greater reality in the model is generally not warranted. Somewhat comparable defenses

TABLE 4-6. Examples of Various Cash Flow Frequencies

Frequency of Cash Flow	Example
Irregular	Repair payments to noncompany personnel*
	Repair costs on a specific property
	Receipts from sale of used plant or equipment
Less frequent than annual	Property insurance payments
Annual	Licensing fees on vehicles
Semiannual	Property tax payments
	Interest payments to bondholders
Quarterly	Dividends paid to stockholders
	FICA payments to U.S. Government
	Income tax payments by corporations
Monthly	Merchandise payments to creditors
	Salaries paid
	Utility service payments
Weekly	Wages paid
Daily	Daily bank deposit of receipts by store
	Receipt by Rent-All agency on a specific item
Continuous	Retail store receipts

*Wages paid to company repair personnel may be quite regular.

could be offered for the continuous cash flow approximation. Our preference for end-of-year rather than continuous cash flow formulation is based on the former's conceptual ease and consequent understandability to a wider range of readers.

The two models tend to represent opposite extremes in the frequency of cash flow. Neither, of course, describes exactly the cash flow which is nonuniform, irregular, or cyclic. In any case, however, the relationship of a continuous cash flow to a lump-sum cash flow is of consequence.

Interest Formulas Relating Continuous and End-of-Year Cash Flows

The relationship of continuous cash flows to end-of-year cash flows can be demonstrated as follows. Note that a, f, g, p, i, n, r, c, and m are as previously defined:

Let a' = the periodic sum (usually annual) in a uniform series of sums continuously flowing during each of n periods

f' = a future sum continuously flowing during the nth period

g' = gradient, the periodic increase in an arithmetically increasing series of sums continuously flowing through each of n periods such that the periodic sums are 0, 1g', 2g', ..., $(n-1)g'$

p' = a present sum continuously flowing over the period just ended

c' = a sum continuously flowing during the first period of a series of sums of $c', c'(1+r), ..., c'(1+r)^{n-1}$

The cash flow diagram for continuous cash flows will be drawn so that the area shown represents cash flow for the year. A continuous cash flow of $a' = \$1$ per year could be thought of as \$1/365 per day, or \$1/52 per week, or still other combinations so that dimensions of the rectangular area are meaningful only in a relative way.

The cash flow diagram below is illustrative and also emphasizes that the continuous flow as defined for the series changing at rate r is not a continuous *function:*

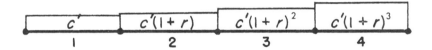

When the cash flows treated are annual, it follows from the definition that:

$$a'/m = \text{size of each receipt or disbursement during each of } n \text{ years}$$
$$f'/m = \text{size of each receipt or disbursement during the } n\text{th year}$$
$$(x-1)\,g'/m = \text{size of each receipt or disbursement during the } x\text{th year}$$
$$p'/m = \text{size of each receipt or disbursement during the year just ended}$$
$$(1+r)^{x-1}\,c'/m = \text{size of each receipt or disbursement during the } x\text{th year}$$

The end-of-year sum equivalent for each year of the continuous flow a' is

$$a = \frac{a'}{m}\left[\frac{(1+i)^m - 1}{i}\right]$$

The end-of-year sum equivalent to the continuous flow f' during the year n is

$$f = \frac{f'}{m}\left[\frac{(1+i)^m - 1}{i}\right]$$

The end-of-year sum equivalent to the continuous flow $g'(x-1)$ during the xth year of a gradient series is

$$g(x-1) = \frac{(x-1)g'}{m}\left[\frac{(1+i)^m - 1}{i}\right]$$

Since the $(x-1)$ term is common to both sides of the equation, the expression is a general one applicable to any and all years of a gradient series.

A cash flow diagram of p and p' may be helpful in emphasizing the timing of p':

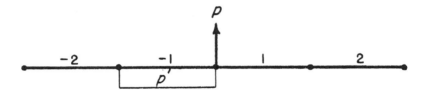

The end-of-year equivalent to the continuous flow p' during the year just ended is

$$p = \frac{p'}{m}\left[\frac{(1+i)^m - 1}{i}\right]$$

The end-of-year equivalent to the continuous flow $c'(1+r)^{x-1}$ during the xth year of a series changing by a rate of r per year is

$$c(1+r)^{x-1} = \frac{c'(1+r)^{x-1}}{m}\left[\frac{(1+i)^m - 1}{i}\right]$$

Since the $(1+r)^{x-1}$ term is common to both sides of the equation, the expression is a general one applicable to any and all years of a series changing by a rate of r per year. See the cash flow diagram appearing with the definition of c' to again note that the *function* as treated is not a continuous one.

Rearranging the terms of the five preceding equations and

substituting,

$$a/a' = f/f' = g/g' = p/p' = c/c' = \left[\frac{(1 + i)^m - 1}{im}\right] = i_a/im$$

$$= \frac{\text{effective annual rate of interest}}{\text{nominal annual rate of interest}}$$

and because compounding is continuous,

$$(a/a') = (f/f') = (g/g') = (p/p') = (c/c')$$

$$= \frac{\text{effective annual rate of interest}}{\text{nominal annual rate of interest}} = (e^{im} - 1)/im$$

$$= i_a/[\ln(1 + i_a)] \tag{4-15}$$

So, for example, at $i_a = 10\%$

$$\frac{a'}{a} = \frac{f'}{f} = \frac{g'}{g} = \frac{p'}{p} = \frac{c'}{c} = \frac{\ln(1 + i_a)}{i_a} = \frac{0.09531}{0.10} = 0.9531$$

Based on this calculation we may state that when $i_a = 10\%$, a lump-sum end-of-year amount of $10,000 is equivalent to $9531 flowing continuously throughout the year. Based on our equation and its reciprocal we can produce Appendix E, which permits us to convert between continuous and lump-sum end-of-year cash flows at various effective rates of interest. These same conversion factors appear on the right-hand side of the tables of Appendix I. Example 5-9 illustrates their application and provides a simple approximation that is helpful in understanding the conversion of continuous cash flows.

For a given interest rate a single factor permits us to convert conveniently from continuous cash flow to equivalent end-of-year cash flow (or vice versa), and additional tables for analysis of continuous cash flows are unnecessary. For example:

$$(p/a') = (p/a)(a/a')$$

This emphasis on the relationship of certain variables seems advisable; it is possible to produce tables of as many as 80 different ratios by pairing of a, a', f, f', g, g', p, p', c, and c' (not including the constant ratios a/a', f/f', g/g', p/p', c/c', or their inversions). Proliferation of tables for 80 different ratios is felt to generate more confusion than convenience; the tables of nine ratios and Equations 4-10, 4-11, 4-12, and 4-15 can be used directly or in combination to produce the same results.

The reader should be cautious in applying the continuous-compounding interest tables of other books because some are based on an integer *nominal* annual interest rate. Too often unlabeled, these tables can be recognized by the equations on which they are based; each equation involves e^{imn}. Because they are based on an integer nominal annual interest rate, such tables are not compatible with tables involving such functions as $(1 + i)^n$.

Compounding Frequency When Not Specified

In this book and in most others dealing with compounding frequency, it may be assumed that compounding frequency is annual ($m = 1$) unless otherwise specified. In Equation 4-13 it should be apparent that when $m = 1$, $i_a = i$. Because they are thus so often interchangeable and are not uniformly defined among the various textbooks which refer to them, the reader is cautioned to use special care in analysis when $m \neq 1$. Note too that when $m = 1$, the number of periods, n, is identical to the number of years, n/m.

Application of Interest Tables

In banks, insurance companies, and other financial institutions, money computations are generally required by law or business practices to be precise to the nearest penny. For a large sum of money this may mean 8-, 10-, or 12-place accuracy. Such accuracy certainly calls for tables and computational methods or devices of comparable precision. Capital expenditures, unlike bank loans, tend to incorporate many estimates; even installed cost of a facility may ultimately prove different from the original estimate, and certainly estimates in the more distant future such as those for salvage, output capacity, life, maintenance costs, operating costs, property taxes, and utilization percentage cannot be expected to be precise. Because we expect the tables in Appendix I to be applied to estimates, the tables are not carried out to the 8-, 10-, or 12-place accuracy which might permit application for the more exacting purposes of the financial institution, and slide rule accuracy is acceptable for virtually all economy analyses.

Examples illustrating use of the interest formulas are given in Chapter 5. In those examples and in the problems beginning with these of Chapter 4, the notation for an interest formula will show the interest rate as a superscript and the number of periods as a subscript, for example, $(p/f)_{10}^{6\%}$ can be read as "the present worth of a future sum ten periods hence with interest at 6% per period." As noted earlier, where geometric growth is involved the presuperscript will be r and the postsuperscript will be i, and $^{4\%}(p/c)_{12}^{8\%}$ can be read as "the present worth of a 12-period series of sums increasing at a rate of 4% per period and being valued at 8% per period.

Problems

4-1. Refer to the compound interest tables of a book such as (1) *Standard Mathematical Tables*, Chemical Rubber Publishing Company, or (2) *Financial Compound Interest and Annuity Tables*, Financial Publishing Company, to find:

   a. *Present worth of a future sum* factor where $n = 40$, $i = 1\frac{1}{4}\%$.

   b. *Future worth of a uniform series* factor where $n = 100$, $i = \frac{3}{4}\%$.

   c. *Capital recovery* factor where $n = 36$, $i = \frac{7}{12}\%$.

4-2. Using either logarithms or slide rule, find:

   a. *Capital recovery* factor when $n = 300$, $i = \frac{5}{12}$ of 1%.

   b. *Present worth of a uniform series* factor where $n = 20$, $i = 17\%$.

   c. *Future worth of a present sum* factor where $n = 20$, $i = 35\%$.

4-3. Using slide rule or log tables, evaluate the following factors:

   a. $(f/p)_{10}^{11\%}$ c. $(g/p)_{10}^{6\%}$

   b. $(g/a)_{2}^{50\%}$ d. $(a/p)_{5}^{16\%}$

4-4. Given $(f/a)_{10}^{17\%} = 22.39$ and the equations of this chapter, find:

   a. $(a/f)_{10}^{17\%}$ d. $(p/f)_{10}^{17\%}$ g. $(a/g)_{10}^{17\%}$

   b. $(a/p)_{10}^{17\%}$ e. $(f/p)_{10}^{17\%}$ h. $(p/g)_{10}^{17\%}$

   c. $(p/a)_{10}^{17\%}$ f. $(f/g)_{10}^{17\%}$

4-5. Using x as a beginning-of-period series rather than a as an end-of-period series, derive interest formulas for:

a. $(x/f)_n^i$ c. $(x/g)_n^i$ e. $(f/x)_n^i$

b. $(x/p)_n^i$ d. $(p/x)_n^i$

4-6. Develop a method for determining the uniform series worth of a present sum (capital recovery) factor when given only a table of:

a. *Present worth of a uniform series* factors.

b. *Uniform series worth of a future sum (sinking-fund)* factors.

c. *Future worth of a present sum* factors.

d. *Future worth of a uniform series* factors.

Check your method by illustrating its application for $i = 10\%$, $n = 5$.

4-7. The population of the United States has grown at the rate of about 3% per year for many years. If this trend were to continue, how long would it take for the population to double?

4-8. Evaluate (a/f), (a/p), (a/g), (p/f), (p/a), (p/g), (f/p), (f/a), and (f/g) using $i = 100\%$ and $n = 4$. Show your result as a fraction rather than converting to the decimal equivalent.

4-9. Evaluate each of the following factors:

a. $(a/g)_\infty^{10\%}$ d. $(p/a)_\infty^{16\%}$ g. $(a/g)_5^{0\%}$

b. $(a/g)_{13}^{\infty\%}$ e. $(p/a)_5^{0\%}$ h. $(p/g)_5^{0\%}$

c. $(p/g)_\infty^{20\%}$ f. $(f/g)_5^{0\%}$

4-10. Prove that:

a. $(a/f)_n^i + i = (a/p)_n^i$

b. $(p/a)_n^i + (p/f)_{n+1}^i = (p/a)_{n+1}^i$

c. $(p/a)_x^i \, (p/f)_y^i = (p/a)_{x+y}^i - (p/a)_y^i$

d. $(f/a)_n^i + (f/p)_n^i = (f/a)_{n+1}^i$

4-11. Show that:

a. $\operatorname*{Lim}_{i \to 0} (a/p) = 1/n$ d. $\operatorname*{Lim}_{i \to 0} (a/g) = (n - 1)/2$

b. $\operatorname*{Lim}_{i \to 0} (f/a) = n$ e. $\operatorname*{Lim}_{i \to \infty} (p/a) = 0$

c. $\operatorname*{Lim}_{i \to 0} (p/g) = (n^2 - n)/2$ f. $\operatorname*{Lim}_{i \to \infty} (a/g) = 0$

4-12. Give an example, different from those listed, of each of the cash flow frequencies listed in Table 4-6.

4-13. a. Using a nominal annual interest rate of 12%, find the effective annual interest rate if compounding is (1) annual, (2) quarterly, (3) monthly, and (4) continuous.

   b. Find the nominal annual interest rate required to produce an effective annual interest rate of 12% if compounding is (1) annual, (2) quarterly, (3) monthly, and (4) continuous.

4-14. Assume $i_a = 6\%$ and a continuous cash flow of \$1 per year for x years:

   a. Convert this to a uniform end-of-year cash flow over the x years.

   b. Convert this to a uniform beginning-of-year cash flow over the x years.

4-15. Using the nine tables of Appendix I and the equations of this chapter, evaluate the following factors:

   a. $(a'/p')_{10}^{6\%}$

   b. $(a'/f')_{6}^{10\%}$

   c. $(f'/p)_{10}^{6\%}$

   d. $(a/p')_{3}^{6\%}$

   e. $(f/g')_{5}^{6\%}$

   f. $(p'/a)_{30}^{10\%}$

   g. $(p/f')_{10}^{10\%}$

   h. $(a'/p)_{20}^{8\%}$

4-16. Using the nine tables of Appendix I and the equations of this chapter, evaluate the following factors:

   a. $^{10\%}(p/c)_{22}^{10\%}$

   b. $^{10\%}(p/c)_{7}^{5\%}$

   c. $^{-20\%}(p/c)_{\infty}^{20\%}$

   d. $^{10\%}(p/c)_{16}^{32\%}$

   e. $^{30\%}(a/c)_{3}^{4\%}$

   f. $^{-4\%}(f/c)_{5}^{20\%}$

5

Application of Interest Formulas

Finding an Equivalent Cash Flow

THIS CHAPTER consists of examples that involve the shifting of cash flows from one date to another through application of the equations developed in Chapter 4. Such shifting is usually necessary to the process of identifying the "best" of a set of competing investment alternatives.

Example 5-1.   Time and Equivalent Cash Flows

  a. *Information:* It is expected that by December 31 of each of five years beginning December 31, 1971, annual amounts of $1000 will be available.

  Objective: Find the sum of money at December 31, 1970, equivalent to the $1000 per year series if $i = 6\%$.

  Analysis: Given $i = 0.06$, $n = 5$, $a = \$1000$, find p, the present worth as of December 31, 1970.

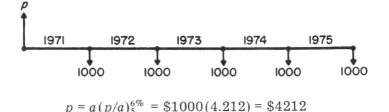

$$p = a(p/a)_5^{6\%} = \$1000(4.212) = \$4212$$

Notice that this is also the result we would show in response to the question, "What present sum would be loaned to a person who agreed to make payments of $1000 at the end of each of five years if $i = 6\%$?" or "What present investment would be required to produce $1000 at the end of each of the next five years?"

b. *Information:* As of December 31, 1970, $4212 is available.

Objective: Find the sum of money at December 31, 1977, equivalent if $i = 6\%$.

Analysis: Given $i = 0.06$, $n = 7$, $p = \$4212$, find f, the future worth as of December 31, 1977.

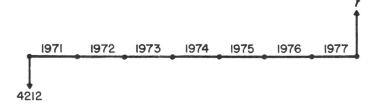

$$f = p(f/p)_7^{6\%} = \$4212(1.504) = \$6335$$

This result is also a proper response to "Using $i = 6\%$, find the amount to which $4212 will accumulate in seven years." or "A present expenditure of $4212 is justified by a saving of _____ seven years hence when $i = 6\%$."

c. *Information:* As of December 31, 1977, $6335 is available.

Objective: Find the annual sum of money for each of

four years beginning December 31, 1978, equivalent if $i = 6\%$.

Analysis: Given $i = 0.06$, $n = 4$, $p = \$6335$, find a, the annual series equivalent for four years.

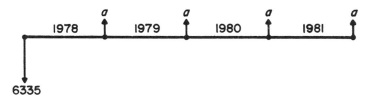

$$a = p(a/p)_4^{6\%} = \$6335(0.28859) = \$1828$$

Our result is also a proper response to "What end-of-year payment is required over each of the next four years to repay a loan of $6335 when $i = 6\%$?" or "What annual end-of-year withdrawals over the next four years will just exhaust a present balance of $6335 with $i = 6\%$?"

d. *Information:* It is expected that by December 31 of each of four years beginning December 31, 1978, annual amounts of $1828 will be available.

Objective: Find the sum of money at December 31, 1981, equivalent to the $1828 per year series if $i = 6\%$.

Analysis: Given $i = 0.06$, $n = 4$, $a = \$1828$, find f, the future worth as of December 31, 1981.

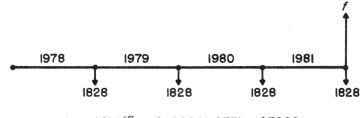

$$f = a(f/a)_4^{6\%} = \$1828(4.375) = \$7998$$

The result is also appropriate to "If $1828 is deposited at the end of each year into an account paying 6% interest, to what size has the account grown after the fourth deposit?" or "End-of-year withdrawals of $1828 per year from a very large account will reduce the amount to which that account would have grown by the end of year four. Use $i = 6\%$, to find the amount of reduction."

e. *Information:* As of December 31, 1981, $7998 is available.

Objective: Find the sum of money at December 31, 1975, equivalent to the $7998 if $i = 6\%$.

Analysis: Given $i = 0.06$, $n = 6$, $f = \$7998$, find p, the present worth as of December 31, 1975.

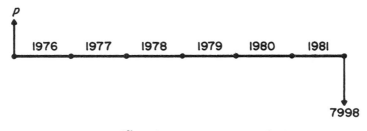

$$p = f(p/f)_6^{6\%} = \$7998(0.7050) = \$5639$$

This is also a correct response to "What present expenditure is warranted for an item that is expected to produce a saving of $7998 six years hence if $i = 6\%$?" or "What present deposit at 6% interest will grow to $7998 in six years?"

f. *Information:* As of December 31, 1975, $5639 is available.

Objective: Find the annual end-of-year sum of money for each of five years beginning December 31, 1971, equivalent if $i = 6\%$.

Analysis: Given $i = 0.06$, $n = 5$, $f = \$5639$, find a, the annual end-of-year series equivalent for five years.

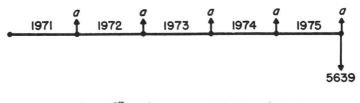

$$a = f(a/f)_5^{6\%} = \$5639(0.17740) = \$1000$$

This response is equally appropriate to "What annual end-of-year deposit will accrue to $5639 after the fifth deposit if $i = 6\%$?" or "A future receipt of $5639 five years hence is equivalent at $i = 6\%$ to an annual end-of-year series of a dollars per year for five years. Find a."

Our example has now gone full circle, having returned to the cash flows of $1000 per period given in Part a.

Example 5-2.   Solving for i

Information: A present loan of $1000 is to be repaid by payments of $400 at the end of each of the next three years.

Objective: Find the interest rate (rate of return) at which the cash flows are equivalent.

Analysis: Given $p = \$1000$, $a = \$400$, $n = 3$, find i.

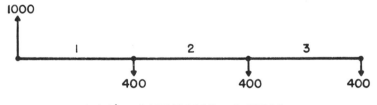

$$(a/p)_3^i = \$400/\$1000 = 0.40000$$

Use the tables to note that

$$(a/p)_3^{8\%} = 0.38803$$

and

$$(a/p)_3^{10\%} = 0.40211$$

By interpolation,

$$i = 8\% + 2\% \, (1197/1408) = 9.7\%$$

or

$$(p/a)_3^i = \$1000/\$400 = 2.500$$

Use the tables to note that

$$(p/a)_3^{8\%} = 2.577$$

and

$$(p/a)_3^{10\%} = 2.487$$

By interpolation,

$$i = 8\% + 2\% \ (77/90) = 9.7\%$$

Example 5-3. Solving for n

Information: End-of-year payments of $300 are made on a present loan of $1000 with interest at 10%.

Objective: Find the number of payments required to repay the loan. Is the last payment a full $300?

Analysis:

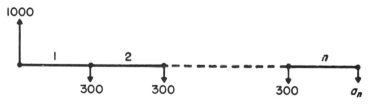

Given p = $1000, a = $300, i = 10%, find n.

$$(a/p)_n^{10\%} = \$300/\$1000 = 0.30000$$

From the tables,

$$(a/p)_4^{10\%} = 0.31547$$

and

$$(a/p)_5^{10\%} = 0.26380$$

From the above it can be seen that four full payments plus a fifth payment less than $300 will repay the loan. The fifth payment is

$$[\$1000 - \$300(p/a)_4^{10\%}]\,(f/p)_5^{10\%} =$$
$$[\$1000 - \$300(3.170)]\ 1.611 = \$78.94$$

Our result can be verified as follows:

End of Year	Principal Plus Interest at 10%	Less Payment	Unpaid Principal
1	$1100.00	$300.00	$800.00
2	880.00	300.00	580.00
3	638.00	300.00	338.00
4	371.80	300.00	71.80
5	78.98	78.98	0

The discrepancy of $0.04 in our result of $78.94 would disappear if we were to use tables of six or more significant digits. A last payment of nonuniform size occurs frequently in such long-term mortgages as on houses, because payments can be made only to the nearest penny even though precise tables indicate payments of fractional amounts. In practice the fractional amount is rounded to the higher penny and the last payment calculated similarly to that of this example.

Example 5-4.   Multifactor Solutions

Information: Withdrawals of $1000 each are desired at the end of years 20 through 23.

Objective: Find the present investment equivalent to the series above if $i = 5\%$.

Analysis:

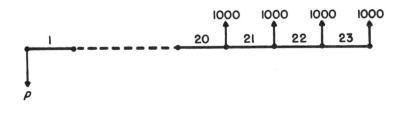

$$p = \$1000\,[(p/f)_{20}^{5\%} + (p/f)_{21}^{5\%} + (p/f)_{22}^{5\%} + (p/f)_{23}^{5\%}]$$
$$= \$1000\,(0.3769 + 0.3589 + 0.3418 + 0.3256)$$
$$= \$1403$$

or

$$p = \$1000\,(p/a)_{4}^{5\%}\,(p/f)_{19}^{5\%}$$
$$= \$1000\,(3.546)(0.3957)$$
$$= \$1403$$

or

$$p = \$1000\,(f/a)_{4}^{5\%}\,(p/f)_{23}^{5\%}$$
$$= \$1000\,(4.310)(0.3256)$$
$$= \$1403$$

or

$$p = \$1000\,[(p/a)_{23}^{5\%} - (p/a)_{19}^{5\%}]$$
$$= \$1000\,(13.489 - 12.085)$$
$$= \$1404$$

Example 5-5. **Patterns of Arithmetic Change: Gradient Formula Applications**

a. *Information:* Repair costs on Equipment A tend to follow the pattern of zero cost the first year, $100 at the end of year two, $200 at the end of year three, $300 at the end of year four, and $400 at the end of year five, the last year of operation.

Objective: Find the annual equivalent end-of-year repair cost if $i = 10\%$.

Analysis:

$$i = 0.10, g = \$100, n = 5, \text{find } a$$
$$a = g(a/g)_5^{10\%} = \$100(1.810) = \$181$$

b. *Information:* Repair costs on Equipment B tend to follow the pattern of $1000 cost at the end of year one, $1100 at the end of year two, $1200 at the end of year three, $1300 at the end of year four, and $1400 at the end of year five, the last year of operation.

Objective: Find the equivalent uniform annual end-of-year cost of repairs if $i = 10\%$.

Analysis:

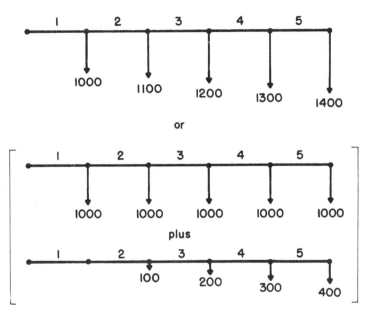

or

plus

Using the latter cash flow diagrams it is apparent that

$$a = \$1000 + \$100(a/g)_5^{10\%}$$
$$= \$1000 + \$100(1.810) = \$1181$$

c. *Information:* Repair costs on Equipment C tend to follow the pattern of $1000 cost at the end of year one, $900 at the end of year two, $800 at the end of year three, $700 at the end of year four, and $600 at the end of year five, the last year of operation.

Objective: Find the equivalent uniform annual end-of-year cost of repairs if $i = 10\%$.

Analysis:

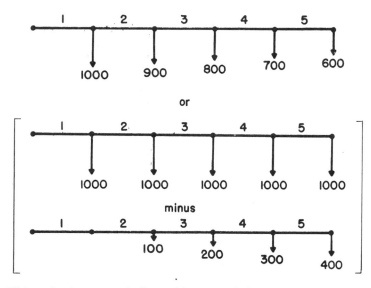

Using the latter cash flow diagrams, it is apparent that

$$a = \$1000 - \$100\,(a/g)_5^{10\%}$$
$$= \$1000 - \$100\,(1.810) = \$819$$

Example 5-6.   Treating a Change in Interest Rate

Information: Ten years ago today Mr. Black made his first annual deposit of $1000 into a fund paying 3% interest compounded annually. Two years ago the interest rate was increased to 4% on all funds.

Objective: Find the current worth of Mr. Black's deposits, the last of which was made today.

Analysis:

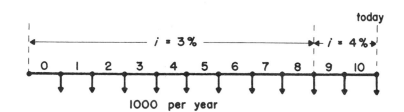

Add 1 to each age interval labeled above and note that there are 11 end-of-year deposits of $1000 each. Worth of fund after the ninth payment has been made:

$$= (f/a)_9^{3\%} \, \$1000$$
$$= (10.159) \, \$1000$$
$$= \$10,159$$

•

Worth of fund after eleventh payment has been made:

$$= \$10,159 \, (f/p)_2^{4\%} + \$1000 \, (f/p)_1^{4\%} + \$1000$$
$$= \$10,159 \, (1.082) + \$1000 \, (1.04) + \$1000$$
$$= \$10,992 + \$1040 + \$1000$$
$$= \$13,032$$

Example 5-7.   Cash Flows Less Frequent than Annual

Information: A series of payments of $1000 each ten years beginning now and lasting forever has been promised.

Objective: Find the present worth of the series of payments if $i = 10\%$.

Analysis:

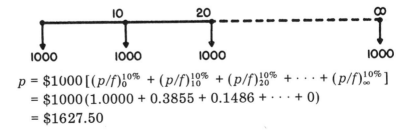

$$p = \$1000 \, [(p/f)_0^{10\%} + (p/f)_{10}^{10\%} + (p/f)_{20}^{10\%} + \cdots + (p/f)_\infty^{10\%}]$$
$$= \$1000 \, (1.0000 + 0.3855 + 0.1486 + \cdots + 0)$$
$$= \$1627.50$$

Or alternatively we might compute an annual equivalent and then take the present worth of the infinite series:

$$p = \$1000 \, (a/p)_{10}^{10\%} \, (p/a)_\infty^{10\%}$$
$$= \$1000 \, (0.16275)(10.0)$$
$$= \$1627.50$$

Or still another solution would be to convert the 10% annual interest to the effective interest rate per ten years:

$$= (1 + 0.10)^{10} - 1 = 2.594 - 1.000 = 1.594 \text{ or } 159.4\%$$

then

$$p = \$1000(p/a)_\infty^{159.4\%} + \$1000$$
$$= \$1000 (1/1.594) + \$1000$$
$$= \$1627.50$$

Note that for problems involving finite values of n this last solution would be much less convenient.

Example 5-8.   Cash Flows More Frequent than Annual

Information: Mr. Green is considering a certain automobile which can be purchased with a $1000 down payment now plus 24 monthly payments of $100 each. The first monthly payment would be due one month after closing the transaction. In a straight cash deal the automobile could be purchased for $3000.

Objective: Find the interest rate (rate of return) at which the two purchase plans (straight cash or monthly payments) would be equivalent.

Analysis: If the monthly payment plan is accepted in lieu of $2000 extra required in the straight cash deal, the cash flow diagram would be:

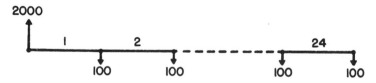

Note that $p = \$2000$, $a = \$100$, $n = 24$, and

$$(a/p)_{24}^i = \$100/\$2000 = 0.05000$$

Since $(a/p)_{24}^{1\%} = 0.04707$ and $(a/p)_{24}^{2\%} = 0.05287$,

$$i = 1\% + 1\% (293/580) = 1.51\% \text{ per month}$$

The nominal annual rate $= 12 (1.51\%) = 18.1\%$
The effective annual rate $= (1 + 0.0151)^{12} - 1$
$$= 0.197 = 19.7\%$$

Example 5-9.   Continuous Cash Flows Subject to Continuous Compounding

Information: It is estimated that a proposed equipment will reduce operating costs by $3000 per year for each of the next three years.   The savings flow continuously so are essentially equivalent to $12 per day for 250 working days per year, or $60 per week for 50 weeks per year, or $250 per month for 12 months per year. Estimated salvage value of the equipment is zero, and the before-tax rate of return required is 20%.

Objective: Find the maximum present expenditure for the proposed equipment justified by the estimated savings in operating costs.

Analysis:

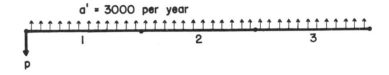

$$p = a' \, (a/a')^{20\%} \, (p/a)_3^{20\%} = \$3000(1.097)(2.106) = \$6931$$

Alternate Analysis: In lieu of the above exact solution, we could use the following approximation which may help one to visualize the conversion from a continuous cash flow to a discrete cash flow, or vice-versa.

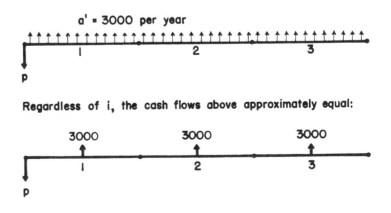

Convert the middle-of-year cash flows to end-of-year flows by allowing for a half-years interest, so for i = 20%:

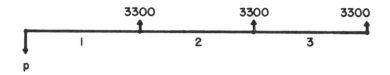

Notice that $3300/3000 = 1.100 = 1 + (i/2) \approx (a/a')^i$. Conversion from continuous cash flow to end-of-year cash flow, or vice-versa, is easily accomplished with this simple approximate factor, $1 + (i/2)$, or its reciprocal. The final result:

$$p = a'(a/a')^{20\%}(p/a)_3^{20\%} \approx \$3000(1.100)(2.106) = \$6950$$

Example 5-10.   Patterns of Geometric Change

a. *Information:* The prospective before-tax cash flows generated by Project A are: $800 at the end of year one, $1200 at the end of year two, $1800 at the end of year three, $2700 at the end of year four.

   Objective: Determine the maximum present expenditure justified if the before-tax rate of return required is 20%.

   Analysis:

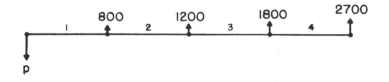

Note that the cash flows increase by 50% each year so $r = 0.50$; $c = \$800$. Since $r > i$, see Equation 4-10:

$$(1 + r)/(1 + i) = 1 + x = 1.50/1.20 = 1.25$$

so $x = 0.25 = 25\%$, and

$$^{50\%}(p/c)_4^{20\%} = [1/(1 + i)]\,(f/a)_4^{25\%}$$
$$= (1/1.20)(5.766)$$
$$= 4.805$$

so $p = \$800\ ^{50\%}(p/c)_4^{20\%} = \$800(4.805) = \$3844.$

b. *Information:* The prospective before-tax cash flows generated by Project B are: $5000 at the end of year one, $4000 at the end of year two, $3200 at the end of year three, $2560 at the end of year four.

Objective: Determine the maximum present expenditure justified if the before-tax rate of return required is 12%.

Analysis:

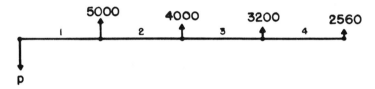

Note that the cash flows decrease by 20% each year so $r = -0.20$; $c = \$5000$. Since $r < i$, see Equation 4-11:

$$(1 + r)/(1 + i) = 1/(1 + x) = 0.80/1.12$$

so $x = 0.40 = 40\%$, and

$$^{-20\%}(p/c)_4^{12\%} = [1/(1 + r)]\,(p/a)_4^{40\%}$$
$$= (1/0.80)(1.849) = 2.311$$

so $p = \$5000\ ^{-20\%}(p/c)_4^{12\%} = \$11{,}555.$

Summary

The examples of this chapter illustrate a number of points which will be helpful in subsequent problems:

1. Interest problems can be phrased in many different ways; the student should familiarize himself with the various phrasings and should note that drawing the cash flow diagram is the first step in solving such problems.

2. A set of cash flows can be shifted from one point in time to another and then moved again; but if the shifting is performed properly, equivalence with the starting point is never lost.
3. There is frequently more than one combination of interest factors which may be applied to produce correct results.
4. Never add together the cash flows of various years without first adjusting for the time value of money.

Problems

5-1. Find the following equivalents using an interest rate of 12%:
   a. An amount on January 1, 1982, equivalent to equal annual amounts of $1000 from January 1, 1975, through January 1, 1982, inclusive.
   b. Equal annual amounts on January 1, 1983, through January 1, 1986, equivalent to amount (a).
   c. An amount on January 1, 1986, equivalent to amount (b).
   d. An amount on January 1, 1974, equivalent to amount (c).
   e. Equal annual amounts on January 1, 1975, through January 1, 1982, inclusive, equivalent to amount (d).
5-2. Find the following equivalents using an interest rate of 8%:
   a. A payment at the end of 20 years equivalent to $100 today.
   b. Equal annual year-end payments during the last 9 of the 20 years equivalent to payment (a).
   c. Payment at the beginning of period (b) equivalent to the nine equal annual year-end payments.
   d. Payment 14 years ago (25 years prior to the beginning of period (b)) equivalent to payment (c).
   e. Equal annual year-end payments during the last 14 years equivalent to payment (d).
   f. Payment today equivalent to payments (e).
5-3. Find how long it takes for a sum of money to double itself at each of the following interest rates: $i = 5\%, i = 10\%, i = 20\%, i = 40\%$.
5-4. A new layout of plant equipment is expected to produce end-of-year savings of $10,000 per year for the next five years. The cost of moving the equipment for the new layout is estimated as $25,000. Find the rate of return (interest) to the nearest 0.1% on the investment in the revised layout.

5-5. Assume in each instance below that $i = 10\%$:
   a. Maintenance costs on a certain equipment are estimated to be $500 at the end of year one, $700 at the end of year two, and to continue to increase by $200 per year. Find the equivalent uniform annual end-of-year cost if $n = 10$.
   b. Certain charges of $2000 in year one, $1800 in year two, and decreasing by $200 per year are expected. Find the equivalent uniform annual end-of-year charge if $n = 10$.
   c. Incomes of $250 at the end of year one, $350 at the end of year two, and increasing by $100 per year are expected. Find the present equivalent of this series of incomes if n is infinite.

5-6. Telephones in service have been increasing at the rate of 6% per year while U.S. population has grown at the rate of 2% per year. Assume that there are currently in service 80 million U.S. telephones and that the current U.S. population is 200 million. Using the growth rates given, estimate the telephones per capita 35 years from now.

5-7. You are considering the purchase of a home and have narrowed your alternatives to two houses which you feel are essentially equal. House A costs $15,000, requires a down payment of $5000, and can be financed by a 20-year 7% loan of $10,000. House B costs $16,000, requires a down payment of $5000, and has an existing $11,000 loan at 5% with 20 years yet to run which the buyer may assume. Assume end-of-year payments (in practice they are actually monthly).
   a. Which house should you buy?
   b. Are there any other factors you should look into?

5-8. You plan to provide a retirement fund for yourself by making end-of-year deposits of $1000 for each of the next 30 years into a fund which pays 4% interest. At the end of each year thereafter, you plan to withdraw $4000. How long will it take to exhaust the fund?

5-9. Why is the following an equality?

$$(p/a)_4^i \, [\$4000 - \$1000(a/g)_4^i \,] = \\ (p/a)_5^i \, [\$4000 - \$1000(a/g)_5^i \,]$$

(Hint: Draw a cash flow diagram for each side of the equation.)

5-10. The United States makes an interest-free loan of $20 million to a foreign nation which agrees to repay the loan with 20 equal end-of-year payments of $1 million each. Interest-free loans have been likened to subsidies. Find the present worth of the "subsidy" assuming the true cost of the money to the U.S. government is 4%.

5-11. The following cash flows occur at the end of years one through seven respectively:

$2000, $3000, $4000, $5000, $4000, $3000, $2000

Using $i = 10\%$, $(p/a)_n^i$ factors and *gradient series factors* only, find the present equivalent.

5-12. Mr. White borrows $20,000 at 6% compounded annually, agreeing to repay the loan in 20 equal annual payments. How much of the principal is still unpaid after he has made the eighth payment? Is his loan 40% "paid off?" Why?

5-13. On January 1, 1970, a corporation borrowed $100,000 at 6% interest, agreeing to repay the loan in equal annual payments of $10,000. The company ran into financial difficulties and was unable to make payments during the first two years. The loan was renegotiated and the company agreed to repay at the rate of $10,000 per year for all future payments, with the first payment on December 31, 1972. By what year should the loan have been repaid? Under the new terms when will the loan be completely repaid?

5-14. Mr. Blue hopes to accumulate $50,000 ten years from now through annual deposits beginning now and ending with a deposit ten years from now:
   a. Find the uniform annual deposit required if $i = 3\%$.
   b. Assume that after seven payments as computed in (a) have been made, the interest rate on all funds is increased to 5%. Find the size of payments required in the remaining years to produce the $50,000.

5-15. A borrower signs a note for $10,000 and agrees to repay the loan with end-of-year payments of $5000 each for the next two years. The lending agency "discounts" the note so the borrower "prepays" $1200 interest and he actually receives $8800 net. Find the actual rate of interest being paid by the borrower.

5-16. Beginning-of-year deposits are made in the hope of accumulating $10,000 by 20 years from today. The deposits begin now and cease with the deposit 20 years from today; the first deposit is $100 and each subsequent deposit is g dollars more than the one preceding. Find g if $i = 5\%$.

5-17. Current per capita use of natural gas in the United States is about 80,000 cubic feet per person per year. If the total consumption of natural gas were to continue increasing at the rate of 7% per year and population at the rate of 2% per year, what would be the annual per capita consumption 35 years from now?

5-18.   a. It is desired to invest a lump sum of money on a boy's sixth birthday to provide him $2000 on each birthday from the eighteenth to the twenty-second, inclusive. If interest of 4% can be obtained on a tax-exempt investment, what lump sum must be invested now?

   b. A man is to receive an annuity of $1000 a year for ten years, with payments beginning on March 1, 1992. He offers to sell the annuity on March 1, 1977. With interest at 5% compounded annually, what is a fair price for the annuity?

5-19.   An investor states that he has just sold a property for a profit of $10,000. He has computed his rate of return to be exactly 10% on the investment he made 20 years ago. What was his original investment? (Note: Neglect income tax and capital gains tax.)

5-20.   BBKLPW Enterprises borrowed $100,000. Of this amount, $50,000 was obtained at a cost of 4%, while the remaining $50,000 was obtained at a cost of 10%. Find the 20 equal end-of-year payments required if the company can apply their payments to the outstanding balance of the debt of their choice. (That is, they are required to repay both loans within the 20 years, but the sequence of repayment is at their discretion.)

5-21.   A 20-year-old student plans to allocate beginning-of-year amounts of $200 per year to some combination of life insurance and investment. He has investigated many alternative insurance and investment programs and has narrowed his choice to three alternatives:

   X. Invest $200 per year.
   Y. Invest $100 per year and pay net premiums of $100 per year on a $10,000 ordinary life insurance policy.
   Z. Pay net premiums of $200 per year on a $10,000 thirty-pay life insurance policy, and thereafter invest $200 per year.
   (Net premium = gross premium - dividend.)

Assume the money is invested so that an after-tax profit rate of 4% is realized.

   a. Compare the alternatives by making a graph on which the vertical axis shows the sum of "life insurance plus investment" and the horizontal axis shows the age of the student (suggested range: 20–80). Which will turn out to be the best of the three alternatives for the person who lives to be 50? 70? When is Z the best alternative? Why?

b. List several types of investments which would yield an after-tax profit rate of 4%.

c. What other alternatives besides X, Y, and Z should be considered?

d. How do "cash value" and "loan value" affect the comparison you made in (a) above?

e. There are various settlement options available in case of death of a policyholder. Option T is a lump-sum payment of $10,000 to the estate. Option U is a uniform end-of-year annual payment of $656.50 for the next 20 years. Suggest a basis on which to make a decision. Show your calculations.

5-22. Using an effective annual interest rate of 8%, find the present worth of the following cash flow streams:

a. $2000 now, $2000 five years from today, and repeating each five years so that service is provided over the next 50 years. There will be ten cash flows of $2000 each.

b. $4000 ten years from today and repeating each ten years forever.

5-23. Find the worth as of January 1, 1984, of deposits of $1000 each *six months* into an account which pays interest of 4% per annum with compounding *quarterly*. The first deposit occurs June 30, 1980, and the last deposit occurs June 30, 1983.

5-24. Find the effective annual interest rate, to the nearest 0.1%, for which there would be equivalence between:

a. A $10,000 present loan and 21 semiannual payments of $815 each. The first payment is due six months from now.

b. Ten deposits of $500 each three years beginning three years from today and one withdrawal of $10,000 thirty years from now.

5-25. You have borrowed $25,000 and are repaying the loan with 35 equal end-of-quarter payments. Immediately after you make payment #14 you decide to liquidate the loan. How much money is required to repay the remaining loan? Assume interest is computed quarterly using a nominal annual rate of 8%.

5-26. Normally, one would think of a checking account as paying no interest and therefore would deposit excess funds in a savings account. In a sense, however, interest is earned if larger balances result in lower service charges on the checking account as shown below:

Monthly Service Charge Schedule on Checking Accounts

Number of Checks	Average Monthly Balance				
	$200 to 299	$300 to 399	$400 to 499	$500 to 599	$600 to 699
21	$1.00	$ .75	$.50	$.25	NC
22	1.05	.80	.55	.30	NC
23	1.10	.85	.60	.35	NC
24	1.15	.90	.65	.40	NC
25	1.20	.95	.70	.45	NC
26	1.25	1.00	.75	.50	$.25
27	1.30	1.05	.80	.55	.30
28	1.35	1.10	.85	.60	.35
29	1.40	1.15	.90	.65	.40
30	1.45	1.20	.95	.70	.45

Based on the preceding table, compute the "interest" earned on funds in the checking account. Show this both as a nominal and effective annual rate.

5-27. A used-car dealer offers you a time payment plan as follows:

Automobile	$1000
6% interest (.06) ($1000)	60
Credit investigation fee	20
	$1080

Monthly payment = $1080/12 = $90 per month (12 months).

With cash, you could actually purchase this car for $900.
a. What monthly rate of interest is actually charged? (Interpolation of tables or use of logarithms may be necessary.)
b. What is the *nominal* annual rate of interest?
c. What is the *effective* annual rate of interest?

5-28. Acquisition of an industrial equipment costing $100,000 is being considered. Salvage value at the end of the equipment's estimated ten-year life is thought to be negligible. If the net savings in manpower costs are treated as continuous cash flows, what minimum weekly savings would be required to justify the expenditure, assuming $i_a = 15\%$?

5-29. Find the end-of-year uniform annual equivalent of the following *continuous* cash flows through the years indicated: $1000 in year one, $1200 in year two, $1400 in year three, $1600 in year four, and $1800 in year five, $i = 15\%$.

5-30. What is the maximum sum which a firm should be willing to pay for an equipment which will save the firm $10,000 per year in labor costs for each of the next five years. Treat the savings of $10,000 as if they are flowing continuously over each of the five years. Assume the equipment will be valueless after the fifth year and that the firm's before-tax rate of return requirement $= i = 25\%$.

5-31. A certain annuity provides for 20 consecutive end-of-year payments of $2000 each, beginning with a payment at the end of this year. Assume inflation occurs at the rate of 2% per year (each year purchasing power of the dollar is 0.98 that of the preceding year) and neglect income taxes. If interest is at 6%, what maximum price would the annuity be worth? Comment on the influence of inflation on the investment desirability.

5-32. Use a before-tax rate of return requirement of 20% to find the end-of-year uniform annual series equivalent to end-of-year sums of $1000, $1500, $2250, . . . , and so forth, increasing at the rate of 50% per year when $n = 10$ years.

III

Methods of Comparing
Alternatives

. . . IT SEEMED suitable to me to describe a general rule for finding which is the most profitable of two or more conditions, and by how much it is more profitable than the other, for in this consists perhaps the principal usefulness of these tables, such because businessmen will daily propose conditions to one another, which frequently neither of the two knows which condition is the best.

SIMON STEVIN in Tables of Interest, 1582

6

Annual Equivalent
and Present Equivalent

The Problem: Comparing Cash Flows

THE TYPICAL INVESTMENT in plant or equipment involves a pattern of a present expenditure followed by annual receipts in excess of annual disbursements and climaxed by disposal of the plant or equipment for some net salvage value (although frequently a receipt, salvage can be negligible or negative, hence a disbursement). This cash flow pattern is shown in Figure 6-1. Even when accurate estimates of such cash flows have been made, inspection alone seldom reveals an answer to the question, "Will it pay?" The difficulty is that cash flows are of different years, and therefore are not directly comparable. We can overcome this difficulty by taking into account the time value of money (interest); treatment of the effect of changes in the purchasing power of money (inflation/deflation) will be deferred. The time value of money is taken into account by converting cash flows to equivalent sums at specific dates. Thus the cash flows might be expressed as either an equivalent annual amount or an equivalent amount as of a certain date. The two most widely used methods are the conversion to (1) equivalent uniform annual amount and (2) equivalent present

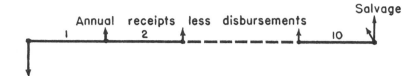

Expenditure

Fig. 6-1. Cash flow diagram of typical plant or equipment investment.

amount; the methods give rise to the labels *annual cost* and *present worth* methods.[1]

Finding equivalent cash flows requires application of the compound interest formula factors. Given the appropriate interest rate, we can convert receipts and disbursements to equivalent sums at specific dates. If the project is to meet the method criteria, equivalent receipts must equal or exceed equivalent disbursements. The investor asking of a prospective investment, "Will it pay?" is saying, "Will the after-tax cash flow be sufficient (at a minimum) to provide repayment of and return on (at some predetermined minimum attractive rate) the investment?"

Before proceeding with the economic evaluation and comparison of alternatives, a number of frequently used terms will be defined or discussed.

Terms Used in Economic Evaluation and Comparison of Alternatives

Minimum Attractive Rate of Return. The minimum attractive rate of return is the effective annual rate of return on investment (after income taxes) which just meets the investor's threshold of acceptability. In Chapters 19 and 20 we will turn to the problem of determining minimum attractive rate of return (*MARR*).

First Cost. First cost of a property is the sum of the costs of purchase, freight in, sales tax, installation, and other such related initial expenditures including preproduction checking. In

---

[1] In his *Tables of Interest* (1582), Simon Stevin wrote: "In-order to set forth this rule [for finding the more profitable of two or more conditions] as shortly as possible I say that it has to be found what is the present value of each proposed condition in respect to a given rate of interest, such in accordance with one of the foregoing examples, the difference between these present values showing by how much one condition is better than the other, which will be clearer from an example."

the case of a building, first cost includes architectural fees, legal fees, permit costs, landscaping costs, property taxes during construction, and interest lost during construction as well as the construction cost itself.[2] Some expenditures, such as for an expanded facility, lead to an expanded need for the items which comprise working capital.[3]

Salvage. Salvage value is the net sum realized from the disposal of a property retired from service. It is the price realized from the disposal of the retired property less the costs of restoration, removal, freight out, and other related disposal expenditures. Salvage is frequently thought of as a receipt to the owner, although it may also represent a disbursement when salvage value is negative. Negative salvage values can occur when it is not possible simply to abandon the property, and certain costs must be incurred in disposal of the property. Salvage ratio is the ratio of salvage to first cost.

Life. Life, or more specifically economic life, of a property is the number of years of service over which the prudent user expects to retain the property in use for its stated purpose. Economic life is that which will produce minimum cost; it can be less than physical life as exemplified by the automobile. The question of how long to retain Machine B really produces an infinitely large set of alternatives; to satisfy this question the reader is asked to assume that the examples and problems which follow show lives which, by preliminary study, have been determined as optimal. In Chapter 17 we will investigate the process of finding economic life.

Annual Sum of Money. Annual sum of money is used in defining operating revenues, cash operating costs, before-tax cash flow, income taxes, and interest on debt capital. It is a time-adjusted amount and is therefore not necessarily identical to the simple average sum.

Example: Assume $i = 10\%$, cash flow = $0, +$100, +$200,

---

[2] Usually partial payments are made to contractors at various stages of building completion; in such cases the owner's payments are made in advance of use. The cost of having an investment in a property prior to its completion is the return foregone on invested funds; this noncash cost is called interest lost during construction (*ILDC*).

[3] Working capital = current assets − current liabilities. Current assets include cash, accounts receivable, and inventory. Current liabilities include accounts payable and taxes accrued.

+$300, +$400, at the end of years one, two, three, four, and five respectively. Here our "annual sum of money" is

$$\$100\,(a/g)_5^{10\%} = \$100(1.810) = \$181$$

while the average is

$$(\$0 + \$100 + \$200 + \$300 + \$400)/5 = \$200$$

Operating Revenue. Operating revenue is the annual sum of money received from the sale and/or rental of goods and/or services including deductions for selling discounts, returned goods, and allowances for uncollectible accounts, but not including excise taxes collected (a very sizable quantity for the liquor and tobacco industries). For decision-making purposes it is not always essential that operating revenues be known; in many cases the alternatives compared have identical effects upon operating revenues.

Cash Operating Cost. Cash operating cost is the annual sum of money required for maintenance, inspection, testing, selling, administration, property taxes, gross receipts taxes, excise tax costs, and FICA, but not including income taxes, interest on debt capital, depreciation expense (and thus not all of the typical "overhead" or "burden" expenses), or costs such as engineering if they have been already allocated to first cost.

Occasionally certain cash operating costs such as property tax are treated separately; in such cases our estimate of remaining cash operating costs must be exclusive of those costs treated separately.

Before-Tax Cash Flow. Before-tax cash flow is the net of operating revenues less cash operating costs.

Depreciation Expense. Depreciation expense is the annual allocation of first cost less salvage over the life of the facility.

Income Tax. Income tax is the annual sum of money required for taxes levied upon net income by municipal, state, federal, and foreign governments.

Interest on Debt Capital. Interest on debt capital is the annual sum of money required for interest (but not principal) due on

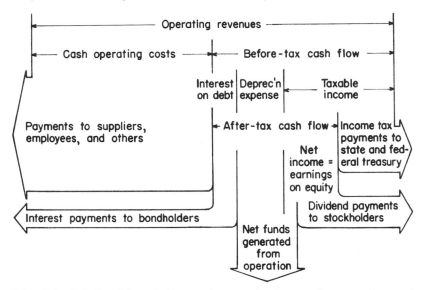

Fig. 6-2. Relationship of the various cash flows of accounting and economy studies.

outstanding borrowed funds. This and other terms just noted are related in Figure 6-2.

Recovery of First Cost and Salvage at the Stipulated Minimum Attractive Rate of Return

In many examples which follow, the levelized annual equivalent of first cost and salvage is computed. This equivalent provides the owner repayment of his investment and return on it during the life of the investment. The combined elements of repayment and return are generally denoted as *capital recovery.* There are many techniques for computing the annual equivalent of first cost and salvage; it is important to distinguish those which are mathematically exact from those which are approximations. (The techniques are exact but are applied to estimates, hence none necessarily produces exact results.) A comparison of the techniques is facilitated by use of the following symbols:

Let B = basis of the property = first cost = installed cost
  V = salvage value at end of the nth year = gross salvage less cost of removal and restoration
  AEC = annual equivalent cost
  i = minimum attractive rate of return
  n = number of years

Exact Techniques. Any of the following exact techniques can be used to determine the annual equivalent of first cost and salvage.

1. The cash flow technique employed in this book is to find annual equivalent of first cost less annual equivalent of salvage:

$$^4AEC = B(a/p)_n^i - V(a/f)_n^i$$

2. Repayment of and return on consumed capital plus return on the residual (salvage) value:

$$^5AEC = (B - V)(a/p)_n^i + Vi$$

which could be rewritten as

$$= B(a/p)_n^i - V[(a/p)_n^i - i]$$

which is equivalent to our first expression.

3. Return on the full investment plus sinking-fund repayment of the consumed capital:

$$^6AEC = Bi + (B - V)(a/f)_n^i$$

which could be rewritten as

$$= B[i + (a/f)_n^i] - V(a/f)_n^i$$

which is equivalent to our first expression.

4. Repayment of and return on first cost less present worth of salvage:

$$^7AEC = [B - V(p/f)_n^i](a/p)_n^i$$

which could be rewritten as

---

[4] *Engineering Economy,* 2nd ed. (New York, American Telephone and Telegraph Company, 1963).

[5] E. L. Grant and W. G. Ireson, *Principles of Engineering Economy,* 4th ed. (New York, Ronald Press, 1960).

[6] P. H. Jeynes, *An Abbreviated Course in Engineering Economics,* Public Service Electric and Gas of New Jersey, 1960; E. P. De Garmo, *Engineering Economy,* 4th ed. (New York, Macmillan, 1967).

[7] W. T. Morris, *The Analysis of Management Decisions,* rev. ed. (Homewood, Ill., Irwin, 1964).

$$= B(a/p)_n^i - V(a/f)_n^i$$

which is our first expression.

5. Return on declining investment plus uniform repayment (= straight-line depreciation) of the consumed capital:

$$[8] AEC = i \left[B - \frac{B - V}{n} (a/g)_n^i \right] + \frac{B - V}{n}$$

since

$$(a/g)_n^i = \frac{1}{i} - \frac{n}{(1 + i)^n - 1} \qquad \text{(See Equation 4-8a)}$$

$$AEC = iB + \frac{B - V}{n} \left[-\frac{i}{i} + \frac{ni}{(1 + i)^n - 1} + 1 \right]$$

$$= iB + (B - V)(a/f)_n^i$$

which was shown in Expression 3 to be equivalent to our first expression.

All the preceding techniques must of course be applied consistently to produce identical results. In Expression 3, the use of a different interest rate for sinking-fund repayment than for return on investment will produce results inconsistent with those of the techniques just described. Such treatment is sometimes based on the assumption that the repayment funds will not be reinvested in the business but will be placed in a special fund or account where the earning rate will be different from the earning rate on the project funds.

Approximations. Several popular approximations are worthy of note because they are convenient and do not require use (and hence explanation) of compound interest tables.

1. *Straight-line depreciation (repayment) plus return on the average investment.*

$$AEC \sim (B - V)[(1/n) + (i/2)] + Vi$$

---

[8]This approach closely resembles the allowed-earnings approach of regulatory commissions on an item or vintage group property, where uniform annual depreciation charges plus an allowed rate of return on a declining rate base are permitted.

and this latter expression is similar to the formula under (2) of the exact methods where

$$AEC = (B - V)(a/p)_n^i + Vi$$

and thus $(1/n) + (i/2)$ is an approximation of $(a/p)_n^i$.

2. *Straight-line depreciation (repayment) plus return on the average investment.* A variation of the formula in (1) above is based on an annual rather than continuous repayment schedule making the average outstanding depreciable investment $(n + 1)/2n$ rather than $(1/2)$ of $(B - V)$:

$$AEC \sim [(B - V)/n] + i(B - V)[(n + 1)/2n] + Vi$$

so with this formula, $1/n + i[(n + 1)/2n]$ is an approximation of $(a/p)_n^i$.

Maximum error of the approximation occurs at $n \longrightarrow \infty$

$$\lim_{n \to \infty} [i(1 + i)^n]/[(1 + i)^n - 1] = i$$

$$\lim_{n \to \infty} (2 + ni + i)/2n = i/2$$

and therefore

$$\lim_{n \to \infty} \text{error} = (i/2)/i = 50\%$$

Figure 6-3 shows the effect of i and n upon the error introduced by the approximation, 2. We can observe there that, if an error of 5% is tolerable, $ni < 100\%$-years will roughly signal the circumstances under which the approximation would be satisfactory. A rough estimate of error size is given by

$$\% \text{ Error} \sim 6\,ni \quad \text{for } ni < 500\%\text{-years}$$

thus when $i = 10\%$, $n = 20$ years,

$$\% \text{ Error} \sim 6(20)(0.10) = 12\%$$

Either approximation, 1 or 2, employs an arithmetic average of investment without adjustment for the interest rate.

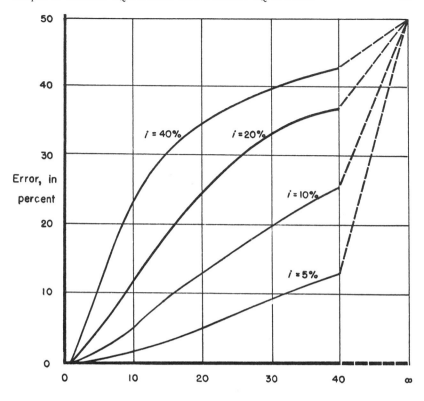

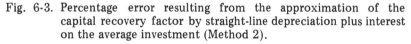

Capital   recovery   period,   in   years

Fig. 6-3. Percentage error resulting from the approximation of the capital recovery factor by straight-line depreciation plus interest on the average investment (Method 2).

As a consequence either approximation understates costs if all three of the following conditions are met: (a) first cost exceeds salvage $(B > V)$, (b) property life exceeds one year $(n > 1)$, (c) minimum attractive rate of return is positive $(i > 0\%)$, and the error increases as i and n increase. If at least one of the following conditions is met, the results will be mathematically exact rather than approximate: (a) first cost equals salvage $(B = V)$, (b) property life equals one year $(n = 1)$, and (c) minimum attractive rate of return is zero $(i = 0\%)$.

Cursory consideration sometimes leads to the view that errors will fall equally upon alternatives; this assumption is dangerous where disparity exists in first cost, salvage, or life of the alternatives considered. Even a relative error of zero seems a poor defense for absolute errors of considerable magnitude.

3. *Straight-line depreciation (repayment) plus return on the full investment.*

$$AEC \sim [(B - V)/n] + iB$$

Interestingly enough, the results are mathematically exact rather than approximate when either (a) $n = \infty$, (b) $B = V$, or (c) $i = 0\%$. The approximation fails to take into account the decreasing nature of the investment and hence leads to overestimation of costs if $B > V$ and n is finite.

The approximations help to emphasize the error involved in a rather inviting analysis. The error is one of double-counting and arises when either depreciation or interest is added to the annual equivalent of first cost less the annual equivalent of salvage. We have just seen that depreciation and interest are an approximation of the annual equivalent of first cost less the annual equivalent of salvage. The double-counting should be apparent; nonetheless it appears all too frequently in practice. Depreciation expense is not a cash flow, and it would be irrelevant to the engineering economy study were it not for its effect on the cash flow item, income tax.

Some Conventions Employed in This Book

Income Taxes. Explicit computation of income taxes is deferred to Chapter 10. In lieu of computing income taxes directly, we will approximate the income tax effect by requiring a before-tax rate of return sufficient to provide for both the after-tax rate of return and income taxes. The combined effective state and federal income tax rate on corporations is about 50%; thus $2 of before-tax return is required to produce $1 of after-tax return to the equity holder. By this same reasoning a before-tax rate of return of 20% should result in an after-tax rate of return of 10%. (Again this is an approximation, as we shall see in Chapter 10.)

To facilitate the approximation:

Let $BTRR$ = before-tax rate of return required
    $MARR$ = minimum attractive rate of return (after taxes)

t = effective tax rate (federal, state, municipal) on taxable income

Then $BTRR \cong MARR/(1 - t)$. So, for example, if the tax rate is 20% and $MARR$ is 8%:

$$BTRR \cong 0.08/(1.00 - 0.20) = 10\%$$

The relationship above is an exact one if there are no interest charges on debt capital (100% equity financing) and no depreciation expense charges (100% salvage values or infinite life). The approximation will be used as a means of implicitly providing for income taxes until Chapter 10 where explicit treatment of income taxes is introduced. In the problems and examples of intervening chapters, income taxes will be included in our studies via the $BTRR$ which provides for both $MARR$ and income taxes. Note that $BTRR = MARR$ if (1) the owner has a tax-exempt status (as for a municipally owned facility) or (2) overall income of the individual or corporation is below the tax threshold.

End-of-Year and Continuous Cash Flows. In this chapter and all subsequent ones, future and annual cash flows will be treated as discrete end-of-year amounts, and present cash flows will be treated as discrete flows now unless otherwise specified.

Mutually Exclusive Alternatives. Most capital expenditure decisions involve choosing the "best" of a number of alternative courses of action. A proposed highway to connect two cities may involve many alternative routes; after an economy study of the routes, the "best" may be selected and remaining route proposals then discarded. The various routes comprise a set of mutually exclusive alternatives. *A set of mutually exclusive alternatives is one in which the selection of one alternative rules out the selection of all others in the set.* Mutually exclusive alternatives tend to be substitutes for each other in filling a given function. Throughout the text, a set of lettered alternatives having either the same or no prefix number are mutually exclusive.

Alternative B and Alternative G are mutually exclusive, or Alternative 3B and Alternative 3G are mutually exclusive.

In a set of *independent* alternatives the acceptance of one has no influence on the acceptance or rejection of other alternatives in that set. An alternative is independent of any other having a different prefix number.

Alternative 5 and Alternative 8 are independent, or Alternative 5B and Alternative 8G are independent.

Example 6-1. Evaluating an Investment Opportunity

Information: Mr. Blue is considering the investment of $100,000 in a venture which will produce estimated revenues of $20,000 at the end of each of the next seven years. His minimum attractive rate of return is 8% and his marginal income tax rate is 20%. Salvage value of the property seven years from now is thought to be negligible. No other costs are involved.

Objective: Analyze the investment to determine an answer to the question, "Will it pay?"

Analysis: Since $MARR$ = 8% and t = 20%,

$$BTRR \cong MARR/(1 - t) = 10\%$$

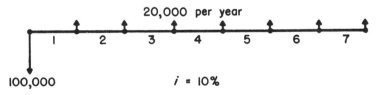

On an annual equivalent basis:

Annual equivalent of revenues $= AER =$ $ 20,000
Annual equivalent of costs
   $= \$100,000(a/p)_7^{10\%}$
   $= \$100,000(0.20541)$ $= AEC =$ 20,541
Annual equivalent of revenues less costs $= AEX = -\$$ 541

On a present equivalent basis:

Present equivalent of revenues
   $= \$20,000(p/a)_7^{10\%} = \$20,000(4.868) = PER =$ $ 97,360
Present equivalent of costs $= PEC =$ 100,000
Present equivalent of revenues less costs $= PEX = -\$$ 2,640

By either method we see that the investment fails to provide revenues which at least match costs at the before-tax rate of return required, so both methods signal rejection of the opportunity.

Example 6-1 illustrates the computation of annual and present equivalents of both revenues and costs. Several points are worthy of mention in this regard:

1. As a shorthand notation throughout the remainder of the text the following mnemonic symbol sets are used:

AEC_B = annual equivalent of costs for Alternative B
AER_B = annual equivalent of revenues for Alternative B
AEX_B = annual equivalent of revenues—costs for Alternative B
PEC_B = present equivalent of costs for Alternative B
PER_B = present equivalent of revenues for Alternative B
PEX_B = present equivalent of revenues—costs for Alternative B
FEC_B = future equivalent of costs for Alternative B
FER_B = future equivalent of revenues for Alternative B
FEX_B = future equivalent of revenues—costs for Alternative B

2. For any equivalent cost, revenue, or combination thereof, the following relationships must exist:

$$AE_B = PE_B(a/p)_n^i = FE_B(a/f)_n^i$$
$$PE_B = AE_B(p/a)_n^i = FE_B(p/f)_n^i$$
$$FE_B = AE_B(f/a)_n^i = PE_B(f/p)_n^i$$

In the example we note, either as a check or as a shortcut computation,

$$AEX_B = PEX_B(a/p)_n^i$$
$$= -\$2640(0.20541)$$
$$= -\$542$$

If our compound interest factors were carried to several more digits, the result here (-$542) would coincide exactly with that of the example (-$541).

It should be apparent that annual, present, or future equivalent comparisons will always signal the same conclusion; the ratio of the equivalents between alternatives is not altered by the method employed:

$$\frac{AEX_A}{AEX_B} = \frac{PEX_A}{PEX_B} = \frac{FEX_A}{FEX_B}$$

and

$$\frac{PEX_A}{AEX_A} = \frac{PEX_B}{AEX_B} = (p/a)_n^i$$

and similarly for other pairings.

3. The technique of finding the excess of the present equivalent of revenues over the present equivalent of costs is referred to by some as the *discounted cash flow technique.*[9] In Example 6-1 the discounted cash flow for Alternative B = $PEX_B = -\$2640$. Being negative, it signals rejection of the proposed alternative.

4. *BTRR* was computed as 10%. If we show the year-by-year declining investment and required return:

Year	Remaining Investment	Return Required	Uniform Repayment
1	$100,000	$10,000.00	$14,286
2	85,714	8,571.40	14,286
3	71,429	7,142.90	14,286
4	57,143	5,714.30	14,286
5	42,857	4,285.70	14,286
6	28,571	2,857.10	14,286
7	14,286	1,428.60	14,286

The equivalent uniform annual total of return on and repayment of investment is

$$\$10,000 - \$1428.60 \, (a/g)_7^{10\%} + \$14,286$$
$$= \$24,286 - \$1428.60 \, (2.622) = \$20,540$$

Again we verify that annual cost exceeds annual revenues by $540.

5. In relating this example to our basic accounting model, we can observe that if (a) the project is approved, (b) deprecia-

---

[9] Joel Dean, *Managerial Economics* (Englewood Cliffs, N.J., Prentice-Hall, 1951).

tion is by the straight-line method, (c) the income tax rate is 20%, (d) financing is entirely by equity capital, and (e) estimated data is exactly correct, then the effect upon the income statement for each of the next seven years will be:

Operating revenues		+$20,000
Cash operating costs	0	
Depreciation expense	+$14,286	
		+ 15,429
Interest on debt capital	0	
Income taxes		
($20,000 − $14,286)(20%) =	+$ 1,143	
Net income = earnings on equity capital		+$ 4,571

This constant return applies to a declining investment; the average investment is approximately

$$[(n + 1)/2n]($100,000) = $57,143$$

and the rate of return is about

$$$4571/$57,143 = 8\%$$

Since income taxes are shown separately, the 8% should be compared to $MARR$, not $BTRR$.

6. The choice of alternatives will be unaffected by whether the analyst compares them by annual equivalent or present equivalent methods; they do not lead to contradictory conclusions. (Nor will any other valid method such as the rate of return method, properly applied, lead to contradictory acceptance-rejection conclusions. In the priority ranking of acceptable projects for capital budgeting purposes there is controversy as to the compatibility of the methods.[10])

7. In Example 6-1 we hope to have answered the question, "Will it pay?" The present and annual equivalent methods permit comparison of revenues and costs so that we can choose between doing nothing and doing the proposed alternative. Next let us compare three or more alternatives with the objective of maximizing profit.

---

[10] Joel Dean proposes ranking projects by rate of return in *Managerial Economics* (Englewood Cliffs, N.J., Prentice-Hall, 1951). H. M. Weingartner rejects this method and favors a ranking based on a present equivalent approach in *Mathematical Programming and the Analysis of Capital Budgeting Problems* (Englewood Cliffs, N.J., Prentice-Hall, 1963). The argument is sustained by other prominent writers.

Example 6-2.   Comparing Four Alternatives (Including Disinvestment)

Information: The following estimates have been made with regard to a service function of the company under various stages of automation. $BTRR = 20\%$.

	Proposal P	Proposal A	Proposal B
Annual revenues	$50,000	$60,000	$70,000
Annual costs of operation, maintenance, and property taxes	54,000	46,000	39,000
Present investment required	0	50,000	100,000
Estimated salvage five years from now	0	0	10,000

Objective: Determine whether Alternative 0, P, A, or B should be recommended. (Alternative 0 is to completely drop the service function. Alternative P is to continue the service function as presently performed.)

Analysis: Although either the annual equivalent or present equivalent approach may be used, it is somewhat simpler to use the former since the majority of cash flows are already on this basis.

	Proposal P	Proposal A	Proposal B
Annual revenues	$50,000	$60,000	$70,000
Annual costs of operation, maintenance, and property taxes	$54,000	$46,000	$39,000
Annual equivalent of first cost and salvage $50,000(a/p)_5^{20\%}$ $100,000(a/p)_5^{20\%}$ $-\$10,000(a/f)_5^{20\%}$	0	16,719	32,094
Annual equivalent cost	$54,000	$62,719	$71,094
Annual equivalent of revenues less costs	$-\$\ 4,000$	$-\$\ 2,719$	$-\$\ 1,094$

Recommend Alternative 0.

Example 6-2 can help us to observe the logic involved in choosing among alternatives.

1. Compare revenues and costs to see if at least one of the proposed alternatives meets the investment criteria. In Example 6-2 the answer is no, and in the absence of other considerations, the best alternative is disinvestment, that is, to drop the service function (accept Alternative 0) and thereby reject Alternatives P, A, and B.

   It is possible, however, that the service function is a mandatory one; perhaps the industrial firm considers its repair service an essential function; perhaps the utility firm is required by its franchise to provide service to a remote area; in still other cases the firm is required by law to provide certain features of employee needs and safety. In such cases Alternative 0 is not available and our task then is one of selecting the alternative which will maximize revenues less costs. In Example 6-2, in the absence of Alternative 0, Alternative B maximizes revenues less costs and would be the recommended alternative.

   The logic of choosing among alternatives is important; perhaps most important is the recognition of those situations where investment is not mandatory. Too often we are tempted to choose the alternative with minimum annual (or present) equivalent cost and examine the problem no further; the end result of such studies would be to produce a "winning" investment from each set of alternatives studied. This would produce capital expenditure levels proportional to the number of economy studies conducted. When they are real options, the alternatives to "do nothing" or "release present investment" (disinvestment) must be investigated; the question of "Will it pay?" cannot be summarily dismissed.

2. Choosing the alternative which minimizes cost may not maximize profit. Only when the revenues of all the various alternatives are identical, can we be certain that the cost-minimizing alternative and the profit-maximizing alternative are one and the same.

3. In any set of mutually exclusive alternatives we should choose the one which maximizes AEX (or PEX), providing that either AEX (or PEX) is positive or the service considered is a mandatory one. Under the latter condition, several relaxations may be made in data required; it is sufficient to know the *difference* in revenues of the alternatives

even if absolute revenues are unknown, and it is likewise sufficient to know the *difference* in costs of the alternatives even if absolute costs are unknown. This is important because many investment decisions involve a choice between alternatives which will produce identical but unknown revenues.

Example 6-3.   Equivalence Studies Involving Alternatives with Known Differences in Costs (and Revenues) but Unknown Levels of Same

Information: Two competing methods of performing a required service are to be compared; selection of one of these two courses of action (methods) is mandatory. $BTRR = 15\%$. Estimates pertaining to the choice:

	Method A	Method B
First cost	$100,000	$200,000
Life	4 years	8 years
Salvage at retirement	0	$10,000
Operating, maintenance, and property tax costs	unknown, say K_1 per year	$10,000 per year less than for A
Revenues	unknown, say K_2 per year	unknown but same as for A

Objective: Find the more economical alternative.

Analysis: Find annual equivalent costs:

	Factor	Method A	Method B
AE_A first cost and salvage = $100,000 $(a/p)_4^{15\%}$	0.35027	$35,027	
AE_B first cost and salvage = $200,000 $(a/p)_8^{15\%}$	0.22285		
$-$$10,000 $(a/f)_8^{15\%}$	0.07285		$43,842
AE operating, maintenance, and property tax costs		K_1	$-$10,000$+K_1$
AEC		$35,027 + K_1$	$33,842 + K_1$

Since annual revenues of A and B are the same, the cost-minimizing Alternative B is also the profit-maximizing (= loss-minimizing) alternative. Note that B is not necessarily a profitable venture; the fact that doing nothing might be better is of no consequence here since the service is mandatory.

As a check or independent solution we could compare the present equivalent costs, being careful to use an *identical service period* (say 8 years; A1 is the original unit, A2 is its replacement).

	Factor	Method A	Method B
PE_{A1} first cost and salvage		$100,000	
PE_{A2} first cost and salvage = $100,000 $(p/f)_4^{15\%}$	0.5718	57,180	
PE_B first cost and salvage = $200,000 - $10,000 $(p/f)_8^{15\%}$	0.3269		196,731
PE operating, maintenance, and property tax costs $K_1 (p/a)_8^{15\%}$	4.487	$4.487K_1$	
$(K_1 - $10,000)(p/a)_8^{15\%}$	4.487		$-44,870 + 4.487K_1$
PEC		$157,180 + 4.487K_1$	$151,861 + 4.487K_1$

Both comparisons indicate that B is the more economic alternative. As noted earlier,

$$PEC_A / AEC_A = PEC_B / AEC_B = 4.487 = (p/a)_8^{15\%}$$

When alternatives with different lives are treated by the annual equivalent method, the results are directly comparable. With the present equivalent method we must compare the alternatives over a common span of time; generally this span will be the least common multiple of lives of the alternatives. In our example we could compare the units over an eight-year span of service; thus an original unit and one replacement unit installed at the end of year four were required for Method A. Had we neglected the replacement unit and its costs, we would have (incorrectly) compared the cost of four years of service from A with the cost of eight years of service from B. Such a comparison would obviously be biased in favor of Alternative A.

Not all sets of alternatives provide a small least common multiple of lives of alternatives. Consider alternatives with lives of 13 and 29 years. It makes little sense to compare these over the least common multiple span of 377 years. It makes still less sense that the difficulty in treating such problems should be dependent upon the relative primeness of the numbers representing lives of the alternatives. This problem will be deferred to a later chapter.

When the need for either perpetual service or service for a long period of time can be satisifed by a facility and its various successors, we can compare either the immediate competing alternatives or the immediate unit and its *replacement chain* of successor units for each of the alternatives.

Valuation Applications

The value of many properties is dependent upon earning capacity. The earning value of a property is the present worth of probable future net earnings as prognosticated on the basis of recent and present expenses and earnings and the business outlook.[11] Examples of properties whose value tends to be primarily a function of earning ability include bonds, stocks, such leased facilities as buildings or equipment, and such rental properties as apartment houses.

Most bonds (1) are issued by a corporate or governmental unit; (2) bear interest semiannually; (3) are redeemable for a maturity value at a specified date, perhaps 20 years hence; and (4) are issued in $1000 denominations. Some bonds are *convertible* into common stock of the firm under specific conditions, or *callable*, that is, they may be repurchased by the issuing organization prior to maturity and according to a specific repurchase schedule. The two basic financial features of the bond are (1) the promise to pay the bondholder semiannual sums of a specified amount and (2) the promise to pay the bondholder the maturity value at a specified date. As a basic approach to the valuation of a bond we can find the present equivalent value of the two promises.

Example 6-4.   Valuation of a Bond

Information: The 3% debenture bond of the Green Corporation was issued eight years ago. Although the bond has a maturity value of $1000, the buyer of eight years ago could purchase the bond for $980 plus brokerage charges of $10. The bond matures 17 years from today and is not callable. Since it is a 3% bond, the semiannual interest payments are $15 each.

---

[11] A. Marston, R. Winfrey, and J. C. Hempstead, *Engineering Valuation and Depreciation*, 2nd ed. (Ames, Iowa State Univ. Press, 1953), p. 8.

Objective: Estimate the present market value of the bond if the current market rate of interest on securities of similar quality is 5%. For simplicity treat the bond interest payments as if they were annual end-of-year amounts and neglect income tax considerations.

Analysis: Find present equivalent revenue:

	Factor	Value
PE interest = $30(p/a)_{17}^{5\%}$	11.274	$338.22
PE maturity = $1000(p/f)_{17}^{5\%}$	0.4363	436.30
PER		$774.52

Example 6-4 shows that the present equivalent approach can be very helpful in the valuation of bonds, even though any specific issue may involve additional facets not treated here. We also note that the original purchaser of the bond who retains it to maturity receives a rate of return slightly less than the rate of return (cost of capital) the corporation pays because of issuing costs, brokerage charges, and the like. This small difference bears an interesting resemblance to engineering concepts of energy losses due to friction. The rate of return for the buyer and cost of capital for the seller would be the same if it were not for "friction."

Prospective rate of return may be quite different from *actual* rate of return. Consider the plight of the party purchasing the bond eight years ago for $1000 and selling the bond today for $775; his actual or realized rate of return before taxes is about 0.2%.

Bond prices tend to move counter to the market rate of interest because the higher interest rates further discount the present equivalent value of the promises of the future. The bond market speculator watches the going rate of interest closely as a guide to action in the bond market.

There is some risk in even the most secure of bond issues; even if there were absolute certainty that interest payments and redemption requirements would be met, there is less than absolute certainty in the day-to-day market and the rate of interest as determined in the market. Fluctuations in the market rate of interest mean fluctuations in bond prices; these fluctuations are more severe for the long-term than for the short-term security.

Example 6-5.   Valuation of Common Stock

Information: The common stock of the Green Corporation paid a quarterly dividend of $1 per share each quarter last year. The company has paid out 100% of earnings as dividends for several years. Based on a study of company growth patterns and future prospects, we have estimated that the current annual dividend rate of $4 will grow by $1 per share per year through the foreseeable future, so the dividend for the coming year is estimated as $5, next years as $6, and so forth. Neglect the cost of commissions to buy or sell, the cost of cashing the dividend, and similar costs whose similarity to friction has already been noted.

Objective: Estimate the present market value of the stock if the going rate of return on securities of similar quality is now 10%. Treat the quarterly dividend payments as if they were annual end-of-year amounts, treat the stock as if it were held forever, and neglect income tax considerations.

Analysis: Find present equivalent revenue:

	Factor	Value
PE dividends = $5 $(p/a)_\infty^{10\%}$	10.000	$ 50
PE growth in dividends = $1 $(p/g)_\infty^{10\%}$	100.000	100
PER		$150

In Example 6-5 the present equivalent approach is extended to the valuation of common stock. Here too there are two basic prospective receipts; the current level of dividends plus the increase (gradient series) in level of dividends. Because we assumed the stock was held forever (as via heirs), it was unnecessary to estimate a third prospective receipt, the resale value of the stock at some future date. It is rather interesting to note that if valuation rates and dividends continue as estimated, the price next year would be

	Factor	Value
PE dividends = $6 $(p/a)_\infty^{10\%}$	10.000	$ 60
PE growth in dividends = $1 $(p/g)_\infty^{10\%}$	100.000	100
PER		$160

and the pattern would be

	End of Year					
	0	1	2	3	4	5
Market price	$150	$160	$170	$180	$190	$200
Earnings per share	4	5	6	7	8	9
Dividends per share	4	5	6	7	8	9
Price to earnings ratio	37.5	32.0	28.3	25.7	23.8	22.2

If our original buyer were to sell his stock according to the "market price" schedule shown, he would still earn a 10% return. Any purchaser buying, then later selling per the schedule, would earn a 10% return on his investment, for example, if purchased now and held for five years:

	Factor	Value
PE dividends $= \$5(p/a)_5^{10\%}$	3.791	$ 18.96
PE growth in dividends $= \$1(p/g)_5^{10\%}$	6.862	6.86
PE resale $= \$200(p/f)_5^{10\%}$	0.6209	124.18
PER		$150.00

The declining price-to-earnings ratio is the result of a declining growth rate (the amount is constant but the rate is declining).

Earnings as well as dividends are an important factor in the valuation of a stock, particularly in the evaluation of cost of capital to the corporation. Three basic financial features of common stock are (1) the prospective dividends (or alternatively, earnings), (2) the prospective growth in dividends (or earnings), and (3) the prospective market price at the time of disposal. Another factor which can be of considerable importance to the prospective buyer is the impact of income taxes; if he is in a high income tax bracket, the prospect of capital gains (taxed at half the ordinary income rate) via price increase may be more attractive than the prospect of generous dividends (taxed at the ordinary income rate).

Capitalized Cost

In engineering economy "capitalized cost" of a project is that special case of present equivalent cost where service is assumed to be provided for an infinite rather than a finite period. The term is probably important, unfortunate, and superfluous. It is important because it is encountered in both the literature of engineering economics and professional engineering examinations and reports; its origination is unfortunate in that accountants had already used the same term to describe a cash outlay which is to be allocated over two or more years rather than "expensed" to the current year. The term is probably superfluous in that it is simply a special case of the present equivalent cost where n is not finite. Why the circumstance of infinite project life should give rise to special terminology is a good question; how a term frequently appearing in the literature of the past can be abandoned is a still better question.

Summary

At the outset of this chapter we noted that comparison of alternatives requires that their cash flows be somehow made comparable. Two methods of comparison have been illustrated: annual equivalent and present equivalent. In Figure 6-4 the factors required for conversion of some common cash flow patterns are given. It is suggested that the reader study through these and use them as a review and reference in his solution of problems. Example 6-6 illustrates many of the cash flows shown in Figure 6-4.

Example 6-6.   An Exercise in Computing Equivalent Cost

Information: A certain proposed facility will require expenditure of $1 million now and $500,000 in 15 years. The facility is expected to have a life of 30 years, after which time it will have an estimated salvage value of $200,000. Annual costs of maintenance are estimated as $5000 for year one, and are expected to increase by $1000 per year throughout the life of the facility. Insurance and property tax costs are estimated to be $30,000 per year for each of the first 15 years and $40,000 per year thereafter. Periodic reconditioning costs of $100,000 at the end of every tenth year including the thirtieth year are expected. The before-tax rate of return required is 20%.

Objective: Find the annual equivalent cost of the facility.

Analysis:

	Factor	Cost
AE first cost $1,000,000 $(a/p)_{30}^{20\%}$	0.20085	\$200,850
AE deferred cost $500,000 $(p/f)_{15}^{20\%}(a/p)_{30}^{20\%}$	0.0649,	6,518
	0.20085	
AE salvage − $200,000 $(a/f)_{30}^{20\%}$	0.00085	−170
AE maintenance $5000 + $1000 $(a/g)_{30}^{20\%}$	4.873	9,873
AE insurance and property tax costs		
$30,000 + $10,000 $(f/a)_{15}^{20\%}(a/f)_{30}^{20\%}$	72.035	30,615
	0.00085	
AE reconditioning $100,000 $(a/f)_{10}^{20\%}$	0.03852	3,852
AEC		\$251,538

A number of points can be observed in Example 6-6:

1. Salvage value is a sizable amount, yet its annual equivalent is virtually negligible. Large values distant in time and discounted at high rates have little significance. In this example a 1% error in estimation of first cost is more serious than a 1000% error in estimation. of salvage. In the example and in general the results of an economy study are more *sensitive* to errors in the estimation of first cost than to errors in the estimation of salvage. By sensitive we mean that our results are quite responsive to changes in a certain parameter such as first cost; our results are *not* very responsive (and hence are not very sensitive) to changes in the parameter, salvage. *Sensitivity studies* are sometimes conducted to determine whether errors of specified size in the estimation of an input variable could change the conclusions of the study. In general, the effect of estimating errors decreases as the number of years distant is increased and also decreases for higher discounting ($BTRR$) rates.
2. Our examples throughout the text are generally carried to the nearest dollar and in a few cases to the nearest cent. Obviously, one must reflect upon the accuracy of the estimates before interpreting results. The digits beyond the second or third are probably not significant; the student is requested however, to carry his calculations to the limits of slide rule accuracy (three or four digits) so that the point being illustrated is not lost in the process of rounding data.

While this chapter has dealt almost exclusively with annual

Type of cash flow	Example	Annual equivalent = x multiplied by:	Present equivalent = x multiplied by:
1. First cost 	Building	$(a/p)_n^i$	1
2. A single deferred cost 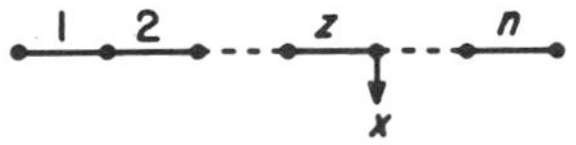	Addition to building	$(p/f)_z^i (a/p)_n^i$	$(p/f)_z^i$
3. Salvage 	Resale of building	$(a/f)_n^i$	$(p/f)_n^i$
4. Annual 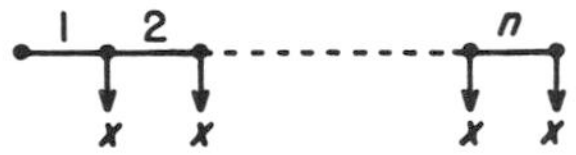	Property tax, maintenance, income tax, insurance	1	$(p/a)_n^i$
5. Deferred annual beginning with year $z+1$ 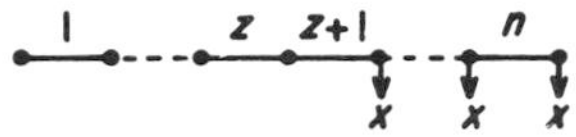	Increased labor cost	$(f/a)_{n-z}^i (a/f)_n^i$ or $(a/p)_n^i\left[(p/a)_n^i - (p/a)_z^i\right]$	$(p/a)_{n-z}^i (p/f)_z^i$ or $(p/a)_n^i - (p/a)_z^i$
6. Arithmetic growth 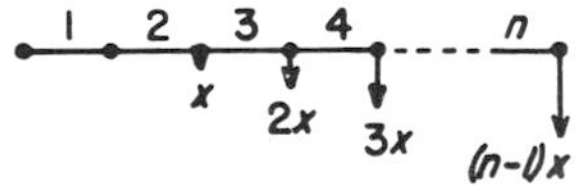	Maintenance	$(a/g)_n^i$	$(p/g)_n^i$
7. Periodic (once each z years beginning with year 0) (n/z is integer >1) 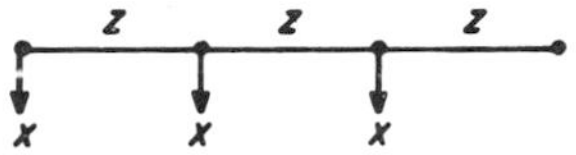	Reroofing if initial roof is considered	$(a/p)_z^i$	$(a/p)_z^i (p/a)_n^i$ or $1 + (p/f)_z^i \cdots + (p/f)_{n-z}^i$
8. Periodic (once each z years beginning with year z) (n/z is integer >1) 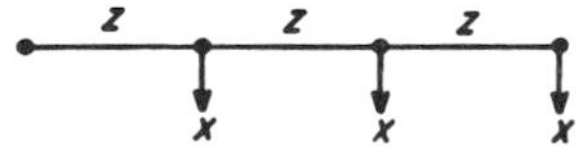	Reroofing	$(a/f)_z^i$	$(a/f)_z^i (p/a)_n^i$ or $(p/f)_z^i + (p/f)_{2z}^i \cdots + (p/f)_n^i$

Fig. 6-4. Conversion of selected types of cash flow to annual or present equivalent.

equivalent and present equivalent methods, costs and revenues of alternative proposals can also be correctly compared on a future equivalent basis.   Although such applications are infrequent, the future equivalent approach can be a very helpful one.

Figures 6-5 and 6-6 show the logic process by which one

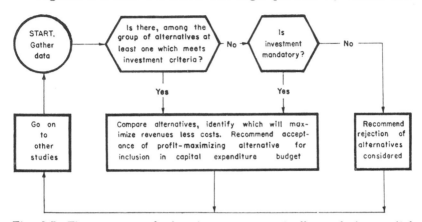

Fig. 6-5. The process of choosing among mutually exclusive capital expenditure alternatives.

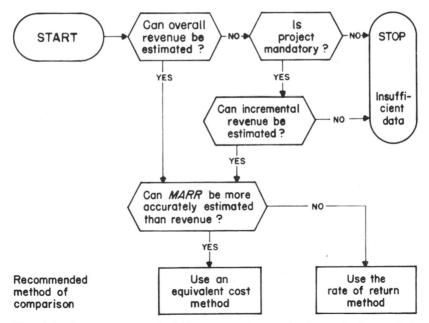

Fig. 6-6. Selecting a method for comparing mutually exclusive capital expenditure alternatives.

chooses from a set of alternative courses of action with regard to capital expenditures, and the method of comparison.

PEX As a Capital Expenditure Criterion

The present equivalent excess of revenues over costs = *PER – PEC = PEX.* Also called the *net present value,*

$$PEX = \sum_{z=0}^{n} (1 + i)^{-z} x_z$$

where i = rate of return required
　n = life of the project, in years
　x_z = the net cash flow in year z

When i and x_z are on a before-tax basis, so is *PEX.*

From a group of mutually exclusive projects select the one with maximum *PEX,* then accept it if either (1) the project function is mandatory or (2) $PEX \geqslant 0$. *AEX* and *FEX* are analogous decision criteria. If in the capital budgeting process the cost of independent projects thus accepted exceeds the funds available, the rate of return required should be revised upward to reflect "opportunity cost." The final value for i should be the one that matches funds required with funds available.

A Caution against Using Payoff Period As a Capital Expenditure Criterion

The payoff period can be defined as the number of years required for the before-tax cash flow from a project to equal zero when the rate of return required is 0%. Prospective payoff period is *NP* such that

$$\sum_{z=0}^{NP} x_z = 0$$

where x_z is the before-tax cash flow in year z.

From a group of mutually exclusive projects one is to select the project with minimum prospective payoff period, then accept it if either (1) the project function is mandatory or (2) prospective payoff period $\leqslant$ the maximum acceptable payoff period. For a group of independent projects one is to rank projects in ascending order of prospective payoff period and then approve projects according to rank until funds available are exhausted. We recommend that this popular criterion be *avoided* because:

1. Cash flows beyond the payoff year are neglected, thus the criterion fails to give weight to cash flows that occur after the date of payout.
2. The timing of cash flows within the payout period is neglected. Project A with end-of-year cash flows of – 100, +50, +50 . . . would be perceived as equivalent to Project B with end-of-year cash flows of – 100, 0, +100 . . . .
3. The "test" is not a uniform one; it discriminates against long-lived projects.
4. Users tend to use shorter and shorter payoff requirements until few investments, if any, can pass the "test."

Problems

6-1. Using the symbols

$$ER_A = \text{equivalent revenue of Alternative A}$$
$$EC_A = \text{equivalent cost of Alternative A}$$

and given Alternatives A and B with the availability of Alternative 0 (disinvestment) unknown, write an expression to specify the circumstances under which A would be the recommended alternative.

6-2. The Bowman Corporation requires a before-tax rate of return of 25%. What is the maximum sum the firm should presently invest in return for estimated pretax cash flow savings of $100 per year for the next ten years?

6-3. Give an example of choices between alternatives where:
   a. Operating revenue will not be affected ($AER_A = AER_B$ . . .).
   b. Operating revenue will be affected.
   c. The "best" of the alternatives is acceptable even if its equivalent costs exceed its equivalent revenues.

6-4. Look in the *Wall Street Journal* for a column headed "Government, Agency and Miscellaneous Securities." The "U.S. Treasury Bills" had maturity dates 90–360 days distant when originally issued. The "U.S. Treasury Notes" originally had maturity dates one to five years distant, and the "Treasury Bonds" had maturity dates more than five years distant when originally issued. List "Yield" and "Discount" figures for maturity dates distant by $1/2$, 1, 2, 4, 8, 16, and 32 years. Is there a pattern?

6-5. This chapter showed five different exact techniques for determining the annual equivalent of first cost and salvage. Show, in formula form, what you consider to be the easiest and most direct method of finding the present equivalent of first cost and salvage.

6-6. Find the present equivalent cost of ten years of service from an equipment of the Van Osdel Company if

First cost = $10,000
Salvage = $1000 (*negative*) at end of year ten
Maintenance, property taxes, and so forth = $2000 in year one, $2200 in year two, and so forth, increasing by $200 per year
Overhaul costs = $3000 at end of year five
Before-tax rate of return requirement = 15%

6-7. A municipally owned water company has proposed rates based on the following costs and submitted these to the city council:

Capital recovery requirements on investment of $100,000 at 4% with a 30-year life = $100,000 $(a/p)_{30}^{4\%}$ = $ 5,783/yr
Operating costs for labor, chemicals, and other direct cash costs   = 15,000/yr
Bond interest (average)   = 2,000/yr
Overhead costs, primarily depreciation   = 3,000/yr
Overhead costs at end of year 15 = $15,000/15   = 1,000/yr
Amortization of bond issue over 20 years = $100,000/20   = 5,000/yr
Local property taxes, state income taxes, and federal income taxes   = 0
        Total $31,783/yr

Please make a more correct analysis and explain very briefly your proposed changes.

6-8. The Revere Service Company has reduced to two the number of alternatives under consideration as the means of providing service to a lakeshore area. Determine which alternative would result in lower annual cost, using the data that follow.

	Land Route	Submarine Route
Length, in miles	10	5
First cost of cable per mile (includes place-ment costs)	$20,000	$34,000
Annual maintenance per mile	500	1,500
Net salvage per mile	6,000	11,000

For both routes, use annual property tax costs of 2% of first cost, a before-tax rate of return requirement of 15%, and a cable life of 20 years.

6-9. You plan to buy one of two bonds; Bond A costs $1000 now and will pay $40 interest at the end of each year of ownership; it can be redeemed at its maturity value of $1000 ten years from now. Bond B is identical except that its maturity date is just two years distant. It should be apparent that the yield on either issue is 4%; the bonds are issued by the same organization. Is there any difference in the risk involved? Explain.

6-10. See the list of "New York Stock Exchange Bonds" in the *Wall Street Journal* for names of some convertible bond issues; these are indicated by "cv" in the data, for example:

High	Low	Bonds	Sales in $1000	High	Low	Close	Net change
310	171	Collins R cv4 s83	140	286	280	282½	−7

Select a convertible bond with a market value of $200 or more and then see the "New York Stock Exchange Transactions" to note the price of the common stock of that same company, for example:

High	Low	Stocks	Divi-dend	Sales in 100's	Open	High	Low	Close	Net change
85¾	46	Collins Rad	.50	427	78⅞	78⅞	74⅞	75½	−2½

Next, see the appropriate *Moody's Manual* (Industrial, Utility, and so forth) to find the specific terms of conversion of the bond into common stock. Do the terms of conversion and market price of the common stock form a basis of value for the convertible bond you selected?

In our illustration here the bond is convertible at $27.50 per share; that is, each $1000 (face value) bond can be exchanged for $1000/$27.50 = 36.36+ shares of stock (cash in lieu of fractional shares). With every $100 of bonds worth 3.636 shares, and valuing these at the days average market of, say $77:

$$\left(\frac{3.636 \text{ common shares}}{\$100 \text{ of cv bond face value}}\right) \left(\frac{\$77 \text{ market value}}{\text{common share}}\right)$$

$$= \frac{\$280 \text{ market value}}{\$100 \text{ of cv bond face value}}$$

This calculation confirms its (inexact) dependence upon market price of the common stock to which it might be converted. Convertible bonds tend to be valued at the higher of their two uses: (1) as bonds and (2) as converted into common stock. If the market price of common stock falls below the conversion price, a valuation "floor" based on retention as a bond would become the basis of value.

6-11. In attempting to price their only product, the Blue Corporation has made estimates of the following:
(1) Annual labor costs
(2) Annual material costs
(3) Annual overhead costs
(4) Annual equivalent of first cost and salvage on all facilities using *BTRR*
(5) Annual depreciation expense
(6) Annual interest on funded debt
(7) Annual dividend payments
(8) Annual income tax payments
(9) Overhaul cost on equipment (will be needed this year)
(10) Allocated costs of selling and administration
(11) Prorated (over total estimated production quantity) costs of engineering and design
It is proposed that these item costs be totaled and then divided by this years estimated production quantity to determine an average cost per unit. What changes would you suggest and why?

6-12. The following costs have been estimated for several insulation alternatives in a factory:

Thickness of Ceiling Insulation	Installed Cost of Insulation	Annual Cost of Heating and Air Conditioning
inches		
0	$ 0	$15000
1	5000	8200
2	5500	7500
3	6000	6900
4	6500	6400
5	7000	6000
6	7500	5700
7	8000	5500
8	8500	5400

Use a life of 20 years and $i = 20\%$ to find the preferred alternative. Make an effort to minimize your computations.

6-13. Which method of comparison (annual or present equivalent) would be easier and more direct to apply if the following conditions existed? (Explain why you had a preference, if any.)

a. Your objective is to identify the preferred alternative, A or B:

End of year	0	1	2	3	4
Cash flow for A	-100	+20	+10	+50	+70
Cash flow for B	-200	+100	+110	+50	+10

b. You were given the following cash flows with the objective of picking the minimum-cost plan of providing the service over the many years ahead:

Plan A costs X dollars each three years
Plan B costs Y dollars each five years
Plan C costs Z dollars each seven years

c. You wished to "tie in" the results of your study with the income statement of the company.
d. You wished to find the maximum expenditure justified now by future cost saving.

6-14. The University Athletic Council is considering an addition to the football stadium. The 10,000 seat addition would cost $500,000, have a life of 25 years, and have negligible salvage value. Added annual costs of ushers, maintenance, ticket personnel, and so forth, are estimated as $5000 per year. Being a part of the university the facility is exempt from both property taxes and income taxes. The minimum attractive rate of return is given as 7%. Assume (1) that annual costs must be recouped through the sale of additional tickets, (2) that there will be five home games each year, and (3) that on the average, 8000 more tickets costing $4 each will be sold for each game if the 10,000-seat addition is constructed. It is expected that revenues will exceed costs; football has traditionally provided an excess of funds that are used to help support certain other sports. With this in mind, compare annual receipts with annual equivalent costs.

6-15. A tax-exempt university organization has authorized the building of certain parking facilities. Total cost of the land and structure is estimated to be $2 million. The life of the structure is estimated as 50 years and salvage at that time is estimated as $200,000. The minimum attractive rate of return is 5%; since the organization is tax exempt, this is also the before-tax rate of return requirement. Costs for operating and maintenance personnel, materials, and power are estimated as $30,000 per year. Capacity of the structure is 640 cars.

a. Find annual equivalent cost for the parking facility.
b. Find present equivalent cost for the parking facility over the next 50 years.
c. Use an estimated utilization of 25% (because of much lower demand during nighttime hours and university vacation periods) to determine an hourly parking cost per vehicle if the facility is to "just pay for itself," including provision for return on investment.
d. Assume that the rate determined in Part c will be charged, so that $AER = AEC$. Find the dollar change in output (AER or AEC or AEX) that would occur if an input estimate is changed by 20%. Do your computation for each of the following input estimates: (1) salvage, (2) life, (3) operating cost, (4) minimum attractive rate of return, (5) first cost, (6) percent utilization, and (7) actual hourly parking charge. Each result is a good indicator of the responsiveness or sensitivity of output results to error in a specific input estimate.

6-16. The membership dues of the Werner professional society are \$30 per year. However, the society offers an incentive to its members to pay their dues for longer periods. The incentive dues are \$30 plus \$20 for each additional year paid for at one time. Dues are payable at the end of the year preceding the period. If a member estimates his personal before-tax rate of return requirement to be 12%, for what period of years should he pay his dues?

6-17. The following estimates have been made for two alternatives; we must choose one of them. Before-tax rate of return required is 25%.

	A	B
First cost	\$50,000	\$70,000
Life	10 years	10 years
Salvage at retirement	0	0
Annual cost of labor, maintenance, property taxes, insurance, and all other out-of-pocket costs	\$15,000	\$10,000

Try to minimize your computations as you determine which course of action to recommend.

6-18. The following estimates of cost apply to equipment Alternatives A and B. The before-tax rate of return required is 20%.

	A	B
First cost	\$50,000	\$20,000
Operating costs for labor, materials, property taxes, insurance, and maintenance	\$5000 at the end of year one and increasing by \$500 per year	\$10,000 at the end of year one and increasing by \$1000 per year
Overhaul costs	\$5000 every five years	none required
Life	20 years	10 years
Salvage value at end of life	\$10,000 if just overhauled	negligible

No future increases in the cost of any of the items above is expected.

a. Compare present equivalent costs using a study period of 20 years.

b. Compare annual equivalent costs. *do First*

c. Compare capitalized costs.

6-19. What is the maximum amount you should pay for:

a. A bond which can be redeemed 20 years hence for $1000 and which will pay interest of $30 at the end of each year. Use $i = 6\%$.

b. One hundred shares of common stock of the Blue Company. It is thought that their current dividend policy of 100% payout and their current dividend rate of $10 per share per quarter will not change appreciably in the years ahead. Treat dividends as end-of-year cash flows. Use $i = 10\%$.

c. A patent and its estimated income potential of $3000 per year for the first ten years, then $2000 per year for the last seven. Use $i = 12\%$.

d. A rental property with prospective pretax net rental receipts of $2000 per year and a prospective resale value ten years hence of $10,000 if $i = 8\%$.

e. Defend the use of different interest rates in (a), (b), (c), and (d) above to represent your personal before-tax rate of return requirement.

6-20. The following cost estimates have been made with regard to two alternative facilities; it has already been determined that the service is an important and necessary one and that our choice is limited to the two facilities below.

	Plan A	Plan B
First cost	$100,000	$60,000
Salvage, at retirement	$10,000	negligible
Economic life	10 years	10 years
Operating costs for labor, material supplies, repair, maintenance, property taxes, and insurance	$10,000 in year one and growing by $1000/yr	$5000 in year one and growing by $1000/yr
Before-tax rate of return requirement	30%	30%

No future increases in the cost of any of the items above is expected. In addition to the costs above, Plan B calls for a second B unit at the end of year four; its salvage value at year ten (when it is six years old) is estimated to be $10,000.

a. Find the present equivalent cost of ten years service for Plan A and Plan B.

b. Find the annual equivalent cost for the second B unit over years five through ten inclusive. Find the annual equivalent cost for the first B unit over years one through ten inclusive. Comment on the effect of period of retention on annual equivalent costs for the B units.

c. Without calculation, speculate upon the relationship of (1) a preference for Plan A or Plan B and (2) the before-tax rate of return requirement which is used.

6-21. Many economy studies involve the comparison of a used equipment presently in service with a new equipment proposed for replacement. If the straight-line depreciation plus return on the average investment method had been used in such studies over many years by a firm, would you expect the firm's replacements, based on this method, to have a consistent difference in timing from replacement timing based on exact methods? Explain.

6-22. A company owns a fleet of trucks and operates its own maintenance shop. A certain type of truck, normally used for five years, has a first cost of $4000 and a salvage value of $1000. Maintenance costs are $500 the first year and increase by $200 each year. Assuming before-tax rate of return required is 20%, find the annual equivalent cost of owning and maintaining the truck.

6-23. Find the uniform annual equivalent cost of the following equipment for a municipality (exempt from income taxes). The equipment, if accepted, will cost $20,000 and financing will be 50% by a bond issue with an interest cost of 4%, and 50% by current revenues of the city. The equipment will cost $2000 to install and will have operating and maintenance expenses estimated at $5000 per year. Estimated salvage value is $4000 at the end of ten years, and we are told straight-line depreciation will amount to $1600 per year. The simple average (not time adjusted) of interest payments to municipal bond-holders is $200 per year. The annual redemption of bonds (related to this equipment) will be $800 per year.

6-24. What is the present worth of the cost of 18 years of service from a machine that has a first cost of $22,000, a life of 18 years, an estimated salvage value of $2000 at the end of life, and annual operation and maintenance costs of $5000? Assume the required before-tax interest rate is 12%.

6-25. You have been asked to compare two alternative bridge designs. The first, a timber structure, has a first cost of $20,000; negligible salvage is expected at its retirement ten years later. Its annual maintenance and other cash costs are estimated as $1000 per year. Similar units with similar costs, are thought to be available in the future. The second design is a steel structure with first cost of $40,000 and negligible salvage expected at its retirement

50 years later. Annual maintenance and other cash costs are estimated as $500 per year for the steel structure. The service need is for a minimum of 50 years and the before-tax rate of return required is 15%.

a. Compare annual costs of the alternatives *using the approximate method*, 2, straight-line depreciation plus return on the average investment.

b. Compare annual equivalent costs of the alternatives using the exact method.

c. Comment on the difference in your results for Parts a and b.

6-26. A machine costing $10,000 has an estimated economic life of ten years and an estimated salvage value of $1000. Maintenance expenses are estimated at zero for each of the first five years and $500 for each of the last five years. Property taxes and insurance are estimated to be $300 the first year and to decrease by $20 per year. Space costs and other costs are estimated as $700 per year. Use a before-tax return requirement of 20% to find the annual equivalent cost of the machine.

6-27. The following estimates apply:

	Equipment A	Equipment B
First cost	$22,000	$11,000
Salvage	$ 2,000	negligible
Life	6 years	4 years
Cash operating costs including maintenance, property tax, and insurance	$ 5,000/yr	$10,000/yr

Use the present equivalent cost method to find the preferred alternative. *BTRR* = 20%.

7

Rate of Return

Rate of Return As a Capital Expenditure Criterion

IF THE RATE OF RETURN *required* (either *BTRR* or *MARR*) for a project is known or specified, we recommend use of the *PEX* criterion given in the preceding chapter. If not, we recommend use of the rate of return, *ROR*, criterion. To apply the *ROR* criterion, we need to compute the rate of return that will be realized if project estimates are realized.

The *prospective*[1] rate of return on an investment is the interest rate at which present equivalent revenues equal present equivalent costs, that is, *PER* = *PEC*, or *AER* = *AEC*, or *FER* = *FEC*, or *AEX* = 0, or *FEX* = 0, or

$$PEX = 0 = \sum_{z=0}^{n} (1 + i)^{-z} x_z$$

---

[1] A more precise term, the *internal rate of return*, emphasizes that we limit ourselves to consideration of project earnings only and disregard the rate of earnings on cash flows (cash throwoff) reinvested in other projects.

where i = rate of return

    n = life of the project, in years

    x_z = the net cash flow in year z

If x_z is expressed on a before-tax basis, the resulting rate of return will also be on a before-tax basis.

If the rate of return required is unknown for a group of mutually exclusive projects, one should use the network diagram procedure of this chapter to reach a conditional decision. If the rate of return required is known, the *PEX* criterion is simpler to apply.

For a group of independent projects, rank them in descending order of prospective rate of return. Then approve projects according to rank until funds available are exhausted. Rate of return studies, properly applied, yield accept-reject decisions that are consistent and compatible with those of the *PEX* criterion.

Example 7-1.   Solving for the Rate of Return

Information: These estimates apply to a proposed project:

$BTCF$ = \$70,000/year = before-tax cash flow

  B = \$250,000 = first cost

  V = \$100,000 = salvage

  n = 5 years = project life

Objectives: Find the prospective before-tax rate of return.

Analysis:

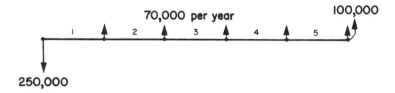

At the solving rate, i, PEX = 0:

$$PEX = -\$250,000 + \$70,000 \, (p/a)_5^i + \$100,000 \, (p/f)_5^i$$

Try i = 15%:

$$PEX = -\$250,000 + \$70,000 \ (3.352)$$
$$+ \$100,000 \ (0.4972) = +\$34,360$$

Try $i = 20\%$:

$$PEX = -\$250,000 + \$70,000 \ (2.991)$$
$$+ \$100,000 \ (0.4019) = -\$440$$

So $i = 15\% + 5\% \ (34,360/34,800) = 19.9\%$. These and other *PEX* values are plotted in Figure 7-1.

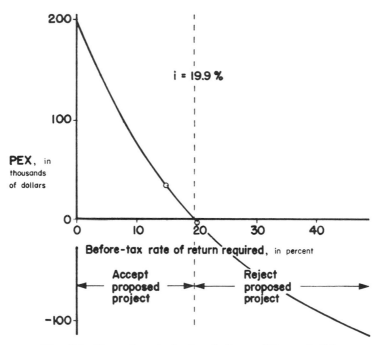

Fig. 7-1. Present equivalent cash flows of Example 7-1.

Since *BTRR* was not specified, we can only state our conclusion in a conditional way:

If *BTRR* is:	Choice
>19.9%	Reject proposed project
≤19.9%	Accept proposed project

Interpolation of Interest Rates

Unless otherwise specified, the problems and examples in this book which call for computation of rate of return are computed via interpolation to the nearest 0.1%. Since the compound interest factors are only approximated by the process of interpolation, a slight error is introduced. To keep this error to a minimum, it is suggested that interpolation be attempted only between adjacent tabled values, for example, between 5% and 6% but not between 4% and 10% since that would introduce a still greater error due to interpolation.

Example 7-2.   Rate of Return Studies Involving Alternatives with Known Differences in Costs (and Revenues) but Unknown Levels of Same (see Example 6-3)

Information: Two competing methods of performing a required service are to be compared; selection of one of these two courses of action (methods) is mandatory. Estimates pertaining to the choice:

	Method A	Method B
First cost	$100,000	$200,000
Life	4 years	8 years
Salvage at retirement	0	$10,000
Operating, maintenance, and property tax costs	Unknown, say K_1 per year	$10,000 per year less than for A
Revenues	Unknown, say K_2 per year	Unknown but same as for A

Assume that eight years of service are required and the replacement unit required for Method A will have the same set of costs as the original unit.

Objective: Find the more economical alternative.

Analysis: The overall rate of return on either of the methods cannot be determined without knowledge of the revenues.

This lack of information is no deterrent here, however. Because the selection of one of the two methods is mandatory, the pertinent question is not, Is the investment worthwhile? but is rather, Is the *extra* investment of B over A worthwhile? An answer to this latter question is facilitated through use of a cash flow diagram.

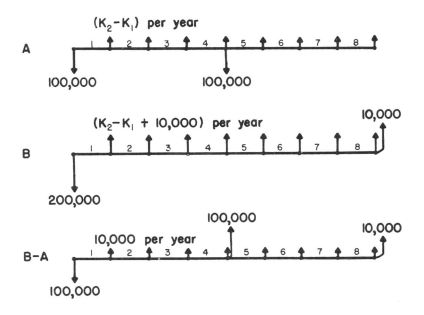

$$PEX = \$0 = -\$100{,}000 + \$10{,}000 \, (p/a)^i_8 + \$100{,}000 \, (p/f)^i_4$$
$$+ \, \$10{,}000 \, (p/f)^i_8$$

At $i = 10\%$: $PEX = -\$100{,}000 + \$10{,}000 \, (5.335)$
$$+ \, \$100{,}000 \, (0.6830) + \$10{,}000 \, (0.4665) = +\$26{,}315$$

Should the next trial be made at a higher rate than 10%? Note that the rate of return would be exactly 10% if the extra cost of Plan B at year zero were $126,315 rather than $100,000. With a present cost of only $100,000 the rate of return must be still higher.

At $i = 15\%$: $PEX = -\$100{,}000 + \$10{,}000 \, (4.487)$
$$+ \, \$100{,}000 \, (0.5718) + \$10{,}000 \, (0.3269) = +\$5319$$

At $i = 20\%$: $PEX = -\$100{,}000 + \$10{,}000 \, (3.837)$
$$+ \, \$100{,}000 \, (0.4823) + \$10{,}000 \, (0.2326) = -\$11{,}074$$

From the calculations we can observe that the rate of return is between 15% and 20%. Since tables for interest rates between 15% and 20% are not provided in this book, we next interpolate:

$$i = 15\% + 5\% \left[5319/(5319 + 11{,}074)\right] = 16.6\%$$

Since this example did not specify *BTRR*, we can only state that the rate of return on the extra investment in Method B is 16.6%; if *BTRR* is equal to or less than 16.6%, the extra investment in B is warranted. This same statement of choice could be shown as a table:

If *BTRR* is:	Choice
>16.6%	A
⩽16.6%	B

In Example 6-3 a *BTRR* of 15% was given and, as we might now expect, Method B was indicated as the more economical of the alternatives. We might observe that while annual equivalent, present equivalent, and rate of return methods are compatible, the information provided is not the same. Perhaps as a matter of compensation for the extra computational effort required, we are provided some insight into the sensitivity (or insensitivity) of our results to small changes in one of the factors (*BTRR*) influencing the acceptance-rejection decision.

Because service was stated to be a mandatory function, we are forced to choose between Method A and Method B. This choice is not influenced by the rate of return on the *overall* investment in A or B; both may be very profitable or unprofitable. In this example it is the rate of return on *incremental* (extra) investment which requires analysis for the acceptance-rejection decision.

Suppose that in Example 7-2 we computed the various pretax rates of return for $K_2 - K_1 = \$25{,}000$ and for $K_2 - K_1 = \$50{,}000$.

$K_2 - K_1$	Overall A	Overall B	Increment B - A
	percent		
$25,000	0.0	8.9	16.6
50,000	35.2	25.3	16.6

When told that you must choose one of these two alternatives and $K_2 - K_1 = \$25,000$, which would you choose if $BTRR = 20\%$? A is the correct response; only if $BTRR \leqslant 16.6\%$ should you choose B. When told that you must choose one of the two alternatives and $K_2 - K_1 = \$50,000$, which would you choose if $BTRR = 15\%$? B is the correct response; only if $BTRR > 16.6\%$ is B unacceptable. Perhaps the clue to the proper interpretation is to recall that our objective is to maximize *profit*, not profit *rate*. Note that the four different overall rates of return of A and B are really *irrelevant* bits of information. The necessary information was already provided in the rate of return on incremental investment analysis.

Mutually Exclusive Multiinvestment Opportunities

The examples and problems which follow deal mainly with the cash flows of the various opportunities and generally do not describe the physical circumstances giving rise to the cash flows. While the brevity and emphasis so achieved seem desirable, it also seems desirable that the reader appreciate the wide applicability of the multiinvestment opportunity which includes such choices as one from a set of:

Different machines to perform a stated function.
Proposed sites for an apartment building (or office building, road, service station, or any of numerous commercial possibilities).
Available pipe sizes for transportation of gas (or water, sewage, gasoline, coal slurry, and the like).
Standard thicknesses of insulation for heating or cooling applications of buildings, pipes, and the like.
Various levels of development of a land area for recreation, hotel, or other uses.
Given levels of pollution control.
Various levels of reduction in smoke output via precipitators.
Possible warehousing or material handling systems consisting of combinations of subsystems.

The sets above involve varying degrees of *separability*. In some cases the increments tend to be physically and financially separable (as with the subsystems of warehousing or material handling system possibilities). In other cases they tend to be *indivisible* (as with the land for a building site). The distinction is important for both capital budgeting purposes and for observing the finality of a decision. Use of one building site in favor of another tends to be an irreversible action; it is difficult to retrieve the opportunity of the rejected site. With a warehousing system, however, when we accept only certain parts of the total system now, we may still be able later to add the rejected (or postponed) part of the system.

Several terms may be of aid in describing our alternatives. A *contingent* alternative is one which requires that some other alternative be accepted first; thus acceptance of an overhead conveyor is contingent upon acceptance of the structure which houses it. The structure is a *prerequisite* of the conveyor system. Since acceptance of the structure is *not* contingent upon acceptance of the conveyor, the relationship of the alternatives is not *symmetric.*

Example 7-3.   **Comparing Three Mutually Exclusive Alternatives by the Rate of Return Method**

Information: We *must* choose one of three alternatives, A, B, or C. In each case the before-tax cash flow will occur for ten years, then cease, but the value of the investment will not diminish with time. Assume that (1) the alternatives are mutually exclusive; (2) the before-tax rate of return requirement, *BTRR*, is not known; and (3) sufficient funds are available for all investments having a prospective rate of return equal to or greater than the unknown *BTRR*. Alternative A costs $2000 and produces a cash flow of -$100/yr; B costs $3000 and produces a cash flow of $150/yr; C costs $4000 and produces a cash flow of $320/yr.

Objective: Determine which alternative will maximize profit (or minimize loss) at various *BTRR* values.

Analysis: The computation of rate of return happens to be especially simple in this problem because the investments do not decline in value.

$$AEX_C = 0 = -\$4000 \, (a/p)_n^i + \$4000 \, (a/f)_n^i + \$320$$
$$0 = -\$4000 \, i + \$320$$
$$i = \$320/\$4000 = 8\%$$

Alter- native	Invest- ment	Annual Before-Tax Cash Flow	Rate of Return on Incremental Investment Compared to:		
			0	A	B
A	$2000	- $100	- 5%		
B	3000	150	5%	25%	
C	4000	320	8%	21%	17%

Recall that one of the three alternatives must be selected; Alternative 0, rejection of all three, is not permitted. The minimum investment which can be made is $2000 for A. Any additional investment would have to satisfy the *BTRR* criterion. So our choice table reads:

If $25\% < BTRR$, choose A

If $17\% < BTRR \leqslant 25\%$, choose B

If $\quad\quad\quad BTRR \leqslant 17\%$, choose C

These choices can be verified by computing *PEX* for various *BTRR* values:

$$PEX_A = -\$2000 + \$2000 \,(p/f)^i_{10} - \$100 \,(p/a)^i_{10}$$
$$PEX_B = -\$3000 + \$3000 \,(p/f)^i_{10} + \$150 \,(p/a)^i_{10}$$
$$PEX_C = -\$4000 + \$4000 \,(p/f)^i_{10} + \$320 \,(p/a)^i_{10}$$

So when $BTRR = i = 25\%$,

$$PEX_A = -\$2000 + \$215 - \$357 = -\$2142$$
$$PEX_B = -\$3000 + \$322 + \$536 = -\$2142$$
$$PEX_C = -\$4000 + \$430 + \$1143 = -\$2427$$

By similar calculations we can produce Figure 7-2. The six circled intersections are break-even points and are the six rates of return listed in the example. The choice table determined by rate of return analysis is identical to one determined according to the criterion of maximizing *PEX* for various *BTRR* rates.

Some Guides to Simplifying the Rate of Return Analysis When Three or More Alternatives Exist

Example 7-3 presents a fairly simple comparison in that:

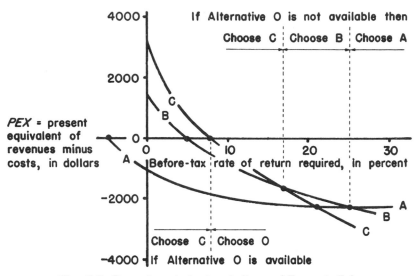

Fig. 7-2. Present equivalent cash flows of Example 7-3.

1. With salvage values estimated to be 100% of first cost, capital repayment and return is simply i; and rate of return can be determined without reference to tables.
2. Rate of return on the overall investment in A is identical to the rate of return on the incremental investment of A over 0.
3. Only three alternatives are involved.

When the "no investment" alternative does not exist and when three or more mutually exclusive alternatives are compared, the number of rates of return on incremental and overall investment is $(x^2 + x)/2$ where x is the number of alternatives available. It is apparent from the expression that where a large number of alternatives are being considered, the number of rates of return to be computed are large; for example, with $x = 7$, a total of 28 rates of return apparently require computation.

    This requirement is only slightly reduced when one of the alternatives is the "no investment" alternative, in which case the number of rates of return to be computed is $(x^2 - x)/2$; when $x = 7$, apparently 21 rates of return require evaluation. If the computational effort required is increased by use of salvage values other than 100% and if the number of alternatives considered is large, the task of comparison could indeed be tedious work. The network diagram and analysis which follow were developed in response to such need.

Network Analysis

When *BTRR* is unknown for a set of mutually exclusive alternatives, the preferred alternative can usually be specified only in a conditional way (for example, choose B if 17% < *BTRR* ≤ 25%). As the number of alternatives considered is increased, the complexity of analysis also increases. As an aid to such analysis, the network diagram described next can be of help in understanding and communicating the responsiveness of the preferred alternative to the variable *BTRR*.

Consider the data of Example 7-3. A network diagram of rate of return relationships can be drawn as an *x*-sided figure to represent the *x* alternatives. Arrange these clockwise in ascending order of investment, connecting all points in the diagram and labeling the rate of return represented by each. Use arrowheads to indicate the direction of ascending investment. Any polygon in the diagram is now subject to interpretation as follows: consider triangle *ABC*, start with the minimum investment point (*A*) and proceed in order of ascending investment to note that incremental investment *B-A* yielding 25% plus incremental

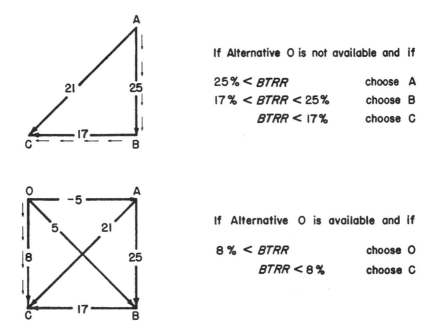

If Alternative O is not available and if

25% < *BTRR*	choose A
17% < *BTRR* < 25%	choose B
BTRR < 17%	choose C

If Alternative O is available and if

| 8% < *BTRR* | choose O |
| *BTRR* < 8% | choose C |

investment *C-B* yielding 17% equals incremental investment *C-A* yielding 21% (and analogous to the resultant in a force diagram).

To construct the choice table, (1) start at the minimum investment alternative using a very large *BTRR* value, then (2) gradually decrease *BTRR* until it permits "escape" to a larger investment, and (3) continue this process until the largest investment is reached. Small arrows have been used to show the "escape" paths.

A second diagram is shown for the condition that Alternative 0 is available. The two network diagrams permit us to make proper choice tables without resorting to the many *PEX* calculations required to produce the compatible but more tediously derived results appearing in Figure 7-2.

Example 7-4.   **Comparing Five Mutually Exclusive Alternatives by the Rate of Return Method**

Information: Five mutually exclusive alternatives are available to us. Investment in one of the five is mandatory; *MARR* itself has not been specified, but the company assures us that any projects showing a prospective rate of return equal to or greater than *MARR* will be accepted. This does not mean that unlimited funds are available; long experience in budgeting of capital expenditures has convinced management that only a limited number and value of projects are capable of meeting the threshold of acceptability, *MARR*. Thus management is confident that the limiting factor at *MARR* will be projects available rather than funds available.

Each of the alternatives below is expected to have an infinite life and produces annual after-tax cash flow savings as shown:

Alternative	First Cost	After-Tax Cash Flow
A	$8000	$920
B	5000	510
C	7000	820
D	6000	640
E	4000	400

Objective: Produce a choice table showing the preferred alternative at the various *MARR* possibilities.

Analysis: First, list the alternatives in order of ascending investment. Second, start with the minimum investment alternative, E, and compute rate of return, *ROR*, on overall and incremental investments (some of these will subsequently prove to be irrelevant):

Alterna-tive	First Cost	After-Tax Cash Flow	Overall ROR	ROR on Incremental Investment over:			
				E	B	D	C
			percent				
E	$4000	$400	10.0				
B	5000	510	10.2	11.0			
D	6000	640	10.7	12.0	13.0		
C	7000	820	11.7	14.0	15.5	18.0	
A	8000	920	11.5	13.0	13.7	14.0	10.0

Use the network diagram to observe that E is the preferred alternative as long as *MARR* exceeds 14% (recall that investment in one of the five is mandatory); for smaller values of *MARR* the extra investment in C will be worthwhile. Note that when *MARR* is small enough for B and D to be satisfactory, C is still better (and therefore C *dominates* B and D). Finally note that when *MARR* is less than 10%, A is preferable to C.

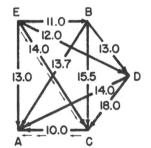

If Alternative O is not available and if

14% < *MARR*	choose E
10% < *MARR* < 14%	choose C
MARR < 10%	choose A

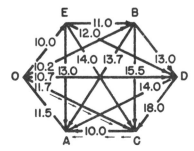

If Alternative O is available and if

11.7% < *MARR*	choose O
10% < *MARR* < 11.7%	choose C
MARR < 10%	choose A

Problems in Which either More than One or No Real Solution to Rate of Return Exists

For nearly all realistic problems in economy studies there exists one and only one solving rate of return (that is, a unique solution). There are, however, a few cash flow patterns which can result in other than one (a nonunique) solving rate of return. By counting the number of changes in the sign of accumulated cash flow, we can find the maximum number of positive real roots *possible*. In Example 7-5, for instance, there *may* be as many as two positive real roots:

End of Year	Net Cash Flow	Accumulated Net Cash Flow
0	-100	-100
		change in sign
1	+230	+130
		change in sign
2	-132	-2

If accumulated year-by-year net cash flows change sign only once, we are assured of a unique solution; with zero changes, a negative rate of return is signaled.

Example 7-5.   **Problems Where either No Solution or Several Solutions Exist to Rate of Return**

Information: An expenditure-receipt-expenditure after-tax cash flow pattern is shown below:

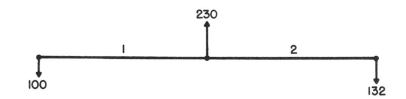

Objective: Find the rate of return.

Analysis: Summation of worths at end of Period 2 =

$$FEX = \$0 = -\$100\,(1 + i)^2 + \$230\,(1 + i) - \$132$$

$$= (1 + i)^2 - 2.30 (1 + i) + 1.32$$
$$= [(1 + i) - 1.1] [(1 + i) - 1.2]$$

so $i = 10\%, 20\%$.

Note that either solution satisfies the equation. Factoring was employed in the solution shown; the quadratic formula could have been applied:

$$1 + i = (- b \pm \sqrt{b^2 - 4ac})/2a$$

where a = coefficient of the first term, in this case, - 100
b = coefficient of the second term, in this case, +230
c = coefficient of the third term, in this case, - 132

As the formula hints and as will be noted shortly, small changes in any one of the three cash flows can result in other dual solutions, a single solution, or imaginary number solutions.

Example 7-5 raises some interesting questions.

1. Who earns the computed rate(s) of return? If we call the expenditure-receipt-expenditure pattern a lending pattern and its opposite a borrowing pattern, was the rate of return earned by borrower or lender? As a matter of fact, which is the borrower's cash flow pattern and which is the lender's?
2. Do slight changes in the cash flow data reveal a "supersensitivity" in the dual rate type of problem? To this question we can respond affirmatively after some calculations which lead to the following analysis:

If the third cash flow amount of $132 had been:	Then the solving rate(s) of return would be:
$130.00	0% and 30%
132.00	10% and 20%
132.25	15%
Over 132.25	imaginary numbers

3. Could there be a cash flow series with more than two solutions? The factoring solution of the example suggests that such a series can be created simply by reversing the factoring process and multiplying by desired solving rates. That is, if we wish to illustrate a cash flow series solving at $i = 10\%$,

20%, and 30%, we simply multiply:

$$[(1 + i) - 1.1] [(1 + i) - 1.2] [(1 + i) - 1.3]$$
$$= (1 + i)^3 - 3.6 (1 + i)^2 + 4.31 (1 + i) - 1.716$$

The above will solve at $i = 0.10$, 0.20, and 0.30. The cash flow series is

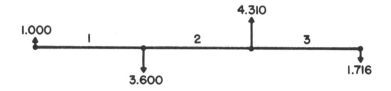

and it should be apparent that in like fashion cash flow series having a larger number of solutions can be constructed. Indeed there should be some series which solve at any rate of interest.

Figure 7-3 and the accompanying data show that with a first cost of $100.20 rather than $100 there would have been just one solving rate of return, 15%. Had the first cost been

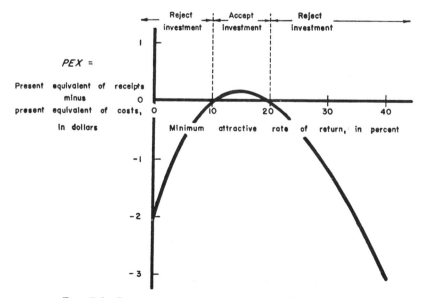

Fig. 7-3. Present equivalent cash flows of Example 7-5.

more than $100.20, there would be no solving rate of return. If we had stopped with our first solution of $i = 10\%$ and then examined the relationship of first cost to rate of return, we might even have concluded that rate of return could be increased by paying $0.20 more (first cost) for the equipment.

Figure 7-3 is based on computations as indicated below:

$$PEX = -\$100 + \$230 \ (p/f)_1^i - \$132 \ (p/f)_2^i$$

So for $i = 5\%$,

$$PEX = -\$100 + \$230 \ (0.9524) - \$132 \ (0.9070) = -\$0.67$$

and similarly at other interest rates.

Situations which could give rise to cash flow patterns resulting in nonunique rates of return include:

1. Incremental cash flows.
2. The purchase of larger oil pumping equipment to hasten the withdrawal of oil from a well.
3. A cash cycle for nuclear fuel when initial costs are followed by receipts from use or sale and finally by the high disposal costs of used nuclear fuel.

The practical problem which involves nonunique rate of return solutions is uncommon. The problems and examples hereafter are limited to those having unique solutions. When nonunique solutions are encountered in practice, an analysis such as portrayed in Figure 7-3 will be helpful in concluding that the project is acceptable if the minimum attractive rate of return is between 10% and 20% inclusive. As noted earlier, the *possibility* of a nonunique solution is signaled when accumulated year-by-year net cash flow changes sign more than once.

Some Cautions in the Interpretation of Interest Rates

In several instances an apparent "earning" or "dividend" is actually a reduction in cost to the recipient. The benefit is in cost, not income, and the federal income tax regulations so treat such matters. Some examples include:

1. The "dividend" on a life insurance policy when it is actually a partial refund of premium based on current mortality ex-

perience (as with U.S. National Service Life Insurance and many commercial life insurance policies).

2. The "dividend" on the share held in a co-op type of firm; when based on purchases made through the co-op, the dividend is simply a reduction in the cost of goods purchased.

3. The investment credit realized through the Investment Credit Act of 1962 and its subsequent revisions provides a direct tax reduction based on the dollar volume of equipment purchased and provides the owner a reduction (via income taxes) in costs, not a return on investment.

4. The dividend paid by a firm with its own stock certificates is referred to as a "stock dividend." The owner's equity in the firm is not changed and tax regulations exclude such stock dividends from taxable income.

There are also some instances in which an interest rate as stated does not actually reflect the true interest rate. These "gimmicks" are apparently a part of our lending system, so a brief description of several of these devices is in order. (A "truth in lending" bill requires, with certain exceptions, that lenders state interest rates on a consistent and uniform basis.)

1. Interest charged on the original balance rather than on the declining balance of a loan causes understatement of the true interest rate. Six percent interest on the original balance is roughly equivalent to 12% interest on the declining balance.

2. Discounting. If you were to borrow $15,000 for the purchase of a home, those funds may be "discounted" so that you receive only $14,700. Effectively the interest rate you are paying is still greater than that stated. If nominal $i = 6\%$ and $n = 25$ end-of-year payments, the true rate is 6.2%.

3. Nominal interest rate. The typical 6% house mortgage actually calls for interest of 0.5% per month which means an *effective* annual rate of 6.2% because of the frequency of compounding.

4. Prepayment of interest. Suppose you were to borrow $1000 at 6% interest. If required to prepay interest, you would receive $940 now and agree to pay $1000 at the end of the year. The effective interest under this circumstance is 6.4%. This device can easily be coupled with that of (1).

5. Required loan insurance. FHA loans require the purchase of loan insurance requiring the payment of monthly premiums.

These premiums increase the FHA loan costs by about 0.5%, from 6% to 6.5% per year, for instance.

6. Administrative charges for credit investigation, handling, and so forth. Especially on small loans these charges can add significantly to the effective interest rate paid by the borrower.
7. "Time price" greater than negotiated cash price. When merchandise purchased on credit costs more than it would at a negotiated cash price, the difference adds significantly to the interest rate actually being paid.
8. Acceleration penalties. Mortgage contracts may or may not contain acceleration penalties; GI mortgages do not, FHA mortgages do. "Acceleration" is prepayment in part or full on the mortgage contract and could be desirable when a mortgagee has an excess of cash or desires to refinance his property at lower interest rates.
9. Compulsory noninterest-bearing escrow account. Some mortgages require monthly deposits toward the annual payment required for property taxes and insurance. If the deposits are held in a noninterest-bearing escrow account, as is usually the case, the effective cost of the loan can be increased by as much as 0.3% (as with an average escrow balance of $500 and an average unpaid loan balance of $10,000 @ 6%).

And finally, we should note instances in which the individual receives more than the stated rate of interest such as:

1. Through savings accounts which state annual nominal interest rates but compute interest payments more frequently than once per year.
2. "Save by the tenth, earn from the first" provisions give the individual something over and above the stated rate of interest.
3. Add-ons such as life insurance coverage, special gifts, and the like.

Summary

When the rate of return comparison is complete, the acceptance-rejection decision will be identical to that indicated by annual or present equivalent comparisons. Additional information, such as sensitivity of the decision to before-tax rate of return required, is observable and may sometimes be helpful to the overall analysis. Certain advantages can be claimed for the rate of return method; primary among these are (1) the broader picture of choice under various values of *BTRR* rather than the narrow one-rate picture portrayed by annual or present equivalent methods (the rate of return method provides a good scale by which sensitivity of conclusions to *BTRR* or *MARR* may be displayed) and (2) subsequent conceptual use of rate of return for establishing a funds priority queue in the overall capital budgeting program of an organization.[2]

Certain disadvantages can also be claimed:

1. The revenue attributable to a particular portion of a system may not be subject to objective and meaningful estimation (for example, What portion of company revenue should be attributed to an insulator on an electric utility power pole?). Knowledge of revenue is generally necessary to determine rate of return.

2. Computation effort is generally greater, especially where the cash flow is irregular or the number of alternatives is large.

3. There are a few unusual and infrequently encountered problems in which more than one or no real solution to rate of return exists.[3]

Problems

7-1. You own a certain property and have narrowed your choices to two:

 A. Dispose of the property now in an "as is" condition, and despite present market conditions, for $11,000.

 B. Retain the property and recondition it at a cost now of $5000. Sell the property a year from now, under

---

[2] Although there are many who advocate use of the ratio of discounted cash flow to investment as a priority device, the businessman's experience is in terms of profit rates, not number ratios. Both the rate of return and ratio of discounted cash flow to investment (an application of the present equivalent technique) are limited in their usefulness as actual priority devices.

[3] It might be countered, however, that the nonunique solution instance revealed by rate of return techniques is obscured by the narrow one-rate picture portrayed by annual or present equivalent methods. If annual or present equivalent methods are used to investigate the effects of varying *BTRR*, they too will produce nonunique responses to such questions as, Is the investment acceptable for *BTRR* values under 20%?

more favorable market conditions, for an estimated $20,000.

Apply rate of return techniques to reach and state a decision rule on a before-income-tax basis.

7-2. Find the pretax rate of return on a $10,000 investment which has zero salvage and which produces annual pretax net receipts of $2000 for each of the next ten years.

7-3. A magazine subscription may be purchased at the following rates: one year, $5; two years, $8; five years, $16.

  a. Find the rate of return on the extra prepayment when a purchaser buys one two-year subscription rather than two one-year subscriptions.

  b. Find the rate of return on the extra prepayment when a purchaser buys one five-year subscription rather than five one-year subscriptions.

  c. Find the rate of return on the extra prepayment when a purchaser buys two five-year subscriptions rather than five two-year subscriptions.

7-4. An investment company offers you a plan in which end-of-year payments of $54.86 for 15 years will "mature" to $1000. What rate of return does this investment offer you?

7-5. We are faced with three mutually exclusive alternatives. A rate of return analysis is made on each alternative so that the following pretax rates of return are known:

Test #1. Rate of return on present equipment.
Test #2. Rate of return on extra investment in proposed equipment.
Test #3. Rate of return on proposed equipment.

These rates are compared with the company's before-tax rate of return requirement of $i\%$. The computed rate of return either "passes" the test (if equal to or greater than $i\%$) or "fails" it (if less than $i\%$). As a result of the tests you must reach one of the following conclusions:

D. Dispose of present equipment.
K. Keep present equipment.
R. Replace present equipment with proposed equipment.
I. An impossible set of test results.

Mark the following test results with the letter D, K, R, or I to indicate your conclusion from the following series of "test" results.

Test #1	Test #2	Test #3	Conclusion
Pass	Fail	Pass	_____ a
Fail	Fail	Fail	_____ b
Pass	Fail	Fail	_____ c
Fail	Fail	Pass	_____ d
Pass	Pass	Pass	_____ e
Fail	Pass	Fail	_____ f
Pass	Pass	Fail	_____ g
Fail	Pass	Pass	_____ h

7-6. Find the before-tax rate of return on an equipment of the Van Osdel Company if:

First cost = $10,000
Salvage = - $1000 (negative) at end of year ten
Maintenance, property taxes, and so forth = $2000 in year one, $2200 in year two, and so forth, increasing by $200 per year
Overhaul costs = $3000 at the end of year five
Revenues = $5000 per year

7-7. Bonds of a certain firm are currently available at a cost, including broker's commission, of $900 each. The bonds mature exactly ten years from now; at maturity the bonds will be redeemed for $1000 each by the firm. Interest payments to bondholders amount to $30 per bond per year; although such payments are generally semiannual, you are requested to treat the interest payment as if it were a single end-of-year amount with the first payment occurring one year from now and the last payment ten years from now.

Find the prospective pretax rate of return to todays buyer who holds his purchase to maturity date.

7-8. The Upsidasium Book Binding Company has asked you to find the before-tax interest rate when $51 is obtained now in exchange for a promise to pay end-of-year amounts of $3.60 at the end of year one, $3.75 at the end of year two, and so forth, growing by $0.15 per year and continuing forever.

7-9. Alternative 0 is the "do nothing" alternative.
Alternative S costs $10,000 and will result in a before-tax cash flow of $13,000 at the end of year one (the overall rate of return is therefore 30%).
Alternative L costs $10,000 and will result in a before-tax cash flow of $15,625 at the end of year two (the overall rate of return is therefore 25%).

Assume that an unlimited supply of capital is available at *BTRR* and that the alternatives are mutually exclusive; use rate of return techniques to respond to the question, In what range of *BTRR* is S the proper choice?

7-10. Some mutually exclusive alternatives are given below. You *must* choose one of them. They are listed in ascending order of investment size. Their lives are identical, and salvage values are negligible. Sufficient funds are available to provide for any alternative meeting the before-tax rate of return requirement. Pretax rates of return are given.

Alternative	Overall ROR	ROR on Incremental Investment over:		
		A	B	C
		percent		
A	10			
B	15	25		
C	12	14	3	
D	16	20	18	32

a. In what range of *BTRR* is A the proper choice?
b. In what range of *BTRR* is B the proper choice?
c. Make the choice table for Alternatives A, B, C, and D for the various possible *BTRR* values.
d. If the zero investment alternative, 0, is added to the set, how would you respond to Part c?
e. If a shortage of funds causes us to choose A instead of D, what is the cost of our investment opportunity foregone, expressed as a rate of return percentage?

7-11. You have developed five *independent* investment alternatives:

Alternative	First Cost	Life, in Years	Pretax Rate of Return
			percent
1C	$10,000	5	25
2A	20,000	20	30
3G	15,000	5	20
4D	40,000	10	40
5D	25,000	5	35

a. Which alternatives would you accept if your *BTRR* were 28%? What total funds are required?

    b. Which alternatives would you accept if your funds available were limited to $65,000? What *BTRR* is inferred by your action?

    c. What problem arises if you have only $22,000 to invest? What are some possible courses of action?

7-12. Seven years ago Mr. Green purchased a convertible debenture by paying $100 for the bond and $30 for the rights required and the handling charges. Since the date of purchase Mr. Green has received end-of-year interest payments of $4 each year. Today he sold his bond for $380. Find his rate of return on investment.

7-13. You have been asked to consider a set of five mutually exclusive alternatives from which you may choose one. These are listed in ascending order of investment size; their lives are identical, as are their salvage ratios. We have been assured that when a *BTRR* is specified, there will be sufficient funds to provide for all investments meeting or surpassing the *BTRR* threshold. Pretax rates of return are:

Alternative	Overall *ROR*	*ROR* on Incremental Investment over:			
		A	B	C	D
A	35%				
B	32%	26%			
C	30%	25%	24%		
D	26%	20%	17%	10%	
E	25%	20%	18%	15%	20%

    a. Without performing any calculations, state which alternative is the proper choice if *BTRR* = 13%. Repeat for *BTRR* = 23%, *BTRR* = 33%, and *BTRR* = 43%.

    b. In what range of *BTRR* values is C the proper choice?

    c. Where x alternatives are involved, a total of $(x^2 + x)/2$ rates of return can be displayed. In this example $(x^2 + x)/2 = 15$; how many of the 15 rates are relevant in responding to Part a of this question?

    d. Are any of the alternatives dominated?

    e. Make a table to show the proper choice for any *BTRR* value.

7-14. On July 11, 1966, the British government issued a new "national development" bond paying annual interest of $5\frac{1}{2}$%. As an incentive to retention of the security, a 2% tax-free bonus was promised to those who held the bond to maturity (July 11, 1971). Find the pretax rate of return to the buyer who retained the security to maturity.

7-15. You have been given three mutually exclusive alternatives to consider; one must be selected. Sufficient funds to cover any alternative meeting the before-tax rate of return requirement are available.

Alterna-tive	Invest-ment	Before-Tax Savings (increased before-tax cash flow)	Life	Salvage
A	$ 70,000	$25,000	6	0
B	100,000	30,400	6	$30,000
0	0	0	0	0

a. Find the relevant rates of return and make a choice table for the various *BTRR* possibilities.
b. Use the annual equivalent method to find the preferred alternative when *BTRR* = 20%.
c. When *BTRR* is known and multialternatives are involved, which method would you recommend for ease of computation, annual equivalent cost or rate of return?
d. Which method (annual equivalent or rate of return) would be most helpful if you hoped to judge the *sensitivity* of your choice to *BTRR*?

7-16. Ten years ago Mr. Blue purchased 100 shares of the Green Corporation at a total cost, including commissions, of $5000. Since that time he has received quarterly dividends which you are asked to treat as if they were end-of-year amounts. Dividends totaled $100 the first year, $130 the second year, and so forth, increasing by $30 per year. Yesterday Mr. Blue sold the stock for a net of $10,000. Find the pretax rate of return on his investment.

7-17. We can compute three different rates of return on two mutually exclusive equipment alternatives as follows:

(1) Rate of return on investment in Equipment A.
(2) Rate of return on incremental investment in Equipment B compared to A.
(3) Rate of return on investment in Equipment B.

The results can then be compared with a company policy of a minimum attractive rate of return of *X*%. The computed rate of return either "passes" the test (if greater than $X\%$) or "fails" it (if less than $X\%$). As a result of the tests you must reach one of the following conclusions:

0. Reject both A and B.
A. Choose A.
B. Choose B.
N. Need results of the other test.

Mark the following test results with 0, A, B, or N to indicate your conclusion from the following series of "test" results.

Test #1	Test #2	Test #3	Conclusion
Pass	Fail		_____ a.
Fail	Fail		_____ b.
	Fail	Fail	_____ c.
	Fail	Pass	_____ d.
Pass	Pass		_____ e.
Fail	Pass		_____ f.
	Pass	Fail	_____ g.
	Pass	Pass	_____ h.
Pass		Fail	_____ i.
Fail		Fail	_____ j.
Pass		Pass	_____ k.
Fail		Pass	_____ l.

$$\left(\begin{array}{c}\text{Investment}\\\text{required for A}\end{array}\right) + \left(\begin{array}{c}\text{incremental}\\\text{investment}\end{array}\right) = \left(\begin{array}{c}\text{investment}\\\text{required for B}\end{array}\right)$$

Note that A and B can be conceived of as:

(1) Two equipment alternatives.
(2) An existing facility which, with an incremental addition, becomes the proposed facility.
(3) An existing equipment which, when traded in with cash "to boot" (incremental investment) provides sufficient funds for the proposed equipment.

7-18. The Blue Corporation has just completed the cycle of obtaining approval from the Securities and Exchange Commission, issuing the necessary information, bearing the costs of printing of bond certificates, and paying the various legal fees involved in a new bond issue. After deduction of these costs and underwriter fees, the company netted $10 million from the issue of 20-year bonds with a maturity value of $11 million and a coupon rate of 5%. Annual handling and clerical costs are estimated as $50,000. Treat the costs and interest payments as if they were end-of-year amounts, and find the actual cost of these borrowed funds expressed as an interest rate.

7-19. The pretax rates of return for three mutually exclusive alternatives are given below. The alternatives are listed in ascending order of investment size and have identical lives and negligible salvage values. Funds are available for any projects which satisfy the before-tax rate of return requirement. Because the service performed is mandatory, the "do nothing" alternative is not permitted.

Alternative	Overall *ROR*	*ROR* on Incremental Investment over:	
		A	B
		percent	
A	-4		
B	6	16	
C	8	14	12

a. Make a choice table for A, B, and C at the various *BTRR* values.

b. If the zero investment alternative, 0, is added to the set, how would you respond to Part a?

7-20. Alternative D. Drop the present activity and dispose of equipment.

Alternative K. Keep the present activity and equipment.

Alternative R. Keep the present activity and replace equipment with updated equipment which earns a pretax rate of return of 35% on the extra investment in the updated equipment.

Assume that there is an unlimited supply of capital and the three alternatives are mutually exclusive; use rate of return concepts to state the circumstances (*BTRR* and pretax *ROR*) under which D is the proper choice. Repeat for K and R.

7-21. A certain manufacturing facility is being considered; the following estimates have been developed:

Cost of land = $200,000 = resale value at end of year ten

Cost of buildings and equipment = $650,000; at the end of ten years it is thought that their value will have declined to $250,000

Working capital required = $50,000

Start-up costs of promotion, research, engineering, training, tooling, and other costs to be recouped over the ten-year life of the project = $100,000

Annual net receipts = operating revenues - operating costs = $200,000 per year

Find the prospective pretax rate of return.

7-22. You have been given three alternatives, A, B, and 0 (do nothing). Use rate of return techniques to produce a rule for choosing the proper alternative when the before-tax cash flows are

End of Year	0	1	2
Cash flow for A	- 1000	+1470	0
Cash flow for B	- 1000	0	+1960

and the before-tax rate of return requirement is unspecified.

7-23. Use the following description of investments and cash flows associated with them to compare pairs of alternatives by *inspection*. When one of the pair is better ($BTRR > 0\%$) by inspection, write, for example, $D > A$ to show D is better than A. Show as many of these as inspection permits.

	Initial	Before-Tax Cash Flow		
Investment	Cost	Year 1	Year 2	Year 3
A	$10,000	$10,000	0	0
B	10,000	5,000	$5,000	$ 5,000
C	10,000	2,000	4,000	12,000
D	10,000	10,000	3,000	3,000
E	10,000	6,000	4,000	5,000
F	10,000	8,000	8,000	2,000

7-24. Five mutually exclusive alternatives are under consideration; each has a life of ten years and zero salvage value. Rate of return on overall investment and on incremental investment is given for each alternative. For what minimum attractive rate of return range is alternative C the "best" alternative? There is an ample supply of both projects and financing at *MARR*. One of the five alternatives must be chosen. They are listed in ascending order of investment size.

Alternative	*ROR* on Overall Investment	*ROR* on Incremental Investment Compared to Alternative:			
		A	B	C	D
		percent			
A	30				
B	28	26			
C	26	24	22		
D	22.5	20	17	12	
E	24	22.5	21.3	21	30

7-25. The owner of a certain property has just received a cash offer of $87,500 on a "now or never" basis. The present owner had not planned to sell the property but had planned to keep it for the next ten years during which time he expected pretax net receipts (before-tax cash flow) of $10,000 per year. After the ten-year period he expected no further receipts or salvage value. Use rate of return techniques to help the owner determine whether he should sell or retain the property. When you find the rate of return, carefully state how it is to be interpreted.

8

Break-even Studies

The Break-even Concept

A BREAK-EVEN point is the value of a selected variable which will produce equivalence in cost and/or revenue. When a certain value of a selected variable results in equivalence between alternatives (the alternatives are equally attractive), the relationship may be expressed as a *break-even* point or *standoff* and the choice as a matter of *indifference*.

Some Reasons for Break-even Studies

Chapter 2 introduced and illustrated *fixed cost* and *variable cost* concepts. Many equipment decisions involve the replacement of man by machine and therefore the replacement of a variable cost (labor) with a fixed cost (investment repayment, return, and related income taxes). Frequently, such a choice is dependent upon sales and production level, with the man system producing lower costs for small output requirements and the machine system resulting in lower costs for large output requirements. Analysis of fixed and variable costs helps to reveal the advantage of either system with the obvious inference that at some level of output the man versus machine costs break even. This analysis also helps to show that the replacement of variable costs by fixed costs results in a profit structure more sensitive to variations in the sales and production levels. Such increased sensitivity results in (1) a reduction in the flexibility to operate economically at reduced sales and production levels and (2) profitability which is more responsive to increased sales and production levels.

Break-even Analysis and Uncertainty

Essentially all data employed in an economy study are uncertain simply because they represent estimates of the future. In some instances we are especially wary of making an estimate because either our information is sketchy or the item is so widely variable. Our very general suggestion will be to (1) examine the parameters and their likely variability; (2) select that one parameter to whose variability the conclusions of the economy study appear most sensitive; and (3) treat that parameter as an unknown variable, solving for its value where the conclusion is a standoff (break-even) situation. Thus, for example, when the parameter to whose variability the conclusions of the economy study appear most sensitive is:

1. *Revenue.* Solve for the revenue required to equal (break even with) the equivalent costs as treated in Chapter 6.[1]
2. *Required rate of return.* Solve for the required rate of return for which choice of alternatives is a standoff (break even) as in Chapter 7.
3. *Sales and production rate.* Solve for the rate required for revenues to break even with costs as illustrated in Figure 2-5.
4. *Equipment life or duration of sales rate.* Solve for the time required for revenues to break even with costs.

Note that uncertainty alone is not the guide to selecting the parameter to be so treated; salvage value generally is rather uncertain, but it will be shown in later chapters that our decisions are seldom sensitive to even large errors in the estimation of salvage.

The usefulness of the break-even concept extends to still other applications, particularly as a means of providing insight

---

[1] As used throughout this book, costs include income taxes and the required rate of return unless specifically noted otherwise.

into such circumstances as:

1. The time period required for the sum of net revenues from the investment to just equal the investment without providing for return on debt or equity capital or income taxes (rate of return = 0 = income taxes). The time period thus computed is called the *payoff period, payout period,* or *payback* period. (In any case the reader is cautioned that the omission of required rate of return and income taxes leaves this method at best as an arithmetically convenient but crude treatment; it provides helpful insights but should not be used for the final acceptance-rejection decision.)
2. The annual sales and production level required for two competing types of equipment of unlike capacity to be equal in equivalent cost.
3. The product price decrease (increase) at which the increased (decreased) revenue just equals the increased (decreased) costs of production, and thus still further change in price is no longer beneficial.
4. The incremental product promotion expenditure which just pays for itself via increased sales.
5. The rate of inflation of prices required for several alternative courses of action to be equivalent.
6. Future income tax rates at which various courses of action would produce similar economy.
7. Future equipment resale values which would result in an indifference as to timing of retirement.

Example 8-1.   Profitability and the Rate of Output

Information: The firm which you organized in Chapter 2 has grown considerably; the following is a projection of the income statement for the year ahead. Administrative costs ($600) are paid to you for services rendered and are independent of the return on investment and production quantity.

Operating revenues (10,000 units @ $0.70 each)	$7000
Operating costs (labor $4100, material $900, administrative costs, $600)	- 5600
Depreciation expense ($7000 - $3000)/4	- 1000
Interest on debt capital	0
Before-tax return on equity (taxable income)	+$ 400

The required before-tax rate of return is 20%; the existing investment consists of investment in working capital and equipment worth $7000 today and with an estimated value of $3000 four years from now. From this information and the methods of Chapter 6, annual equivalent costs could be estimated as follows:

Before-tax cash flow required for equipment and
   working capital = $7000 $(a/p)_4^{20\%}$ - $3000 $(a/f)_4^{20\%}$ = $2145
Operating costs for labor, material, and
   administrative costs   5600
Annual equivalent costs $\overline{\$7745}$

Comparing the costs and revenues, we find a deficit of $745; if there is no hope of improving revenues, efficiencies, and so forth, we might well consider disposing of, rather than updating of, this operation.

    Since the number of units presently produced is 10,000, the equation for total costs (fixed plus variable, with administrative expenses of $600 having been determined as a fixed cost) of producing q units by present methods is

Total cost = fixed cost + variable cost
$$= \$2145 + \$600 + [(\$4100 + \$900)/10,000]\, q$$
$$= \$2745 + \$0.50q$$

where q = the number of units produced.

    In reviewing the situation, you note the availability of an Automatic Assembler machine costing $6000; the machine will mechanize certain operations now being performed by hand. The machine is estimated to be useful for about four years and at that time its net salvage is expected to be zero. The machine is attractive from two standpoints: (1) It will save an estimated $2500 per year in labor cost and (2) it will double the theoretical output capacity of productive facilities (presently 20,000 units). Sales levels (10,000) have never actually reached present production capacity although the trend does point to increased sales.

a. *Objective:* Find annual equivalent cost and use this as a basis for acceptance or rejection of the machine.

   Analysis:

    AE first cost and salvage = $6000 $(a/p)_4^{20\%}$
$$= \$6000\,(0.38629) = \$2318$$

The estimated savings of $2500 in labor costs exceed the annual equivalent cost of $2318, so the proposed equipment is preferable to the present equipment. It is not yet certain, however, that the proposed equipment is superior to the disinvestment alternative.

b. *Objective:* Show the effect that acquisition and use of the Automatic Assembler would have upon the income statement and total-cost equation.

Analysis: The effects of use of the Automatic Assembler upon the annual income statement are indicated below:

	Before	After	Net Change
Operating revenues (sales)	$7000	$7000	0
Operating costs (labor, materials, administrative costs)	5600	3100	- $2500
Depreciation expense	1000	2500	+ 1500
Interest on debt capital	0	0	0
Before-tax return on equity = (taxable income) =	$ 400	$1400	+$1000

and the new equation for total costs including the 20% before-tax rate of return is

Total cost

$$= \$2745 + \$6000 \, (a/p)_4^{20\%} + \frac{\$4100 + \$900 - \$2500}{10,000} q$$

$$= \$5063 + \$0.25q$$

where q = the number of units produced.

c. *Objective:* On the basis of data already given, find the sales and production level at which the present and proposed methods break even.

Analysis: Several break-even points are shown in Figure 8-1:

q_1 = the requested break-even point where costs (includ-

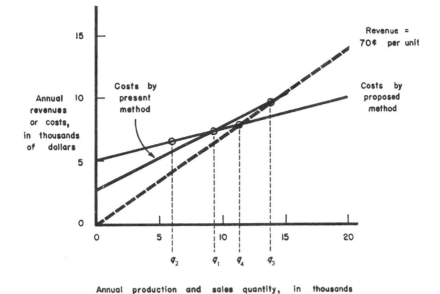

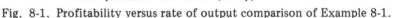

Annual production and sales quantity, in thousands

Fig. 8-1. Profitability versus rate of output comparison of Example 8-1.

ing the 20% pretax return) of the two methods break even.

$$\$2745 + \$0.50q_1 = \$5063 + \$0.25q_1$$
$$q_1 = 9272$$

q_2 = the break-even point where costs, exclusive of required pretax profit, of the two methods break even (thus the rate of return would be 0%).

$$\$1600 + \$0.50q_2 = \$3100 + \$0.25q_2$$
$$q_2 = 6000$$

q_3 = the break-even point where revenues and costs (including the 20% pretax return) of the present method are equal.

$$\$2745 + \$0.50q_3 = \$0.70q_3$$
$$q_3 = 13,725$$

q_4 = the break-even point where revenues and costs (including the 20% pretax return) of the proposed method are equal.

$$\$5063 + \$0.25q_4 = \$0.70q_4$$
$$q_4 = 11{,}251$$

The preceding data show that for any production and sales level above 9272 units per year the proposed method is superior to the present method, but they also show that our required rate of return will not be satisfied by the proposed method unless production and sales levels can be increased to 11,251 units per year. Acceptability of the Automatic Assembler is still uncertain.

d. *Objective:* On the basis of data already given, find the time required for the proposed equipment to pay for itself at 0% interest.

Analysis: The time period required above is best known as the *payoff period,* the ratio of the expenditure to the net annual cash receipts or savings. In this case

Payoff period = $6000/$2500 = 2.4 years

e. *Objective:* Using the fixed and variable costs computed earlier for the proposed equipment, find the average cost per unit at the following annual output levels: 10,000, 15,000, 20,000, 25,000, and 30,000.

Analysis: Total costs are $5063 + $0.25 per unit of output.

Annual Output	Total Cost	Average Cost per Unit	Incremental Cost per Unit
10,000	$ 7,563	$0.76	$0.25
15,000	8,813	0.59	0.25
20,000	10,063	0.50	0.25
25,000	11,313	0.45	0.25
30,000	12,563	0.42	

f. *Objective:* Assume the Automatic Assembler has been purchased and installed. In an attempt to enhance present levels of profit one might examine the effect of a change in the selling price, presently $0.70 per unit. Assume that a market study has been conducted and the following estimates of the relation of sales volume to prices have been made:

If product price is:	Then estimated sales are:
$0.40	30,625
0.50	19,600
0.60	13,611
Present level....0.70.........10,000.....
0.80	7,656
0.90	6,049

Find the price at which profit is maximized.

Analysis:

A	B	C	D	E
Price	Sales level	Revenues = (col. A) (col. B)	Costs = $5063 + ($0.25) (col. B)	Extra profit* = col. C - col. D
$0.40	30,625	$12,250	$12,719	- $ 469
0.50	19,600	9,800	9,963	- 163
0.60	13,611	8,167	8,466	- 299
0.70	10,000	7,000	7,563	- 563
0.80	7,656	6,125	6,977	- 852
0.90	6,049	5,444	6,575	- 1131

*Extra profit = pretax profits in excess of the required before-tax rate of return of 20%.

Based on the data above, reducing the price to $0.50 per unit will increase profit, presumably to a maximum for the choices given and neglecting the possibility of some in-between price such as $0.53.

g. *Objective:* After the preceding price reduction to $0.50 per unit was made, the sales level did increase to an annual rate of about 19,600 units. Up to this point no funds had been used for product promotion (for example, advertising). A study was conducted and the following estimates of the relation of sales volume to promotion expenditures were made:

If $ _____ are spent for product promotion:	Then the present sales level would be increased to that shown below:
0	19,600
$1000	26,000
2000	29,200
3000	29,200
4000	26,000

Find the promotional expenditure level at which profit is maximized. Neglect possible cumulative or lagged benefits from the present advertising outlay and neglect the interdependence of price and promotion expenditures.

Analysis: The *marginal cost* per additional unit is $0.25, while the *marginal revenue* per additional unit is $0.50. This means that up to $0.25 per unit could profitably be spent in product promotion. For each $1000 of promotion, sales must increase at least 4000 units to justify the expenditure. By inspection, $1000 of promotion appears to maximize profits (again, from the choices given). This conclusion can be verified as below:

A	B	C	D	E
			Costs = $5063 +	
		Revenues =	($0.25)	
Promo-	Sales	(col. B)	(col. B)	Extra profit =
tion	level	($0.50)	+ col. A	col. C - col. D
0	19,600	$ 9,800	$ 9,963	-$ 163
$1000	26,000	13,000	12,563	+ 437
2000	29,200	14,600	14,363	+ 237
3000	29,200	14,600	15,363	- 763
4000	26,000	13,000	15,563	- 2563

Through (1) acquisition of new equipment, (2) reduction of selling price from $0.70 to $0.50 per unit, and (3) expenditure of $1000 for product promotion, the difference between annual equivalent revenues and annual equivalent costs (required to produce a before-tax rate of return of 20%) is changed from a deficit of $745 to an excess of $437 (in accounting terminology the effect is an increase in pre-tax profits from $400 to $2400 per year), while sales increase from 10,000 to 26,000 units per year.

TABLE 8-1. Estimated Profitability for a Certain Product at Various Levels of Price and Promotion

	4000	-6344	-2563	-841	-216	-126
Promotion	3000	-3117	- 763	+105	+261	+ 96
expenditure,	2000	-1063	+ 237	+510	+360	+ 48
in dollars	1000	- 180	+ 437	+376	+ 86	-267
	0	- 469	- 163	-299	-563	-852
		0.40	0.50	0.60	0.70	0.80
			Price, in dollars			

Note: The profit figure shown in the table is the pretax profit in excess of the required 20% before-tax rate of return.

Isoplanar Views and Contour Maps

In Example 8-1 we treated promotion expense and product price as if they were two independent factors. In general, however, additional promotion and price discounting tend to be partly substitutive factors, that is, they are interdependent. If promotion and price are interdependent, then optimizing each on a "one at a time" basis may not optimize profit.

To explore the possible interdependence and to enhance our optimization procedure, we might perform market studies to reveal a more complete picture than that already found in Example 8-1 and shown in bold figures in Table 8-1. The additional estimates are shown as light figures in the same table.

The nature of the relationship of price, promotion, and profit might be still further clarified through graphic representation. One method of graphically portraying the relationship of three variables on a two-dimensional surface is to hold constant one of the variables (for example, price); this is the basis upon which the data of Table 8-1 were estimated. Table 8-1 is presented graphically as Figure 8-2.

Several terms used in the graph require definition. An *isoplane* is a cutting plane parallel to an axis (it therefore represents a fixed quantity of one input or output variable). An *isoplanar view* is a cross-sectional view of an isoplane. An *isoprice* line is one (as from an isoplanar view) on which price is constant. In general, isolines are characterized by equal magnitudes of some variable. Thus we also show *isopromotion* and *isoprofit* lines. We can also include such terms as *isodesirability* = *isoquants* = *indifference curves.* [2]

---

[2] Further exposition of isoquant analysis is provided by D. V. Heebink in "Isoquants and Investment Decisions," *The Engineering Economist,* Vol. 7 (Summer 1962) and Vol. 8 (Fall 1962).

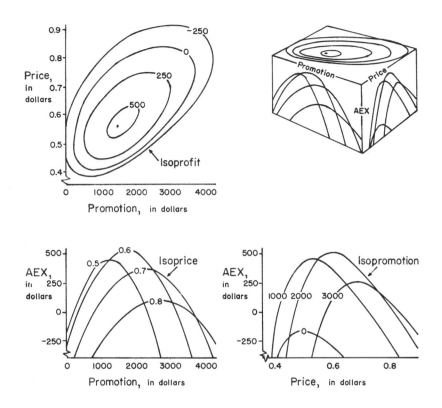

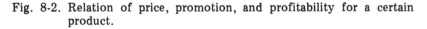

AEX = annual profit in excess of that produced at the required rate of return

Fig. 8-2. Relation of price, promotion, and profitability for a certain product.

A *contour map* is a graphic display having lines connecting points of equal desirability (isoquants, in this case isoprofits, as illustrated in the price versus promotion graph). The contour map can also be termed a *profit hill* where we seek a peak; had its dimension been cost instead of profit, we could seek the saddle (low) point of the *cost bowl*. The contour map draws its name from the similar and analogous topographic elevation map. Figure 8-2 aids us in a number of important observations:

1. It is clear that optimizing price and then optimizing promotion has failed to maximize profit (though only slightly). The graph suggests further investigation of prices between $0.50 and $0.60 and promotional outlays between $1000 and $2000. Further study indicates that a price of $0.56 and

a promotional outlay of $1500 should produce sales of 22,903 units and a profit of $537.[3]
2. Price cuts and promotional outlays tend to be partly substitutive for one another. This should be no surprise, however, since price cuts are, in a sense, a form of promotional effort.
3. If a product is underpriced, underpromotion may be necessary to maximize profit at that price. If a product is overpriced, overpromotion may be necessary to maximize profit at that price.
4. The penalties (lost profits) from overpricing are not as severe as the penalties of underpricing; because of this typical but not universal circumstance, the firm may, under conditions of uncertainty, choose to slightly overprice the product.

The contour map can be of considerable aid in the optimizing process as long as no more than three variables are involved. Tabulation and further tabulation (iterative sectioning by isoplanes) can also be helpful. Where the output variables (for example, cost, profit) can be expressed by an equation, differentiation may be helpful (but not always, for example, when the differentiation is too cumbersome).

Table 8-2 illustrates a means by which the search for the optimum price and promotional outlay could be pursued.[4] It also illustrates the use of incremental analysis since *marginal revenue* and *marginal cost* concepts can be applied to determine the best of the alternatives given.

Optimum price for the choices given can be obtained (1) by selecting the one which maximizes "extra profit" or (2) by accepting successive price reductions as long as marginal revenue exceeds marginal cost (they can be compared on either a total or a per unit basis).

---

[3] Based on the equations developed in the further study

$$\text{Profit} = (\text{price} - \$0.25)\,\text{quantity} - \$5063 - \text{promotion}$$

where

$$\text{Quantity} = \frac{4900 + 2\,(\text{promotion}) - \text{promotion}^2/[10,000\,(\text{price})^2]}{\text{price}^2}$$

[4] The method shown is not the most efficient in determining the exact optimum. More efficient generally are the method of par-tangents, or the method of steepest ascent, or differentiation, perhaps coupled with a computer. The point here, however, is an illustration of the relationship of marginal analysis with the profit-maximizing process.

TABLE 8-2. **Marginal Revenue and Marginal Cost in the Analysis of Profitability**

Promotion Outlay	Selling Price per Unit	Sales in Units per Year	Total Revenue	Marginal Revenue Total	Marginal Revenue Per Unit	Total Cost	Marginal Cost Per Unit	Marginal Cost Total	Extra Profit
$1500	$0.60	20,208	$12,125			$11,615			$510
				$341	26.5¢ ◄— accept —► 25¢			$322	
1500	0.58	21,496	12,466			11,937			529
				360	25.6¢ ◄— accept —► 25¢			352	
1500	0.56	22,903	12,826			12,289			537
				375	24.3¢ ◄— reject —► 25¢			386	
1500	0.54	24,446	13,201			12,675			526

Changes in Input Factors Related to Rising Levels of Output

As production levels increase, there is an observable tendency to substitute fixed cost inputs (for example, equipment) for variable cost inputs (for example, labor). When equipment inputs partly replace labor inputs, the production function is said to be more *capital intensive,* a characteristic which distinguishes some firms in the United States from their counterparts elsewhere in the world. This capital intensive nature is, in part, attributable to (1) relatively high production levels, (2) a relatively abundant supply of investment funds, and (3) a relative scarcity of manpower.

In time it is likely that there will be changes in the prices or required quantities of input factors, the demand, and the responsiveness of demand to price or promotional policies. For this reason it is essential to recognize that any present "solution" will be subject to future reviews and change.

Beginning with the data of Chapter 2 and continuing through Chapter 8, a number of changes in production quantity, method, selling price, promotional outlay, and equipment are involved. These changes are detailed in Table 8-3.

Note that sales levels responded to price reductions. The price reductions are limited by costs; in Situation 2, price had fallen below the level at which costs, including the pretax rate of return, could be recouped. Per unit costs generally do fall as production quantities rise; the relationship here can be roughly described by the equation:

$$\text{Cost} = 11.97q^{-0.305}$$

where q = the sales and production quantity in units per year. The equation produces the predicted costs shown below and in Figure 8-3.

TABLE 8-3. Changing Input Factors Related to Rising Levels of Output

Situation	Selling Price per Unit	Production and Sales Quantity per Year	Fixed Costs per Unit at That Output	Variable Costs per Unit	Total Costs per Unit at That Output
1. See Chapter 2	$1.45	1,000	$0.69	$0.75	$1.44
2. "Present method" as first described in Example 8-1	0.70	10,000	0.27	0.50	0.77
3. Latest method noted after the conclusion of Example 8-1 and with the help of the contour map	0.56	22,903	0.29	0.25	0.54

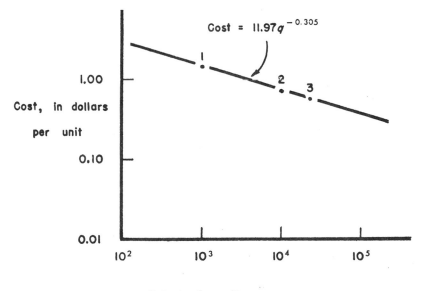

Fig. 8-3. Evolution of cost per unit versus rate of output under changing production facilities.

Production and Sales Quantity per Year	Predicted Total Cost per Unit at That Output
1,000	$1.46
10,000	0.73
22,903	0.56

Figure 8-4 shows the relationship among the various costs of production. Annual equivalent cost of repayment and pretax return on investment for the existing equipment plus the Automatic Assembler had already been computed as $4463; it can be verified in

$$\$13{,}000 \ (a/p)_4^{20\%} - \$3{,}000 \ (a/f)_4^{20\%} = \$4463$$

To separate this annual equivalent cost into its two parts, we need only to find the repayment of ($13,000 – $3,000)/4 = $2500 and by subtraction find the pretax return as $1963. This latter figure can be verified by noting that the pretax return needs to be earned on declining balances of $13,000, $10,500, $8000, and $5500 at the end of years one through four respectively; this means returns of $2600, $2100, $1600, and $1100 are required in those years. Their uniform equivalent is

$$\$2600 - \$500 \ (a/g)_4^{20\%} = \$1963$$

Using this information, we can find the production and sales quantity which would produce a 0% return:

Revenues = costs with a 0% return
$$\$0.56q_1 = \$2500 + \$600 + \$1500 + \$0.25q_1$$
$$q_1 = 14{,}839 \text{ units}$$

or the quantity which would produce a pretax return of 20%:

$$\$0.56q_2 = \$4463 + \$600 + \$1500 + \$0.25q_2$$
$$q_2 = 21{,}171 \text{ units}$$

or the quantity which would produce a pretax return of 40%; since

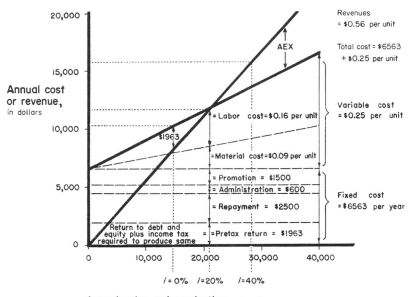

Fig. 8-4. Variation of costs and revenues with the level of production for the proposed method of Example 8-1.

$$\$13,000\,(a/p)_4^{40\%} - \$3000\,(a/f)_4^{40\%} = \$6608$$
$$\$0.56q_3 = \$6608 + \$600 + \$1500 + \$0.25q_3$$
$$q_3 = 28,090 \text{ units}$$

Utilization of Capacity

Although the Automatic Assembler increased production capacity to 40,000 units per year, the actual utilization amounts to only $22,903/40,000 = 57\%$ of that capacity. Since additional units of production have to bear the extra costs of labor and materials only, it is apparent that efforts to find new uses and new markets for the product are worthwhile, as are efforts to use the facilities for other products. In retrospect, we might also question whether a less costly Automatic Assembler of smaller capacity might have been available and whether future growth in sales will soon absorb the unused capacity. One indicator of the relative significance of plant and equipment costs is the ratio of these costs to the sales dollars. In our example, about $\$4463/\$12,826 \sim 35\%$ of the sales dollar is required to meet plant and equipment costs. (A still higher percentage would be reflected if this simplified example included the property taxes and insurance costs resulting from the investment.)   Still another helpful indicator is the ratio of annual sales dollars to invested capital. In our example the invested capital goes from $13,000 now to $3000 four years from now, an average of $8,000, so the ratio is $12,826/$8000 = 1.60. The *Fortune 500* reports the following industry medians for "sales per dollar of invested capital":[5]

Merchandising firms	4.91
Aircraft and parts	4.16
Food and beverage	3.29
Apparel	3.29
Soaps, cosmetics	2.85
Appliances, electronics	2.58
Motor vehicles and parts	2.54
Rubber	2.53
Metal products	2.40
Office machinery (includes computers)	2.38
Publishing and printing	2.21
All Industry	2.14
Textiles	2.09
Farm and industrial machinery	2.04
Shipbuilding and railroad equipment	2.00
Chemicals	1.72
Paper and wood products	1.70
Pharmaceuticals	1.59
Metal manufacturing	1.58
Glass, cement, gypsum, concrete	1.49
Tobacco	1.34
Petroleum refining	1.27
Mining	1.13
Transportation companies	0.77
Utilities	0.64

Two terms which are helpful in describing capacity characteristics are *capacity factor* and *demand factor:*

---

[5] See current *Fortune* directory for most recent data. Invested capital is end-of-year net worth. For merchandising firms, transportation companies, and utilities the figure reported is the average of the 50 largest firms in that group.

    Capacity factor = (average demand)/(rated capacity)
    Demand factor = (average demand)/(maximum demand)

In Figure 8-5 the varying demands upon the generating system
of an electric utility are shown. It can be seen that maximum
demand is 4 megawatts; by careful inspection we can find the
average demand is 2 megawatts and thus for the data given, the
capacity factor is 40% and the demand factor is 50%.

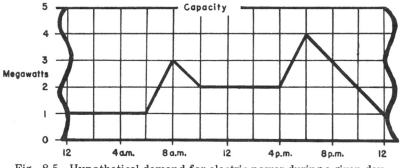

Fig. 8-5. Hypothetical demand for electric power during a given day.

    We should also note that capacity is generally composed of
similar but not identical units. In the trucking firm, the airline,
the electric power company, or most any type of firm there is a
subtle but continuous *demotion* of equipment from first-line
service to standby service. Terborgh shows this clearly with his
studies of numerous industrial equipment items, each of which
is operated to a lesser extent each year.[6] For a group of similar
equipment items of various ages the demands tend to fall first
on the newest and lastly on the oldest of the equipment items
which make up the total capacity capability. Because of the
wide disparity in ages and value of the equipment which com-
prises capacity, we should note that oldest units of capacity
tend to have both low value and high operating costs, while the
reverse is true for the newer units. The penalties of unused
capacity are therefore not necessarily linear nor are they always
as great as inferred by consideration of new units only. Figure
8-6 is an attempt to emphasize this point.

    Suppose that a trucking company owns five trucks which are
similar in size but vary in age as indicated below:

---

    [6] See G. Terborgh, *Dynamic Equipment Policy* (New York, McGraw-
Hill, Inc., 1949).

Truck Number	Age in Years	Annual Equivalent Costs	Miles Driven per Year	Total Costs per Mile
1	1	$5000 + $0.12 per mile	50,000	$0.22
2	2	4000 + 0.15 per mile	40,000	0.25
3	3	3000 + 0.18 per mile	30,000	0.28
4	4	2000 + 0.21 per mile	20,000	0.31
5	5	1000 + 0.24 per mile	10,000	0.34

Because of these cost characteristics, the company follows the policy of assigning work to the lowest numbered (lowest-cost) truck available. As a slight modification of this policy, short-haul jobs are sometimes assigned to high-numbered trucks to maintain availability of low-cost trucks for long haul and to permit service work on the lower-numbered units.

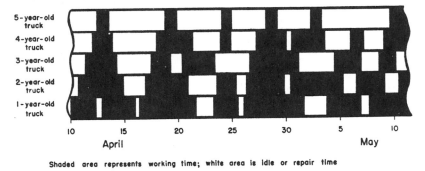

Shaded area represents working time; white area is idle or repair time

Fig. 8-6. Hypothetical demand for trucks during a given month.

These policies can be observed in Figure 8-6. Note that demotion of high-numbered units has really taken on two forms: (1) reduced utilization and (2) reduced category of utilization (primarily short-haul rather than long-haul).

It is nonetheless apparent that idle capacity is not costless. Primary among the causes of our problems is the apparent need to provide capacity which will meet peak rather than average rates of demand. Both the trucking company and the electric utility had peak demands far in excess of average demand; capacity needs could be reduced if only demand were a nonfluctuating function. The department store in the Christmas season or the farmer's use of corn-drying equipment also illustrate extreme seasonal peaks of demand. In like fashion, storm sewer systems experience the peaks and valleys of demand with heavy rainfall or its lack.

In treating this problem, at times of peak demand the manager might simply choose to "turn away" some of that demand. This certainly is observable in those universities where student applications exceed acceptances in ratios of 4 to 1 or perhaps even 10 to 1; on the other hand, the franchise of the public utility rules out the right to "turn away" customers, and the consequences of the storm sewer "turning away" water must be reckoned with. On some intermediate ground are the laundromat, barber shop, and repair shop where customers will tolerate a certain amount of waiting for service. Waiting line (queuing) studies can be excellent tools of analysis for such problems.

When production capacity is already strained, production schedule "stretch-outs" may also be practiced in helping to ease the strain and vice-versa.

Another possibility in treating the problems of widely fluctuating demand is the storage of excess output in times of low demand to help meet peak demand. The electric utility can store excess generated energy with the pumped storage of water in elevated reservoirs (potential energy). Because storage space and inventory investment are not without cost, the manufacturer of large and costly goods may find the storage possibility is of very limited help. Gas distributing utilities are sometimes able to use natural underground "containers" for the storage of gas pumped to them in the summer and stored for use in the winter months.

Other measures which can help to smooth the peaks and valleys of demand include pricing practices where "bargain rates" are offered for the off-peak hours of the bowling alley, theater, electric utility energy, or off-peak day travel by airlines. Similar seasonal bargains by the department store or by dealers in farm implements or automotive equipment all help to increase the utilization of existing facilities. The success or failure of such measures is dependent upon the responsiveness of consumers to the opportunity; there is even the danger that, if overresponsive, a new and higher peak demand of another type or at another hour will have been created. Somewhat analogous to these measures is the sale to foreign markets ("dumping") of extra production at bargain rates which little more than cover incremental production costs.

Additional possibilities include the multishift operation of a factory or the addition of evening hour operation for the supermarket. A multiplicity of product lines, customer types, areas serviced, and seasonal characteristics can be helpful, as illustrated in the manufacture of ice skates and roller skates or in the provision of electric power to both industrial (primarily daytime) and residential (primarily evening time) consumers.

The manager is frequently left with the problem of doing the best he can with the uncertain future growth in sales and the restriction that economic incremental additions to capacity seldom are available in sizes which exactly fit the needs of the firm. The problem of determining the size of the most economic increment of added capacity is deferred to Chapter 15.

Still other options are open to the manager; where surplus capacity is available, he may well consider "pulling in" some of the components presently purchased or the subunits presently subcontracted. He may consider the opposite action where capacity is already strained and thus reduce the apparent peaks and valleys of demand by letting these fall over numerous suppliers whose diversified products and outlets may cushion the change (but not so for the supplier with few products and outlets). This brings us to the make or buy decision.

Make versus Buy Decisions

While problems of either insufficient or excess capacity frequently lead to the question of whether the firm should procure a given item internally or externally, we should recognize that many other circumstances, such as cost reduction studies, can also lead to this same question. We will amplify the economic considerations by way of an example; before we do, however, it is well to note that some factors not treated in the example can be of great importance in the final analysis. Before taking action we should examine effects of the proposed decision upon such factors as:

1. The relationship, both legal and ethical, with the present or proposed supplier.
2. The controllability for scheduling and engineering design purposes.
3. The probable need for and cost of liaison with the supplier.

Example 8-2.   A Make versus Buy Decision

Information: Recent cost studies of 3 of the 100 items manufactured by the Green Corporation show the following costs per unit. An outside supplier has bid on each of the three items; the bid prices for A, B, and C respectively are $8.25, $6.75, and $6.50.

		Item A	Item B	Item C
Variable costs	direct labor	$4.00	$2.00	$3.00
	direct material	1.00	5.00	2.00
Fixed costs	overhead:	3.00	1.00	3.00
	allocated space rent,	2.50	0.50	0.20
	administration, insurance, property taxes, repayment and pretax return on investment	0.50	0.50	2.80
Total costs		$8.00	$8.00	$8.00

a. *Objective:* Assume that Green Corporation will not actually alter any of its overhead costs if any of the three items are purchased. Determine which, if any, of the items should be purchased.

   Analysis: If overhead costs will continue regardless of the make or buy decision, the only real savings will be in the variable costs of labor and material; on this basis, Green Corporation should continue manufacture of Items A and C. Purchase of Item B will reduce costs by $0.25 per unit and appears advisable.

b. *Objective:* Assume that Green Corporation is badly in need of space and their alternative is either to purchase some items presently being manufactured, releasing the needed space, or to rent additional space at a cost double the present per square foot cost. Overhead costs other than space are not expected to decrease if the items are purchased from an outside supplier, and it is felt that the production which will fill any vacated space should not be charged with those costs. Find which, if any, of the items should be purchased. Assume the supplier is willing to contract for the work on a short-term renewable basis.

   Analysis: The cost studies can be adjusted to reflect the urgency (dearness) of space needs by doubling the floor-space cost shown; in the marginal sense this is the alternative cost of such space. On this basis the per unit comparison becomes:

	Item A	Item B	Item C
Costs which will be eliminated if part is purchased. Space cost is doubled to reflect its alternative cost	$10.00	$8.00	$5.40
Total cost to buy	8.25	6.75	6.50
Savings by buying	$ 1.75	$1.25	-$0.90

That A requires considerable space was inferred by the basic cost study. Releasing this space for other use when space capacity is very strained has made A and B very likely candidates for subcontracting. Notice that C should apparently still be procured internally.

Example 8-3. **Short-Run Output Decisions Based upon Product Price**

Information: Past data plus estimates of future cost have produced the following output versus cost information for the F & V Corporation:

	temporary shutdown	80%	100%	120%
Output as a percent of designed capacity				
Output in units per month	0	80,000	100,000	120,000
Total cost in dollars per month (includes required pretax return)	$30,000	$270,000	$320,000	$390,000

Assume that F & V is one of many producers of similar products and is unable to influence market prices by withholding goods from the market.

Objectives:
a. Assume the current market price for the product is $4 per unit and find how many units should be produced this month.
b. Find how much extra pretax profit could be earned if the selling price of the product were $4 per unit.

c. Find the price level at which F & V should cut back from an output rate of 120% to one of 100%.
d. Find the price level at which F & V can no longer meet the required pretax return of 20%.
e. Find the price level at which, in the short run, F & V should shut down production.

Analysis: Additional information can be computed from the given data.

Output as a percent of designed capacity	0%	80%	100%	120%
Output in units per month	0	80,000	100,000	120,000
Total cost in dollars per month (includes required pretax return)	$30,000	$270,000	$320,000	$390,000
Average cost per unit		$3.38	$3.20	$3.25
Incremental cost per unit		$3.00	$2.50	$3.50
Extra profit per unit if selling price is $4		$0.62	$0.80	$0.75

a. From the data above we can observe that when the market price is $4 the plant should produce 120,000 units. Our objective is to maximize profit rather than profit rate; it is better to have extra profit per unit of $0.75 on 120,000 units than $0.80 on 100,000 units.
b. The extra pretax profit is $0.75 (120,000) = $90,000.
c. The simplest way to find the answer is by looking at the incremental cost of the last 20,000 units, $3.50. At any lower price the incremental units will not pay for themselves. You may wish to check profits for 100% and 120% output with prices of $3.49 and $3.51 to verify this point.
d. If the selling price drops below $3.20 per unit, F & V can no longer make the required pretax rate of return.
e. If shut down, F & V will lose $30,000 per month; smaller losses will be incurred for any selling price, x, exceeding

the value in the equation below:

$$\text{Deficit if shut down} = \text{deficit if operating at 100\% of capacity}$$
$$-\$30,000 = 100,000x - \$320,000$$
$$x = \$2.90 \text{ per unit}$$

So in the short run the firm should continue to operate at 100% of capacity if its product continues to sell for any price in excess of $2.90 per unit. If prices continue at low levels in the long run, the firm should seek cost reduction possibilities; and if no long-term cures are in sight, it should consider disinvestment.

Problems

8-1. In manufacturing a newly designed product, a company will need an estimated 10,000 units per year of Component 7C11. If manufactured, estimated labor costs of $0.20 per unit and material costs of $0.10 per unit will be incurred. In addition, a machine costing $10,000 with an estimated life of ten years will be required. The machine can produce Component 7C11 at the rate of one unit per 0.25 minutes. The before-tax rate of return requirement is 20%. Other costs for space, administration, property taxes, and insurance will also be incurred.
   a. What immediate uneconomic situation do you foresee and what can the firm do about it?
   b. Would the situation be any different if production needs had been 100,000 units per year or 1000 units per year?

8-2. Past policy on a certain item of vehicular equipment has been to replace the equipment every five years. In your opinion the replacement should take place more frequently. You have gathered the following information:

Equipment cost new = $10,000
Salvage value at end of year five = $1000
Operating and maintenance costs = $2000 in year one, $2500 in year two, and increasing each year by $500
Before-tax rate of return requirement = 20% = $BTRR$

   a. What salvage values at the end of year one, two, three, and four would result in your being indifferent to timing of replacement?
   b. If you subsequently find that salvage value at the end of year six is still $1000, what would you conclude?

8-3. This problem has insufficient facts for a decision. List the *minimum* additional facts needed and give the best answer you can, for example, "If the . . . is . . . , then . . . is . . . ."

A company is currently using three pieces of equipment which cost $10,000 each several years ago. They can be replaced by one unit of new equipment that costs $20,000 and which is expected to save $4000 per year in labor and maintenance for the next five years. Both present and proposed equipment will last another five years and both are expected to have negligible value by that time. Assume that the company must choose to use either the present or proposed equipment. Should the new equipment be purchased?

8-4. A certain building is soon to be constructed; an addition to the building is planned for some time in the future as required by the expanding needs. If structural provisions costing $50,000 are made now for the second unit, an estimated $150,000 in construction costs will be saved when the second unit is added.

a. Assume that the building is a county-owned structure and that $BTRR = MARR = 6\%$. The time at which the second unit will be required is unclear. Make an analysis and a recommendation based on the period of deferral of the second unit.

b. Assume that the building is a privately owned structure and that $BTRR = 20\%$. The time at which the second unit will be required is unclear. Make an analysis and a recommendation based on the period of deferral of the second unit.

c. What generalization might you make regarding the comparisons of (a) and (b) above, including comments on sensitivity, $BTRR$, and income tax rates?

8-5. The Blue Manufacturing Company is interested in purchasing a patent which you own. You have already decided to sell the patent to the company and now wish to choose between the alternative payment plans they have offered. Plan A calls for cash of $10,000 now; Plan B calls for a royalty payment of $0.10 per item sold. Assume that the items are sold uniformly over each of the next 17 years and $BTRR$ is 12%. What minimum annual quantity of items sold is necessary for Plan B to be preferable from your point of view?

8-6. Blue Corporation has fixed costs of $1 million per year and variable costs of $5 per unit of output. Green Corporation is highly mechanized and has fixed costs of $1,800,000 per year and variable costs of $3 per unit.

Output capacity for each company is 400,000 units per year.

 a. If the selling price is $9 per unit, at what percent of output capacity must Green Corporation manufacture and sell to break even?

 b. At what percent of capacity are the total costs of the two companies equal? How is this break-even point affected by selling price?

 c. Assuming that fixed costs are not alterable in the short run, holding production as inventory is not practical, and the market is large enough so the firm cannot influence price (as by withholding output), how low can the selling price go before Green Corporation should shut down production (in the short run)?

 d. For which of the two firms is profit more sensitive to changes in the sales and output level?

8-7. This problem has insufficient facts for a decision. List the *minimum* additional facts needed and give the best answer you can, for example, "If the . . . is . . . , then . . . is . . . ."

A company must choose between a small plant costing $1 million and a large plant costing $1,500,000. The earnings of both plants are computed, and it is found that the small plant would yield an after-tax return of 30% and the larger plant an after-tax return of 25%. Both have lives of 20 years and negligible salvage. Which plant should be chosen?

8-8. The lining of a plating tank must be replaced every three years at a cost of approximately $2000. A new lining material has been developed that is more resistant to the corrosive effects of the plating liquid and will cost approximately $4000. If $BTRR = 20\%$ and annual property taxes and insurance amount to about 4% of the first cost, how long must the new lining last to be more economical than the present one?

8-9. Two years ago the POI Company purchased certain equipment costing $12,000. At the time of purchase, equipment life was estimated to be 12 years and salvage value was estimated to be negligible. Currently the depreciated book value of the equipment is $8000. Its serial number is 8XBZ5209. Since the announcement of the availability of technologically improved equipment, the market value of the present used equipment has dropped drastically. The improved equipment will reduce operating, maintenance, property tax, and insurance costs by a total of $5000 per year. It is now expected that eight more years of service can be obtained from either the improved or present equipment and the salvage value at that time will be negligible for the present equipment and $1000 for the improved equipment. $BTRR = 20\%$.

Neglect any possible "loss on disposal" and "investment credit" tax considerations; assume revenues will be unaffected by your choice.

Find the maximum *extra* sum which POI can afford to pay if they trade in the present equipment on the improved equipment.

8-10. Alternative A requires an investment only half that of B. On the other hand, B has a life expectancy twice that of A. If all other costs (maintenance, labor, and so forth) appear to be equal, which alternative, if either, is better?

8-11. Gene Green is 20 years old and single; he is weighing two alternatives: (1) Pay a net premium of $300 per year for a $20,000 ordinary life insurance policy or (2) invest $300 per year. Gene thinks he can invest his funds to earn an after-tax interest rate of 6%.

a. How long would it take for his investment to grow to $20,000?

b. What does this mean?

8-12. A small electric utility company is considering the acquisition of a truck-mounted power-driven posthole auger for use in the installation of power poles. The equipment, including the truck, has an estimated first cost of $10,000 with negligible salvage value after ten years of use. The equipment can dig a hole in 0.3 hours and requires two operators; each earns gross wages, including all fringe benefits, of $4 per hour. Cost for fuel and lubricants is an estimated $1 per hour of operation. Costs for licensing and insurance are expected to average $200 per year.

If the equipment is rejected, the company will continue to employ hand labor costing $3 per hour and averaging 10.0 man-hours per hole dug.

a. Find the number of holes per year required for the hand versus machine methods to break even if a before-tax rate of return of 15% is required.

b. Repeat Part a for a 0% return.

8-13. The following data have been estimated for Machine A:

First cost $= K_1$
Salvage $= K_2$
Life $= K_3$
Operating cost per year $= K_4$
Before-tax rate of return requirement $= K_5$
Output per year (50% of capacity) $= K_6$

Exactly identical data have also been estimated for a completely different equipment, Machine B. As far as we have been able to discover, the only difference between the machines is that A is expected to be retired after K_3 years because it is obsolete, while Machine B is expected to be retired after K_3 years because it is worn out. Are investment costs (repayment plus pretax return) *fixed costs* for both machines? Why?

8-14. Make a profitability analysis similar to Table 8-2 and based on the equation

$$\text{Quantity} = \frac{4900 + 2(\text{promotion}) - \dfrac{\text{promotion}^2}{10,000(\text{price})^2}}{\text{price}^2}$$

Use a selling price of $0.56 and show profitability information for promotion outlays of $1300, $1500, and $1700. Which of the three maximizes profit?

8-15. A small company which rents tools and equipment is considering the purchase of a special cutting tool which can be mounted in a $1/2$-inch electric drill to saw a 4-inch diameter hole. The tool costs $10 and its life is estimated at five years with zero salvage. Annual costs of storing, insurance, and property taxes are estimated as 10% of first cost regardless of amount of use. The cost of cleaning, handling, and sharpening is estimated to average $0.50 each time the tool is used. It is thought that the tool would rent for $1.50 per half-day. The required before-tax rate of return on investments is 20%.

   a. Graph the annual fixed plus variable costs versus number of rentals. Repeat for revenue versus number of rentals.

   b. How many half-day rentals per year are required to meet the stipulated rate of return? Express this both as a number and a percent of the maximum possible number of rentals per year.

8-16. Two alternative fixtures for aiding a certain assembly operation are being considered. Fixture A costs $300 and will save about $0.30 of operator time per unit produced, while Fixture B costs $600 and will save about $0.40 per unit. $BTRR = 20\%$. Annual property taxes and insurance total 3% of first cost. The life of either fixture is three years, and salvage of either is estimated as negligible at that time.

   a. Find the annual production rate required for Fixture B to repay its costs, including pretax return.

   b. At what minimum annual production rate do you recommend Fixture B?

   c. Defend the difference in your responses to (a) and (b).

8-17. a. Use the appropriate *Moody's Manual* to obtain data for estimation of the percent of sales dollar required for annual plant and equipment costs. For the latter, include depreciation expense; interest on debt capital; return on equity capital; federal, state, and local income taxes; and state and local property taxes. Do this for one company of each of the following types:

(1) An electric utility.
(2) A chemical company.
(3) An electronics company.
(4) A tobacco company.
(5) A food processing company.

b. Also show the investment dollars or asset dollars per employee for each of the companies selected.

8-18. a. Refer to Example 8-1. Find the before-tax rate of return on the overall investment for the data given with sales of 22,903 per year, promotion of $1500, selling price of $0.56 per unit, and using the Automatic Assembler.

b. Refer to Figure 8-4. Where would dividends on common and preferred stock be included? Where would interest on debt capital be included? Where would the bond-redemption or sinking-fund payments be included?

8-19. Certain waste materials from a plant are currently hauled away for disposal at a cost of $3 per ton. If processing equipment costing $20,000 is acquired, the present 5000 tons per year of waste can be converted into useful material. The processing equipment has an estimated life of ten years and negligible salvage value at that time; annual operating costs are estimated as $6000. The before-tax rate of return requirement is 20%.

a. What is the minimum price the company can afford to charge for the converted waste material?

b. What is the minimum price the company can afford to charge for the converted waste material if demand soars to 10,000 tons per year and the company is forced to buy 5000 tons additional waste material at a cost of $7 per ton to meet demand?

8-20. The Green Company has a certain property costing $1000 and having a life of ten years. A chemically treated property of the same type costs $1200; the treatment will prolong the life of the property by an undetermined amount of time. Assume the before-tax rate of return requirement is 20% and that annual property taxes and insurance average 3% of first cost. How many years must the treatment prolong equipment life if it is to meet the economic criteria? Assume replacement units will be required over many years into the future.

8-21. The incremental cost of providing a certain product is $6 per unit. A current market demand analysis indicates the following relationship of price and demand.

If price is ____ per unit:	An estimated ____ units per month can be sold:
$18	8,000
17	9,000
16	10,000
15	11,000
14	12,000
13	13,000

a. How should the product be priced to maximize profit?
b. If additional units could be "dumped" in a foreign market without danger of their returning to upset national price and demand structures, how low could such units be priced?
c. Defend your having responded to (a) and (b) without knowing the total cost per unit.

8-22. The Miscellaneous Manufacturing Company finds the following breakdown:

Department	Floor Space Used by Department, Sq. Ft.	HP of Machinery	Direct Labor Personnel	Machinery First Cost	Life, Years
A	2000	50	24	$ 15,000	5
B	1500	25	8	19,000	5
C	900	10	6	75,000	10
D	600	15	2	127,000	10
Total	5000	100	40	$236,000	

Factory expense items are as follows:

Depreciation of building	$10,000
Heat, light, fire insurance	2,000
General (allocate by number of men)	15,000
Power	2,400
Superintendent's salary	15,000

Wages (including all fringe benefits) in Departments A and B average $6250 per man per year. In Departments C and D wages average $8000 per man per year. Salvage value of machinery is negligible.

a. Find the overhead rate (ratio of dollars of overhead to dollars of direct labor) for the factory without departmentalizing costs.

b. Find the overhead rate in Department B by departmentalizing costs. Why bother departmentalizing?

c. Find the factory cost for a product manufactured in the plant assuming direct material costs of $1.50 per unit, and direct labor costs of $2.00 per unit (all in department D).

8-23. You are in the process of trying to determine whether to continue manufacture of a certain part or to buy the part from a manufacturer who has offered to make the part for $0.10 each. Your own factory costs have been estimated as:

Labor and material	$0.02
Overhead	0.18
	$0.20

It is apparent that overhead costs are a crucial factor in your decision.

Why might your decision be influenced by whether or not many other similar simultaneous make or buy decisions are being made by your company?

8-24. Make a demand curve analogous to Figures 8-5 and 8-6 for one of the following. Whether the horizontal scale is in hours, days, or weeks depends upon the facility you study.

a. University classroom space.

b. A bowling alley.

c. A service station.

d. A highway or street.

e. A laundromat.

f. A barbershop.

g. Other as assigned or selected.

What suggestions do you have to improve the utilization of existing facilities? What types of resistance do you foresee to such changes?

8-25. An industrial firm has used 14,400 kwh this month. Its maximum demand, as defined above, has been 200 kw. A schedule of electric power rates for a certain community follows.

RESIDENTIAL:

Rate: (Net) per month per meter

Energy charge: For first 25 KwH ____6.0c per KwH
 For next 125 KwH __3.2c per KwH
 For next 850 KwH __2.0c per KwH
 All in excess of
  1,000 KwH ____1.0c per KwH
 Minimum: $1.50 per month

COMMERCIAL: A rate available for general commercial and miscellaneous power uses where consumption of energy does not exceed 10,000 KwH in any month during any calendar year.

Rate: (Net) per month per meter

Energy charge: For the first 25 KwH __6.0c per KwH
 For the next 375 KwH _4.0c per KwH
 For the next 3,600 KwH 3.0c per KwH
 All in excess of
  4,000 KwH ____1.5c per KwH
 Minimum: $1.50 per month

GENERAL POWER: A rate available for service supplied to any commercial or industrial customer whose consumption in any month during the calendar year exceeds 10,000 KwH. A customer who exceeds 10,000 KwH per month in any one month may elect to receive power under this rate. A customer who exceeds 10,000 KwH in any three months or who exceeds 12,000 KwH in any one month during a calendar year shall be required to receive power under this rate at the option of the supplier. A customer who elects at his own option to receive power under this rate may not return to the commercial service rate except at the option of the supplier.

Rate: (Net) per month per meter

Kw is rate of flow. 1 Kw for 1 hour is 1 KwH.

Demand Charge: For the first 30 Kw of maximum demand per month _$2.50 per KwH
 For all maximum demand per month in excess of 30 Kw _$1.25 per Kw

Energy charge: For the first 100 KwH per Kw of maximum demand per month _____2.0c per KwH
 For the next 200 KwH per Kw of maximum demand per month _____1.2c per KwH
 All in excess of 300 KwH per Kw of maximum demand per month _____0.5c per KwH
Minimum Charge: The minimum monthly bill shall be the demand charge for the month.

Determination of Maximum Demand: The maximum demand shall be either the highest integrated Kw load during any thirty minute period occurring during the billing month for which the determination is made, or 75 per cent of the highest maximum demand which has occurred in the preceding eleven months, whichever is greater.

Water Heating: 1.0c per KwH with a minimum monthly charge of $1.00.

a. Using the rates given, find the firm's electric power cost for this month.

b. Find the *demand factor*, as defined in this chapter, for the firm.

c. If shifting the hours of certain electric oven work could be accomplished at a cost of $40 per month and if the shift would reduce maximum demand to 150 kw, would the change be justifiable?

d. A residential consumer is analyzing the operating costs of certain equipment; without the equipment his electric consumption has averaged 1100 kwh per month as follows: 1500, 1500, 1300, 1100, 900, 800, 900, 1100, 800, 900, 1100, 1300. The equipment will use about 150 kwh each month. Estimate annual costs of electric power for operation of the equipment.

8-26. You have been asked to compare the costs of alternative methods of hand-drying equipment.

A. Paper towel.

B. Roller cloth towel service.

C. Push-button electric dryers.

 Gather the necessary data and make estimates where necessary to compare the economics of the three methods. Carefully show in detail the basis of your estimates. Graph your comparison by placing cost on the vertical axis and use in times per day on the horizontal axis. Note break-even points, if any.

8-27. The owner-manager of a medium-sized motel feels that to maintain his "share of the market" he must provide a swimming pool and do so without increasing the rates on rooms. Apparently the pool will add to costs but not receipts. Upon what economic basis should the owner-manager make his accept-reject decision on the swimming pool addition?

8-28. Gene Green is contemplating the purchase of a used light airplane for $10,000. Fixed cash operating costs will be incurred for insurance, hangar rental, and registration fees. Variable cash operating costs will be incurred for fuel, lubrication, maintenance, repair, and periodic engine overhaul. The following estimates apply:

Cash operating cost	If used for self and no more than one co-owner	If used for instruction or if more than one co-owner
Fixed	$800 per year	$1000 per year
Variable	$8 per flight hour	$9 per flight hour

Personal flying time of Green = 200 flight hours per year

Potential revenue from use of plane and instruction of pilot by Green = $20 per flight hour

Resale value of airplane five years from now = $5000

Before-tax rate of return required by Green = 10% (This is also the pretax return he can earn if he invests funds elsewhere)

a. Find *AEC* for Green if he is sole owner and sole user of the airplane.

b. Green is considering the possibility of giving flying lessons to help finance his personal flying. He holds a license which permits him to give flying lessons, but if he does so, cash operating costs will increase. Green could earn $7 per hour elsewhere if he were not busy giving flying lessons. How many flight hours of instruction per year would be required just to cover the increased cash operating costs?

c. In lieu of giving flying lessons to help finance his personal flying, Green considers the possibility of reducing costs by co-ownership of the airplane. Green would look for a potential co-owner with similar flying skills and personal flying time per year. Find Green's *AEC* for this possibility. Since his *AEC* is less than in Part a, did he get something for nothing? Explain your response.

d. Find the average cost per mile flown by Green. Although cruising speed of the airplane is 140 mph, takeoffs, landings, and the like reduce the overall average speed to 120 mph.

8-29. For utilization of its idle capacity, the Energetic Electric Company is investigating the following engineering proposal for its economic advisability.

During evening and weekend hours when the demand for power is slack, water will be pumped up Mighty Mountain by pumps powered by electricity from steam-powered generators elsewhere in the electric utility system. The water will be stored in a reservoir until demand for power reaches a peak. Then it will be sent downhill to spin the two generators of the plant for electricity production.

The plant will produce only 2 kwh of electricity for each 3 kwh it burns to pump the water uphill. But the cost of producing the 3 kwh will be relatively low because the electric utility can use equipment that otherwise would be idle at the time. "In effect the only cost will be for fuel," says J. W. McPlan, president of the utility.

Since the facility will not need coal storage and handling equipment of steam lines and boilers, it will cost only $50 million to build the proposed plant compared with an estimated $67 million for a steam-power plant with the same capacity (350,000 kw). The proposed facility, because of automation, will require only about 15 maintenance employees compared to about 90 persons needed to maintain and operate a comparable steam plant; this is the basis for the estimated saving of $600,000 per year in maintenance costs.

Power men say that interest in pumped storage plants has been spurred by the development in the last ten years of versatile equipment that can both pump and generate and thus greatly cut capital costs. Prior to this advance, separate sets of equipment were needed for the pumping and generating cycles.

Initially, the electric utility will fill a 395-acre reservoir at the foot of Mighty Mountain with water from the east fork of the West River. This reservoir is connected by a tunnel with a 54-acre reservoir at the mountain's summit, 800 feet above. A powerhouse containing two 160-ton waterwheels and the combination pump-generators sits just above the lower reservoir.

When the generators are working as pumps, they will lift 17,000 gallons of water per second from the lower pool and force it through the tunnel to the top. The task is comparable to raising 5 million tons, or about 100 battleships, half again as high as the 555-foot Washington Monument, according to company officials.

To generate power, the upper reservoir will empty its 1.4 billion gallons into a yawning drain which feeds the tunnel leading to the powerhouse where generators will turn at 200 rpm. Because water is recirculated, the plant would not be as dependent upon rainfall to renew the water supply as the conventional hydroelectric plant.

a. Compare the annual cost of alternatives on a pretax basis if the investment is to be recovered over 30 years with either the steam or hydroelectric alternatives and if a before-tax return requirement of 15% is appropriate. The following data apply:

	Steam	Hydro-electric
Cost per kwh for coal	$0.002	$0.000
Cost per kwh for operating expenses other than coal, maintenance, depreciation, taxes	0.003	0.003
Cost per kwh for water lift (3/2) ($0.002)	0.000	0.003
Variable cost per kwh	$0.005	$0.006˙

Base your annual cost comparison on an average utilization which is 5% of the rated capacity.

b. Defend the use of so low a utilization estimate.

c. Is there anything wrong with assuming a 50% utilization? Defend your response on engineering grounds.

d. Explain the use of the fraction ($3/2$) in the computation of cost for water lift. Comment upon the resulting figure of $0.003. Do you agree?

IV

Income Tax Considerations

THE PUBLIC revenues are a portion
that each subject gives of his property,
in order to secure the remainder . . . .
This nation is fond of liberty, because
this liberty is real, and it is possible for
it, in its defense, to sacrifice its wealth,
its ease, its interest, and to support the
burden of the heaviest taxes, even such
as a despotic prince durst not lay upon
his subjects.

CHARLES DE MONTESQUIEU in *The Spirit of Laws, 1748*

9

Depreciation and Depletion

Intuitive Decisions: A Caution

IF YOU were choosing between two alternatives, A and B, identical in all respects except that A has a salvage value and B does not, which would you choose? The choice seems obvious, yet if we take into account depreciation and income taxes, as we do in Example 9-1, our comparison may provide a surprise.

Example 9-1. Comparison of Two Alternatives, Identical except for Estimated Net Salvage

Information: The following estimates have been made:

	Alternative A	Alternative B
First cost	$120,000	Same as A
Salvage	$ 60,000	0
Life	30 years	Same as A
Financing	100% by equity	Same as A
Income tax rate	50%	Same as A
Before-tax cash flow	$22,000/yr	Same as A
Depreciation method	Straight-line	Same as A
MARR	10%	Same as A

Objective: Compare the two alternatives.

Analysis:

	Alternative A	Alternative B
Before-tax cash flow	$22,000	$22,000
Less: Depreciation expense	2,000	4,000
Equals: Taxable income	$20,000	$18,000
Less: Income tax @ 50% rate	10,000	9,000
Equals: Reported net income	$10,000	$ 9,000

At this point we might be tempted still to prefer Alternative A on the basis of greater "reported" annual net income. Resisting this temptation, we might subtract income taxes from the before-tax cash flow to find that after-tax cash flow is $12,000 per year for A and $13,000 per year for B (the same numerical result is produced by adding together depreciation expense and reported net income). Calculations reveal a rate of return for A of 9.7%, while return for B is 10.2%, and verifies that B is the better alternative. This result is something of a surprise in the light of prior knowledge and conceptions; even the accountant's reported net income seems to indicate an advantage of A. The tricky result is due to depreciation and income taxes; up to this chapter we have been well aware of A's advantage of salvage and could express it in an annual form:

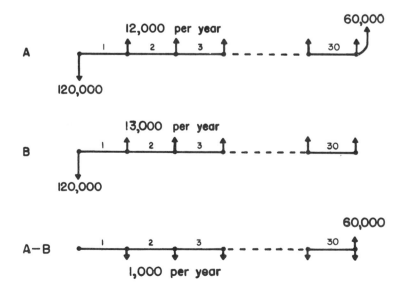

$$A\text{'s advantage} = \$60,000 \, (a/f)_{30}^{10\%} = \$365/\text{yr}$$

but note that on income taxes

$$B\text{'s advantage} = \$10,000 - \$9,000 = \$1000/\text{yr}$$

In other words, we have found that B's advantage of greater depreciation charges (and therefore smaller income tax expenses) more than offsets A's very distant salvage value. A's advantage becomes more important for small values of n and i. Several break-even points where tax savings are just offset by salvage foregone are:

i	Break-even Point
percent	*years*
6	22.4
8	17.2
10	14.0

Thus, when $i = 8\%$ and $n = 17.2$, (1) the presence (or absence) of an estimated salvage value has virtually no effect on the economy of the alternative, and (2) an error in estimating the salvage value produces a net error of zero in the economy analysis. Unfortunately the same consolation cannot be offered the accountant; capital gains or losses affecting income taxes will result when properties are sold for an amount different from their book value.

Depreciation, a noncash expense, affects cash flow via its effect upon income taxes; it can be a decisive decision factor.[1]

Depreciation and Its Meanings

Depreciation can mean (1) an allocation of cost, (2) a cost of operation, (3) a change in value, or (4) a change in the state of physical condition of a property. Within each of these meanings, variations in measurement are possible; for example, Is change in value to be measured by the difference between (1) original cost and present salvage value, (2) present-day cost of the item (or its updated model equivalent) and present salvage value, or (3) some other pair of values? Furthermore, if (2) is specified, there is still the problem of recognizing that a new

---

[1] See also Example 10-1 and Table 10-3.

model may incorporate improvements making it similar rather than identical to the asset in question.

It is important to recognize that estimates of depreciation for the economy analysis may be quite different from those used for managerial control, reporting, or pricing purposes. Our primary application is the economy analysis where tax depreciation is an allowable expense in computing taxable income and thus affects the cash disbursement for income taxes. The material which follows is therefore limited to depreciation for income tax purposes.

Items Not Depreciable for Tax Purposes

Items that require a present cash outlay and produce benefits extending over two or more years into the future (one definition of a capital expenditure) are not necessarily depreciable for income tax purposes, for example:

1. Items which do not diminish in value (theoretically) with time, such as salvage value residual, working capital, and land.
2. Items which do not represent a cash outlay, specifically, interest lost during construction (*ILDC*).
3. Items which are expensed, such as rearrangement costs, research costs, advertising costs, training costs, start-up computer programming costs, and so forth. Expensing a property means allocating its full cost to the year of acquisition. It is, effectively, immediate depreciation.

Income Tax and Depreciation Policies of Government

For income tax purposes our federal government treats depreciation as an allocation of cost. The cost base used is the price actually paid at the time a property is acquired and includes transportation and installation costs. Proponents of replacement-cost depreciation (also called price-level depreciation) argue that such treatment neglects the problems of inflation and recoups past cost, but not current cost.

Capital expenditures by privately owned firms are determined by management but are influenced by governmental income tax and depreciation policies. The changing nature of these policies is evident:

Year	Change
1940	Accelerated amortization ruling permitted five-year write-off of cost under special circumstances on property acquired after June 10, 1940.
1953	Liberalized depreciation generally permitted use of a depreciation method which resulted in write-off of about two-thirds of cost in the first half of property life. Applicable to property acquired after January 1, 1954.
1958	Additional first-year depreciation permitted an extra 20% of cost of a property to be written off in the first year of a property life under certain circumstances and limitations.
1962	"Guideline" depreciation gave corporations more freedom in estimating property life; burden of proof remains upon the corporation but a three-year "test" period is used.
1962	Investment credit—a percentage of the first cost of qualifying properties is allowed as a direct deduction from federal income taxes.
1964	Clarification and further liberalization—cost base of a depreciable property need *not* be reduced by the amount of the investment credit.
1964	Reduction of personal and corporate income tax rates.
1965	Excise tax rate reductions.
1966	Excise tax rate increases. Investment credit and liberalized depreciation suspended in October.
1967	Investment credit and liberalized depreciation reinstated in March.
1969	Tax Reform Act effective January 1, 1970; investment credit suspended in April.
1971	Permission to reduce depreciable life of newly acquired asset by 20%. Investment credit reinstated. Approval of ASY depreciation.

Estimates of Life for Tax Depreciation Purposes

The 1971 depreciation liberalization permits the taxpayer to use a property life that is between 80% and 120% of the *Guidelines* life.[2] A sampling of guideline lives follows:

_____

[2] See *Depreciation Guidelines and Rules*, U.S. Treasury Department, 1962.

Description	Depreciable Life, in Years
Buildings, Office, or Factory	45
Manufacturing	
Cement	20
Electrical equipment	12
Electronic equipment	8
Glass	14
Metal refining (Nonferrous)	14
Petroleum refining	16
Plastics	11
Printing	11
Pulp and paper	16
Rubber	14
Textile	14
Tobacco	15
Utilities	
Electric transmission and distribution	30
Gas distribution	35
Pipelines	22
Water	50

Group versus Item Depreciation

One of the options available to a company in its determination of depreciation policy is the choice between accounting for each item of property via separate item accounts and accounting for groups of items. The possible grouping may be as general as a single overall composite account for the company or may be in detail according to any one or more characteristics such as (1) similarity of items within an account, (2) date of installation, (3) average life, (4) mortality characteristics, or (5) geographic location. Companies having a substantial number of similar items, such as the water company with its numerous water meters, generally have chosen to use group accounting.

Unless otherwise specified the discussion and examples involve item accounting. Group accounting and item accounting involve important differences; retirements are treated differently and the allocations and accruals are related to mortality characteristics of the property group. Gain or loss on disposal charges are not generally allowed where group accounting procedures are applied.

Notation Used with Depreciation Formulas

The following notation is used in Table 9-1 and subsequent material.

Let ASY = adjusted SYD depreciation method begun in 1971
B_t = tax depreciation basis of property
B_z = portion of installed cost unallocated at age z
DC_z = accrued (cumulative) depreciation allocation at end of year z
DDB = double-declining-balance method of depreciation
D_z = depreciation allocation at end of year z

TABLE 9-1. Comparison of Selected Tax Depreciation Methods

Method	D_z, Depreciation Allocation at End of Year z*	DC_z, Accrued (Cumulative) Depreciation Allocation at End of Year z*
SL	$(B - V)(1/n)$	$(B - V)(z/n)$
SYD†	$(B - V)\left[\dfrac{n + 1 - z}{\frac{1}{2}(n^2 + n)}\right]$	$(B - V)\left[\dfrac{2nz - z^2 + z}{n^2 + n}\right]$
ASY†	$B\left[\dfrac{n + 1 - z}{\frac{1}{2}(n^2 + n)}\right]$	$B\left[\dfrac{2nz - z^2 + z}{n^2 + n}\right]$
1SY†	Use DDB in year one, SYD thereafter	
1AS†	Use DDB in year one, ASY thereafter	
DDB†	$B\left(\dfrac{2}{n}\right)\left(\dfrac{n - 2}{n}\right)^{z-1}$	$B\left[1 - \left(\dfrac{n - 2}{n}\right)^z\right]$
150‡	$B\left(\dfrac{1.5}{n}\right)\left(\dfrac{n - 1.5}{n}\right)^{z-1}$	$B\left[1 - \left(\dfrac{n - 1.5}{n}\right)^z\right]$
125‡	$B\left(\dfrac{1.25}{n}\right)\left(\dfrac{n - 1.25}{n}\right)^{z-1}$	$B\left[1 - \left(\dfrac{n - 1.25}{n}\right)^z\right]$
SF	$(B - V)\left[\dfrac{i(1 + i)^{z-1}}{(1 + i)^n - 1}\right]$	$(B - V)\left[\dfrac{(1 + i)^z - 1}{(1 + i)^n - 1}\right]$

*Accrued depreciation must never exceed $B - V$. Formulas apply only so long as neither termination nor method switch occurs. The subscript t has been omitted from n and B for the usual case that $B_t = B$ and $n_t = n$.

†Not allowed when $n < 3$. When DDB *is* allowed, a subsequent switch at any year to any of the other methods listed is allowed. To maximize *PED* when $V > 0$, 1AS is optimal; when $0 > V/B > - 1/n$, 1SY is optimal; when $- 1/n > V/B$, SYD is optimal. Relevant equations are given in Appendix A.

‡Method switch limited to SL. For nonresidential business real estate the taxpayer may be limited to 150, 125, or SL.

i = interest rate, as a decimal ratio
n_t = tax life of the property, in years
SF = sinking-fund depreciation method
SL = straight-line depreciation method
SYD = sum-of-the-years-digits depreciation method
V = estimated net salvage value of the property
z = attained age of the property, in years
1AS = use DDB in year one, ASY thereafter
1SY = use DDB in year one, SYD thereafter
125 = 125% declining-balance depreciation method
150 = 150% declining-balance depreciation method

Example 9-2.   Computation of Depreciation by Several Methods

Information: An asset has been acquired at a cost of $250,-000 and is expected to be used for five years, then sold for an estimated net of $100,000. *MARR* = 10%.

Objective: Find (1) annual depreciation allocations, (2) accrued depreciation allocations at each age, and (3) portion of installed cost unallocated at each age using the methods listed.
a. Straight-line.
b. Sum-of-the-years-digits.
c. Double-declining-balance.
d. Sinking-fund; use i = 10%.

Analysis:
a. By the straight-line method, depreciation allocation for any year

$$= (B - V)/n = (\$250,000 - \$100,000)/5 = \$30,000$$

and in table form:

Age	Depreciation Allocation	Accrued Depreciation at Age Shown	Installed Cost Unallocated at Age Shown
years			
1	$30,000	$ 30,000	$220,000
2	30,000	60,000	190,000
3	30,000	90,000	160,000
4	30,000	120,000	130,000
5	30,000	150,000	100,000

b. By the sum-of-the-years-digits method the denominator is $1 + 2 + 3 + 4 + 5 = 15$, and the numerator of 5 decreases by one each year so that the first-year allocation is (5/15) ($250,000 - $100,000), the second-year allocation is (4/15) ($250,000 - $100,000), and so forth.

Age	Depreciation Allocation	Accrued Depreciation at Age Shown	Installed Cost Unallocated at Age Shown
years			
1	$50,000	$ 50,000	$200,000
2	40,000	90,000	160,000
3	30,000	120,000	130,000
4	20,000	140,000	110,000
5	10,000	150,000	100,000

Note that identical results will be produced if the rate applied is

(Remaining life)/(sum of digits of remaining years)

and the base is the undepreciated portion of the depreciable base (unallocated cost less salvage); thus the first-year allocation is (5/15) ($250,000 - $100,000) = $50,000, the second-year allocation is (4/10) ($100,000) = $40,000, the third-year allocation is (3/6) ($60,000) = $30,000, and so forth. This alternative treatment is used for group depreciation computations especially for non-integer values of n.

c. By the double-declining-balance method the constant rate, $2/n$ is 40%, so the first-year allocation is 40% of ($250,000) = $100,000. The second-year allocation would be 40% ($250,000 - $100,000) = $60,000, but this violates the constraint that accrued depreciation must never exceed the depreciable base, $B - V$; therefore, only $50,000 is permitted.

Age	Depreciation Allocation	Accrued Depreciation at Age Shown	Installed Cost Unallocated at Age Shown
years			
1	$100,000	$100,000	$150,000
2	50,000	150,000	100,000
3	0	150,000	100,000
4	0	150,000	100,000
5	0	150,000	100,000

d. By the sinking-fund method the first-year allocation is ($250,000 - $100,000)$(a/f)_5^{10\%}$ = $150,000(0.16380) = $24,570. In year two the allocation is $24,570(1.10) = $27,027 and in year three is $27,027(1.10) = $29,729.

Age	Depreciation Allocation	Accrued Depreciation at Age Shown	Installed Cost Unallocated at Age Shown
years			
1	$24,570	$ 24,570	$225,430
2	27,027	51,597	198,403
3	29,729	81,326	168,674
4	32,702	114,028	135,972
5	35,972	150,000	100,000

Five-Year Write-off

In a few cases, where the use of an asset is deemed important to national interests, tax laws permit a five-year write-off of the asset. Under certain circumstances pollution control facilities, railroad rolling stock, and coal mine safety equipment can qualify. In such cases no estimate of life is needed for computation of depreciation rates. We might observe, however, that the advantage of the short write-off is greatest where estimated life is considerably more than five years. This is not an additional method, it is simply a special case of the straight-line method.

Choosing a Depreciation Method

If income tax rates were constant over the period of study, then total income taxes would also be identical regardless of the depreciation method employed. Note that in Example 9-2 each of the four depreciation methods resulted in a total allocation of $150,000. Although the total of income taxes over the years can be unaffected by choice of method, the *timing* of those income taxes is affected. Given a nonnegative annual series of taxable incomes and the choice of paying a dollar of income tax now versus a dollar of income tax later, the time value of money $(i > 0\%)$ encourages us to choose the postponement. The postponement is achieved by depreciation methods which yield higher depreciation allocations in the early years of an asset.

When the time value of money, i, and the income tax rate, t, are positive and time-invariant, an objective of maximizing

present equivalent depreciation allocations, *PED*, is identical to
an objective of minimizing the present equivalent of income tax
payments. To illustrate, consider Example 9-2 where $n = 5$,
$V/B = 0.4$, and $MARR = 10\%$. Then

PED_{SL} = present equivalent depreciation allocations by the
straight-line method
= $30,000 $(p/a)_5^{10\%}$ = $113,730

PED_{SYD} = $50,000 $(p/a)_5^{10\%}$ - $10,000 $(p/g)_5^{10\%}$ = $120,930

PED_{DDB} = $100,000 $(p/f)_1^{10\%}$ + $50,000 $(p/f)_2^{10\%}$ = $132,230

PED_{SF} = $24,570 $(p/f)_1^{10\%}$ + \cdots + $35,972 $(p/f)_5^{10\%}$
= $111,682

And for the methods given above:

Maximum PED/B = $132,230/$250,000 = 0.5289

It can be shown that to maximize *PED* one should use the
shortest tax life allowed, and if choosing from the methods
listed in Table 9-1,

1. If $V/B > 0$, use DDB for first year and ASY thereafter. (Call
   this the 1AS method; if $1 - V/B \geqslant (1 - 2/n_t)^3$, the results
   from DDB, without switch, are identical to the results of
   1AS.)
2. If $0 \geqslant V/B > -1/n_t$, use DDB for first year and SYD thereafter. (Call this the 1SY method).
3. If $-1/n_t \geqslant V/B$, use SYD.

Table 9-2 is based on application of these rules with *PED*
evaluated at $i = 10\%$ for selected values of V/B and n.

For *real-world* studies one could generally use Table 9-2, or
one like it, to determine *PED* for the engineering economy
study. One could thus incorporate both the maximum permissible shortening of life and the *PED*-maximizing method.

For *classroom* studies one can assume that economic life
(n_e) and tax life (n_t) are identical and that depreciation is by
the straight-line method, unless otherwise specified.

Note that *PED* is computed using tax life but that the an-
nualized effect, *AED*, should be spread over the *economic* life
of the property. For example, if $n_e \neq n_t$, *AED* for the straight-
line method is *not* simply $(B - V)/n$.

TABLE 9-2. Present Equivalent Depreciation Allocations As a Percent of First Cost = 100% (*PED/B*)

Amount shown is maximum available from methods of Table 9-1, evaluated at i = 10%. To maximize PED, use the shortest tax life allowed.

Tax Life, in years	Ratio of Salvage Value to First Cost of Property = V/B						
	−0.2	−0.1	0.0	0.1	0.2	0.3	0.4
3	103.35	95.33	87.32	79.81	71.63	63.36	54.55
4	99.88	92.11	84.33	77.39	69.87	61.98	53.72
5	96.74	89.19	81.64	75.19	68.22	60.71	52.89
6	93.99	86.49	79.16	73.20	66.58	59.52	52.01
7	91.35	83.95	76.83	71.36	65.16	58.36	51.12
8	88.84	81.55	74.63	69.53	63.66	57.25	50.42
9	86.43	79.28	72.55	67.81	62.32	56.20	49.52
10	84.12	77.11	70.56	66.19	60.98	55.14	48.77
11	81.91	75.08	68.67	64.61	59.70	54.18	48.04
12	79.79	73.14	66.86	63.09	58.48	53.17	47.26
13	77.76	71.28	65.13	61.64	57.27	52.27	46.59
14	75.81	69.49	63.48	60.24	56.15	51.32	45.87
15	73.94	67.78	61.90	58.88	55.01	50.46	45.18
16	72.14	66.13	60.38	57.58	53.95	49.56	44.56
17	70.42	64.55	58.92	56.33	52.88	48.73	43.87
18	68.76	63.03	57.52	55.11	51.87	47.90	43.24
19	67.17	61.57	56.18	53.95	50.88	47.10	42.63
20	65.64	60.17	54.89	52.82	49.93	46.32	42.00
21	64.16	58.82	53.65	51.72	49.00	45.56	41.43
22	62.74	57.51	52.45	50.67	48.10	44.83	40.84
23	61.38	56.26	51.30	49.65	47.23	44.10	40.26
24	60.06	55.06	50.20	48.66	46.37	43.41	39.72
25	58.79	53.89	49.13	47.70	45.55	42.70	39.16
26	57.57	52.77	48.11	46.78	44.74	42.05	38.62
27	56.39	51.69	47.12	45.88	43.97	41.38	38.11
28	55.25	50.65	46.16	45.02	43.21	40.75	37.58
29	54.15	49.64	45.24	44.18	42.47	40.12	37.09
30	53.09	48.67	44.35	43.36	41.75	39.51	36.60
31	52.07	47.73	43.49	42.57	41.05	38.92	36.11
32	51.08	46.82	42.66	41.81	40.38	38.34	35.65
33	50.12	45.94	41.86	41.07	39.71	37.78	35.18
34	49.19	45.09	41.08	40.35	39.07	37.22	34.72
35	48.30	44.27	40.33	39.65	38.44	36.69	34.28
40	44.23	40.54	36.92	36.45	35.55	34.16	32.19
45	40.74	37.34	34.01	33.68	33.01	31.91	30.28
50	37.73	34.58	31.49	31.26	30.76	29.89	28.55
55	35.11	32.18	29.30	29.14	28.76	28.08	26.97
60	32.81	30.07	27.38	27.27	26.98	26.44	25.53
Tax Life, in years	SYD		ISY	IAS			
	Tax Depreciation Method						

Ultimately the choice between depreciation methods for income tax purposes can be an important one which must be based on all pertinent factors including:

1. Expected future trend of income tax rates as affected by legislation and variability of company earnings.
2. Presence or absence of carry-forward losses.
3. Effects of choice upon present and future prices of goods or services produced. For investor-owned public utility companies the choice of depreciation methods is partly contingent upon regulatory rate-determination policies.

Unit-of-Production Method

The preceding methods permit the taxpayer to base his depreciation computation on elapsed time on the theory that retirement of property is mainly a function of time. When retirement tends to be more a function of use, depreciation may be based on the unit of production method *under appropriate circumstances,*[3] that is:

... the method must be confined to those items in the property account whose useful lives are determined by the factors of wear and tear or where the extent of use or the rate of production measures the rate of exhaustion of the property. . . . For most property it is not possible to obtain this information with any reasonable degree of accuracy and, therefore, the method is not considered an acceptable one for general application to the machinery account of industrial concerns or to the property of those companies exploiting a natural resource with reserves sufficient to extend operations beyond the physical life of the original plant.[4]

The unit-of-production method requires an estimate of total lifetime use or production; the asset cost is then allocated equally over the units produced rather than the years of production. Thus

Depreciation rate per unit of production

$$= (B - V)/(\text{estimated lifetime production})$$

---

[3] See *Bulletin #6182,* U.S. Treasury Department, 1954, p. 12.
[4] See *Bulletin F,* U.S. Treasury Department, 1942, p. 5.

Example 9-3. **Computation of Depreciation by the Unit-of-Production Method**

Information: As in Example 9-2, an asset has been acquired at a cost of $250,000 and is expected to be used for five years, then sold for an estimated $100,000.

Additional information includes: Estimated lifetime production of asset = 100,000 units, and estimated annual production is 10,000 units in year one; 20,000 units in year two; 30,000 units in year three; 30,000 units in year four; and 10,000 units in year five.

Objective: On the basis of the information above, estimate (1) annual depreciation allocations, (2) accrued depreciation at each age, and (3) portion of installed cost unallocated at each age using the unit-of-production method.

Analysis:

Depreciation rate per unit of production

$$= (B - V)/\text{estimated lifetime production}$$

$$= (\$250,000 - \$100,000)/100,000$$

$$= \$1.50/\text{unit}$$

Thus first-year depreciation would be 10,000 units @ $1.50 = $15,000. In similar fashion the other data is:

Age, in Years	Depreciation Allocation	Accrued Depreciation at Age Shown	Installed Cost Unallocated at Age Shown
1	$15,000	$ 15,000	$235,000
2	30,000	45,000	205,000
3	45,000	90,000	160,000
4	45,000	135,000	115,000
5	15,000	150,000	100,000

Retirement of Property

Property retirement is the withdrawal of an asset from its primary service function. Withdrawal can mean physical removal, abandonment, or reassignment to a secondary service function. Reassignment is exemplified by the old building whose use is downgraded from office space to warehouse space or by the old electric generator downgraded from regular service to standby service.

The managerial decision to retire a property is usually based on one or more of the following factors.

1. *Unsatisfactory functional characteristics of the asset.* The question of retirement is eventually raised by the wear and tear of usage; deterioration of time; accident; disaster; or inadequacy of the asset to meet such updated requirements as those for safety, output quality, and output capacity. Retirement is appropriate if analysis shows that repair (including repair of appearance), reconditioning, modernizing, or modification is uneconomic.

2. *Unsatisfactory economic characteristics of the asset.* Through technological change, improved assets currently available may have reduced operating costs to an extent that the old asset is uneconomic.

3. *Termination of need.* Abandonment of the local enterprise, termination of a contract, or withdrawal from a certain phase of the business, can lead to retirement of an asset.

4. *Requirements for maintenance policy, style, or governmental authority.* Maintenance policy based on study may indicate economic replacement prior to failure; group replacement at specified intervals may also be part of a well-planned economic replacement schedule. Style is subject to the same hazards of obsolescence as function and may also be a factor in the retirement decision. Governmental authority may lead to retirement of an asset to make way for highways, transit systems, defense sites, or flood control projects; assets retired may include structures, underground services, and related facilities.

Other factors can also affect the timing of retirement; certainly the availability of funds with which to acquire a replacement, the urgency of the need, and even the availability of equipment or installation personnel (as in wartime periods of emergency) can influence the timing of retirement. Still other influential factors include tax laws as they affect retirement timing via regulation of income tax rates, depreciation rates and disposal adjustments, and expected future price trends on new and used equipment.

Gain or Loss on Disposal

When property is retired from its primary service function but retained for use in a secondary function, the cost base for income tax purposes is generally the portion of installed cost unallocated at the date of retirement. Under these circumstances there is no gain or loss on disposal (although the depreciation rate for the property may change).

When a property is disposed of, the selling price less cost of removal may not coincide with installed cost less accumulated depreciation. In such cases accounting entries showing a gain or loss on disposal or adjustment of the cost basis of the replacement property are allowable when item accounting (but not group accounting) is employed. Under some circumstances the profitable sale of business property held more than six months is taxed as long-term capital gain and thus is taxed at a lesser rate than that applying to ordinary income. Recapture provisions of income tax laws limit the extent to which that lesser rate can be applied.

The gains and losses of a given year must be offset against each other so that a single final net gain or loss for the year is the base upon which income tax is computed. This treatment tends to reduce or eliminate the amount of gain upon which the favorable tax rate is applied. It is generally to the taxpayer's advantage, therefore, if losses tend to be grouped in one year, gains in the next, and so forth, in alternation; thus it is apparent that tax laws can influence the timing of property retirement.

Example 9-4.   Determining Gain or Loss on Disposal

Information: A property acquired for $250,000 was originally estimated to have a life of five years and $100,000 salvage value. The company uses straight-line depreciation.

Objective: Compute the "gain on disposal" or "loss on disposal" if the property is sold for $120,000 at the end of year three.

Analysis: Unallocated cost of the property at the date of sale = $250,000 - $90,000 = $160,000. "Loss on disposal" = $160,000 - $120,000 = $40,000.

Depletion

Depletion is the expiration of an asset caused by its conversion into a salable product. Assets subject to depletion are such natural resources as oil and natural gas reserves, metal and mineral mines, orchards, vineyards, fisheries, and timber. Not all natural resources are subject to depletion, for example, soil fertility and urban land.

Properties subject to depletion are sometimes referred to as wasting assets. Many are exhaustible, extractive, destructible, nonreproducible, and irreplaceable; most are quantitatively withdrawn in service and, unlike depreciable assets, are not subject to physical deterioration, wear and tear, or obsolescence. Timber properties provide exception to many of the preceding descriptive terms; other exceptions occur to an extent that the terms can be used only to describe roughly, rather than to define, properties subject to depletion.

As an accounting device, depletion is similar to depreciation; both are periodic noncash expenses charged to income, reflecting the decline in value of a property. Unlike depreciation, depletion accruals are not necessarily limited to the cost base of the property. It is argued that depletion helps to provide capital for future survey, exploration,[5] acquisition, development, and compensation for risk involved in development of properties which may prove to be uneconomical; in this sense high depletion allowances may encourage prospecting and development of our national resources. Opponents of the depletion allowance criticize it as arbitrary and excessive.

Two methods of computation are permitted by the Internal Revenue Code, the cost method and the percentage method. Their applicability is portrayed in Figure 9-1.

The cost method is similar to the unit-of-production method of computing depreciation. Annual depletion charges are the product of (1) cost or other basis of the property and (2) the ratio of units sold during the year to the estimated total number of units available. Write-off by the cost method limits the accumulated depletion to the cost or other basis of the property. Allowable depletion is never less than that computed by means of the cost method.

In the percentage method, annual depletion charges are com-

---

[5] See *Federal Tax Regulations*, two volumes (St. Paul, Minn., West Publishing Co.), annual. Sections 615 and 616 of the code permit limited deductions for exploration and/or development expenditures for ore or other minerals, but not for oil or gas.

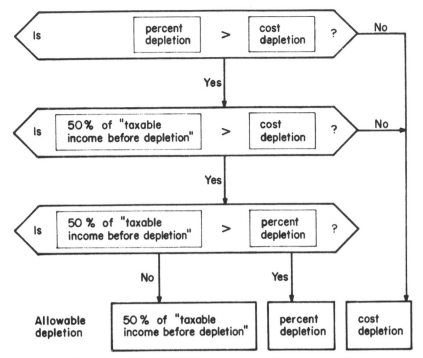

Fig. 9-1. Flow chart for determination of allowable depletion.

puted as a percentage of gross income. The applicable percentage and other limitations are prescribed in the current Internal Revenue Code; the percentage method is applicable to mine or well properties, but not timber properties. The Internal Revenue Code of 1954 as later amended provides:

SEC. 613 PERCENTAGE DEPLETION
(a) GENERAL RULE—In the case of the mines, wells and other natural deposits listed in subsection (b), the allowance for depletion under section 611 shall be the percentage, specified in subsection (b), of the gross income from the property excluding from such gross income an amount equal to any rents or royalties paid or incurred by the taxpayer in respect to the property. Such allowance shall not exceed 50 percent of the taxpayer's taxable income from the property (computed without allowance for depletion). In no case shall the allowance for depletion under section 611 be less than it would be if computed without reference to this section.
(b) PERCENTAGE DEPLETION RATES—The mines, wells,

and other natural deposits, and the percentages, referred to in subsection (a) are as follows:

(1) 22 percent
  (A) oil and gas wells
  (B) sulphur and uranium; and
  (C) if from deposits in the United States—anorthosite, clay, latertite, and nephelite syenite (to the extent that alumina and aluminum compounds are extracted therefrom), asbestos, bauxite, celestite, chromite, corundum, fluorspar, graphite, ilmenite, kyanite, mica, olivine, quartz crystals (radio grade), rutile, block steatite talc, zircon, and ores of the following metals: antimony, beryllium, bismuth, cadmium, cobalt, columbium, lead, lithium, manganese, mercury, molybdenum, nickel, platinum and platinum group metals, tantalum, thorium, tin, titanium, tungsten, vanadium, and zinc.

(2) 15 percent (if from deposits in the United States)
  (A) gold, silver, copper, and iron ore; and
  (B) oil shale (except shale described in paragraph 5).

(3) 14 percent
  (A) metal mines (if paragraph (1C) or (2A) does not apply), rock asphalt, and vermiculite; and
  (B) if paragraph (1C), (5), or (6B) does not apply, ball clay, bentonite, china clay, sagger clay, and clay used or sold for use for purposes dependent on its refractory properties.

(4) 10 percent—asbestos (if paragraph (1C) does not apply), brucite, coal, lignite, perlite, sodium chloride, and wollastonite.

(5) 7½ percent—clay and shale used or sold for use in the manufacture of sewer pipe or brick and clay, shale, and slate used or sold for use as sintered or burned lightweight aggregates.

(6) 5 percent
  (A) gravel, peat, pumice, sand, scoria, shale (except shale described in paragraph (2B) or (5)), and stone, except stone described in paragraph (7);
  (B) clay used, or sold for use, in the manufacture of drainage and roofing tile, flower pots, and kindred products; and
  (C) (if from brine wells), bromine, calcium chloride, and magnesium chloride.

(7) 14 percent—all other minerals (including but not limited to), aplite, barite, borax, calcium carbonates,

diatomaceous earth, dolomite, feldspar, fullers earth, garnet, gilsonite, granite, limestone, magnesite, magnesium carbonates, marble, phosphate rock, potash, quartzite, slate, soapstone, stone (used or sold for use by the mine owner or operator as dimension stone or ornamental stone), thenardite, tripoli, trona, and (if paragraph (1C) does not apply) bauxite, flake graphite, fluorspar, lepidolite, mica, spodumene, and talc (including pyrophyllite), except that, unless sold on bid in direct competition with a bona fide bid to sell a mineral listed in paragraph (3), the percentage shall be 5 percent for any such other mineral (other than slate to which paragraph (5) applies) when used, or sold for use, by the mine owner or operator as riprap, ballast, road material, rubble, concrete aggregates, or for similar purposes. For purposes of this paragraph, the term "all other minerals" does not include:

(A) soil, sod, dirt, turf, water, or mosses; or

(B) minerals from sea water, the air, or similar inexhaustible sources.

Example 9-5.   Computation of Allowable Depletion

Information: An antimony mine has a cost basis of $200,000; percentage depletion has been elected. Estimated salvage is zero. No other investment or depreciable property is involved; financing is by equity capital. Withdrawal this year of 10% of the ore originally estimated to be present has resulted in deductible expenses of $40,000, while gross income has amounted to $100,000. Section 613 permits use of a depletion rate of 22%. Computations are as follows:

Percentage depletion: (22%) ($100,000) = $22,000
Cost depletion: (10%) ($200,000) = $20,000
50% of taxable income before depletion:
                    (50%) ($100,000 − $40,000) = $30,000

Objective: Find allowable depletion.

Analysis: By reference to the limitations of Section 613, the allowance found is $22,000.

Problems

9-1. The following estimates have been made with regard to a certain asset:

Installed cost = $100,000
Net salvage = $16,000 at the end of year 20

Find the depreciation allocation for the second year if the method used is:
a. Straight-line.
b. Sum-of-the-years-digits.
c. Double-declining-balance.
d. 10% sinking-fund.
e. Five-year write-off.
f. Unit-of-production if total production of 500,000 units of product are expected to be produced at the rate of 10,000 units per year for the first five years and the last five years, and 40,000 units per year for the middle ten years.

9-2. The following estimates have been made for a certain equipment item:

Installed cost = $100,000
Net salvage = $18,000 at the end of year 40

Find the accrued depreciation allocations at age ten years by the following methods:
a. Straight-line.
b. Sum-of-the-years-digits.
c. Double-declining-balance.
d. 6% sinking-fund.
e. Unit-of-production if total production of 1 million units of product are expected to be produced at the rate of 10,000 units per year for the first 5 years, 40,000 units per year for the next 20 years, and 10,000 per year for the last 15 years.

9-3. The following estimates have been made for a certain equipment item:

Installed cost = $110,000 $i = 10\%$
Salvage = $0 at the end of year 10

Find annual equivalent depreciation expense for each of the following methods:

     a. Straight-line.
     b. Sum-of-the-years-digits.
     c. Five-year write-off.

9-4. A "Certificate of Necessity" permitting a five-year write-off of a certain property has been issued. Installed cost is $100,000 and estimated salvage is zero, 20 years from now. Actual depreciation charges would be $20,000 per year for five years; express this as an annual equivalent depreciation charge over the 20-year period using $i = 10\%$ and compare your result to the SYD and SL method equivalents.

9-5. The following information applies to the property of a corporation:

    (1) Effective income tax rate is 50% on ordinary income.
    (2) Tax law recapture provisions require that the gain on this property be taxed as ordinary income.
    (3) Income for the corporation involved is enough so that the income tax bracket will not be affected by the gain or loss involved.
    (4) Property in question has an installed cost of $100,000 and is now five years old.
    (5) Property was originally estimated to have a life of ten years and zero net salvage.

    a. Find the capital gains tax if the property is sold now for $58,000 and if the depreciation method used has been:

      (1) Straight-line.
      (2) Sum-of-the-years-digits.

    b. Since the capital gains tax was higher when SYD method is used, does this mean the method is less advisable for such circumstances?

9-6. If minimum attractive rate of return = 10% and estimated life = 15 years, find the uniform annual equivalent depreciation expense over the 15 years. Express your result as a percent of installed cost less net salvage for the method(s) specified:
    a. Straight-line.
    b. Sum-of-the-years-digits.
    c. 10% sinking-fund.
    d. Five-year write-off.

9-7. Assume that any of the depreciation methods listed in Table 9-1 are allowed and that the taxpayer wishes to maximize present equivalent depreciation expense. Use the rules given in this chapter to identify the optimal depreciation method for each of the properties below.

Prop- erty	First Cost	Estimated Net Salvage	Economic Life, in Years	Tax Life, in Years
A	$100,000	$ 11,000	5	5
B	100,000	- 11,000	10	8
C	100,000	- 11,000	20	16
D	100,000	11,000	40	32

9-8. Two years ago Blue Corporation purchased an equipment costing $100,000; estimated equipment life was 20 years and estimated salvage was zero. DDB depreciation has been used. Today the company sold the equipment for $41,000. Find the gain or loss on disposal.

9-9. Distinguish between *depreciation* and *depletion*.

9-10. Find allowed depletion for each of the properties below.

Depletion Computed by:

Property	Percentage method	Cost method	50% of taxable income before allowance for depletion
A	$10,000	$ 8,000	$ 6,000
B	20,000	15,000	30,000
C	30,000	40,000	50,000
D	40,000	60,000	50,000
E	50,000	20,000	30,000
F	60,000	80,000	40,000

9-11. Use symbols as follows:

R = Gross income or revenues.

$M + D + I$ = Deductible expenses except depletion.

D_q = Percentage depletion rate (expressed as a decimal ratio) allowed for property in question.

Formulate an expression to show the ratio $(M + D + I)/R$ necessary for the following limitation to take effect: *Allowance shall not exceed 50 percent of the taxpayer's taxable income from the property (computed without allowance for depletion).*

9-12. Find allowed depletion for each of the properties below.

Depletion Computed by:

Property	Percentage method	Cost method	50% of taxable income before allowance for depletion
G	$2000	$4000	$6000
H	6000	2000	4000
I	4000	6000	2000
J	2000	6000	4000
K	4000	2000	6000
L	6000	4000	2000

9-13. Compute depletion allowance for the following properties assuming that the percentage method has been used in past years.

Property	Depletion Rate	Natural Resource	Amount Removed This Year	Cost Basis	Gross Income	Deductible Expenses
	percent		*percent*			
A	22	Sulphur	5	$300,000	$100,000	$ 60,000
B	22	Oil	10	150,000	100,000	20,000
C	10	Sodium chloride	10	400,000	100,000	40,000
D	5	Gravel	4	300,000	200,000	170,000

9-14. A mine costing $1 million has an estimated 200,000 tons of recoverable ore with a current market of $60 per ton. Assume operating expenses will be about $10 per ton removed and the applicable depletion rate is 15%; compute depletion charges per ton by the:
a. Cost method.
b. Percentage method.

9-15. *Depreciation reserve* for an item of property is the accumulation of *depreciation expense* charges over the past years. Inspection of the annual report of almost any firm, however, reveals that the difference between last years and this years depreciation reserve is less than this years depreciation expense. Explain why.

9-16. A small manufacturing concern has incurred the following costs. Write *capitalize* or *expense* by each to indicate the appropriate treatment.
a. Equipment maintenance $ 400
b. Employee training costs   800
c. Paper punch for office   15

d.	Filing cabinet	80
e.	Rearrangement of facilities	1600
f.	Purchase of a used lathe	2200
g.	Trucking charge for a used lathe	90
h.	Installation cost for a used lathe	300
i.	Case of fluorescent lamp bulbs	120
j.	Advertising costs	1000
k.	Automatic feed added to punch press	1400
l.	Stamping die for a special order	1000
m.	New battery for truck	25

9-17. It would appear that the criterion of maximizing profit or the requirement of meeting a predetermined before-tax rate of return requirement could also be expressed in a policy which states that investments which improve reported *earnings per share* will be accepted, while those which do not will be rejected. What are the shortcomings of this criterion?

9-18. Federal income tax regulations permit the computation of depreciation charges by several different methods. These include (1) straight-line, (2) sum-of-the-years-digits, (3) double-declining-balance, and (4) sinking-fund methods. Assume the following data for an equipment:

Installed cost (includes transportation and installation costs) = $B = \$100,000$

Net salvage value (market value at the end of service life less cost of removal) = $V = \$10,000$

Probable service life, in years = $n = 5$ years

Interest rate per period = $i = 10\%$

Let z = age of property, in years

   D_z = depreciation allocation for the zth year

   DC_z = accrued depreciation allocation to age z

   B_z = portion of installed cost unallocated at age z

Using the data above, compute D_z, DC_z, and B_z by all four methods. Show your results in a table similar to that given below:

Straight-Line				Sum-of-the-Years-Digits			
Year	D_z	DC_z	B_z	Year	D_z	DC_z	B_z
1				1			
2				2			
3				3			
4				4			
5				5			

Double-Declining-Balance   Sinking-Fund

Year	D_z	DC_z	B_z	Year	D_z	DC_z	B_z
1				1			
2				2			
3				3			
4				4			
5				5			

9-19. You have purchased an automatic screw machine for $12,000. It has an estimated service life of ten years and an estimated salvage value of $1000.
   a. Find depreciation expense for the second year if the double-declining-balance method is used each year.
   b. Find depreciation reserve at the end of the third year if the sum-of-the-years-digits method is used each year.
   c. The following data comes from this years annual report:

12-31-X1 Depreciation reserve	$5,584,030
19X2 Depreciation expense	1,219,007
Subtotal	$6,803,037
12-31-X2 Depreciation reserve	5,949,326
"Difference"	$ 853,711

   Explain what this "difference" is and specify how it arises.
9-20. The following data applies to a certain property:

   First cost = $100,000
   Life = 10 years
   Salvage = negative $10,000 (removal cost exceeds gross salvage)
   $i = 10\%$

   a. Find the annual equivalent depreciation expense if SL depreciation is used.
   b. Repeat for SYD depreciation.
   c. Find *PED* by the SYD method and compare your result with Table 9-2.
9-21. An unidentified organization has an office copying machine which was accidentally damaged; the machine will cost $200 to repair or can be sold "as is" to the supplier of these machines for $100. The copying machine was purchased a year ago for $600; at that time it was thought the life would be six years. The salesman has

used the following comparison to suggest that a replacement unit be rented (no long-term commitment is required) and the damaged unit turned in.

Cost of Owning		Cost of Renting	
Depreciation		Rental costs	
$600/6	= $100	12($25/month)	= $300
Repair	200	Less trade-in	− 100
Property tax and			
insurance	25		
	$325		$200

Make your own comparison of alternatives commenting
on the changes you suggest.

I0

Income Taxes

Objections to the Before-Tax Approach

IN THE EXAMPLES AND PROBLEMS of Chapters 1 through 8 the before-tax rate of return (required or prospective) had to provide for both income taxes and after-tax rate of return; thus income taxes were *implicitly* treated. In this chapter methods of *explicitly* computing income taxes will be introduced.

The main objection to the use of a before-tax rate of return requirement (approximately double the after-tax rate of return requirement when income taxes are at a 50% rate) is that the method makes no allowance for the rate of write-off. In each set of circumstances illustrated in Example 10-2, the before-tax approach indicates an annual revenue requirement of $30,291; under the after-tax approach, those requirements are shown to vary from $18,619 to $37,239. Example 10-3 portrays circumstances under which the after-tax rate of return on composite capital varies from 8.6% to 22%; in each case the before-tax approach indicates a pretax rate of return of 20%, thus failing to discern among varied circumstances. Additional objections to the before-tax approach include the following:

1. Because the before-tax approach makes no allowance for the rate of write-off, it fails to provide any guidelines for choice of a depreciation method (liberalized versus straight-line) nor does it give a proper basis for the accept-reject decision where liberalized depreciation is employed.
2. Even with straight-line depreciation the approach is inexact. Its use results in understatement of prospective rate of return or overstatement of revenue requirements for properties which are depreciable.
3. In allocating funds to projects (capital budgeting) the after-tax approach may be the only alternative when comparing projects subject to depletion against those subject to depreciation. In any case, the ranking of projects may be different on a before-tax basis than on an after-tax basis.
4. As usually treated in practice, the before-tax approach does not take into account the firm's financial structure (debt-equity relationship).

Because of the preceding objections the before-tax approach should be limited to use as:

1. An intermediate teaching and learning device for inclusion of income taxes prior to explicit treatment of same.
2. An exact approach if income taxes are not applicable (as with certain tax-exempt organizations of government).

Example 10-1.   Comparison of Two Alternatives, Identical except for Estimated Life

Information: The following estimates have been made:

	Alternative B	Alternative C
First cost	$120,000	Same as B
Salvage	Zero	Same as B
Life	30 years	60 years
Financing	100% by equity	Same as B
Income tax rate	50%	Same as B
Before-tax cash flow	$22,000/yr	Same as B
Depreciation method	Straight-line	Same as B
MARR	10%	Same as B

Objective: Compare the two alternatives.

Analysis:

	Alternative B	Alternative C
Before-tax cash flow	$22,000	$22,000
Less: Depreciation expense	4,000	2,000
Equals: Taxable income	$18,000	$20,000
Less: Income tax @ 50% rate	9,000	10,000
Equals: Reported net income	$ 9,000	$10,000

    Again, as in Example 9-1, we are tempted to choose the alternative with greater "reported" annual net income. Resisting this temptation, we might subtract income taxes from the before-tax cash flow to find the after-tax cash flow is $13,000 per year for B and $12,000 per year for C (the same numerical result is produced by adding together depreciation expense and reported net income). Calculations reveal a rate of return for C of 10.0%, while B's return is 10.2%, and thus verify that B is the better of the alternatives. This result is something of a surprise in the light of prior knowledge and conceptions. The tricky result is due to depreciation and income taxes; until this chapter we have been well aware of C's advantage of longer life and could express it in an annual form:

$$\text{C's advantage} = \left(\begin{array}{c}\text{annual equivalent of savings}\\ \text{due to longer life}\end{array}\right)$$

$$= \$120,000\,[(a/p)_{30}^{10\%} - (a/p)_{60}^{10\%}\,]$$

$$= \$690/\text{yr}$$

and

$$\text{B's advantage} = \left(\begin{array}{c}\text{annual equivalent of income tax}\\ \text{savings via greater depreciation}\end{array}\right)$$

$$= \$120,000\,(1/30 - 1/60)(50\% \text{ tax rate})$$

$$= \$1000/\text{yr}$$

In other words, we have found that B's advantage of greater depreciation charges (and therefore smaller income tax expenses) more than offsets C's very distant extra years of service. C's advantage becomes more important for small values of n and i.

    Break-even points where tax savings are just offset by

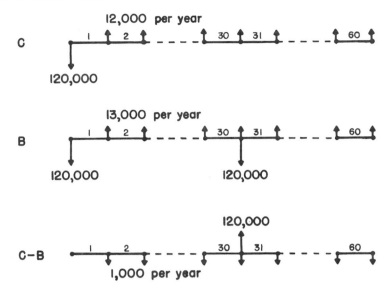

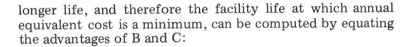

longer life, and therefore the facility life at which annual equivalent cost is a minimum, can be computed by equating the advantages of B and C:

$$(a/p)_n^i - (a/p)_{n+1}^i = \left(\frac{1}{n} - \frac{1}{n+1}\right) 0.50$$

Break-even Point	Approximate i
years	*percent*
50	6
40	8
20	17
10	41

Thus, when $i = 8\%$ and $n = 40$, (1) a small error in estimated equipment life produces no error in the computed annual equivalent cost and (2) additional years of estimated equipment life beyond age 40 actually increase the annual equivalent cost. Because of income taxes optimal facility life is less than infinite.

Examples 9-1 and 10-1 both emphasize the need for careful analysis where income taxes are involved; both produced results different from what may have been expected. These same effects are illustrated later in Table 10-3.

Cash Flow and the Impact of Income Taxes

Graphic treatment of a probabilistic before-tax cash flow stream and the resulting after-tax cash flow stream (see Figs. 10-1 and 10-2) suggests that, in a sense, the government is a partner in the business, sharing in both gains and losses, and as such (via income taxes) reduces the variability (risk?) of prospective cash flow. Furthermore, the deductibility of expenditures (either currently or prorated via depreciation charges over a number of accounting periods) in effect introduces the government as part financier (via tax savings) of the expenditure; the investment credit act further supports the analogy. Carry-backward and carry-forward provisions of our federal tax laws also tend to reduce variability in corporate cash flow.

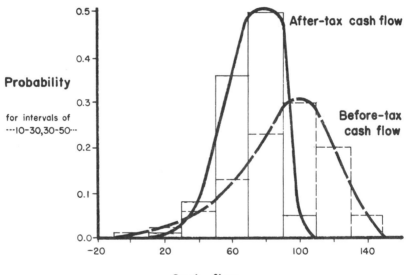

For example, if depreciation expense = 50, debt ratio = 0, marginal income tax rate = 50%, and before-tax cash flows given apply, then after-tax cash flows will be:

Before-tax cash flow	0	20	40	60	80	100	120	140
Probability	0.01	0.02	0.06	0.13	0.23	0.30	0.20	0.05
After-tax cash flow	25	35	45	55	65	75	85	95

Fig. 10-1. Income taxes tend to reduce variability of cash flow for a project.

Corporate Flow of Cash to the U.S. Treasury

The nomenclature, taxation, and relationship of various cash flows are illustrated next:

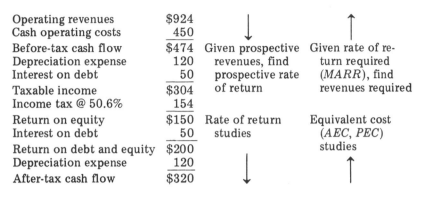

Operating revenues	$924		
Cash operating costs	450		
Before-tax cash flow	$474	Given prospective	Given rate of re-
Depreciation expense	120	revenues, find	turn required
Interest on debt	50	prospective rate	(*MARR*), find
Taxable income	$304	of return	revenues required
Income tax @ 50.6%	154		
Return on equity	$150	Rate of return	Equivalent cost
Interest on debt	50	studies	(*AEC, PEC*)
Return on debt and equity	$200		studies
Depreciation expense	120		
After-tax cash flow	$320		

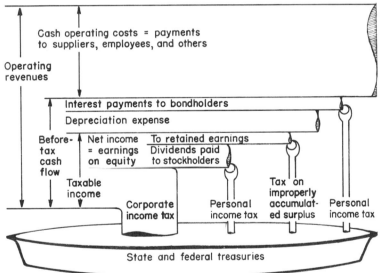

Fig. 10-2. The corporate flow of cash to state and federal treasuries.

Tax Rates on Corporate Income

The U.S. federal tax on a corporation's taxable income (the excess of receipts over such allowable expenses as labor, materials, bond interest, and depreciation) consists of (1) a 22%

"normal tax" on all taxable income, and (2) a 26% "surtax" on all taxable income in excess of $25,000.

Much has been said about double taxation of corporate dividends; first, they are taxed at the rate of 48% as corporate income, then are taxed at the various income tax rates of the stockholders. It has been suggested that corporate dividends be made an allowable deduction in the determination of corporate income tax,[1] just as debt interest already is, in an effort to promote the formation of new capital funds and to regularize investment in capital expenditures. As a minor concession to such arguments, present tax laws exempt the first $100 of dividend income from taxes. (Married couples filing joint returns and having appropriately split ownership of stock may receive up to $200 in tax-free dividend income.) In the case of a closely held corporation, with owners in a tax bracket averaging 50%, each dollar of corporate income will be decreased by $0.48 for federal taxes and by another $0.26 for personal income taxes, thus the effective rate of tax is 74%. Since high tax rates discourage the payment of dividends, the tax laws attempt to protect against the possibility that earnings of the closely held corporation will be held in excess of needs in an effort to avoid income taxes. The protection is in the form of a tax on "undistributed profits." The current annual tax rate on such undistributed profits is 27.5% of the first $100,000 and 38.5% of any further excess ($100,000 is allowed before the tax is applied).

Tax Rates on Individual Income

Federal tax rates on individual income are given in Table 10-1. The rates are graduated, the higher rates applying to the higher levels of income. Because they increase at the higher levels of income, they are said to be *progressive* rates. Based in part on concepts of "ability to pay" and "equal and proportional sacrifice," they reflect the conscious effort of taxing bodies to reduce inequalities of income and wealth. Sales taxes, which tend to fall with an impact that is less than proportional to wealth, are said to be regressive; the combined effect of all types of taxes in the system needs to be considered in developing a tax system and structure.

---

[1] See National Bureau of Economic Research, *Regularization of Business Investments* (Princeton, N.J., Princeton Univ. Press, 1954), p. 227, for this proposal by E. W. Morehouse.

TABLE 10-1. Federal Tax Rates on Individual Income

Taxable Income (after deductions and exemptions) for Married Couple Filing Joint Return (in thousands of dollars)	Rates applicable from 1965 to 1974 excluding surtax of 1968–1970 Amount of Tax
Under 1	14% of taxable income
1 to 2	$ 140 plus 15% of excess over $ 1,000
2 to 3	290 plus 16% of excess over 2,000
3 to 4	450 plus 17% of excess over 3,000
4 to 8	620 plus 19% of excess over 4,000
8 to 12	1,380 plus 22% of excess over 8,000
12 to 16	2,260 plus 25% of excess over 12,000
16 to 20	3,260 plus 28% of excess over 16,000
20 to 24	4,380 plus 32% of excess over 20,000
24 to 28	5,660 plus 36% of excess over 24,000
28 to 32	7,100 plus 39% of excess over 28,000
32 to 36	8,660 plus 42% of excess over 32,000
36 to 40	10,340 plus 45% of excess over 36,000
40 to 44	12,140 plus 48% of excess over 40,000
44 to 52	14,060 plus 50% of excess over 44,000
52 to 64	18,060 plus 53% of excess over 52,000
64 to 76	24,420 plus 55% of excess over 64,000
76 to 88	31,020 plus 58% of excess over 76,000
88 to 100	37,980 plus 60% of excess over 88,000
100 to 120	45,180 plus 62% of excess over 100,000
120 to 140	57,580 plus 64% of excess over 120,000
140 to 160	70,380 plus 66% of excess over 140,000
160 to 180	83,580 plus 68% of excess over 160,000
180 to 200	97,180 plus 69% of excess over 180,000
over 200	110,980 plus 70% of excess over 200,000

Tax Rates on Increments of Income

When an economy study deals with the effects of a project and its related cash flows upon an existing business, we are concerned with incremental effects of accepting or rejecting the project. Among these incremental effects we may look for answers to such questions as:

If the project is accepted, how much will costs be reduced?
If the project is accepted, how much will taxable income and income taxes be increased?

Since most economy studies are based on such incremental costs, incremental cost savings, and incremental revenues, nearly all economy studies use a tax rate applicable to incremental income rather than an average tax rate applicable to overall income.

Combining Federal, State, and Municipal Income Taxes into a Single Effective Tax Rate

Most states and a few municipalities levy a tax both on corporate and on individual income. The deductibility of various expenses, including federal income tax, varies widely. State rates tend to be much lower than federal ones. Persons making economy studies are urged to incorporate in them the estimated future tax rates of states and municipalities.

It is usually convenient to work with a single composite rate which effectively combines federal, state, and municipal tax rates as follows:

Let incremental federal tax rate = 48%
        incremental state tax rate =   5%

Then, if federal income tax *is not* deductible in computing state income tax:

	percent
Income subject to tax	100
State tax = 5% of 100	5
Income subject to federal tax	95
Federal tax = 48% of 95	45.6
Combined state and federal rate = 5% + 45.6% =	50.6

Or, if federal income tax *is* deductible in computing state income tax:

	percent
Income subject to tax	100
Let federal tax be	x
Income subject to state tax	$100 - x$
State tax 5% $(100 - x)$	$5 - 0.05x$
Income subject to federal tax = $100 - 5 + 0.05x =$	$95 + 0.05x$
Federal tax 48% $(95 + 0.05x) = x$ =	46.72
State tax 5% $(100 - 46.72)$ =	2.66
Combined state and federal rate =	49.38

From these two illustrations it is apparent that the convenient approximation of a 50% tax rate is also quite realistic. The examples and problems of this chapter frequently depart from this convenient device to avoid the possible ambiguity arising when the tax and net income figures are identical.

Concessions to Fluctuating Income: Carry-backward and Carry-forward Provisions

Imagine two firms, A and B, similar in many respects except that the taxable income of B is much more variable than that of A:

	Year 1	Year 2	Year 3	Year 4	Total
Firm A	$100,000	$100,000	$100,000	$100,000	$400,000
Firm B	500,000	-300,000	500,000	-300,000	400,000

If current tax rates (22% on all taxable income plus 26% on taxable income in excess of $25,000) were applied without adjustment of taxable income for variability, the resultant taxes would be:

	Year 1	Year 2	Year 3	Year 4	Total
Firm A	$ 41,500	$ 41,500	$ 41,500	$ 41,500	$166,000
Firm B	233,500	0	233,500	0	467,000

Not only is the inequity apparent but the tax rate for Firm B is *more than 100% of the four-year total taxable income!*

In recognition of the unequal burden which could be caused by variability of taxable income, federal tax regulations permit the corporation to carry backward three years and carry forward up to five years any operating or net capital losses. This particular provision has varied from time to time; from 1933 to 1939 no carry backward or carry forward was permitted due to cash needs of the federal government.

An interesting by-product of this tax-law feature has been the mutual attractiveness between firms having substantial unused tax-loss carry forward and firms having substantial profits. Indeed, the firm having a substantial unused tax-loss carry forward may be valued primarily on the basis of potential tax savings rather than on the basis of future profitable operations. Merger for purely taxsaving reasons can be grounds for denying the saving, so the attractiveness is not without qualification.

For operating losses of individuals the carry back is three years and carry forward is five years. Capital losses of individuals are limited to $1000 per year; there is no carry back, but carry forward is permitted without limit. *Income averaging* procedures are available to individual taxpayers whose year-to-year taxable income fluctuates markedly.

Investment Credit

Major provisions of the Investment Credit Act[2] are:

1. U.S. corporations are granted a direct tax reduction equal to the dollar expenditure for "qualifying property" multiplied by an "appropriate rate." This tax reduction has no effect upon the depreciation base.
2. "Qualifying property" includes most of the plant and equipment expenditures of the corporation except buildings and land. Certain building equipment items such as elevators, escalators, and air conditioners are included.
3. "Appropriate rate" depends upon whether the corporation is an industrial or utility firm and upon the life of the qualifying property. The rates are:

	$3 \leqslant n < 5$	$5 \leqslant n < 7$	$7 \leqslant n$
	percent		
Industrials	$2\frac{1}{3}$	$4\frac{2}{3}$	7
Utilities	$1\frac{1}{3}$	$2\frac{2}{3}$	4

4. The excess of investment credit over $25,000 must not exceed 25% of the excess of the precredit tax liability over $25,000. Thus if the precredit tax liability is $105,000, the investment credit must not exceed $45,000 = [$25,000 + 25% ($105,000 - $25,000)]. Excess credits may, however, be carried forward or backward.

Gains and Losses on Disposal of Business Real Estate and Depreciable Property

If sale of an unneeded item of depreciable property gives rise to a substantial gain which would be taxed at high rates, the

---

[2] With some exceptions, benefits of the Investment Credit Act were suspended from September 1966 to March 1967 and from April 1969 to April 1971.

incentive to convert unneeded property to cash is diminished by the prospect of conversion of much of the cash to a tax liability. High tax rates can, in this sense, be a barrier to optimal allocation of resources. Because of this, in 1942 a more favorable treatment of gains on disposal of business real estate and depreciable property was allowed; specifically, a tax rate ceiling of the lesser of 25% or half the taxpayer's marginal tax rate was prescribed. This ceiling has since been modified; gains subject to "recapture" provisions are now taxable as ordinary income, and the 25% limit is increased under some circumstances.

Certain definitions, rules, and details are essential to economy study applications regarding business real estate and depreciable property:

1. Short-term gains and losses are those from assets owned for six months or less. Long-term gains and losses are those from assets owned for more than six months.
2. Losses and gains for the tax year must be offset against each other prior to computation of tax.
3. Long-term losses are fully deductible for the corporate taxpayer, thus permitting savings at a rate identical to the ordinary income tax rate. For the individual taxpayer long-term losses are only half deductible.
4. Because of (2) and (3) it is sometimes good tax strategy for the firm to retire or sell "loss" properties in one year and "gain" properties the next and to continue such alternation over future years.
5. Provisions similar to those for business real estate and depreciable properties are applicable to such capital assets as stocks, bonds, and various securities.
6. The gain or loss is the difference between remaining book value of the property and actual net salvage value realized.
7. Gains or losses on disposal for income tax purposes are not recognized on properties accounted for by group depreciation methods.
8. When one depreciable asset is traded for another of equal or greater cost, the gain or loss is to be incorporated in the book value of the new asset, thus the tax impact, if any, is deferred.
9. Any portion of gain subject to recapture provisions is taxed as ordinary income. For details of recapture provisions, see a current tax reference.[3]

---

[3] Such as *Federal Tax Course* (New York, Prentice-Hall) annual; *Federal Tax Regulations*, 2 vols. (St. Paul, Minn., West Publishing Co.) annual.

Gross Receipts Taxes

When taxes are levied on gross receipts of the firm, they are applied without permitting deduction of the expenses allowed in the computation of income taxes. The amount of the tax is simply the gross receipts tax rate multiplied by gross receipts. When revenues are unknown and the objective of the economy study is revenue requirements, the various revenue requirements are added together. Then gross receipts required will be

$$\frac{\text{Sum of all other revenue requirements}}{1 - \text{gross receipts tax rate}}$$

and the tax itself is

$$\left(\frac{\text{Gross receipts tax rate}}{1 - \text{gross receipts tax rate}}\right)(\text{sum of all other revenue requirements})$$

Tax-exempt Securities

It has long been the policy in our tax system that one branch of government will not tax the income from an instrument of another governing body; thus the federal government exempts from taxation the income from certain bonds of states and municipalities. Examples of such bonds include school district bonds, municipal water revenue bonds, general obligation bonds of the city, university dormitory bonds, state toll road bonds, and so forth. Because the interest payments to holders of such bonds are not taxed by the federal government, they are called tax-exempt.

Tax-exempt securities are especially attractive to persons in high income tax brackets. To a person whose effective income tax rate is 60%, the after-tax yield from a newly issued tax-exempt bond with a 4% coupon rate is comparable to that of a taxable bond with a 10% coupon rate.

Limitation on the Use of Income Tax Information in This Book

The specific income tax regulations of the federal government occupy many times the volume of this entire book. Thorough treatment involving examples and interpretations would require still more space. The detail given here is obviously not intended for computation of actual tax liability; however, it satisfies the needs of the economy study by permitting computation of estimated future income tax liability and acquaints the reader with those characteristics of income tax regulations which are both major in importance and frequent in occurrence.

TABLE 10–2. Sample Data Relating Cash Flows and Repayment of Investment When Debt Capital Is Involved.

End of Year	z	1	2	3
Required operating revenues	C_z	$22,226	$22,226	$22,226
Cash operating costs	M_z	3,000	3,000	3,000
Before-tax cash flow	$C_z - M_z$	19,226	19,226	19,226
Interest on debt	$B_{z-1}r_d i_d$	2,000	1,556	1,058
Depreciation expense	D_z	10,000	10,000	10,000
Taxable income	$C_z - M_z - D_z - B_{z-1}r_d i_d$	7,226	7,670	8,168
Income tax	$t(C_z - M_z - D_z - B_{z-1}r_d i_d)$	4,336	4,602	4,901
After-tax cash flow	$(1 - t)(C_z - M_z)$			
	$+ tD_z + B_{z-1}tr_d i_d$	14,890	14,624	14,325
Required return on debt and equity	$B_{z-1}i_c$	6,000	4,666	3,173
Repayment	$(1 - t)(C_z - M_z) + tD_z$			
	$- B_{z-1}(i_c - tr_d i_d)$	8,890	9,958	11,152
Investment remaining after repayment	B_z	31,110	21,152	10,000

Assumptions:

$B_0 = B = \$40,000 =$ first cost
$\qquad =$ net remaining value at end of year zero
$B_3 = V = \$10,000 =$ salvage
$\qquad =$ net remaining value at end of year three
$n = 3$ years = project life

For income tax purposes depreciation is computed by the straight-line method, so $D_z = (B - V)/n = \$10,000$

$C_z = \$22,226$ per year = required operating revenues
$M_z = \$3,000$ per year = cash operating costs
$\quad t = 60\% =$ combined effective income tax rate
$r_d = 50\% =$ debt ratio = proportion of total financing by debt capital; the remainder of financing is by equity capital.
$i_d = 10\% =$ rate of return required on debt capital.
$i_e = 20\% =$ rate of return required on equity capital.

Because $i_c = r_d i_d + (1 - r_d) i_e$ it follows that:

$\quad i_c = 15\% =$ rate of return required on composite capital

Formulating the Relationship between "Operating Revenues" and "Return on Debt and Equity"

The data assumed for Table 10-2 consist of ten independent variables. The symbols given there permit us to formulate the relationship between "operating revenues" and "return on debt and equity."

Observe that

$$B_z = B_{z-1} - [(1 - t)(C_z - M_z) + tD_z - B_{z-1}(i_c - tr_d i_d)]$$

Simplify

$$B_z = B_{z-1}(1 + i_c - tr_d i_d) - [(1 - t)(C_z - M_z) + tD_z]$$

Let $i_c - tr_d i_d = i_a$, then, for example,

$$B_1 = B_0(1 + i_a) - [(1 - t)(C_1 - M_1) + tD_1]$$
$$B_2 = B_1(1 + i_a) - [(1 - t)(C_2 - M_2) + tD_2]$$

Substitute for B_1:

$$B_2 = B_0(1 + i_a)^2 - [(1 - t)(C_1 - M_1) + tD_1](1 + i_a)$$
$$- [(1 - t)(C_2 - M_2) + tD_2]$$

$$B_3 = B_0(1 + i_a)^3 - [(1 - t)(C_1 - M_1) + tD_1](1 + i_a)^2$$
$$- [(1 - t)(C_2 - M_2) + tD_2](1 + i_a)$$
$$- [(1 - t)(C_3 - M_3) + tD_3]$$

Or in general,

$$B_n = B_0(1 + i_a)^n - [(1 - t)(C_1 - M_1) + tD_1](1 + i_a)^{n-1} \cdots$$
$$- [(1 - t)(C_n - M_n) + tD_n]$$

Multiply by $(1 + i_a)^{-n}$ which is identical to $(p/f)_n^{i_a}$; note that $B_n = V$ and that $B_0 = B$.

$$V(p/f)_n^{i_a} = B - \sum_{z=1}^{n} \{[(1 - t)(C_z - M_z) + tD_z](p/f)_z^{i_a}\} \quad (10\text{-}1)$$

$$\sum_{z=1}^{z=n} C_z(p/f)_z^{i_a} = \begin{pmatrix} \text{present equivalent} \\ \text{cost or revenue} \\ \text{required} \end{pmatrix} = PEC$$

$$\sum_{z=1}^{z=n} M_z \, (p/f)_z^{ia} = \begin{pmatrix} \text{present equivalent} \\ \text{cash operating cost} \end{pmatrix} = PEM$$

$$\sum_{z=1}^{z=n} D_z \, (p/f)_z^{ia} = \begin{pmatrix} \text{present equivalent} \\ \text{depreciation expense} \end{pmatrix} = PED$$

Substituting into Equation 10-1 and rearranging terms,

$$V(p/f)_n^{ia} = B - (1 - t)\,PEC + (1 - t)\,PEM - t\,(PED)$$

Rearranging terms,

$$PEC = PEM + [B - V(p/f)_n^{ia} - t(PED)]/(1 - t) \quad (10\text{-}2)$$

Multiply by $(a/p)_n^{ia}$: AFTER TAX

$$AEC = AEM + [B(a/p)_n^{ia} - V(a/f)_n^{ia} - t(AED)]/(1 - t) \quad (10\text{-}3)$$

where $i_a = i_c - t r_d i_d$ = minimum attractive rate of return. Note that

BEFORE TAX

$$[B(a/p)_n^{ia} - V(a/f)_n^{ia} - t(AED)]/(1 - t) = AE(BTCFR) \quad (10\text{-}3a)$$

where $BTCFR$ = before-tax cash flow required. If straight-line depreciation is used,

$$AEC = AEM + [B(a/p)_n^{ia} - V(a/f)_n^{ia} - t(B - V)/n]/(1 - t)$$

If $r = 0$, then $i_a = i_c = i_e \approx BTRR(1 - t)$, so

icon 239

$$AEC = AEM + [B(a/p)_n^{i} - V(a/f)_n^{i} - t(AED)]/(1 - t)$$

If $t = 0$, then $i_a = i_c \approx BTRR$, so

$$AEC = AEM + B(a/p)_n^{i} - V(a/f)_n^{i}$$

which is the model developed in Chapter 6. Recall that

$$PEX = (PER - PEC)(1 - t)$$

Substitute PEC from Equation 10-2 into the equation above to yield

$$PEX = (PER - PEM)(1 - t) - B + V(p/f)_n^{ia} + t(PED) \quad (10\text{-}4)$$

Alternate equations include

$$PEX = \left[PE\left(\begin{array}{c}\text{before-tax cash}\\ \text{flow available}\end{array}\right) - PE\left(\begin{array}{c}\text{before-tax cash}\\ \text{flow required}\end{array}\right) \right] (1 - t)$$

$$(10\text{-}4a)$$

$$PEX = \sum_{z=0}^{n} (1 + i)^{-z} x_z \qquad (10\text{-}4b)$$

where i = after-tax rate of return required
$\quad n$ = life of the project, in years
$\quad x_z$ = after-tax cash flow in year z

Similar to Equation 10-4 we could show that

$$AEX = (AER - AEM)(1 - t) - B(a/p)_n^{ia} + V(a/f)_n^{ia} + t(AED)$$

$$(10\text{-}5)$$

Given the Rate of Return Required, Find the Revenues Required

Let operating revenues, C_z, be unknown; and let B, V, n, D_z, M_z, t, r_d, i_d, i_e, and i_c be as specified in Table 10-2. Note that

$$i_a = i_c - tr_d i_d = 15\% - 60\%(50\%)(10\%) = 12\%$$

and substitute into Equation 10-3:

$$AEC = \$3000 + [\,\$40,000(a/p)_3^{12\%} - \$10,000(a/f)_3^{12\%}$$
$$- 0.60(\$10,000)\,]/0.40$$

$$= \$22,226$$

Given the Prospective Revenues, Find the Prospective Rate of Return

Let operating revenues be \$26,193 per year; let i_e and i_c be unknown; and let B, V, n, D_z, M_z, t, r_d, and i_d be as specified in Table 10-2. Apply Equation 10-5 to solve for i_a.

$$AEX = 0 = (\$26,193 - \$3000)(0.40) - \$40,000(a/p)_3^{i_a}$$
$$+ \$10,000 \, (a/f)_3^{i_a} + \$6000$$
$$= \$15,277 - \$40,000 \, (a/p)_3^{i_a} + \$10,000 \, (a/f)_3^{i_a}$$

at $i_a = 15\%$: $AEX = +\$638$
at $i_a = 20\%$: $AEX = -\$965$
so $i_a = 15\% + 5\%(638/1603) = 17.0\%$

Since $i_a = i_c - tr_d i_d$,

$$0.170 = i_c - 0.030$$
$$i_c = 20.0\%$$

Since $i_c = ri_d + (1 - r_d) \, i_e$,

$$0.200 = 0.050 + 0.5 \, i_e$$
$$i_e = 30.0\%$$

Additional illustrations of finding revenues required or prospective after-tax rate of return appear in Examples 10-2 through 10-7.

Relating $BTRR$, $MARR$, i_a, i_b, i_c, i_d, i_e, r_d, and t

Appendix A provides a description of each of the eight symbols given. Equations which relate them are

$$i_c = r_d i_d + (1 - r_d) \, i_e$$
$$i_a = i_c - tr_d i_d = (1 - t) \, r_d i_d + (1 - r_d) \, i_e$$
$$i_b = BTRR \approx r_d i_d + (1 - r_d) \, i_e/(1 - t) = i_a/(1 - t)$$

Many books and articles neglect the possibility of debt-financing and therefore do not distinguish between rates of return that are identical only when $r_d = 0$. In other reading where $r_d > 0$ and where the base to which $MARR$ applies is not given, one should be very careful in its interpretation. Minimum attractive rate of return, $MARR$, could refer to

$MARR$ on debt capital, i_d
$MARR$ on equity capital, i_e
$MARR$ on composite capital, i_c
$MARR$ on tax-sheltered composite capital, i_a

In this book it can always be interpreted as $MARR$ on tax-sheltered composite capital, i_a. The distinction is most important when $r_d > 0$ as in some of the problems and examples of Chapters 10 and 19. Note that

when $r_d = 0$: $i_a = i_c = i_e \approx BTRR\,(1 - t)$
$\qquad\qquad = MARR$ on tax-sheltered composite capital
$\qquad\qquad = MARR$ on composite capital
$\qquad\qquad = MARR$ on equity capital
when $t = 0$: $i_a = i_c \approx BTRR$
$\qquad\qquad = MARR$ on tax-sheltered composite capital
$\qquad\qquad = MARR$ on composite capital

In deriving Equations 10-2 and 10-3, it has been shown that the rate of return required or realized is i_a. A given rate i_a, is representative of *many* combinations of i_d, i_e, r_d, and t as tabled next. Because a single rate, i_a, can be used to describe any of these rate of return requirements, it is the most general of the rate of return requirements we could impose upon an investment.

					Resulting Value of:	
					Approximate	
t	r_d	i_d	i_e	i_c	BTRR	i_a
			percent			
0	0	—	10	10	10	10
0	50	8	12	10	10	10
0	100	10	—	10	10	10
20	0	—	10	10	12.5	10
20	50	8	13.6	10.8	12.5	10
20	100	12.5	—	12.5	12.5	10
50	0	—	10	10	20	10
50	50	8	16	12	20	10
50	100	20	—	20	20	10

Impact of Life and Salvage Value When Income Tax Considerations are Explicitly Treated

For any specified firm t, r_d, i_d, i_e, and the depreciation method employed tend to be fixed for a year or more at a time.

Under these circumstances the firm may choose to tabulate the before-tax cash flow requirement for various B, V, and n values as in Table 10-3. Such a table is conveniently and directly applicable to that firm's revenue requirement studies (as illustrated in Problems 10-22, 10-24, 10-26, and 10-28). Based on Equation 10-3, we can write the following equation for use in computing the tabled data:

$$BTCFR/B = [(a/p)_n^{ia} - (V/B)(a/f)_n^{ia} - t(AED/B)]/(1 - t)$$

when $t = 0.50$,

$$BTCFR/B = 2(a/p)_n^{ia} - (2V/B)(a/f)_n^{ia} - AED/B$$

Observe that at $n \approx 14$ years, $BTCFR = 20.00\%$ of first cost and is independent of salvage. For larger values of n the presence of salvage has a detrimental effect upon $BTCFR$. Also observe that if $B > V$, $BTCFR$ is a minimum at $n \approx 32$ years. Additional years of property life have a detrimental effect upon $BTCFR$. Recall Examples 9-1 and 10-1 which emphasize that, due to income tax considerations, neither greater salvage nor greater property life is necessarily beneficial. The two n values noted will decrease if i_a is increased, if t is increased, or if depreciation is increased (computed by a method more liberal than SL).

Table 10-3 is based on a tax rate, $t = 50\%$, and the tax-sheltered rate of return requirement, $i_a = 10\%$. When $t = 50\%$,

TABLE 10-3. Annual Before-Tax Cash Flow Requirements Expressed as a Percent of First Cost of Property

Life of Property	Ratio of Salvage Value to First Cost of Property = V/B				
	0.0	0.1	0.2	0.4	1.0
years					
5	32.76	31.48	30.21	27.66	20.00
10	22.55	22.30	22.04	21.53	20.00
15	19.63	19.66	19.70	19.78	20.00
20	18.49	18.64	18.79	19.10	20.00
30	17.88	18.09	18.31	18.73	20.00
40	17.95	18.16	18.36	18.77	20.00
60	18.40	18.56	18.72	19.04	20.00
∞	20.00	20.00	20.00	20.00	20.00

Note: Assumptions:
$i_a = 10\%$; $t = 50\%$ SL depreciation; $B > 0$
Tax life = economic life   No investment credit

the rate $i_a = 10\%$ will result from many different combinations of r_d, i_d, and i_e values, for example:

| | | | | Resulting Value of: | |
| | | | | Approximate | |
r_d	i_d	i_e	i_c	BTRR	i_a
			percent		
0	—	10	10	20	10
30	6	13	10.9	20	10
50	8	16	12	20	10
60	10	17.5	13	20	10
100	20	—	20	20	10

Format for Revenue Requirement (Equivalent Cost) Studies

When financing is entirely by equity capital ($r_d = 0$), explicit display of the income tax requirement causes no extra calculation. Results can be displayed:

$$ATCFR = B(a/p)_n^i - V(a/f)_n^i \qquad = \qquad K_1$$

$$\text{Income tax} = \left(\frac{t}{1-t}\right)(K_1 - AED) = \qquad K_2$$

$$AEM \qquad\qquad\qquad\qquad = \qquad \underline{K_3}$$

$$AEC \qquad\qquad\qquad\qquad = K_1 + K_2 + K_3$$

When $r_d > 0$, it is more convenient to continue with the format of Equations 10-2 and 10-3.

Example 10-2. Revenue Requirements for a Project under Various Write-off Conditions

a. *Information:* These estimates apply to a proposed project:

$B = \$100,000 = $ first cost

$V = \$4000 = $ salvage

$n = 15$ years $=$ project life

$i_e = 16\% = $ rate of return required on equity capital

$i_d = 8\% = $ rate of return required on debt capital

r_d = 50% = debt ratio = proportion of total financing by debt capital; the remainder of financing is by equity capital.

t = 50% = effective tax rate on taxable income

M_z = \$5000 + \$1000$(z - 1)$ = cash operating costs for the zth year

For income tax purposes depreciation is computed by the straight-line method, so $D_z = (B - V)/n$ = \$6400 per year. Because $i_c = r_d i_d + (1 - r_d) i_e$, it follows that i_c = 12% = rate of return required on composite capital. Because $i_a = i_c - t r_d i_d$, it follows that i_a = 10% = discount rate at which revenue requirements must be computed. If we were to use the before-tax approach with $BTRR$ = 20%, annual equivalent revenue requirements = \$5000 + \$1000 $(a/g)_{15}^{20\%}$ + \$100,000 $(a/p)_{15}^{20\%}$ - \$4000$(a/f)_{15}^{20\%}$ = \$30,291.

Objective: Find the annual equivalent revenue requirement, AEC.

Analysis: Substitute into Equation 10-3:

$$AEC = \$5000 + \$1000(a/g)_{15}^{10\%} + [\$100,000(a/p)_{15}^{10\%}$$
$$- \$4000(a/f)_{15}^{10\%} - 0.50(\$6400)]/0.50 = \$29,921$$

Or alternatively,

$$BTCFR = [\$100,000(a/p)_{15}^{10\%} - \$4000(a/f)_{15}^{10\%}$$
$$- 0.50(\$6400)]/0.50 = \$19,642$$

$$AEM = \$5000 + \$1000(a/g)_{15}^{10\%} \qquad = \quad 10,279$$

$$AEC \qquad\qquad\qquad\qquad\qquad\qquad = \$29,921$$

b. *Information:* Same as Part a except that for income tax purposes depreciation is computed by the sum-of-the-years-digits method.

Objective: Same as Part a.

Analysis: D_1 = \$12,000; D_2 = \$11,200; . . .

$$AED = \$12,000 - \$800(a/g)_{15}^{10\%} = \$7777$$

$$BTCFR = [\$100,000(a/p)_{15}^{10\%} - \$4000(a/f)_{15}^{10\%}$$
$$- 0.50(\$7777)]/0.50 = \$18,265$$

$$AEM = \$5000 + \$1000(a/g)_{15}^{10\%} \qquad\qquad = \quad 10,279$$

$$AEC \qquad\qquad\qquad\qquad\qquad\qquad\qquad\qquad = \$28,544$$

Compare this result with those of Examples 10-4b and 10-5b.

c. *Information:* Same as Part a except that investment credit applies. The company is an industrial firm.

Objective: Same as Part a.

Analysis: Investment credit = 0.07($100,000) = $7000.

$$BTCFR = [\$93,000(a/p)_{15}^{10\%} - \$4000(a/f)_{15}^{10\%}$$
$$- 0.50(\$6400)]/0.50 = \$17,801$$

$$AEM = \$5000 + \$1000(a/g)_{15}^{10\%} \qquad\qquad = \quad 10,279$$

$$AEC \qquad\qquad\qquad\qquad\qquad\qquad\qquad\qquad = \$28,080$$

d. *Information:* Same as Part a except that for income tax purposes the property qualifies for a five-year write-off.

Objective: Same as Part a.

Analysis: $(B - V)/5 = \$19,200.$

$$AED = \$19,200(p/a)_{5}^{10\%} (a/p)_{15}^{10\%} = \$9569$$
$$BTCFR = [\$100,000(a/p)_{15}^{10\%} - \$4000(a/f)_{15}^{10\%}$$
$$- 0.50(\$9569)]/0.50 = \$16,473$$

$$AEM = \$5000 + \$1000(a/g)_{15}^{10\%} \qquad\qquad = \quad 10,279$$

$$AEC \qquad\qquad\qquad\qquad\qquad\qquad\qquad\qquad = \$26,752$$

e. *Information:* The investment is in research, advertising, training, computer programming "start-up" costs, or rearrangement of facilities and can therefore be written off immediately (expensed). $M_z = 0$; $V = 0$; $B = \$141,625$; n, i_d, r_d, t, and i_e are as specified in Part a. Annual revenue requirements computed by the before-tax approach using a 20% rate = $\$141,625 (a/p)_{15}^{20\%} = \$30,291.$

Objective: Same as Part a.

Analysis: $AED = \$141,625 (a/p)_{15}^{10\%}$.

$$AEC = [\$141,625 (a/p)_{15}^{10\%}$$
$$- 0.50(\$141,625) (a/p)_{15}^{10\%}]/0.50 = \$18,619$$

f. *Information:* The investment is in "interest lost during construction" (*ILDC*); tax laws do not permit write-off. $M_z = 0$; $V = 0$; $B = \$141,625$ as in Part e; n, i_d, r_d, t, and i_e are as specified in Part a. Annual revenue requirements computed by the before-tax approach using a 20% rate = \$30,291.

Objective: Same as Part a.

Analysis:

$$AEC = [\$141,625 (a/p)_{15}^{10\%} - 0.50(0)]/0.50 = \$37,239$$

g. *Information:* The investment is in land or working capital; no write-off is permitted. $B = V = \$106,660$. M_z, n, i_d, r_d, t, and i_e are as specified in Part a. Annual revenue requirements computed by the before-tax approach using a 20% rate = \$30,291.

Objective: Same as Part a.

Analysis:

$$BTCFR = [\$106,660 (a/p)_{15}^{10\%}$$
$$- \$106,660 (a/f)_{15}^{10\%}]/0.50 \qquad = \$21,332$$
$$AEM = \$5000 + \$1000 (a/g)_{15}^{10\%} \qquad = \underline{10,279}$$
$$AEC \qquad = \$31,611$$

Example 10-3   After-Tax Rates of Return for a Project under Various Write-off Conditions

a. *Information:* These estimates apply to a proposed project:

$B = \$100,000 = $ first cost
$V = \$4000 = $ salvage

n = 15 years = project life

i_d = 8% = rate of return required on debt capital

r_d = 50% = debt ratio = proportion of total financing by debt capital; the remainder of financing is by equity capital.

t = 50% = effective tax rate on taxable income

M_z = \$5000 + \$1000$(z - 1)$ = cash operating costs for the zth year

For income tax purposes, depreciation is computed by the straight-line method, so

D_z = $(B - V)/n$ = \$6400 per year

AER = annual equivalent revenue

= sufficient to produce a pretax rate of return of 20%

= \$5000 + \$1000$(a/g)_{15}^{20\%}$ + \$100,000$(a/p)_{15}^{20\%}$
$$- \$4000 (a/f)_{15}^{20\%}$$

= \$30,291

Objective: Find the prospective after-tax rate of return on composite capital, i_c.

Analysis: Apply Equation 10-5:

$$AEX = 0 = (AER - AEM)(1 - t) - B(a/p)_{15}^{i_a} + V(a/f)_{15}^{i_a} + t(AED)$$

$$= [\$30{,}291 - \$5000 - \$1000(a/g)_{15}^{i_a}]0.50$$
$$- \$100{,}000(a/p)_{15}^{i_a} + \$4000(a/f)_{15}^{i_a} + 0.50(\$6400)$$
$$= \$15{,}846 - \$500(a/g)_{15}^{i_a} - \$100{,}000(a/p)_{15}^{i_a} + \$4000(a/f)_{15}^{i_a}$$

Compare the *AEX* equation and results with those of Example 10-4a.

At i_a = 10%: AEX = +\$185

At i_a = 12%: AEX = -\$1219

so i_a = 10% + 2% (185/1404) = 10.3%

Since $i_a = i_c - tr_d i_d$,

$$0.103 = i_c - 0.020$$
$$i_c = 12.3\%$$

Since $i_c = r_d i_d + (1 - r_d) i_e$,

$$0.123 = 0.040 + 0.500\, i_e$$
$$i_e = 16.6\%$$

b. *Information:* Same as Part a except that for income tax purposes depreciation is computed by the sum-of-the-years-digits method.

Objective: Same as Part a.

Analysis: $D_1 = \$12,000; D_2 = \$11,200; \ldots$
$$AED = \$12,000 - \$800\,(a/g)_{15}^{i_a}$$

$$AEX = 0 = [\,\$30,291 - \$5000 - \$1000\,(a/g)_{15}^{i_a}\,]\,0.50$$
$$- \$100,000\,(a/p)_{15}^{i_a} + \$4000\,(a/f)_{15}^{i_a}$$
$$+ 0.50\,[\,\$12,000 - \$800\,(a/g)_{15}^{i_a}\,]$$
$$= \$18,646 - \$900\,(a/g)_{15}^{i_a} - \$100,000\,(a/p)_{15}^{i_a}$$
$$+ \$4000\,(a/f)_{15}^{i_a}$$

At $i_a = 10\%$: $AEX = +\$874$
At $i_a = 12\%$: $AEX = -\$411$
so $i_a = 10\% + 2\%\,(874/1285) = 11.4\%$

Since $i_a = i_c - tr_d i_d$,

$$0.114 = i_c - 0.020$$
$$i_c = 13.4\%$$

c. *Information:* Same as Part a except that investment credit applies. The company is an industrial firm.

Objective: Same as Part a.

Analysis: Investment credit = $0.07\,(\$100,000) = \$7000$.

$$AEX = 0 = [\,\$30,291 - \$5000 - \$1000\,(a/g)_{15}^{i_a}\,]\,0.50$$
$$- \$93,000\,(a/p)_{15}^{i_a} + \$4000\,(a/f)_{15}^{i_a} + 0.50\,(\$6400)$$
$$= \$15,846 - \$500\,(a/g)_{15}^{i_a} - \$93,000\,(a/p)_{15}^{i_a}$$
$$+ \$4000\,(a/f)_{15}^{i_a}$$

At $i_a = 10\%$: $AEX = +\$1106$

At i_a = 12%: AEX = - \$191

so i_a = 10% + 2%(1106/1297) = 11.7%

Since $i_a = i_c - tr_d i_d$,

$$0.117 = i_c - 0.020$$
$$i_c = 13.7\%$$

d. *Information:* Same as Part a except that for income tax purposes the property qualifies for a five-year write-off.

Objective: Same as Part a.

Analysis: $(B - V)/5 = \$19{,}200.$

$AED = \$19{,}200(p/a)_5^{i_a} \, (a/p)_{15}^{i_a}$

$AEX = 0 = [\$30{,}291 - \$5000 - \$1000(a/g)_{15}^{i_a}]0.50$
$\qquad - \$100{,}000(a/p)_{15}^{i_a} + 4000(a/f)_{15}^{i_a}$
$\qquad\qquad + 0.50[\$19{,}200(p/a)_5^{i_a} (a/p)_{15}^{i_a}]$

$\quad = \$12{,}646 - 500(a/g)_{15}^{i_a} - \$100{,}000(a/p)_{15}^{i_a}$
$\qquad\qquad + \$4000(a/f)_{15}^{i_a} + \$9600(p/a)_5^{i_a} (a/p)_{15}^{i_a}$

At i_a = 12%: AEX = +\$662

At i_a = 15%: AEX = - \$1152

so i_a = 12% + 3%(662/1814) = 13.1%

Since $i_a = i_c - tr_d i_d$,

$$0.131 = i_c - 0.020$$
$$i_c = 15.1\%$$

e. *Information:* The investment is in research, advertising, training, computer programming "start-up" costs, or rearrangement of facilities and can therefore be written off immediately (expensed). $M_z = 0$; $V = 0$; $B = \$141{,}625$; n, i_d, r_d, t, and AER are as specified in Part a. Note that

AER = sufficient to produce a pretax rate
of return of 20%

$\qquad = \$141{,}625(a/p)_{15}^{20\%} = \$30{,}921$

Objective: Same as Part a.

Analysis: The annual equivalent of the immediate write-off is

$$AED = \$141{,}625\,(a/p)_{15}^{i_a}$$

Apply Equation 10-5:

$$AEX = 0 = \$30{,}291\,(0.50) - \$141{,}625\,(a/p)_{15}^{i_a}$$
$$+ 0.50(\$141{,}625)(a/p)_{15}^{i_a}$$
$$= \$15{,}146 - \$70{,}813\,(a/p)_{15}^{i_a}$$

At i_a = 20%: AEX = 0
so i_a = 20.0%

Since $i_a = i_c - tr_d\,i_d$,

$$0.200 = i_c - 0.020$$
$$i_c = 22.0\%$$

f. *Information:* The investment is in "interest lost during construction" (*ILDC*); tax laws do not permit write-off. M_z = 0, V = 0; B = \$141,625 as in Part e; n, i_d, r_d, t, and AER are as specified in Part a. As in all parts AER is sufficient to produce a pretax rate of return of 20%.

Objective: Same as Part a.

Analysis:

$$AEX = 0 = \$30{,}291\,(0.50) - \$141{,}625\,(a/p)_{15}^{i_a}$$
$$= \$15{,}146 - \$141{,}625\,(a/p)_{15}^{i_a}$$

At i_a = 6%: AEX = +\$564
At i_a = 7%: AEX = - \$404
so i_a = 6% + 1%(564/968) = 6.6%

Since $i_a = i_c - tr_d\,i_d$,

$$0.066 = i_c - 0.020$$
$$i_c = 8.6\%$$

g. *Information:* The investment is in land or working capital; no write-off is permitted. $B = V = \$106,660$. M_z, n, i_d, r_d, t, and AER are as specified in Part a. AER is sufficient to produce a pretax rate of return of 20%.

Objective: Same as Part a.

Analysis:

$$AEX = 0 = [\$30,291 - \$5000 - \$1000\,(a/g)_{15}^{i_a}]\,0.50$$
$$- \$106,660\,(a/p)_{15}^{i_a} + \$106,660\,(a/f)_{15}^{i_a}$$

$$= \$12,646 - \$500\,(a/g)_{15}^{i_a} - \$106,660\,i_a$$

At $i_a = 8\%$: $AEX = +\$1316$
At $i_a = 10\%$: $AEX = -\$660$
so $i_a = 8\% + 2\%\,(1316/1976) = 9.3\%$

Since $i_a = i_c - tr_d i_d$,

$$0.093 = i_c - 0.020$$
$$i_c = 11.3\%$$

The Cash Flow Table As a Sometime Alternate to Equations 10-2 and 10-3.

The cash flow table provides an alternate method of taking an after-tax approach to problems in which (1) after-tax rate of return is sought, *and the debt ratio is zero,* or (2) revenue requirements are sought and the repayment pattern is known. These two situations are illustrated in Example 10-4. Such solutions tend to be less efficient than those employing Equations 10-2 or 10-3, especially when the resultant cash flow pattern is neither a uniform nor gradient series.

Example 10—4.   After-Tax Analyses Using Cash Flow Tables

a. *Information:* These estimates apply to a proposed project:

$B = \$100,000 = $ first cost
$V = \quad \$4,000 \quad = $ salvage
$n = \quad 15$ years $ = $ project life

For income tax purposes depreciation is computed by the SL method.

M_z = $5,000 + $1,000 $(z - 1)$ = cash operating costs for the zth year

Realized operating revenues = $30,291 per year
 t = 50% = combined effective income tax rate
 r_d = 0 = debt ratio; that is, all financing is by equity capital

Objective: Find the prospective after-tax rate of return.

Analysis: The preceding information permits us to complete a cash flow table as follows:

End of Year	1	2	15
Operating revenues	$30,291	$30,291	$30,291
Cash operating cost	5,000	6,000	19,000
Before-tax cash flow	25,291	24,291	11,291
Depreciation expense	6,400	6,400	6,400
Taxable income	18,891	17,891	4,891
Income tax	9,445	8,945	2,445
After-tax cash flow	15,846	15,346 decreasing by $500/yr	8,846

After-tax first cost = $100,000; after-tax salvage = $4000

Then, based on the after-tax cash flows, we can write an equation:

$$AEX = 0 = \$15,846 - \$500\,(a/g)_{15}^i - \$100,000\,(a/p)_{15}^i$$
$$+ \$4000\,(a/f)_{15}^i$$

At i = 10%: AEX = + $185
At i = 12%: AEX = - $1219
so i = 10% + 2% (185/1404) = 10.3%

Compare the AEX equation and results with those of Example 10–3a.

b. *Information:* The data of Example 10–2b apply. Repayment is to be uniform, that is, $6400 each year.

Objective: Find the year-by-year revenue requirements, then convert them to an annual equivalent revenue requirement, AEC.

Analysis:

End of Year	1	2	3
Investment remaining before repayment	$100,000	$93,600	$87,200
Repayment	6,400	6,400	6,400
Return on debt and equity	12,000	11,232	10,464
After-tax cash flow	18,400	17,632	16,864
Income tax	2,400	2,688	2,976
Depreciation expense	12,000	11,200	10,400
Interest on debt	4,000	3,744	3,488
Before-tax cash flow	20,800	20,320	19,840
Cash operating costs	5,000	6,000	7,000
Required operating revenues	25,800	26,320	26,840 increasing by $520/yr

To complete the table we need to observe, perhaps by reference to Figure 10-2, that

$$\left(\begin{array}{c}\text{Before-tax}\\\text{cash flow}\end{array}\right) - \left(\begin{array}{c}\text{income}\\\text{tax}\end{array}\right) = \left(\begin{array}{c}\text{after-tax}\\\text{cash flow}\end{array}\right)$$

$$\left(\begin{array}{c}\text{Before-tax}\\\text{cash flow}\end{array}\right) - t\left(\begin{array}{c}\text{before-tax}\\\text{cash flow}\end{array} - \begin{array}{c}\text{interest}\\\text{on debt}\end{array} - \begin{array}{c}\text{depreciation}\\\text{expense}\end{array}\right)$$
$$= \left(\begin{array}{c}\text{after-tax}\\\text{cash flow}\end{array}\right)$$

$$\left(\begin{array}{c}\text{Before-tax}\\\text{cash flow}\end{array}\right) = \frac{1}{1-t}\left(\begin{array}{c}\text{after-tax}\\\text{cash flow}\end{array}\right)$$
$$- \frac{t}{1-t}\left(\begin{array}{c}\text{interest}\\\text{on debt}\end{array} + \begin{array}{c}\text{depreciation}\\\text{expense}\end{array}\right)$$

Since $t = 0.5$,

$$\left(\begin{array}{c}\text{Before tax}\\\text{cash flow}\end{array}\right) = 2\left(\begin{array}{c}\text{after-tax}\\\text{cash flow}\end{array}\right) - \left(\begin{array}{c}\text{interest}\\\text{on debt}\end{array} + \begin{array}{c}\text{depreciation}\\\text{expense}\end{array}\right)$$

The annual equivalent revenue requirement

$$AEC = \$25,800 + \$520\,(a/g)_{15}^{10\%} = \$28,545$$

Compare this result with those of Example 10-2b and 10-5b.

Example 10—5.   Relationship between Revenue Requirements and "Regulatory Allowed Revenue"

a. *Information:* The data of Example 10-2a apply to an investor-owned utility company. Straight-line depreciation is used for both income tax and regulatory purposes.

   Objective: Find annual revenue requirements by the "allowed revenue" approach of the regulatory commission.

   Analysis:

End of Year	1	2	3
Investment remaining before repayment (rate base)	$100,000	$93,600	$87,200
Repayment	6,400	6,400	6,400
*Interest on debt	4,000	3,744	3,488
*Return on equity	8,000	7,488	6,976
*Income tax	8,000	7,488	6,976
*Depreciation expense	6,400	6,400	6,400
*Cash operating costs	5,000	6,000	7,000
Required operating revenues = sum of * items	31,400	31,120	30,840

   Levelized annual required operating revenue = $31,400 - $280 $(a/g)_{15}^{10\%}$ = $29,922 as was determined in Example 10-2a.

b. *Information:* The data of Example 10-2b apply to an investor-owned utility company. The company uses SYD depreciation for income tax purposes. Under such circumstances the regulatory commission follows a "normalize" procedure in which the commission allows revenues based on (1) SL depreciation or recovery of investment, (2) income taxes deferred as well as those paid, and (3) a rate base reduced by both accumulated deferred income tax liability and accrued SL depreciation.

   Objective: Find annual revenue requirements by the "allowed revenue" approach of the regulatory commission.

Analysis:

End of Year	z	1	2	3
Investment remaining before repayment (rate base)	B_{z-1}	$100,000	$90,800	$82,000
Repayment (includes deferred income tax)	D_{zr} + $(D_{zt} - D_{zr})t$	9,200	8,800	8,400
*Interest on debt	$B_{z-1}\, r_d i_d$	4,000	3,632	3,280
*Return on equity	$B_{z-1}\,(1 - r_d) i_e$	8,000	7,264	6,560
*Income tax paid		5,200	4,864	4,560
*Income tax deferred	$(D_{zt} - D_{zr})t$	2,800	2,400	2,000
Depreciation expense: for income tax	D_{zt}	12,000	11,200	10,400
*for regulation	D_{zr}	6,400	6,400	6,400
*Cash operating costs	M_z	5,000	6,000	7,000
Required operating revenues = sum of * items	C_z	31,400	30,560	29,800

Since income tax paid $= t(C_z - M_z - D_{zt} - B_{z-1} r_d i_d)$ and income tax deferred $= t(D_{zt} - D_{zr})$,

$$C_z = B_{z-1}\,[(1 - r_d) i_e + r_d i_d]$$
$$+ t(C_z - M_z - D_{zt} - B_{z-1} r_d i_d + D_{zt} - D_{zr})$$
$$+ D_{zr} + M_z$$

$$(1 - t)C_z = B_{z-1}\,(i_c - tr_d i_d) + M_z\,(1 - t) + D_{zr}\,(1 - t)$$

$$C_z = B_{z-1}\,(i_c - tr_d i_d)/(1 - t) + M_z + D_{zr}$$

So income tax paid

$$= t\,[B_{z-1}\,(i_c - tr_d i_d)/(1 - t) + D_{zr} - D_{zt} - B_{z-1} r_d i_d]$$
$$= [t/(1 - t)]\,B_{z-1}\,(i_c - r_d i_d) - t(D_{zt} - D_{zr})$$
$$= [t/(1 - t)]\,B_{z-1}\,(1 - r_d) i_e - t(D_{zt} - D_{zr})$$

The revenue requirement decreases by $840, then $760, then $680, and so forth. Using $i_a = 10\%$, it can be shown that the annual equivalent revenue requirement = $28,545, as in Examples 10-2b, 10-4b, and 10-5c.

Although the levelized annual revenue requirement, *AEC*, produced under the regulatory "normalize" policy is identical to that produced under the "flow through" policy, the *pattern* of revenue requirements is significantly different.   When a utility changes its tax depreciation method from SL to SYD or DDB, the "normalize" policy tends to provide an orderly transition in revenues allowed, while the "flow through" policy tends to tip rates to favor the present consumer at the expense of the future consumer.

c. *Information:* The data of Example 10-2b apply to an investor-owned utility company.   The company uses the SYD depreciation method for income tax purposes. Under such circumstances the regulatory commission follows a "flow through" procedure in which the commission allows revenues based on (1) SL depreciation or recovery of investment, (2) income taxes paid but not those deferred, and (3) a rate base reduced by accumulated SL depreciation.

Objective: Find annual revenue requirements by the "allowed revenue" approach of the regulatory commission.

Analysis:

End of Year	z	1	2	3
Investment remaining before repayment (rate base)	B_{z-1}	$100,000	$93,600	$87,200
Repayment	D_{zr}	6,400	6,400	6,400
*Interest on debt	$B_{z-1}r_d i_d$	4,000	3,744	3,488
*Return on equity	$B_{z-1}(1-r_d)i_e$	8,000	7,488	6,976
*Income tax paid		2,400	2,688	2,976
Depreciation expense: for income tax	D_{zt}	12,000	11,200	10,400
*for regulation	D_{zr}	6,400	6,400	6,400
*Cash operating costs	M_z	5,000	6,000	7,000
Required operating revenues = sum of * items	C_z	25,800	26,320	26,840

Since income tax paid = $t(C_z - M_z - D_{zt} - B_{z-1}\,r_d i_d)$,

$$C_z = B_{z-1}\,i_c + t(C_z - M_z - D_{zt} - B_{z-1}\,r_d i_d)$$
$$+ D_{zr} + M_z$$
$$(1-t)C_z = B_{z-1}\,(i_c - tr_d i_d) + M_z\,(1-t) + D_{zr} - tD_{zt}$$
$$C_z = [B_{z-1}\,(i_c - tr_d i_d) + D_{zr} - tD_{zt}]/(1-t) + M_z$$

So income tax paid

$$= [t/(1 - t)] \; [B_{z-1} \; (i_c - tr_d i_d) + D_{zr} - tD_{zt} - D_{zt} \; (1 - t)$$
$$- B_{z-1} \; r_d i_d \; (1 - t)]$$

$$= [t/(1 - t)] \; [B_{z-1} \; (i_c - r_d i_d) + D_{zr} - D_{zt}]$$

$$= [t/(1 - t)] \; [B_{z-1} \; (1 - r_d) i_e + D_{zr} - D_{zt}]$$

The revenue requirement increases by \$520 each year so when i_a = 10%, the annual equivalent revenue requirement = AEC = \$25,800 + \$520 $(a/g)_{15}^{10\%}$ = \$28,545, as in Examples 10–2b, 10–4b, and 10–5b.

Example 10-6.   Finding the Present Equivalent Revenue Requirement

Information: The data of Table 10-2 apply except that the required operating revenue, C, is unknown.

Objective: Find the present equivalent revenue requirement, *PEC.* Prove the result by completing a cash flow table similar to Table 10-2.

Analysis: Using Equation 10-2,

$$PEC = \$3000 (p/a)_3^{12\%} + [\$40,000 - \$10,000 (p/f)_3^{12\%}$$
$$0.6 (\$10,000)(p/a)_3^{12\%} \;] / 0.4$$

$$= \$53,381$$

End of Year	0	1	2	3
Operating revenues	\$53,381	\$ 0	\$ 0	\$ 0
Cash operating costs	. . .	3,000	3,000	3,000
Before-tax cash flow	53,381	- 3,000	- 3,000	- 3,000
Interest on debt	. . .	932	804	661
Depreciation expense	. . .	10,000	10,000	10,000
Taxable income	53,381	- 13,932	- 13,804	- 13,661
Income tax	32,028	- 8,359	- 8,283	- 8,197
After-tax cash flow	21,353	5,359	5,283	5,197
Required return on debt and equity	. . .	2,797	2,413	1,982
Repayment	21,353	2,562	2,870	3,215
Investment remaining after repayment	18,647	16,085	13,215	10,000

Example 10-7.   Year-by-Year Analysis of Income Tax When Depletion Charges Are Involved

Information: Bruce Blue is considering investment of $120,000 to acquire, drill, and develop an oil well having estimated reserves of 200,000 barrels. An additional $80,000 will be required for equipment needed to operate the well. That equipment is expected to have zero salvage and will be depreciated by the unit-of-production method. No debt capital is involved. Mr. Blue's combined effective state and federal income tax rate is 60%. Cost depletion = $120,000/200,000 barrels = $0.60 per barrel. Other estimates:

End of Year	1	2	3
Barrels of oil withdrawn	100,000	70,000	30,000
Gross income = operating revenues	$350,000	$240,000	$80,000
Cash operating costs	153,000	112,000	45,000
Before-tax cash flow	197,000	128,000	35,000
Depreciation expense	40,000	28,000	12,000
Taxable income before deducting depletion	157,000	100,000	23,000
50% of above	78,500	50,000*	11,500
Percent depletion @ 22%	77,000*	52,800	17,600
Cost depletion @ 60¢/bbl	60,000	42,000	18,000*
Taxable income	80,000	50,000	5,000
Income tax @ 60%	48,000	30,000	3,000
After-tax cash flow	149,000	98,000	32,000

   * = Allowable depletion.

Objective: Find Mr. Blue's prospective after-tax rate of return.

Analysis: Using Equation 10-4b and the after-tax cash flows tabled above,

$$PEX = 0 = -\$200,000 + \$149,000(p/f)_1^i + \$98,000(p/f)_2^i + \$32,000(p/f)_3^i$$

At $i = 20\%$: $PEX = +\$10,842$
At $i = 25\%$: $PEX = -\$1696$
so $i = 20\% + 5\% (10,842/12,538) = 24.3\%$

Problems

In these problems and those of subsequent chapters the following assumptions prevail unless otherwise specified:

1. *Basis for tax depreciation, B_t, acquisition, B, retention, B_v, and regulation B_r, are identical.*
2. *Life for tax depreciation, n_t, acquisition or retention, n, and regulation are identical.*
3. *Investment credit percentage is zero.*
4. *The debt ratio is zero; that is, financing is entirely by equity capital.*
5. *The rate of return, required or prospective, is positive and time invariant.*
6. *Item accounting, not group accounting is used, so gain or loss on disposal will be allowed for income tax purposes.*
7. *Cash flows are end of year.*
8. *Compounding frequency is annual.*
9. *The after-tax approach introduced in this chapter is to be used.*
10. *The life given for a facility is certain. It is not subject to dispersion. Dispersion aspects are treated in Chapter 14. The life given is the optimal (economic) life of the facility.*

10-1. Consider the manner in which annual equivalent costs or present equivalent costs are computed and respond to the writer who states, "With our current tax rate of 48%, any increased outlays for wages, contributions to charitable organizations, or advertising now cost the corporation 52 cents per dollar."

10-2. The Green Corporation is subject to an income tax rate of 50%. Find the year-by-year net income (return on equity) if year-by-year taxable income over the first four years of operation follows the pattern shown:
   a. $100,000; $100,000; $100,000; $100,000.
   b. $500,000; - $300,000; $500,000; - $300,000. Assume that losses are carried backward and the tax-saving adjustment is immediate.
   c. - $200,000; - $200,000; - $200,000; $1,000,000. Assume that losses are carried forward to the profitable year and the tax savings are realized at the end of that year.

10-3. Assume investment credit is in effect and show the investment credit rate, if any, which applies.

Company	Item Purchased	Tax Life
a. Ohio Bell Telephone	heavy-duty vehicles	6 years
b. J. C. Penney	land for future use	. . .
c. ABC Electric & Gas	generating facilities	30 years
d. United Airlines	office building	40 years
e. Standard Oil	computer equipment	6 years
f. General Motors	fabricating dies	2 years
g. Eastman Kodak	processing equipment	10 years

10-4. a. Briefly state your objections to treating the effect of investment credit identically for studies that are (1) for choosing among various investment alternatives and (2) for determining the price to charge per unit of good or service.

b. Briefly state your objections to the statement, "Investment credit rates are greater for industrial firms than for utility firms and thus discriminate against the utility firms."

10-5. The owner of a certain property has just received a cash offer of $87,500 on a "now or never" basis. The present owner had not planned to sell the property but had planned to keep it for the next ten years during which time he expected pretax net receipts (before-tax cash flow) of $10,000 per year. After the ten-year period, he expected no further receipts or salvage value. Suppose you were asked to compute the after-tax rate of return he will earn in the years ahead if he retains the property and if he has no debt financing. What additional information would you ask for?

10-6. Find the combined effective tax rate on income if the federal rate is 48%; the state rate is 8%; and in computing state income taxes, only 50% of the total federal income taxes paid can be deducted. State income taxes paid are fully deductible in determining income subject to federal income tax.

10-7. Mr. Green is considering the purchase of a bond which pays interest of $20 each six months; treat these receipts as if they were end-of-year sums of $40 each. The bond matures nine years from today and will be redeemable for $1000 at that time. The cost of the bond today is $720. Mr. Green has a marginal effective combined state and federal income tax rate of 40%. Assume that any capital gain will be taxed at half the ordinary rate.

a. Write an equation that would permit you to find his prospective after-tax rate of return. Simplify your result.

   b. Find Mr. Green's annual effective after-tax rate of
      return.

10-8. See alternatives A, B, and C in Example 9-1 and in
      Example 10-1. Compare "*ATCF* available" with "*ATCF*
      required" for each.

10-9. Galloway Corporation manufactures electrical equip-
      ment of various types.   According to depreciation
      Guidelines the entire depreciable property of such com-
      panies can be written off as if it has a composite life of
      12 years. On this basis a question has been raised as to
      whether the income tax portion of annual equivalent
      cost for all proposed depreciable properties should be
      based on the composite 12-year depreciable life rather
      than on the estimated ownership life of each individual
      property. Take a stand and defend your position.

10-10. Write an equation which would enable you to find the
      prospective after-tax rate of return on a project to which
      the following estimates apply:

      Effective combined state and federal tax rate = 50%
      First cost = $100,000
      Salvage = $10,000
      Operating receipts = $25,000/yr
      Property taxes, insurance, operating costs, and mainte-
        nance = $6000/yr
      Project life = 10 years
      Depreciation method = straight-line

10-11. Find the after-tax rate of return on equity capital in:
      a. Example 10-3e.
      b. Example 10-3f.

10-12. Construct a cash flow table similar to the one given in
      Example 10-4a for:
      a. Problem 10-17a.
      b. Problem 10-13b.
      c. Problem 10-13c.
      d. Problem 10-36.
      e. Problem 10-13e.
      f. Problem 10-10.

10-13. It has been proposed that Green Corporation spend
      $120,000 today in exchange for the prospect of future
      savings in operating costs of $26,000 per year over the
      next 15 years; future salvage value is estimated to be
      negligible. Combined effective state and federal income
      tax rate = 50%.

      a. Find the before-tax rate of return, as in Chapter 7.

    b. Find the after-tax rate of return if SL depreciation is used.

    c. Find the after-tax rate of return if SYD depreciation is used.

    d. Find the after-tax rate of return if a 7% investment credit applies and SYD depreciation is used.

    e. Find the after-tax rate of return if the $120,000 is expensed.

    f. Find the after-tax rate of return if the $120,000 is expensed but the actual taxsaving effect upon cash occurs one year later.

10-14. Bruce Blue owns and operates his own unincorporated business and is subject to federal tax on his personal income only. His taxable income over the years ahead is estimated to be $18,000 per year. He is considering an investment which would cost $100,000 now, would be entirely financed by him, and would result in before-tax net receipts of $20,000 per year for each of the next ten years. The investment is expected to have no salvage value after the ten years, and depreciation is by the straight-line method. Compute his prospective after-tax rate of return on the investment. Mr. and Mrs. Blue file a joint income tax return.

10-15. A certain manufacturing facility is being considered; the following estimates have been developed:

Cost of land = $200,000 = resale value at end of year ten

Cost of buildings and equipment = $650,000; at the end of ten years it is thought that their value will have declined to $250,000

Working capital required = $50,000

Start-up costs of promotion, research, engineering, training, tooling, and other costs to be written off now and recouped over the ten-year life of the project = $100,000

Annual net receipts = operating revenues less cash operating costs = $200,000 per year

Financing is entirely by equity capital

Combined state and federal income tax rate = 50%

Depreciation is computed on a straight-line basis

Find the prospective after-tax rate of return.

10-16. See the before-tax cash flow data given in Figure 10-1. Refer to Figure 10-2, then tabulate the calculations that would permit you to verify the after-tax cash flow and interval probability data given in Figure 10-1.

10-17. Nineteen years ago Mr. Green purchased an apartment building costing $140,000. Annual revenues have been about $24,000 per year, while costs of insurance, property tax, advertising, maintenance, and operation have averaged $8000 per year. He has just had a cash offer of $112,000 for the building. Over all these past 19 years his effective combined state and federal income tax rate has been about 50%. He has no debt financing, and he has used SL depreciation on the building, assuming a 25-year life and $40,000 salvage value. He has no offsetting capital losses this year.

    a. Find his historic after-tax rate of return earned over the past 19 years. Assume that he sells the property now and the gain is taxed at a 25% rate.

    b. Find his prospective after-tax rate of return earned over the next six years if he retains the property then sells it six years hence for $40,000.

    c. Which rate of return (from Part a or from Part b) is more relevant to his decision?

10-18. Refer to Table 10-2. Assume the values given there for B, V, n, D_z, M_z, t, r_d, and i_d. Assume operating revenues are $20,000 each year. Find the after-tax rate of return on equity capital.

10-19. Refer to Example 10-3a. Assume the values given there for B, V, n, D_z, M_z, t, r_d, and i_d. Assume operating revenues are $21,671 + $1000\ (z- 1)$ in the zth year.

    a. Find the after-tax rate of return on equity capital.

    b. Find the uniform annual operating revenue which would produce identical after-tax rates of return. (**Hint:** Use the rate of return i_a computed in Part a).

10-20. a. Construct a cash flow table similar to the one given in Example 10-4b assuming that B, V, n, D_z, M_z, t, r_d, i_d, and i_e are as specified in Table 10-2; "repayment" is uniform; and "required operating revenues" are unknown.

    b. Find the levelized annual equivalent required operating revenue, AEC, based on the year-by-year requirements determined in Part a.

10-21. Table 10-2 is based on data where $i_a = 12\%$ and $t = 60\%$. Although the table further stipulates that $i_e = 20\%$, $r = 50\%$, and $i_d = 10\%$, many other combinations of those three variables will result in $i_a = 12\%$ and therefore identical $BTCFR$ and identical AEC. Find one such combination.

10-22. Green Corporation is considering a property for which:

$$B = \$50,000 = \text{first cost} \qquad t = 50\% \qquad i_e = 13\%$$

$V = \$0$ = salvage $\qquad r_d = 30\% \qquad i_d = 6\%$
$n = 5$ years = life
$M = \$5000$ per year = cash operating cost
SL depreciation will be used for income tax purposes

a. Find the annual equivalent cost, AEC.
b. Use Table 10-3 to check your result in Part a.
c. Recompute AEC assuming that first cost is expensed.

10-23. Find the present equivalent cost = PEC = present equivalent revenue requirement for a property for which:

$B = \$100,000$ = first cost $\qquad t = 50\% \qquad i_e = 12\%$
$V = \$10,000$ = salvage $\qquad r_d = 25\% \qquad i_d = 8\%$
$n = 10$ years = life
$M = \$7000$ per year = cash operating costs
SL depreciation will be used for income tax purposes

10-24. Green Corporation is considering a property for which:

$B = \$110,000$ = first cost $\qquad t = 50\% \qquad i_e = 16\%$
$V = \$0$ = salvage $\qquad r_d = 50\% \qquad i_d = 8\%$
$n = 10$ years = life
$M = 10\%$ of first cost = $0.1B$ per year = cash operating cost
SYD depreciation will be used for income tax purposes

Find the annual equivalent cost = AEC = annual equivalent revenue requirement.

10-25. A small investor-owned water utility has a plant worth $1 million in current dollars after consideration of depreciation. Composite average remaining life of the depreciable plant is 25 years and straight-line depreciation will be used. Of the $1 million current plant value, $200,000 is nondepreciable plant in the form of land and residual salvage values. Financing is 40% debt, 60% equity; $i_d = 5\%$, $i_e = 10\%$. Effective combined state and federal income tax is 50%. In addition to the cost of the plant and income taxes there are administrative salary costs of $50,000 per year and operating costs (electricity, labor, chemicals, and so forth, not including depreciation) of $100 per million gallons (mg). Assume annual sales of 500 mg and compute annual equivalent revenue requirements per mg.

10-26. Green Corporation is considering a property for which:

$B = \$100,000$ = first cost $\qquad t = 50\% \qquad i_e = 13\%$
$V = \$40,000$ = salvage $\qquad r_d = 30\% \qquad i_d = 6\%$

$n = 20$ years = economic life
$M = \$10,000$ per year = cash operating cost

Find the annual equivalent cost, AEC, using:
a. The before-tax approach of Chapter 6 and the appropriate $BTRR$.
b. The after-tax approach and SL depreciation. Check your result using Table 10-3.
c. The after-tax approach and SYD depreciation.
d. The after-tax approach, the shortest tax life allowed (16 years), and 1AS depreciation. Refer to Table 9-2 to find PED, then convert to AED using the economic life of 20 years.

10-27. Refer to Table 10-2. Assume the values given there for B, V, n, D, M, and t. Assume $r_d = 1.0$, $i_d = 30\%$, and that C is unknown. Find the levelized annual equivalent required operating revenue, AEC. Prove your result by completing a cash flow table similar to Table 10-2.

10-28. Green Corporation is considering five mutually exclusive alternatives.

Alternative	First Cost	Annual Before-Tax Cash Flows
0	$ 0	$ 0
A	100,000	40,000
B	200,000	75,000
C	300,000	105,000
D	400,000	130,000

The cash flow will cease after year five and salvage will be negligible. SL depreciation is used, $t = 50\%$, and $MARR = 10\%$. Carefully choose whether to compare by AEX, PEX, ROR or other procedure, then find the preferred alternative.

10-29. Equipment 007-X has just been retired after only 15 years of service; at the time of its acquisition useful life of the equipment was estimated to be 60 years. Other data:

First cost = $120,000
Estimated salvage = actual salvage = $0
Estimated cash operating costs = $6000 per year
Minimum attractive rate of return = 10%
Combined state and federal income tax rate = $t = 50\%$
Depreciation method is SL, the undepreciated balance is written off at the time of retirement, and "loss on disposal" is not offset by gains from other retirements.

Compare annual revenue requirements for the estimated 60-year life with those for the actual 15-year life.

10-30. Determine annual equivalent cost for each of two competing plans. Include investment credit in your comparison. $MARR = 10\%$, $t = 50\%$; the company is an industrial and uses SL depreciation.

Plan A. Continue a present accounting center operation which results in end-of-year ad valorem taxes of $160 and annual costs of $80,000 for maintenance and operation.

Plan B. Invest $300,000 today in business machines having an economic life of five years and a salvage value of $60,000. End-of-year ad valorem taxes are estimated as $8000, and annual costs for maintenance and operation are an estimated $29,000.

10-31. Estimates appropriate to the acquisition of a certain facility are given below:

Land cost = $13,000; resale value ten years hence = $25,000

Building cost = $20,000; resale value ten years hence = $10,000

Equipment cost = $30,000; resale value ten years hence = $0

Working capital required during ownership but available whenever ownership is relinquished = $18,000

Rearrangements, preproduction checking, debugging, employee training, and other costs connected with startup = $11,000 (expensed)

Operation, maintenance, property taxes, and insurance = $10,000 per year

The company is an industrial firm and uses straight-line depreciation. Investment credit is to be taken where it is appropriate to do so. Profit on sale of the land ten years hence will be taxed as a capital gain; use a capital gains tax rate of 25%. $MARR = 10\%$, $t = 54.55\%$. Find the present equivalent cost for ten years of service. (Suggestion: Combine the various first-cost figures before multiplication by the appropriate factors.)

10-32. Compute the prospective after-tax rate of return on the antimony mine of Example 9-5. Assume (1) investment is equivalent to the cost basis shown, (2) investment life equals the reciprocal of fraction removed this year, (3) an income tax rate of 50% applies, and (4) because financing is 100% equity and no depreciation charges are involved, before-tax cash flow = taxable income + depletion = gross income – deductible expenses.

10-33. a. Compute the prospective after-tax rate of return on Property A of Problem 9-13. Assume (1) investment is equivalent to the cost basis shown, (2) investment life equals the reciprocal of the fraction removed this year, (3) an income tax rate of 50% applies, and (4) because financing is 100% equity and no depreciation charges are involved, before-tax cash flow = taxable income + depletion = gross income – deductible expenses.
    b. Repeat Part a for Property B.
    c. Repeat Part a for Property C.
    d. Repeat Part a for Property D.

10-34. A mine investment of $100,000 is expected to produce before-tax cash flow of $20,212 for each of the next 25 years. Depletion allowances based on the appropriate percentage are expected to average $10,000 per year.
    a. Find the prospective before-tax rate of return.
    b. Find the prospective after-tax rate of return assuming (1) an income tax rate of 50% applies and (2) because financing is 100% equity and no depreciation charges are involved, before-tax cash flow = taxable income + depletion.

10-35. Green Corporation invested funds in Property B of Problem 9-13 and also invested $100,000 in each of three other wells which proved unproductive. The unproductive wells were expensed for income tax purposes so they represent an after-tax ($t = 50\%$) investment of $50,000 each. Continue with the four assumptions of Problem 10-33. Find the rate of return on the combined investment in the one producing and three nonproducing wells.

10-36. A prospector acquired a mine containing 500,000 tons of ore at a cost of $400,000. His costs, exclusive of depletion and income taxes, are $240,000 per year when ore is removed at the rate of 100,000 tons per year. The ore is subject to a percentage depletion allowance of 22% and sells for $5 per ton. Financing is 100% equity, and a combined effective state and federal income tax rate of 60% applies. Salvage is negligible. No depreciation charges are involved. Find the prospector's prospective after-tax rate of return.

I I

Analysis of Projects
Financed by Public Funds

Pertinence of the Study of Governmental Expenditures

OUR CONCERN for economy in governmental expenditures arises from our role as citizen and taxpayer, although our involvement may arise from our role as employee, consultant, or competitor of the governmental activity; the involvement may be one of design, construction, operation, or management. Government spending is shown in Table 11-1; the magnitude of spending is obviously worthy of our concern. As in preceding chapters our attention will be directed to that portion of spending which relates to capital expenditures.

Some Types of Governmental Expenditures

Capital expenditures by various divisions of government are applied to a vast array of purposes; some of these are listed below:

NATURAL RESOURCES
  Flood control
  Watershed and drainage

TABLE 11-1. Relationship of Government Purchases of Goods and Services, Gross Private Domestic Investment, and Gross National Product for Selected Years

Year	Gross National Product	Gross Private Domestic Investment	Governmental Purchases of Goods and Services	
			Federal	State and local
	billions of dollars			
1929	103.1	16.2	1.3	7.2
1934	65.1	3.3	3.0	6.8
1939	90.5	9.3	5.1	8.2
1944	210.1	7.1	89.0	7.5
1949	256.5	35.7	20.1	17.7
1954	364.8	51.7	47.4	27.4
1959	483.7	75.3	53.7	43.3
1964	632.4	94.0	65.2	63.5
1969	931.4	139.8	101.3	110.8

Source: *Economic Report of the President,* USGPO, 1971, p. 197.

Forest management
Irrigation and water supply
Sewage and garbage disposal
Pollution control (of air and water)
Navigation, rivers, harbors, and locks
Conservation of fish and wildlife
Research (as for desalinization)

ECONOMIC SERVICES
Communication—comsat
Emergency air landing fields
Power—Rural Electrification Association, Tennessee Valley Authority, and other
Postal
Transportation—transit systems
Highways, turnpikes, bridges, roads, streets
U.S. Mint
Research—Bureau of Standards, Bureau of Mines, Department of Agriculture, Copyright Office, Patent Office, weather research
Regulatory—Federal Communications Commission (FCC), Federal Power Commission (FPC), Interstate Commerce Commission (ICC), Securities and Exchange Commission (SEC), and others
Hospitals, veterans homes, rehabilitation centers
Housing loan programs

Subsidies
Social security and medicare
Printing office

PROTECTION
Armed forces
Crime control and Federal Bureau of Investigation
Police and fire
Traffic control
Judicial system
Air beacons, lighthouses, and Coast Guard
Research (as of interplanetary space)

CULTURAL DEVELOPMENT
Armed forces academies
Parks and camping grounds
Scenic and recreational areas
Historic and wildlife preserves
Education, elementary through university

Criteria for Acceptance or Rejection of Governmental Projects

As the long but only partial listing of uses of governmental funds is considered, many possible decision criteria applicable to analyses of expenditures for such purposes become apparent, for example:

Maximization of the ratio of annual benefits to annual cost of those benefits.
Minimization of present value sacrificed if the program were cut.[1]
Maximization of the excess of benefits over costs on an annual or present equivalent basis.
Minimization of combined annual cost to the public user and the agency supplier of the facility when benefits are fixed. (These last two criteria, appropriately defined, are identical.)

Ultimately involved in many decisions regarding governmental expenditures is the question of "general welfare" of the public. Divergent views of "general welfare" necessarily lead to disagreement as to the appropriateness of any given expenditure. Thus a project seemingly justifiable on an economic basis may

---

[1] For this and many additional possible criteria see R. N. McKean, *Efficiency in Government through Systems Analysis,* 2nd ed. (New York, Wiley 1964).

be unacceptable because we consider it to be in competition with investor-owned services. Conversely, such projects as a historic preserve, wildlife refuge, or recreation site may fail to meet economic criteria, but nonetheless may be accepted as being in the best interests of the public.

Overriding criteria may also be encountered; the timing of some expenditures may be more influenced by wars and recessions than by the merits of the individual projects.

Example 11-1.   **Tax Considerations (in Addition to Others) in Comparing the Costs of Governmental Projects**

Information: A municipality is considering the merits of a municipally owned versus an investor-owned electric utility service. A consultant was hired to estimate costs of the municipally owned facility; included in his report were the following estimates, all expressed as a cost per kilowatt-hour (kwh).

1. Cost per kwh for fuel, wages, and other cash costs   = $0.0100
2. Cost per kwh for depreciation of equipment   = 0.0018
3. Cost per kwh for 4% interest on the half financed by bonds   = 0.0008
4. Cost per kwh for 6% return on the half financed by appropriation   = 0.0012
5. Cost per kwh for property taxes   = 0.0014
6. Cost per kwh for state income taxes   = 0.0002
7. Cost per kwh for federal income taxes   = 0.0020

In his report the consultant states that items 4, 5, 6, and 7 have been included for possible comparative purposes and notes that the municipally owned utility would not be required to pay 4, a return on the taxpayer's investment; 5, property taxes; 6, state income taxes; or 7, federal income taxes.

Objective: Find the cost per kwh of the municipally owned facility for comparison with that of an investor-owned utility.

Analysis:
Fuel, wages, and other cash costs are incurred in the production of electric power and should be included.

Depreciation of equipment, interest on bonds, and *return on appropriation funds* or after-tax cash flow required should be included so that a proper provision for return on and repayment of investment is accomplished. Funds required for repayment of bonds should not be added; to do so would be double-counting since repayment of the entire investment is already provided for. The market interest rate on bonds should be adjusted upward to include flotation, administration, underwriting, legal, and other costs. We suggest that public funds taken from private citizens via taxes should be invested with an efficiency (rate of return) at least equal to that which the private citizen might have earned on those funds; for example, cash rather than credit purchases of furniture and automobiles (8–15%), or acceleration of payments on same or on house mortgages (6%), or in conservative investments (5–7%).

Property taxes. The selection of a city-owned versus investor-owned utility has no effect on the property tax revenue needs of the community; if the city-owned facility does not pay property taxes, the property tax (millage) rate will have to be somewhat higher for those who have taxable property. With municipal ownership the choice is apparently one of paying property taxes (for example, Detroit Street Railways which pays taxes to the city, county, and state) or equivalent (perhaps free electric power to the city for street lighting) or not doing so and thus increasing taxes for property owners, while providing relatively lower rates for utility customers and thus shifting some of the consumers' burden to the property owners.

State income taxes. As with property taxes the selection of a municipally owned versus an investor-owned utility has no effect on the income tax needs of the state; if the city-owned facility does not pay state income taxes, the difference will have to be made up somewhere in the revenue system of the state. With municipal ownership the choice is apparently one of paying state income taxes or not doing so and thus increasing the burden of the state's taxpayers, while providing relatively lower rates for utility customers and thus shifting some of the city consumer burden to the taxpayers of the state. The latter is certainly tempting from a local point of view; the political considerations are obvious. If we examine the choice from a broader point of view, say that of citizens of the state, the supposed savings on state income taxes appear to be "phantom" savings.

Federal income taxes. Again the question of municipally owned versus investor-owned has no bearing on revenue needs of the federal government. The real choice is one of paying federal income taxes or not doing so and thus increasing the burden to the other taxpayers of the federal government, while providing relatively lower rates for utility customers and thus shifting some of the city consumer burden to the taxpayer of the federal government. When a portion of the burden of 20,000 consumers can be spread over 100 million taxpayers who are remote, anonymous, unaware, and apparently 5000 times (= 100,000,000/ 20,000) more able to carry such a trivial burden, the appeal of such a shift must be both obvious and great. Yet if we adopt a less provincial view and consider the choice from the viewpoint of citizens of this country, the supposed savings on federal income taxes appear to be "phantom" savings.

The choice between a municipally owned and investor-owned facility *does not* affect the revenue needs of city, state, or federal governments; it *can* affect the allocation of that burden among the citizens of the political divisions considered.

Point of View

Consider the following points of view and the possible objectives of each and consider how these might lead to differing conclusions with regard to such economic decisions as are involved in Example 11-1.

1. A city council member.
2. A property owner in that city.
3. A taxpayer in that state.
4. A taxpayer in some other state.

It is difficult to say whether viewpoint is economic or political in nature; perhaps it has economic impact and is determined in a political manner (for example, voting). An agency specifying the viewpoint from which a study is to be made may be making value judgments not within its defined scope. The analyst who assumes or defines viewpoint may be expressing his personal views rather than producing an objective, unbiased study.

With these cautions in mind we suggest:

1. Where federal projects are involved, especially those with "national" costs and "local" benefits, it is essential that our point of view be national rather than local;[2] such a view admittedly may be both difficult and unpopular. It does not, however, follow that the pricing of services must be based on this total cost. Even if costs exceed revenues, the project may be in the public interest and the difference between costs and revenues may be an acceptable amount of federal aid. Only if we properly estimate and identify the costs can legislators or voters be in a position to appraise properly the alternative courses of action and to make the value judgment, Is this amount of federal aid acceptable?, required in the acceptance-rejection decision.

2. The analysis of municipal, county, or state projects that provide products or services elsewhere available from commercial (privately owned) facilities should permit cost comparisons from all the necessary viewpoints (perhaps city, county, state, and federal). While a national viewpoint is theoretically desirable, the specification of viewpoint is not the analyst's prerogative, and he therefore may not be in a position to provide an unqualified recommended course of action. In such instances we recommend that he separate and clearly label (as in Example 11-1) the controversial tax shifts and their estimated impact so that the voters or their representatives can appraise properly the alternative courses of action. Again, the pricing of services may be based on a different viewpoint than the acceptance-rejection decision. Analogous comments apply to county and state projects.

Little extension of these concepts is necessary for inclusion of studies involving cooperatively owned facilities or those owned or financed through the Rural Electrification Administration. This is not to plead for or against governmental or cooperative ownership, but to point out some very real pitfalls in the analysis of publicly owned facilities. The pitfalls must be avoided if we are to produce objective and unbiased analyses.

---

[2] That is, "federal income taxes foregone" should be included where the product or service might be obtained from commercial sources. The estimate of tax revenue foregone should be based on what the commercial source would have paid, not on what the government might have paid. The view is supported in Bureau of the Budget *Bulletin A-76*, dated March 3, 1966.

Project impact and an appropriate viewpoint of it can extend beyond national or world borders. There certainly are projects which are viewed best from an international viewpoint or even a universe viewpoint. We recommend that the viewpoint of the analyst tend toward the broad rather than provincial and that at a minimum the circumstances of "apparent savings" be clearly elicited.

Financing Considerations in Public Projects

Several distinguishing characteristics in the financing of public projects warrant our attention: The three main sources of funds are (1) taxes; (2) borrowing (via bonds, notes, bills, and so forth); and (3) internal funds as generated by the postal system, a city-owned water system, or a state-owned turnpike. Typical types of tax funds include sales taxes, gasoline taxes, income taxes, vehicle licensing fees, and property taxes.

Frequently there are limits on the maximum bonded indebtedness. Though it is periodically revised, the limit on the national debt may have some value as a deterrent to spending; in most cities, counties, and states the law specifies the maximum bonded indebtedness permitted. That limit is frequently based on the assessed value of property within the specific area. In many cases the sale of bonds must be approved in advance by two-thirds of the voters. Unlike industrial projects, many city, county, and state governmental projects are financed entirely by debt capital (bonds). Because interest payments on city, county, and state bonds are generally exempt from federal income tax, they are quite attractive to investors in high-income tax brackets. For this reason, tax-exempt securities generally sell at a lower interest cost (yield) than their industrial counterparts.

With few exceptions federal projects are financed from a common pool of tax funds and borrowed funds. Various forms of federal aid may be involved in the financing of a project. One form is direct cost sharing which includes the 100% payment of the financial needs of projects for such purposes as flood control, navigation, or historic facilities of national significance. Cost sharing also includes interstate highway projects where, in many cases, the federal government pays 90% of the costs. Circumstances in which 50% federal participation is available are not uncommon. The investment credit act described in Chapter 10 has analogous aspects; the federal government pays (via income tax forgiveness) a percentage of the cost of qualifying investments.

A second form of aid is the low-interest-rate loan. Certain irrigation projects, university buildings, foreign aid, and other projects may qualify for interest-free loans. Again, in an analogous sense, the industrial firm has been somewhat similarly aided when applying the five-year write-off of facilities or when applying liberalized depreciation. In either case the relative effect is that of an interest-free loan (via deferral of income tax payment requirements). Other loans at rates below that representing the cost of money to the federal government can be noted: the Rural Electrification Administration; the Tennessee Valley Authority; and many state colleges, universities, and their students are among the recipients of such loans.

A third form of aid is the indirect grant or subsidy used to encourage development of airmail, maintenance of modern merchant ships, price stability, and the like.

Still other forms of aid may be involved in our study, as for government-insured loans or government-financed projects aimed at stabilizing a depressed economy.

Highway Projects As an Illustration of Benefit-Cost Concepts

To produce user benefits, a project must result in user costs that are less than project users *would be willing to pay.* To be economically acceptable, a project must produce user benefits which meet or exceed the cost of providing those benefits. Figure 11-1 illustrates this criterion for evaluating public projects. To illustrate the application of these concepts, we next consider the benefits which might be generated by various types of highway improvement projects.

Possible benefits of highway projects include cost savings from (1) reduced travel distance, as with shorter routes; (2) reduced travel time, as with shorter routes, higher speeds, or elimination of stops; (3) reduced vehicle wear and tear, as with improved road surfaces, reduced grade, or reduced curvature; (4) reduced congestion, as with additional lanes; (5) reduced accident costs, as with improved guard rails, restricted access, or elimination of intersections.[3] The magnitude of possible savings

---

[3] For specific data on road-user costs see *Road-User Benefit Analyses for Highway Improvements,* American Association of State Highway Officials, Washington, D.C., 1960.

Line-Haul Trucking Costs, Highway Research Board Bulletin 301, Washington, D.C., 1960.

If each sum below is measured on the same basis, say present value, then the fundamental equation for evaluating a public project proposal is:

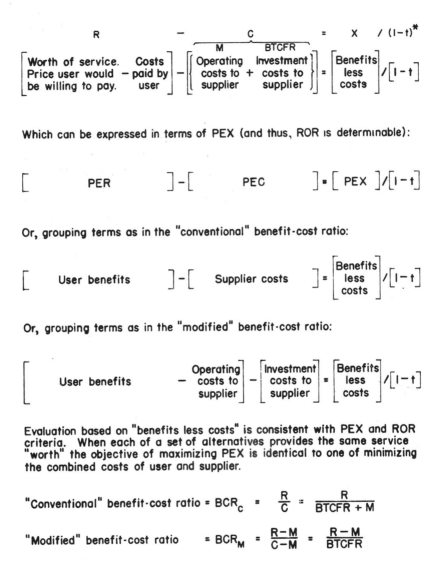

Which can be expressed in terms of PEX (and thus, ROR is determinable):

$$\left[\quad PER \quad\right] - \left[\quad PEC \quad\right] = \left[\ PEX\ \right]\bigl/\left[1-t\right]$$

Or, grouping terms as in the "conventional" benefit-cost ratio:

$$\left[\quad \text{User benefits} \quad\right] - \left[\ \text{Supplier costs}\ \right] = \begin{bmatrix}\text{Benefits}\\\text{less}\\\text{costs}\end{bmatrix}\bigl/\left[1-t\right]$$

Or, grouping terms as in the "modified" benefit-cost ratio:

$$\begin{bmatrix}\\\text{User benefits}\\\\\end{bmatrix} - \begin{bmatrix}\text{Operating}\\\text{costs to}\\\text{supplier}\end{bmatrix} - \begin{bmatrix}\text{Investment}\\\text{costs to}\\\text{supplier}\end{bmatrix} = \begin{bmatrix}\text{Benefits}\\\text{less}\\\text{costs}\end{bmatrix}\bigl/\left[1-t\right]$$

Evaluation based on "benefits less costs" is consistent with PEX and ROR criteria. When each of a set of alternatives provides the same service "worth" the objective of maximizing PEX is identical to one of minimizing the combined costs of user and supplier.

"Conventional" benefit-cost ratio = BCR_C = $\dfrac{R}{C}$ = $\dfrac{R}{BTCFR + M}$

"Modified" benefit-cost ratio   = BCR_M = $\dfrac{R-M}{C-M}$ = $\dfrac{R-M}{BTCFR}$

* In many public projects the appropriate tax rate t is zero.

Fig. 11-1. Relating benefits and costs to *BCR*, *PEX*, and *ROR* criteria.

TABLE 11-2. Proposed Highway Improvements and Possible Road-User Benefits

Proposal	Possible Benefits				
	Time	Dis-tance	Wear and tear	Reduced con-gestion	Im-proved safety
Eliminate stop with an overpass	+		+		+
Restricted rather than un-restricted access		−			+
Provision of a shorter route	+	+			
Improve surface	+		+		
Improve alignment	+	+	+		
Improve congested condi-tion	+		+	+	+
"Flatten" grades—improve sight distance			+		+
Replace intersection with cloverleaf	+	−			+

+ = positive benefit
− = negative benefit

of the five benefit categories tends to be in the preceding order. The listing in Table 11-2 suggests some of the possible benefits connected with various proposals. Note too, that the road user can experience negative as well as positive benefits.

The department of government which provides a highway facility must consider not only costs of the right-of-way, construction, and subsequent maintenance and resurfacing but also the less obvious costs of survey, design, and inspection of construction. In general, the basis of the property, B, for any type of governmental project should include:

Amounts paid directly to the supplier or contractor.
Transportation charges.
Bid evaluation expense.
Costs of contract negotiating, awarding, and managing.
Costs of materials furnished by the division of government.
Costs due to incentive or premium provisions in the contract.
Costs necessary for preparing and maintaining an existing unit for standby service.
Interest lost during construction (interest on the payments made prior to use of the facility).

Wage and salary costs, usually a part of the annual cost estimate, should include prorated fringe benefit costs for retirement, insurance, health and medical plans, vacation, travel expense, leave, termination, gift, separation, or other such allowances.

Other costs to be recognized are the cost of risk of damage or loss (as with automobiles where an additional $0.01 per mile per vehicle cost for risk or for accidents may be appropriate). Allocated indirect costs of supervision, personnel, accounting, and other applicable services should be included where appropriate.

Example 11-2.   Comparing Costs and Benefits of Proposed Highway Alternatives

Information: In considering alternative routes to connect City A and City B we have narrowed the possibilities to two, Route S and Route L, for which the following data has been estimated:

Route	Length in Miles	Cost of Right-of-Way	Cost of Bridges Culverts, Drainage, and Grading	Cost of Paving	Annual Cost of Maintenance
S	10	$200,000	$1,000,000	$ 800,000	$30,000
L	12	250,000	500,000	1,000,000	20,000

	Life	Salvage
Right-of-way	infinite	same as first cost
Bridges, culverts, drainage, and grading	40 years	negligible
Paving	20 years	negligible

MARR = 6%. The cost estimates above include an allowance for the survey, inspection, design, and related highway department costs based on average annual costs per dollar of construction work. Estimates pertaining to road-user costs include:

Distance costs = $0.06 per passenger car mile and $0.20 per truck mile. (Includes incremental costs but not cost of license, depreciation, or driver wages. The estimate is dependent upon average speeds and highway characteristics such as grade.)

Time costs = $0.10 per commercial vehicle mile and $0.00 per noncommercial vehicle mile, based on average hourly wage, including fringe benefits, and dependent upon the average speed of vehicles.

Wear and tear, congestion, and safety benefits are felt to slightly favor Route S but are neglected here.

Traffic is estimated as follows:

Noncommercial passenger cars = 400,000 vehicles/yr
Commercial passenger cars   = 100,000 vehicles/yr
Commercial trucks   = 50,000 vehicles/yr

Objective: Compare Route L and Route S. State which is the better of the two alternatives.

Analysis:

Annual Highway Department Costs		Factor	Route S	Route L
Right-of-way	$\$\ 200{,}000(a/p)_\infty^{6\%}$	0.0600	$\$\ 12{,}000$	
	$250{,}000(a/p)_\infty^{6\%}$	0.0600		$\$\ 15{,}000$
Bridges, culverts,	$1{,}000{,}000(a/p)_{40}^{6\%}$	0.06646	66,460	
drainage, grading	$500{,}000(a/p)_{40}^{6\%}$	0.06646		33,230
Paving	$800{,}000(a/p)_{20}^{6\%}$	0.08718	69,744	
	$1{,}000{,}000(a/p)_{20}^{6\%}$	0.08718		87,180
Maintenance			30,000	20,000
Annual equivalent highway department costs =			$\$178{,}204$	$\$155{,}410$
Annual equivalent highway department cost difference for S - L =				$\$22{,}794$

Annual Road-User Costs	Route S	Route L
Distance costs = \$0.06 (500,000) (length in miles)	$\$300{,}000$	$\$360{,}000$
Distance costs = \$0.20 (50,000) (length in miles)	100,000	120,000
Time costs = \$0.10 (150,000) (length in miles)	150,000	180,000
Annual equivalent road-user costs =	$\$550{,}000$	$\$660{,}000$
Annual equivalent road-user cost difference (benefit or savings) for L − S =		$\$110{,}000$

Thus, additional expenditures by the highway department, equivalent to $22,794 per year, could save road users an estimated $110,000 per year. This increment analysis shows Route S is clearly superior to Route L. The combined costs to highway department and road user are $728,204 for Route S, and $815,410 for Route L. AEX_{S-L} = $87,206. If the "do-nothing" alternative has been ruled out, or if its combined cost is greater than that of Route S, we should choose S.

Example 11-3.   **Comparing the Conventional Benefit-Cost Ratio Criterion with Others**

Information: Four alternative proposals for a governmental facility are being considered. Each requires an investment of $100,000 and each has a life of 20 years. A minimum attractive rate of return of 5% is used to compute equivalence.

Alternative	User Benefits (decreased user costs)	Increased Governmental Operating Costs	User Benefits less Governmental Operating Costs
A	$23,856	$7880	$15,976
B	15,904	− 72*	15,976
C	11,928	−4048*	15,976
D	8,836	−7140*	15,976

*The negative sign indicates that these costs decrease as a result of the project.

Objective: For each alternative find (1) the conventional benefit-cost ratio (2) the modified benefit-cost ratio, and (3) the rate of return on the investment.

Analysis: Refer to Figure 11-1, then note that

$$BTCFR = \text{investment costs to supplier}$$
$$= (B)(a/p)_{20}^{5\%}$$
$$= (\$100,000)(0.08024)$$
$$= \$8024$$

And for alternative A the conventional benefit-cost ratio

$$= \$23,856/(\$8024 + \$7880) = 1.50$$

and similarly for alternatives B, C, and D.

Modified benefit-cost ratio

$$= (\$23,856 - \$7880)/\$8024 = 1.99$$

The amount by which annual benefits exceed annual costs

$$= \$23,856 - \$7880 - \$8024 = \$7952$$

Rate of return can be computed by equating

$$\$23,856 - \$7880 - \$100,000 \, (a/p)^i_{20} = 0$$

and substituting as appropriate for alternatives A, B, C, and D. Results of the calculations are summarized below:

Alternative	Conventional Benefit-Cost Ratio	Modified Benefit-Cost Ratio	Annual Benefit less Annual Cost	Rate of Return
				percent
A	1.50	1.99	$7,952	15
B	2.00	1.99	7,952	15
C	3.00	1.99	7,952	15
D	10.00	1.99	7,952	15

In Example 11-3 no incremental analysis is necessary since each alternative involves the same investment over the same period of time and the cash flow is identical (the distribution between the facility user and the facility provider does vary, however). There is no need, therefore, to restrict the illustration to either mutually exclusive or nonmutually exclusive alternatives. Neither restraint would diminish the bias introduced by the conventional benefit-cost ratio. Grouping annual benefits to the *users* of the public facility in the numerator and net annual cost effects to *operators* of the public facility in the denominator of the benefit-cost ratio seems to be a logical procedure advocated by highly competent sources (including AASHO)[4] and used quite widely in practice.

It has been shown that the conventionally computed ratio can introduce a bias into the comparison of alternatives. If we agree that public funds should be allocated to their various purposes to maximize the long-run gains (such as reduced cost to users of the facility and decreased maintenance costs to operators of the facility) of such investments, it follows that a criterion which satisfactorily measures the desirability of alternatives is mandatory. The danger involved in continued use of the conventional benefit-cost ratio method is the possibility that some high-yield projects will be delayed or denied because funds have been exhausted in some low-yield projects. The

---

[4] See *Road-User Benefit Analyses for Highway Improvements*, American Association of State Highway Officials, Washington, D.C., 1960. In particular note pages 27–28.

conventional benefit-cost ratio is not recommended for the comparison of either mutually exclusive or nonmutually exclusive alternatives.

Large-Scale Long-Range Governmental Project

The economic analysis of a federal project proposal is frequently printed as a U.S. "House Document." An overview of one of these is offered with the purpose of acquainting the reader with some of the characteristics of federal projects. Distinguishing characteristics include the tendency to be long term, large scale, complex, multipurpose, multiagency involved, and diversely viewed by a public having dissimilar views. In this latter respect the government agency has an involvement quite different from that of the nongovernmental investor.

Example 11-4.   **Provision of an Integral Third Power Plant at Grand Coulee Dam**

Information: The proposed project involves construction of a third power plant at Grand Coulee Dam;[5] it includes a visitor center and would increase the present 2.0 million kilowatt (kw) capacity to 5.6 million kw through the installation of 12 turbine generator units of 300,000 kw each to be installed in pairs with the final pair in operation by 1983.

When completed, the facility would be larger than any single hydroelectric development in the world today. The international significance of the project is illustrated by the Columbia River Treaty between the United States and Canada which provides for cooperative development of the Columbia River. The broad impact of such projects is apparent in a listing of the governmental organizations which compiled data and information specifically for the report:

Bureau of Reclamation
Bonneville Power Administration
National Park Service
Fish and Wildlife Service (Department of Interior)
Corps of Engineers (Department of the Army)
Public Health Service (Department of Health, Education, and Welfare)

---

[5] This example is based on the project (located on the Columbia River in the central part of the State of Washington) as reported in 89th Congress, 1st session, House Document #142, Reference Number 19173.

and those whose data and information was applicable to portions of the report:

U.S. Geological Survey
U.S. Weather Bureau
Federal Power Commission

and other groups to whom the project has significance:

Water Resource Board
Soil Conservation Service
Watershed District

Objective: (As indicated by the analysis contained in the report.) Treat the proposed project as an incremental addition to the Columbia Basin Project and find the benefit to cost ratio.

Analysis: (As found in the report.)

Annual equivalent benefits:

Power	$ 43,838,000
Flood control	1,470,000
Recreation	389,000
Navigation	0
Fish and wildlife	0
Control of water temperature	0
	$ 45,697,000

Net federal investment:

Total project cost	$364,310,000
Less: Investigational costs	289,000
Plus: Interest during construction	36,585,000
	$400,606,000

Annual equivalent costs:

Net federal investment	$12,529,000
Equivalent annual operating costs	1,563,000
	$14,092,000

Conventional benefit to cost ratio $= \dfrac{\$45,697,000}{\$14,092,000} = 3.24$

Basis for Analysis and Comments: Power benefits (annual) were based on capacity benefits valued at $12 per kw and energy benefits valued at $0.0016 per kw; the basis was provided by the Federal Power Commission. In a 1954 study of this project, capacity benefits had been valued at $3.67 per kw. Elsewhere in the report (Page 47) the annual equivalent of revenues is shown as $30,740,000, considerably less than the estimated benefit. The benefit is roughly equivalent to $0.01 per kwh of average annual energy production. The estimates are based on the assumption that successive stages of increased capacity will be utilized fully; the existing intertie with the Pacific Southwest area is cited as a means of marketing surplus power. Flood control benefits (annual) realized via the provision of Canadian storage and the third power plant are based on controlling peak discharges of up to 800,000 cubic feet per second (cfps). Since 1858 there had been 13 years in which peak discharge had exceeded this figure, the maximum being 1,240,000 cfps in 1894.

Recreation benefits (annual) are based on estimates of annual recreation visitor numbers of 450,000 growing to 600,000 per year by the year 2000. The National Park Service has estimated a benefit value of $0.65 per visitor day. Whether the cost of guides has been (1) excluded, (2) included via the use of a *net* benefit of $0.65, or (3) included via the annual operating costs is not clear. For the sake of comparison a benefit value of $5 per fisherman-day used on the Skagit River project is noted. Navigation benefits were estimated as negligible since adequate pondage for reregulation of releases from Grand Coulee was already available. Fish and wildlife benefits were estimated as negligible, although the project might provide some opportunity to control release of cold water from storage and hence achieve a small amount of downstream temperature control. Because storage behind Grand Coulee Dam is relatively small in comparison with flow through the reservoir, the degree of water temperature control is limited, and therefore the temperature control benefits were estimated as negligible.

Federal costs include interest during construction, a cost which is based on the concept that a considerable sum of money is required for payments in advance of the actual completion of facilities. In finding equivalent annual costs, an interest rate of $3\frac{1}{8}$ % and a project life of 100 years were used. Because of the stage of development (and payments), the conversion of present to annual equivalent costs involves data not given in the report and an exact reconciliation of the conversion is not offered. Income taxes foregone (those which would result under private ownership) were not computed.

The benefit-cost ratio computed in the conventional manner is 3.24, the modified benefit-cost ratio

$$= \frac{\$45,697,000 - \$1,563,000}{\$14,092,000 - \$1,563,000} = 3.52$$

In either case, the benefit-cost ratio seems to indicate that the project is very desirable and not very sensitive to minor variations in the assumptions or estimates. The vulnerability of the conclusions is apparent, however, as certain questions are considered. The questions are noted here only to help the reader recognize that certain issues of significant economic consequence must be answered before project acceptability can be judged.

Questions	Approximate Effect
1. Is it reasonable to use an interest rate of 6% instead of 3⅛% on the grounds that the taxpayer can apply his funds to accounts costing or investments earning such an interest rate or that because of a limited supply of funds the marginal investment opportunity of the government will be at the higher rate?	If yes, then $AEC =$ annual equivalent cost is increased by $11–$13 million.
2. Should an allowance for property taxes foregone by the local area (compared to the alternative of privately owned facilities) be included?	If yes, then AEC is increased by $8–$10 million (if property tax rate is 2.5%).
3. Should an allowance for state and federal income taxes foregone (compared to the alternative of privately owned facilities) be included?	If yes, then AEC is increased by $11–$24 million.

Another question which has a bearing on the final decision is, Will the proposed facility change the existing proportions of public and private power in the region served?

Cautions in the Estimation of Project Benefits

Project benefits are not always what they seem. To illustrate this, consider an irrigation project which might upgrade the crop potential of a semiarid area. The project would seem to offer increased income potential for the landowners and farmers and increased market value for the land involved. In fact, however, the latter is simply the result of the former; to count both is double-counting. If the upgraded crop output can cause a depressed market price for a farm commodity already in oversupply or an increased federal burden, as when the commodity price is already being supported in the marketplace, the remaining benefit is still overstated; localized benefits may be entirely offset by undesirable consequences at the national level. Obviously it is essential to consider the impact of a project upon the nation (point of view); it should also be apparent that such circumstances illustrate the possible conflict between area objectives and national objectives.

Price or cost do not necessarily measure *value* to society. For example, if we attach a value of zero to clean water, we are saying that it is free to use or abuse.

The *distribution* of project benefits and costs is important. A proposed city freeway project with a composite benefit-cost ratio of 2.0 may produce a benefit-cost ratio of 3.0 for suburbanites and -2.0 for the central city poor.

Multipurpose Projects

Many governmental projects involve a number of purposes and benefits such as:

Flood control	Water quality control	Irrigation
Hydroelectric power	Fish and wildlife	Navigation
Recreation (tourists)	Land stabilization	Drainage
Water supply		

Two important cautions follow. First, the incremental cost associated with any particular benefit should be separately treated whenever possible; that is, each of the various investment elements should prove itself economic; a good flood control project should not be used to "carry" a navigation project which is uneconomic. Second, we should note that these multipurposes are not entirely free from conflicting objectives; consider the various and conflicting desired water levels for each of the above purposes.

Example 11-5.   Applying Incremental Investment Concepts in the Interpretation of Benefit-Cost Ratios for Multialternative Projects

Information: The following benefits and costs have been estimated for a certain public project:

Alternative	Investment Required $= B$	Life, in Years $= n$	Annual User Benefits less Supplier Operating Costs	Annual Investment Costs $= B(a/p)_{20}^{8\%}$	Modified Benefit-Cost Ratio
E	$ 981,800	20	$400,000	$100,000	4.0
F	1,178,160	20	432,000	120,000	3.6
G	1,374,520	20	434,000	140,000	3.1
H	1,570,880	20	480,000	160,000	3.0

A minimum attractive rate of return of 8% was used in computing annual equivalent costs, $(a/p)_{20}^{8\%} = 0.10185$. Salvage of each alternative is negligible.

Objective: Determine which of the alternatives should be recommended.

Analysis: The annual equivalent excess of benefits minus costs:

$$AEX_E = \$300,000 \qquad AEX_G = \$294,000$$
$$AEX_F = \$312,000 \qquad AEX_H = \$320,000$$

This data is sufficient for us to conclude that H is the best of the alternatives. (H maximizes AEX and thus minimizes the combined annual cost to the public user and agency supplier of the facility.)

Alternative Analyses Using Modified Benefit-Cost Ratio: Compute the ratios for the various incremental investments and then reason as follows:

Compare E and O: 4.0 is an acceptable ratio, so E is better than O.
Compare F and E: 32,000/20,000 = 1.6 (increment ratio), so F is better than E
Compare G and F: 2,000/20,000 = 0.1 (increment ratio), so F is better than G
Compare H and F: 48,000/40,000 = 1.2 (increment ratio), so H is best of the set

Or use a network diagram similar to those of Chapter 7.

Start with the minimum investment, 0. Move to a greater investment as soon as a decreasing minimum acceptable benefit-cost ratio permits it (take the path of "first escape"). Proceed in this manner so long as (1) the ratio required $\geqslant$ 1 and (2) a still greater investment exists. The logic "path" thus followed is marked below; the final choice is Investment H.

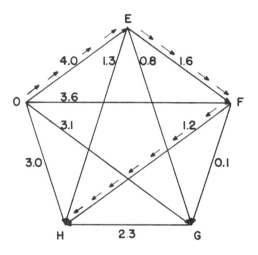

Defining the Threshold of Acceptability

When the cost of worthwhile public projects exceeds the funds available for financing, *capital rationing* causes us to postpone projects which are worthwhile but somehow less urgent than certain others. In response to capital rationing, some agencies define a minimum acceptable (cutoff) benefit-cost ratio, say 2.0, on the basis that there is sufficient capital to finance projects which surmount such a hurdle. There are at least two serious difficulties with this practice. First, the hurdle really presents dual criteria since some specified minimum attractive rate of return *and* minimum benefit-cost ratio must be met. Second, the hurdle is not a uniform one. In Example 11-3 the modified benefit-cost ratio of 1.99 corresponds to a rate of return of 15% while in Example 11-4 the modified benefit-cost ratio of 3.52 corresponds to a rate of return of about 11%. When funds available are not sufficient to finance the acceptable projects, the minimum attractive rate of return should be revised upward to reflect the *opportunity* cost of capital. This procedure is illustrated in Example 11-6.

Example 11-6.   Benefit-Cost Analyses Involving Capital Rationing

Information: Several *independent* projects are being considered. In each case estimated salvage is zero.

Project	Investment Required	Project Life, in Years	User Benefits less Supplier Operating Costs
1	$1,000,000	5	$300,000
2	1,0C0,000	10	$180,000
3	1,000,000	20	$110,000

Objective: Develop a conditional response to the question of which projects should be accepted under varying degrees of capital rationing.

Analysis: The modified benefit-cost ratio at selected interest rates:

Project	at 5%	at 10%	at 15%
1	1.30	1.14	1.01
2	1.39	1.11	0.90
3	1.37	0.94	0.69

Then, accepting all projects with a benefit-cost ratio $\geq$ 1.0,

Funds Available	Example i	Projects Accepted
$ 0	20%	0
1,000,000	15%	1
2,000,000	10%	1, 2
3,000,000	5%	1, 2, 3

When studying the projects at $i = 5\%$, we found that Project 2 had a higher benefit-cost ratio than did Project 1. But we must not use such data to say that Project 2 is more urgent than Project 1. All we can say is that all three projects are acceptable because the benefit-cost ratio exceeds one. If resources are inadequate to fund all three projects, the *opportunity* cost of capital is greater than 5% and the study rate should be revised.

Alternate Analysis: The "trial-and-error" approach above can be avoided by computing rate of return for each project, then displaying the conditional response. The rates of return are, respectively, 15.2%, 12.4%, and 9.0%.

"Study" Rate or Cutoff	Projects Accepted	Funds Required
15.2% or more	0	$ 0
12.4%–15.2%	1	1,000,000
9.0%–12.4%	1,2	2,000,000
9.0% or less	1,2,3	3,000,000

Modified Benefit-Cost Ratio As a Capital Expenditure Criterion

Examples 11-5 and 11-6 illustrate that the modified benefit-cost ratio criterion, properly applied, leads to the same accept-reject decisions and the same capital-rationed funding decisions as would be reached by use of *PEX* or *ROR* criterion. Unfortunately, the modified benefit-cost ratio criterion is sometimes inefficient and too often misapplied or misinterpreted. We therefore recommend that public project evaluation at least be verified by the *PEX* or the *ROR* criterion.

Problems

11-1. Compare and contrast the role of the industrial firm with that of a governmental unit with respect to:
    a. The tendency to defer capital expenditures for projects whose benefits or savings are rather distant in time.
    b. The "treasurer's problem" of obtaining large sums of money for financing of projects whose cost may be spread over several years; part of the problem is that the funds temporarily in excess are generally put into short-term securities earning a fairly low rate of interest. This in turn means the timing of the acquisition of funds is quite critical. Still other aspects of the "treasurer's problem" include appropriation procedures and philosophies.
    c. The problem of obtaining acceptance of the proposed capital expenditure.
11-2. One principle of a fair tax system is that taxes should fall upon taxpayers according to benefits received, yet there are some taxpayers who have never used such municipal facilities as the city library, schools, or airport. Are such persons being fairly taxed?

11-3. What are the monetary benefits of:
   a. A city park.
   b. A city library.
   c. A new fire station.
   d. A parking lot at a high school.
   e. Landscaping of public buildings or roads.
   f. A municipal band and bandshell.
   g. A city-owned retirement home.
   h. A municipal zoo.
   i. A municipal museum.

11-4. Assume the following estimates are appropriate to a a group of independent investment opportunities. Income taxes do not apply.

Alternative	Facility-User Savings/Yr	Maintenance and Operating Savings/Yr	Investment	Life, in Years
2J	$ 1,883	$ 7,000	$100,000	30
3K	6,608	4,000	100,000	30
7L	24,414	-12,000	100,000	30

   a. Compute the conventional benefit-cost ratio using a minimum attractive rate of return of 6%.
   b. Compare the three alternatives by the rate of return method.

11-5. Complete the statements which follow the data below (income taxes do not apply).

Alternative	Facility-User Savings/Yr	Maintenance and Operating Savings/Yr	Annualized Investment Cost
E	-$10,000	$40,000	$10,000
F	0	30,000	10,000
G	22,000	8,000	10,000
H	40,000	-10,000	10,000

   a. Use of the conventional benefit-cost ratio would introduce a bias favoring Alternative ____ .
   b. The conventional benefit-cost ratio for Alternative F is ____ .
   c. The conventional method of finding benefit-cost ratio would introduce a bias which acts most adversely toward Alternative ____ ; although Alternative(s) ____ are also unfavorably evaluated by the method.

d. If Alternative I were introduced and were to be evaluated identically by both conventional and modified benefit-cost ratios, the data would have to be:

Alternative	Facility-User Savings/Yr	Maintenance and Operating Savings/Yr	Annualized Investment Cost

11-6. It has been proposed that a certain gravel road be replaced by asphaltic pavement. Annual maintenance costs have averaged $1400 per mile and are expected to drop to $600 per mile for the proposed paving. Annual traffic is 100,000 vehicles.

Principal benefits of the new paving are twofold; the improved surface will result in *time* savings by virtue of the higher average speeds and reduced *wear and tear*. There may also be some *safety* benefits, but these have not been quantitatively estimated. Total road-user costs of $10,000 per mile per year will be reduced by $2000 if the asphaltic paving is used.

The new paving is expected to cost about $20,000 per mile and has an estimated life of 20 years. An interest rate of 5% is to be used; income taxes do not apply.

Evaluate the proposed project by finding:

a. Both types of benefit-cost ratio.

b. The amount by which annual benefits exceed annual costs.

c. Rate of return.

d. The amount by which present equivalent benefits exceed present equivalent costs, using $n = 20$.

e. Find the number of years the paving must last to pay for itself including the 5% return (break even).

11-7. After a detailed study of the wearing qualities of many different paints, four paints have been selected on the basis of durability. It is desired now to select one of these four paints as "best" for a certain highway application.

	A	B	C	D
Expected life, years	2	3	4	5
Cost per gallon	$6	$8	$10	$14

From further study you determine that a gallon of paint will cover approximately 0.40 miles (of 4-inch wide strip). You have estimated the per mile cost of painting as $15 per mile of *painted strip.* This includes *all* operation cost of the painting equipment including depreciation, plus the cost of the operators and of traveling from one painted strip to another.

Compare annual equivalent cost using an interest rate of 6%; income taxes do not apply.

11-8. The annual equivalent of Highway Department costs for a certain project will be $100,000 per year. These costs begin in the same year that the road is available to the road user. Prospective savings of the proposed facility have been estimated as $0.20 per vehicle per year. The present traffic load of 300,000 vehicles per year will grow by 50,000 each year for the next 20 years, then remain a constant 1,300,000 per year.

a. Annualize project benefits over the first 20 years, then compare benefits with costs to see whether the project is acceptable. *MARR* = 6%; income taxes do not apply.

b. Compare prospective savings (benefits) with costs on a year-by-year basis to find the optimal starting date of the project.

11-9. Two cities named Remote and Unreachable are located only one mile apart. Because a sheer cliff separates the two cities, the road traveler must actually drive 50 miles to get from one city to the other. The economic feasibility of a tunnel is explored; annual equivalent costs are found to be $10,000. Average road-user costs for time, distance, and so forth, are estimated as $0.20 per mile and the traffic without the tunnel is estimated as 1000 vehicles per year; the traffic is estimated as 100,000 vehicles per year if the tunnel is dug. On this basis road-user costs were computed as follows:

Annual Road-User Costs	Without Tunnel	With Tunnel
1000($0.20)(50)	$10,000	
100,000($0.20)(1)		$20,000

The tunnel is obviously an improvement, yet the increased traffic results in increased road-user costs. Suggest a better way to analyze road-user costs. Income taxes do not apply.

11-10. Assume the following estimates are appropriate to a group of mutually exclusive alternatives and that income taxes do not apply.

Alternative	Facility-User Savings/Yr	Maintenance and Operating Savings/Yr	Present Investment	Life, in years
M	$ 8,000	$ 6,059	$100,000	30
N	12,000	12,118	200,000	30
Q	20,000	14,177	300,000	30

Using a minimum attractive rate of return of 7%, compute for each alternative:

a. The conventional benefit-cost ratio.

b. The amount by which annual benefits exceed annual costs.

11-11. *A Tunnel under the English Channel.*[6] Nearly 100 years ago Napoleon III listened to a French mining engineer's proposal to dig a hole to England. Napoleon was interested, but when war broke out with England the scheme was dropped. A generation later Thome de Gamond, a Frenchman, proved the idea a practical one when his test borings indicated that a floor of chalk extended across the channel. Chalk is ideal for tunneling since it is soft, impermeable, and capable of withstanding the pressure of the sea above without the aid of compressed air. A successful start was made in 1880, but was stopped when Britain's military leaders became concerned over possible loss of her "perfect insularity."

According to engineers, there are two practical methods of constructing the tunnel; either will cost about half the $800 million cost of a bridge over the 23-mile distance.

One method is boring; modern equipment can bore a tunnel 22 feet in diameter at the rate of 50 feet per day. With equipment working from both ends of the tunnel, the boring could be completed in about four years.

A second method, called "build and sink," would involve the use of seagoing towers to lower big steel tubes into a trench on the ocean floor where they would be riveted together.

In either case, trains carrying passengers and their autos would provide the new means of transportation. Assume the following hypothetical estimates:

First cost = $400 million
Life of the tunnel = infinite (periodic major rebuilding

---

[6] Based on an article appearing in the *Des Moines Register.*

costs have been included in the annual equivalent operating and maintenance costs)

Annual equivalent operating and maintenance costs = $10 million

Minimum attractive rate of return = 6%

Estimated passenger revenue = 12 million passengers per year (excludes train-fare revenue which is applicable separately to those excluded costs) at $1.50 per person

Estimated vehicle revenue = 800,000 vehicles per year at $10 per vehicle.

    a. Analyze the economic practicality of the tunnel based on the hypothetical estimates.

    b. Comment on other factors which might influence your decision.

11-12. a. Under what conditions will the conventional and modified benefit-cost ratios be identical?

    b. Under what conditions will the conventional benefit-cost ratio be zero?

    c. Would it ever be possible to justify a project which increases facility-user costs?

11-13. Four independent highway projects are under consideration. They are identical in all respects (first cost, life, and so forth) except for the benefit differences noted below:

Road 1C saves 100 miles of driving for 1000 vehicles per year.

Road 2B saves 10 miles of driving for 10,000 vehicles per year.

Road 3G saves 1 mile of driving for 100,000 vehicles per year.

Road 4A saves 0.1 miles of driving for 1,000,000 vehicles per year.

Are these four road projects equally beneficial? Defend your response and comment upon financing of such projects.

11-14. Consider governmental projects and financing and state under what circumstances it is *not* appropriate to equate:

    a. Project life and life of bonds used in financing.

    b. Project risk and risk of bonds used in financing.

11-15. a. Estimate the annual tax revenues from all applicable sources per mile of highway using a traffic load of 1 million vehicles per year.

    b. Repeat Part a for a highway for which toll revenues are also appropriate to include.

c. Discuss the relationship of traffic load as an economic criterion in the design of a highway.

11-16. *Mt. Blanc Tunnel.* One of the most dramatic current examples of highway facility decisions is Europe's Mt. Blanc tunnel joining Chamonix, France, and Courmayeur, Italy. The tunnel is 7.2 miles long and cost $50 million.

The tunnel reduces the Paris to Rome mileage by 125 miles; Geneva and Turin, 197 miles apart in the summer and 491 in the winter when the snow causes a detour to the Riviera and around the Alps, will now be 168 miles apart all year long. The new tunnel can accommodate up to 500 vehicles per hour. Cost of the tunnel is to be partly recouped through a toll fee. Income taxes do not apply.

a. What is the maximum toll fee a person traveling from Paris to Rome apparently should be willing to pay if his time and distance total costs are valued at 10 cents per mile?

b. If a toll fee of $3 were charged, the cost per mile of tunnel traveled would be about 42 cents. For a car averaging 21.6 miles per gallon (as for many of the European cars), this is the equivalent of a gasoline tax (on the 7.2 miles) of $9 per gallon! Do you think the toll is unfair? Why?

c. Who would carry the burden of costs if no toll fee were charged?

d. Assume that 1 million vehicles per year are expected to use the tunnel, that the average saving per vehicle is $10, $i = 6\%$, operating and maintenance costs are $1 million per year, and the life of the tunnel is infinite. Find the annual benefits and costs.

e. How long would it take to recoup the $50-million investment if the average vehicle fee is $3?

11-17. In Example 11-2, Route S was indicated as the preferred alternative. The annual cost of Route S was $178,204, yet the prospective revenues from road-user gasoline taxes (both state and federal) seems to be much less:

$$\frac{500{,}000 \text{ cars}}{\text{year}} \left| \frac{10 \text{ miles}}{} \right| \frac{\text{gallon}}{10 \text{ miles}} \left| \frac{\$0.10 \text{ tax}}{\text{gallon}} \right. = \$50{,}000/\text{yr}$$

$$\frac{50{,}000 \text{ trucks}}{\text{year}} \left| \frac{10 \text{ miles}}{} \right| \frac{\text{gallon}}{5 \text{ miles}} \left| \frac{\$0.10 \text{ tax}}{\text{gallon}} \right. = \$10{,}000/\text{yr}$$

Assume that revenues from state sales taxes on vehicles and the licensing fees for trucks and cars together produce another $60,000 per year of revenue. It would seem that approval of projects which result in revenue less than the annual costs of the projects will accentuate the problems of capital rationing. Income taxes do not apply.

Suggest a means by which the financing problem might be somewhat alleviated.

11-18. *Ocean-Tide Power Plant.*[7] The dream of harnessing the power of ocean tides may become a reality in the near future. Below the Cherbourg Peninsula of France such a project is already nearing completion. It has been reported that the Soviet Union has plans for utilizing tides of the Arctic Sea to produce 5 million kilowatts of power. Perhaps one of the most inviting locations is the Bay of Fundy near the Nova Scotia coast where the rise and fall of tides is the greatest in the world, ranging from 40 to 50 feet in height. Passamaquoddy Bay, an arm of the Bay of Fundy, experiences an average rise and fall of tides of 18 feet; the topography of the area is well suited to such a project as indicated in the accompanying figures. This American project was first conceived in 1919 and has been strongly advocated in the ensuing years. In 1935 our nation launched initial development work of the project with an appropriation of $7 million. Further appropriations were blocked by

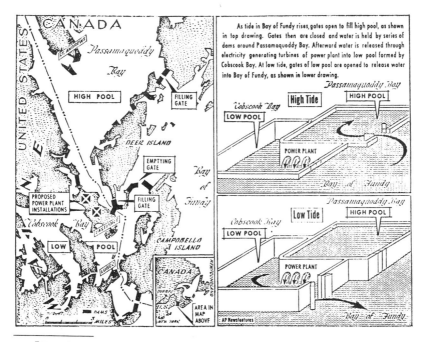

As tide in Bay of Fundy rises, gates open to fill high pool, as shown in top drawing. Gates then are closed and water is held by series of dams around Passamaquoddy Bay. Afterward water is released through electricity generating turbines of power plant into low pool formed by Cobscook Bay. At low tide, gates of low pool are opened to release water into Bay of Fundy, as shown in lower drawing.

---

[7] Based on an article appearing in the *Des Moines Register.*

those who, tongue in cheek, referred to the project as "moondoggling."

The project was enthusiastically supported by President Franklin D. Roosevelt whose summer home on Campobello Island permitted him first-hand observation of the surging tides. The island is in a key position for helping to form one of the huge reservoirs necessary for the hydroelectric development.

Current plans call for a total of 7½ miles of ocean dams constructed at various places to form two pools. Passamaquoddy Bay would trap the incoming tidewaters and provide a storage pool for the water to be dropped later through turbines into a low pool formed by adjacent Cobscook Bay. At low tides the water in Cobscook Bay would be released. The project would take an estimated 15 years for completion at which time power needs of northeastern United States and adjacent areas of Canada would be far greater than now. The facility could produce one million kw during a peak one-hour period daily and lesser amounts at other hours. Project cost has been estimated as $1 billion.

President John F. Kennedy told a group of New England congressmen: "Harnessing the energy of the tides is an exciting technological undertaking . . . . Each day over a million kilowatts of power surge in and out of the Passamaquoddy Bay. Man needs only to exercise his engineering ingenuity to convert the ocean's surge into a national asset."

a. Assume the following hypothetical data, then examine the economic aspects of the project.

Cost of all dams, generating facilities, and so forth = $1 billion

Life of project = 60 years

Estimated salvage = negligible

Operating, maintenance, and other cash costs per year = $10 million

Average price charged per kwh = $0.02

Average kwh energy sales per year = 2 billion kwh

Minimum attractive rate of return = 6%

Income taxes = not applicable

Interest lost during construction: Base on uniform payments over 15-year period from first payment to initial operation of the facility.

b. If a project fails to "pay its way" must we reject it?

11-19. Consider the statements below, then discuss why you agree or disagree:

Highway safety is an extremely important factor in highway planning. Because it is difficult to place a dollar value on safety, we unfortunately find ourselves in the position of justifying projects on the basis of reduced time or distance or similar factors which can be measured with greater objectivity. In general, we are pressed to justify projects on some basis other than safety.

11-20. The municipal water company has raised the question of how much money and effort should be spent in detecting and repairing leaks in the water system. As a preliminary to a detailed study the following estimates have been made:

Leakage Rate in Cubic Feet per Minute	Estimated Number of Leaks That Size	Annualized Cost of Locating and Repairing All Leaks That Size
0.1	1000	$250,000
1.0	100	50,000
10.	10	10,000
100.	1	2,000

Water is sold for $5 per 1000 cubic feet; note however, that revenue will not be enhanced by the reduction of leaks. The incremental cost of water (amount paid for chemicals and pumping) amounts to about $1 per 1000 cubic feet. If a significant reduction in leakage can be achieved, the city may be able to delay the expensive drilling of additional well capacity.
   a. Make an economic analysis to help determine an optimum level of effort in reducing system leakage.
   b. Justify the value of water you use in your analysis.

11-21. See the diagram and data on the next page.[8]   Assume the following additional data:

Project life = 15 years
i = 7%
Cost per fatal accident = $20,000
Cost per injury accident: Note that the number of persons injured *before* the change was 54 in 24 accidents, an average of 2.25; on an *after* basis the persons injured were 8 in 8 accidents. Because of this difference, assume a cost per injury accident of $4000 on the before basis and $2000 on the after basis.

---

[8]These data are from the U.S. Bureau of Public Roads and were published in the *Engineer*, Engineers Joint Council (Summer 1966), p. 4.

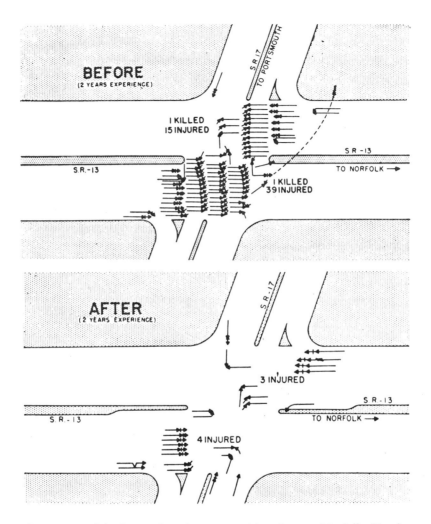

Correction of high accident location achieved near Norfolk, Va. by installation of left turn lanes and new traffic signal system, including left-turn arrow. Project cost $16,700, according to U.S. Bureau of Public Roads. After two years experience before and after, the result:

BEFORE		AFTER
159.4	Accident Rate	26.2
80	Total Accidents	24
2	Fatal Accidents	0
24	Injury Accidents	8
54	Property Damage Accidents	16
2	Persons Killed	0
54	Persons Injured	8
$33,500	Amount of Property Damage	$8,600

Maintenance and operating costs are unaffected by the change. Find the benefit-cost ratio.

11-22. The City of Greensville is considering the acquisition of a water-softening facility for their municipally owned water system. It is felt that the softened water will eliminate the need for "home" softening units and will reduce the amount of soaps and detergents required. This saving has been estimated as $10 per person per year. The average consumption of water per person in Greensville is a steady 50,000 gallons per person per year, and the population seems to be stabilized at 10,000. Other benefits from the water-softening such as reduced fuel costs for water heating, reduced maintenance costs on plumbing, and reduced wear and tear on fabrics and on cooking utensils have an estimated value of $0.05 per 1,000 gallons. Other estimates:

Cost for chemicals, labor, maintenance, repair, and power for the softening plant = $0.15/1000 gallons

First cost of water softening facility = $50,000 per 100 million gallons per year of capacity.

Salvage of facility = negligible at the end of its 30-year life

Minimum attractive rate of return = 8%

Income taxes do not apply

Find the (modified) benefit-cost ratio.

11-23. *Delaware Bay Superport.* The Maritime Commission has asked you to investigate the economic feasibility of constructing a huge floating platform off Delaware Bay. Such a "Superport" would be capable of serving "super-tanker" ships that cannot dock at conventional ports. Supertankers require water depths of about 100 feet, twice that available in most East Coast and Gulf Coast harbors. Current studies indicate that (1) the cost of dredging conventional harbors to the required depth is prohibitive and (2) the docking of supertankers in conventional harbors would present grave ecological dangers.

Type of Ship and Deadweight	Capacity in Barrels	Transportation Cost for Crude Oil Delivered from Persian Gulf to East Coast
Conventional 50,000 ton	375,000	$2.00 per barrel
Supertanker 400,000 ton	3,000,000	$1.40 per barrel

The proposed superport cost for the floating platform, shore storage tanks, connecting pipelines, and related facilities is $400,000,000. Estimated cash operating costs consist of (1) a fixed component of $4,568,000 per year and (2) a variable component of $0.07 per barrel processed. Assume U.S. imports of crude oil amount to 2 billion barrels per year; it is conservatively estimated that 5% of that amount would be shipped via supertankers to the Delaware Bay superport. Those 100,000,000 barrels per year would require about 33 supertanker deliveries per year. Each supertanker round trip takes about two months, so six supertankers might be in constant round-trip use at the superport.

Ecological hazards associated with the handling of crude oil by conventional ships in conventional ports represents a subtle transportation cost "extra" estimated to be $0.10 per barrel over and above the direct costs of transportation. For supertanker and superport, reduced dangers are recognized via the estimated transportation cost "extra" of $0.05 per barrel.

Estimated net salvage of the superport is zero at the end of its 30-year life. The minimum attractive rate of return is 10%. Transportation benefits will partly accrue to the public via relatively lower prices paid for gasoline than if deliveries were limited to conventional ships. It is also felt that public interests are served via greater ecological protection. Some benefits also accrue to U.S. oil refiners, so the federal government is proposing a sharing of construction costs with the refiner group.

a. Find the amount by which annual benefits exceed annual costs.

b. Find the modified benefit-cost ratio

11-24. Find the optimal date for construction of a public project facility to which the following estimates apply:

First cost = $10,000   Project life is infinite
Cash operating costs = 0 $MARR = 50\%$
Project benefits = $0 at the end of this year; $2000 at the end of next year; and $2000 $(z - 1)$ at the end of year z. Thus, if the facility is constructed three years from now, its first year of use (year four) will produce end-of-year benefits of $6000; the next year of use will produce end-of-year benefits of $8000; and so forth.

V

Multioutcome Considerations: Risk and Uncertainty

ONLY ONE thing is certain—that is,
nothing is certain. If this
statement is true, it is also false.

ANCIENT PARADOX

1 2

Multioutcome Estimates and the Cost of Risk

Probability

PROBABILITY is the long-run relative frequency with which an event occurs.[1] It is the ratio of the number of occurrences to the number of opportunities for the occurrence. A probability distribution is a list or graphic portrayal of the relative frequency with which each possible outcome occurs. The probability that a flipped coin will come up "heads" is 0.5, while the probability that a card drawn from a deck of cards will be a heart is 0.25; and there is a 0.1 probability that the last digit of your telephone number is 7. But the probability that your last name begins with S is not 1 in 26—why? Probability is a very useful concept in treating capital expenditures which necessarily deal with an uncertain future having nonunique possible outcomes.

Frequency Distribution

Suppose that we wish to estimate the life of a certain item for which we have some historical experience. That experience

---

[1] This "classical" definition is not without its objectors.

can be arrayed in a tabular form called a frequency distribution. For convenience in recording, presentation, and analysis, data can be grouped into classes of service of ½ to 1½ months, and so forth; such grouping means our reported data is actually a *grouped frequency distribution* as shown in Table 12-1.

TABLE 12-1. Frequency Distribution of Service-Life Data

	Observed Life	Number of Units	Unit-Months of Service Rendered
	months		
	0–½	0	0
	½–1½	1	1
	1½–2½	2	4
	2½–3½	4	12
Observed	3½–4½	8	32
data	4½–5½	11	55
	5½–6½	14	84
	6½–7½	8	56
	7½–8½	2	16
		Total 50	260

A total of 260 months of service was rendered by the 50 units, thus a *mean* of 5.2 months of service was rendered. The most popular of the lives observed was 6 months and is called the *mode*. The middle value, called the *median*, is 5 months; 25 units had equal or longer lives, and 25 units had equal or shorter lives. If we estimate item life with a single estimate, we prefer use of the mean (= average = expected value) of 5.2 months (because in some cases, such as normally distributed data, it will be the least variable linear unbiased estimate and because its use will sometimes lead to simpler mathematics). If our estimate is to be multivalued, then we should attempt to include some measure of the variation of actual lives from this mean; several possibilities can be pursued: (1) Show the actual data as in the table; (2) approximate the actual data, for example, by a theoretical probability distribution such as the normal, exponential, uniform, Poisson, binomial, or other (see Figure 12-1); or (3) describe variability.

Our original data can be portrayed graphically as a *frequency histogram* (see Figure 12-2). After plotting the points, we can treat the data as *continuous* and connect the points by the line shown or treat it as *discrete* by representing each tabled life by a vertical bar, thus producing the stair-step representation also shown. We might also note that the curve is not *symmetric*

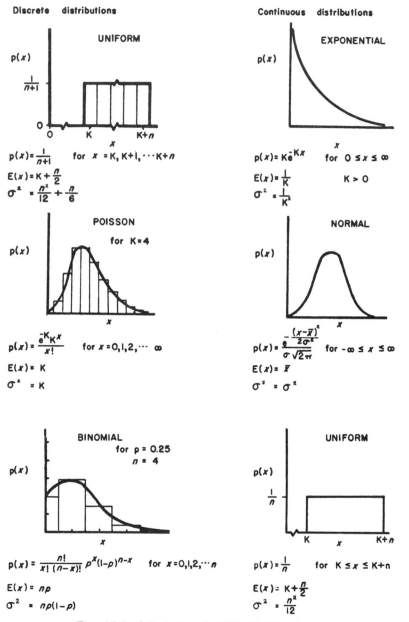

Discrete distributions

UNIFORM

$p(x)$

$\frac{1}{n+1}$

0

O K K+n

x

$p(x) = \frac{1}{n+1}$ for x = K, K+1, $\cdots$ K+n

$E(x) = K + \frac{n}{2}$

$\sigma^2 = \frac{n^2}{12} + \frac{n}{6}$

POISSON

for K = 4

$p(x)$

x

$p(x) = \frac{e^{-K}K^x}{x!}$ for x = 0,1,2,$\cdots$ ∞

$E(x) = K$

$\sigma^2 = K$

BINOMIAL

for p = 0.25

n = 4

$p(x)$

x

$p(x) = \frac{n!}{x!(n-x)!} p^x(1-p)^{n-x}$ for x = 0,1,2,$\cdots$ n

$E(x) = np$

$\sigma^2 = np(1-p)$

Continuous distributions

EXPONENTIAL

$p(x)$

x

$p(x) = Ke^{-Kx}$ for $0 \leq x \leq \infty$

$E(x) = \frac{1}{K}$ K > 0

$\sigma^2 = \frac{1}{K^2}$

NORMAL

$p(x)$

x

$p(x) = \frac{e^{-\frac{(x-\bar{x})^2}{2\sigma^2}}}{\sigma\sqrt{2\pi}}$ for $-\infty \leq x \leq \infty$

$E(x) = \bar{x}$

$\sigma^2 = \sigma^2$

UNIFORM

$p(x)$

$\frac{1}{n}$

K x K+n

$p(x) = \frac{1}{n}$ for $K \leq x \leq K+n$

$E(x) = K + \frac{n}{2}$

$\sigma^2 = \frac{n^2}{12}$

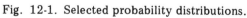

Fig. 12-1. Selected probability distributions.

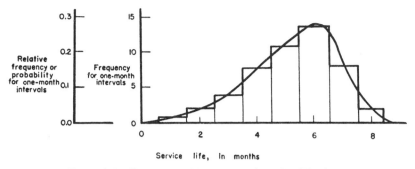

Fig. 12-2. Frequency histogram of service-life data.

about a vertical axis; this is a clue that mean and median values
are probably dissimilar.

Risk

Because the future outcome of a present course of action is
uncertain, we say the course of action involves some *risk,* the
possibility that one of the less desirable outcomes will be realized.   Less desirable outcomes include (1) damage from such
acts of nature as floods, hurricanes, windstorms, and lightning;
(2) losses from such acts of man as highway accidents, fires, and
faulty design; and (3) an undesirable variation of actual result
from predicted result.   In this sense, all risk is the result of our
inability to control the future.   Although the possible variation
of the actual course of events from the predicted course of
events is a type of risk; *variability in itself does not necessarily
cost anything* (as when many similar independent investments
will be subject to similar variability and as a group the "average"
result is realized even though individual investments varied from
the average).

Uncertainty

Capital expenditure decisions deal with fairly well-defined
present cash outlays compared with less definite prospective
receipts of cash.   Because such decisions are based on a predicted but unknown future course of events, they are decisions
made under conditions of *uncertainty.*   Because the decisions involve various possible outcomes, some less desirable than others,
they are decisions made under conditions of *risk.*   All capital expenditure decisions are subject to the conditions of risk and un-

certainty. The treatment of information and estimates of varying reliability can be emphasized by classifying decisions as:

1. Decisions based on single-valued estimates of future outcomes.
2. Decisions based on multivalued estimates of future outcomes coupled with estimates of the probability of those outcomes.
3. Decisions based on multivalued estimates of future outcomes under circumstances where estimates of the *probability* of those outcomes cannot be made with even remote confidence.

Progression through the sequence above suggests a reduction in available information.[2]

Risk-reducing Capital Expenditures

To a limited extent man can control the *frequency* and/or *severity* of possible outcomes. Expenditures which tend to reduce frequency and/or severity of an undesirable outcome are called risk reducing. Examples include dams to control floods; controlled-access roads for highway safety; monitoring for control of water pollution; windbreaks, lightning rods, and construction methods as protection from storm; warning devices, sprinkler systems, fire doors, and the like, for protection from fire; operator safety via mechanical guards, and safety consciousness programs; and in many instances insurance is a risk-reducing device.

In each instance above the pertinent question is how much risk reduction to purchase. There is a trade-off between expenditure and risk reduction which in its optimal state produces minimum total cost (or maximum profit if revenues or benefits are affected). This point is illustrated in Example 12-1.

---

[2] And further suggests the classification of decisions as those under certainty, risk, and uncertainty. We prefer not to use these classifications, however, because: (1) All capital expenditure decisions are subject to conditions of risk and uncertainty. (2) While it is true that a decision under assumed certainty requires only a single-valued estimate of the future, the application of a single-valued estimate does *not* require the assumption of certainty (for example, when expected value is used). (3) Separability of risk and uncertainty is questionable, especially when the separation is based on ability or inability to estimate the probability of an outcome when it is assumed that the outcome itself is estimable.

Example 12-1.   Optimizing the Level of Risk Reduction

Information: Floods can be costly from the standpoint of their danger both to life and to property. Absolute protection from flood may, however, be uneconomic. Given the task of finding the optimum level of protection from flood, we have produced the following list of alternatives and estimates:

Design	Will protect against floods with flow rates less than ___ cfps	And the probability that a still greater flow will occur during the year is ___ , p	And the average cost of that still greater flow is ___ , X	And the cost of that design is ___ , Y
0	1,000	0.100	$350,000	0
A	2,000	0.050	400,000	$100,000
B	3,000	0.010	550,000	200,000
C	4,000	0.005	700,000	300,000
D	5,000	0.001	1,000,000	400,000

Floods occur no more than once per year. Each of the proposed structures has an estimated life of 40 years. This is a governmental project; assume a minimum attractive rate of return of 8% and neglect income taxes. Salvage is estimated to be negligible for all designs.

Objective: Determine the optimum level of flood protection.

Analysis:

		Annual Cost	
Design	Flood risk $= (p)X$	Capital recovery $= Y(a/p)_{40}^{8\%}$ $= 0.08386Y$	Total
0	0.100($350,000)	+ $0	= $35,000
A	0.050(400,000)	+ 0.08386(100,000)=	28,386
B	0.010(550,000)	+ 0.08386(200,000)=	22,272
C	0.005(700,000)	+ 0.08386(300,000)=	28,658
D	0.001(1,000,000)	+ 0.08386(400,000)=	34,544

The analysis signals Alternative B as the preferred choice.

Expected Value

Expected value of an outcome, $E(x)$, is the average of possible outcomes weighted according to the frequency with which they occur. Consider an investment opportunity which has the probabilistic outcomes shown:

Outcome, x	Probability of Occurrence, p
-$ 80,000	0.75
320,000	0.25

$$E(x) = -\$80{,}000(0.75) + \$320{,}000(0.25) = +\$20{,}000$$

When the outcome function is discrete,

$$E(x) = \sum_{j=1}^{k} x_j\, p_j$$

And when the outcome function is continuous,

$$E(x) = \int x\, p(x)dx$$

For two probabilistic investments considered jointly,

$$E(x_1 + x_2) = E(x_1) + E(x_2)$$

For a probabilistic investment selected K times,

$$E(Kx) = K\, E(x)$$

Maximizing expected value is often an appropriate investment decision criterion. It is most appropriate when a large number of similar sized investments with independent outcomes will be undertaken.

Variability

Variability is the tendency toward scattering or dispersion of data; for example, outdoor temperatures have more variability than indoor temperatures.

One measure of variability is *variance*, σ^2. When the outcome function is discrete,

$$\sigma^2 = \sum_{j=1}^{k} [x_j - E(x)]^2 \, p(x_j)$$

And when the outcome function is continuous,

$$\sigma^2 = \int [x - E(x)]^2 \, p(x)dx$$

Thus for the probabilistic investment opportunity just considered,

$$= (-80,000 - 20,000)^2 \, (0.75) + (320,000 - 20,000)^2 \, (0.25)$$
$$= 30,000,000,000 \text{ or } 3.0 \times 10^{10}$$

Another measure, the *standard deviation*, is the square root of the variance so

$$\sigma = 173,200 \text{ or } 1.732 \times 10^5$$

Range is the algebraic difference between extreme values of the outcomes, so

$$= \$320,000 - (-\$80,000) = \$400,000$$

Cost of Variability and/or Possible Loss

Under some circumstances the cost penalty of a parameter being x units less than its expected value is different from the cost penalty caused by it being x units greater than expected value. Where such circumstances apply, we can refer to variability as leading to *unequal cost penalties*. Examples include:

1. The manufacturing process where the cost of having removed too little material is less than that for having removed too much material.
2. The product order situation where the cost of running out of stock (lost profits, a shutdown on the production line, and so forth) may be different from the cost of carrying too much stock (obsolescence, spoilage, interest on inventory, storage costs, and so forth).
3. Equipment life where the lost receipts from equipment with

a life x years less than average are not exactly offset by the extra receipts from an equipment with a life x years more than average.

The variability and/or loss resulting from the multivalued future of a parameter of an alternative may have a cost. In general, variability and/or possible loss is treated by use of expected value; in our earlier illustration the expected value of the two outcomes was

$$-\$80,000(0.75) + \$320,000(0.25) = +\$20,000$$

When expected value is the sole criterion, the decision maker is saying that he does not object to the variability and/or possible loss the opportunity involves. Such a view is quite appropriate if he expects to accept many similar opportunities whose outcomes are independent of each other and if the variability involves equal cost penalties.

On the other hand, sole use of expected value is inadvisable (1) when unequal cost penalties of variability are involved or (2) when marginal utility of money is not uniform and the number of similar opportunities with independent outcomes is small or when the number of similar opportunities with dependent outcomes is large. Discussion of (2) is deferred to Chapter 13.

Single-valued versus Multivalued Estimates

Virtually all the examples and problems treated in previous chapters have involved only single-valued estimates of economy study parameters such as first cost, economic life, salvage, rate of return, operating costs, future income tax rates, and the like. Single-valued estimates are not necessarily evidence of either naiveté or the assumption of certainty regarding the future for at least four reasons:[3]

1. Such treatment is a part of a very logical procession in the building of a mathematical model. Starting with very simple models of repayment, we proceed to relax restrictions by considering properties involving such complications as unequal operating costs, unequal lives, unequal capacity, un-

---

[3] Helpful comments on the task of estimating and the effect of errors are provided by G. A. Taylor, *Managerial and Engineering Economy* (Princeton, N.J., Van Nostrand, 1964), pp. 333–45.

known or nonarbitrary rate of return specification, or explicit treatment of income taxes. Our model still falls short of reality by not yet explicitly treating inflation, dispersion of lives, risk, uncertainty, obsolescence, and an inexhaustible supply of still other considerations. Because the time and money which can be economically spent in the search for and analysis of alternatives has a finite limit, we adopt models which are inexact replicas of reality. The penalties of an over-zealous search or model may be as great as those of inadequate effort.

2. Although a study which omits consideration of variability and/or possible loss actually estimates it to be zero, it is quite possible that after considering the variety of outcomes and their impact, our estimate of the net dollar effect of that variety is zero.

3. Variability and/or possible loss may have been treated implicitly (a) if the minimum attractive rate of return employed exceeds that ordinarily required with the differential representing full compensation for variability and/or loss (or risk or uncertainty) or (b) if the range of outcome is neglected because it is small or its effect is trivial (for example, salvage under most circumstances).

4. Variability and/or possible loss may have been treated explicitly (a) if expected values were employed as the single-valued estimates of parameters with equal cost penalties or (b) if an unlikely, but possible, serious outcome (for example, fire) has been insured against and its cost (insurance premium) is explicitly treated.

Single-valued estimates ought to be used most of the time in most studies, especially in those studies where (1) the likelihood of an outcome other than the single-valued estimate appears to be negligible; or (2) the application of multivalued estimates is not practical, as when the decision involves a small nonrepetitive dollar value; or (3) the single-valued estimate is expected value of the multivalued function, excepting for circumstances when uniform weighting of outcomes is not appropriate (for example, dispersion of lives, nonuniform marginal utility of money, or annually fluctuating operating costs with a trend).

Single-valued Estimates of All but One Parameter

By varying just one parameter at a time and observing the effect on acceptability as indicated by such measures as rate of return or before-tax cash flow requirement of a project, we can provide some helpful indication of the responsiveness of our computed results to the parameter being varied. This procedure is generally referred to as a sensitivity analysis. Sensitivity analysis can help to answer such questions as:

1. What accuracy is required in estimating Parameter X?
2. For what range of Parameter X does the study decision remain unaltered?
3. Are the penalties of underestimation equal to the penalties of overestimation?

Example 12–2.   Sensitivity Analysis

Information: These estimates apply to a proposed project.

i_a = 10%   Depreciation is by straight-line method
                              First cost = $100,000
                              Salvage = $40,000 for all values of n
t = 50%   Property life = 15 years
                              Operating costs = $8000 per year

Objectives: Using the data given,
a. Compute annual equivalent costs (AEC).
b. Then recompute AEC for various salvage values.
c. Recompute AEC for various operating costs.
d. Recompute AEC for various property lives.

Analysis:

a. $ATCFR = \$100,000(a/p)_{15}^{10\%} - \$40,000(a/f)_{15}^{10\%} = \$11,888$

Income tax = 1.00 ($11,888 - $4000)	= 7,888
Operating costs	= 8,000
AEC	= $27,776

By referring to Table 10-3, we can note the $BTCF$ requirement, then add operating costs of $8000 per year to produce the following tabulations:

b.
Salvage	$0	$10,000	$20,000	$40,000	$100,000
AEC	$27,630	$27,660	$27,700	$27,780	$ 28,000

c.
Operating costs	$0	$ 5,000	$10,000	$15,000	$ 20,000
AEC	$19,780	$24,780	$29,780	$34,780	$ 39,780

d. Life   5   10   15   20   30
  AEC $35,660 $29,530 $27,780 $27,100 $ 26,730

  Life   40   60   ∞
  AEC $26,770 $27,040 $28,000

These data are the basis of Figure 12-3. It is apparent that:

1. Errors in the estimation of salvage value are of trivial consequence in their impact on revenue requirements, thus the requirements are very insensitive to errors in estimating salvage. Despite uncertainty, multivalued estimates of salvage are not warranted.
2. Errors in the estimation of operating cost are of considerable consequence in their impact on revenue requirements, thus the requirements are quite sensitive to errors in estimating operating cost. Where considerable un-

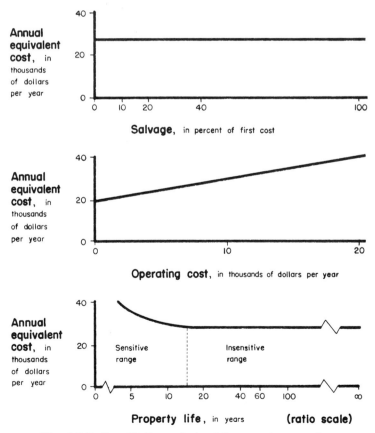

Fig. 12-3. Sensitivity analysis of three cost parameters.

certainty exists, multivalued estimates of operating cost are appropriate.

3. Revenue requirements are sensitive to errors in estimation of property life for $n < 15$ but insensitive for $n > 15$.

Multivalued Estimates of Several Parameters

Simultaneous variation of more than one parameter sometimes offers several advantages:[4]

1. It is more realistic than "one-at-a-time" variation, especially in its output of a multivalued set of rate of return possibilities or a multivalued set of revenue requirement possibilities.
2. It may discriminate better and signal the catastrophic result which could occur when multiplicative rather than compensating variations are involved.

The disadvantage of this approach is primarily its unwieldiness. If two different parameters are trivalued, there are nine sets of probabilities and outcomes to be computed. For still larger numbers of parameters and parameter values, the computational effort grows rapidly. A partial answer to the difficulty is afforded by the computer. By employing simulation techniques such as Monte Carlo, the computer can be used to generate combinations of input parameters and sum the various outcomes.

Example 12-3.   Expressing Outcome As a Multivalued Function

Information: These estimates apply to a proposed project.

Company minimum attractive rate of return = 12%
Duration of project = 15 years
Estimated salvage = \$0
$ATCF$ = after-tax cash flow
      = operating revenues - operating costs - income taxes

---

[4] See D. B. Hertz, "Risk Analysis in Capital Investment," *Harvard Business Review* (Jan.–Feb. 1964); J. F. Magee, "Decision Trees for Decision Making," *Harvard Business Review* (July–Aug. 1964); J. F. Magee, "How to Use Decision Trees in Capital Investment," *Harvard Business Review* (Sept.–Oct. 1964); F. S. Hillier and D. V. Heebink, "Evaluating Risky Capital Investments," *California Management Review* (Winter 1965).

Assume that parameter values given below are independent of each other (for reasons of simplicity, not realism).

Probability	ATCF	Probability	First Cost
	dollars/yr		
0.5	$10,000	0.3	$136,220
0.5	20,000	0.7	68,110

Objective: Using the data given, compute the multivalued outcome and consider the question of acceptability.

Analysis: Alternative outcomes from a decision on the proposed project are illustrated in the *decision tree* below, and in the table that follows.

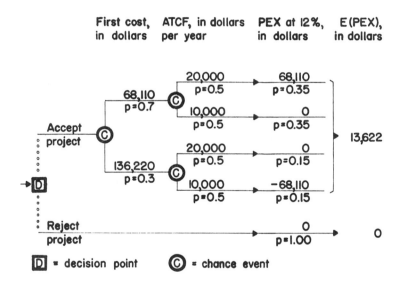

First cost, in dollars	ATCF, in dollars per year	PEX at 12%, in dollars	E(PEX), in dollars

D = decision point   C = chance event

| | | Probability | Computed |
ATCF	First Cost	of Same	*PEX i = 12%
dollars/yr			
$10,000	$136,220	(0.5)(0.3) = 0.15	− $68,110
10,000	68,110	(0.5)(0.7) = 0.35	0
20,000	136,220	(0.5)(0.3) = 0.15	0
20,000	68,110	(0.5)(0.7) = 0.35	+ 68,110
15,000	88,543	expected values	+$13,622

*PEX = present equivalent receipts less present equivalent costs
     = $10,000 $(p/a)_{15}^{12\%}$ − $136,220 = −$68,110

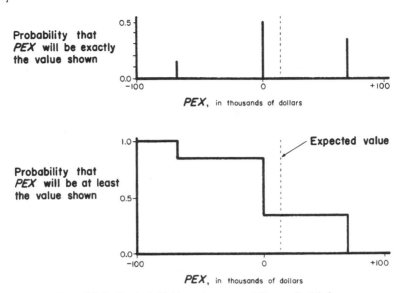

Fig. 12-4. Probabilistic outcomes of Example 12-3.

Because the expected value of *PEX* = +$13,622, we might conclude that the investment opportunity is a satisfactory one. Unlike the problems considered earlier, however, there is an obvious element of risk; that risk is portrayed in Figure 12-4.

The question of acceptability might now be restated as: Should the firm accept the risk of a *PEX* deficit of $68,110 with a probability of 0.15 when there is a probability of 0.35 that a *PEX* surplus of $68,110 will be realized? The answer depends upon the relative weighting placed upon the outcomes. If a uniform weighting is applied, the decision is one based on expected value, which in this example is positive and therefore acceptable. If weighting of outcomes is not uniform, we might choose to reject the opportunity. Such a weighting scheme is illustrated below and in Figure 12-5 and will be considered further in Chapter 13.

Outcome, *PEX*	Probability	Weighting	Weighted Outcome
- $68,110	0.15	3	- 30,650
0	0.50	2	0
+ 68,110	0.35	1	23,839
	weighted value		- 6,811

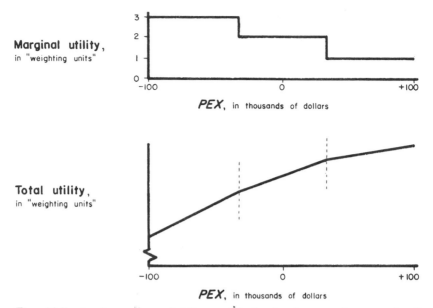

Marginal utility, in "weighting units"

PEX, in thousands of dollars

Total utility, in "weighting units"

PEX, in thousands of dollars

Fig. 12-5. A scheme for weighting multivalued outcomes: An example of declining utility for money based on Example 12-3.

Monte Carlo Simulation

The analysis of Example 12-3 is too unwieldy for treatment of problems where many parameters have multivalued estimates. The obstacle of unwieldiness has been alleviated by the digital computer which can serve to (1) generate simulated experience, (2) compute rate of return or a similar measure of acceptability, and (3) repeat the preceding through many cycles as illustrated in Figure 12-6.

$E(PEX)$ As a Capital Expenditure Criterion

$$E(PEX) = \sum_{z=0}^{n} (1 + i)^{-z} E(x_z) = \sum_{j=1}^{k} \sum_{z-0}^{n} (1 + i)^{-z} x_{jz}\, p_j$$

where x_{jz} = a set of probabilistic cash flows over each of the next n years. $j = 1, 2, \ldots k$; z = year 0, year 1, ... year n

p_j = the probability that set x_{jz} occurs

If all cash flows occur at the end of year zero, as in a bet

1. Make multivalued estimates of input parameters for the proposed facility.

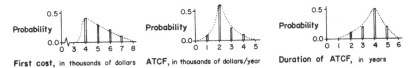

First cost, in thousands of dollars   ATCF, in thousands of dollars/year   Duration of ATCF, in years

Discrete distributions are illustrated; for simulation of a continuous probability distribution, the dials below can be calibrated in thousandths.

2. Calibrate each dial according to the independent input parameter probabilities.

First cost, in thousands of dollars   ATCF, in thousands of dollars/year   Duration of ATCF, in years

3. "Draw" one estimate for each input parameter by spinning the dial (or by generating a uniformly distributed random digit from a table or computer).

4. Find the outcome — PEX or rate of return — for that one set of input estimates. Repeat steps 3 and 4 — perhaps as many as 1000 times.   MARR = 10%.

5. List resulting outcomes and count frequency of occurrence.   Display results.

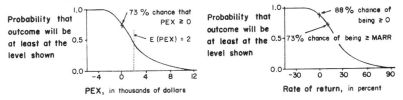

6. Repeat steps 1 through 5 for each alternative, then compare results.

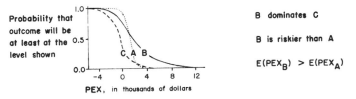

Fig. 12-6. Monte Carlo simulation of multivalued inputs.

$(z = n = 0)$, the equation above simplifies to

$$E(PEX) = \sum_{j=1}^{k} x_j p_j$$

The procedure for application is the same as for the *PEX* criterion.

$E(i)$ As a Capital Expenditure Criterion

The expected rate of return is i such that $E(PEX) = 0$. Be careful to note that $E(i) \neq (p_1 i_1 + p_2 i_2 \cdots + p_j i_j)/j$. The procedure for application is the same as for the rate of return criterion.

Problems

12-1. Repair costs have been recorded as follows:

$ 75	$125	$225	$400
100	75	100	50
25	275	50	25
1750	25	75	150
150	100	50	175

On the basis of the data above:
a. Find the expected value of the cost to repair.
b. Find the probability that a given repair will cost less than $200.
c. Show the data in graph form with "Cost to Repair" as the horizontal axis, and "Probability That Repair Cost Will Not Exceed $_" as the vertical axis.

12-2. Give some specific examples where variation from expected value leads to unequal cost penalties.

12-3. The Blue Corporation has just purchased a new dupulator machine and is now questioning whether to retire the old dupulator or keep it for standby service. Total annual costs to retain it are an estimated $1200 per year; this provides for the cost of space, property taxes, investment costs of return, repayment, income taxes, and so forth. The primary advantage of retaining it is that production can be continued (no workers idled) when the new dupulator is "down" for repairs; this advantage is valued at $30 per hour of use. Repairs average about eight hours. There are about 2000 working hours per year. The estimated repair frequency is:

Number of Times "Down" for Repair per Year	Probability
0–2	0.10
3–5	0.20
6–8	0.40
9–11	0.20
12–14	0.10

Should the old dupulator be retained if repairs follow the grouped frequency distribution given?

12-4. You are at a fair gambling wheel which has the numbers 1 through 100 on it; each number has an equal chance of being the result. You plan to make just one bet and have decided to choose between:

Bet A. Bet that the outcome will be an even number; if it is, you win $10; if not, you lose $10.

Bet B. Bet that the outcome will be the number 13; if it is, you win $990; if not, you lose $10.

Bet C. Bet that the outcome will not be the number 17; if it is not, you win $10; if it is, you lose $990.

a. Find the expected value of each of the bets.
b. Which bet gives you the highest probability of winning?
c. Does Bet B involve more risk than Bet A? Why?
d. Does Bet C involve more risk than Bet B? Why?
e. Find the variance of each of the bets.

12-5. Blue Corporation is involved in a patent suit which is thought to have three possible outcomes:

A. If the case is settled out of court now, Blue will receive $80,000 (after taxes).

B. If the case is continued and won, the present after-tax value to Blue is estimated as $500,000. The probability of winning is p.

C. If the case is continued but lost, the present after-tax cost to Blue is estimated as $200,000. The probability of losing is $(1 - p)$.

Find the minimum probability of winning necessary if Blue is to pass up the settlement.

12-6. Studies of the relationship of rainfall and yield from several different types of hybrid corn have been studied with the following hypothetical results:

Inches of Rainfall for Year	Yield (in net after-tax dollars per acre for seed type)			Rainfall Probability for Area XY
	A	B	C	
0–15	$25	$15	$10	0.1
15–25	45	40	36	0.2
25–35	55	62	60	0.4
over 35	60	68	75	0.3

The rainfall probabilities for Area XY have also been noted above. Which seed do you recommend for Area XY?

12-7. A utility company provides service to a remote area. Windstorms, snow, and lightning frequently lead to the need for repairs. Because the area is remote, repairs are rather costly, averaging $1000 each. The probability of damage during a years time is:

Number of Times Repair Is Required	Probability of Exactly That Number of Repairs
0	0.4
1	0.3
2	0.2
3	0.1

It is estimated that relocating and reworking of the line could reduce the probabilities to:

Number of Times Repair Is Required	Probability of Exactly That Number of Repairs
0	0.9
1	0.1

Assume a life of 20 years, zero salvage value, $MARR = 20\%$, $t = 0$, and that operating costs other than those for repairs are unaffected by the proposed change.

How much could the company afford to pay for the relocated and reworked line?

12-8. Industry and government standards require that a certain material sold as having a 10 mil thickness must have at least a thickness of 10.0 mils (no maximum is specified). Certain equipment is available which will monitor and control production so that the present variability in thickness of the product is decreased. The present equipment controls thickness to ± 0.3 mils, while the new equipment controls thickness to ± 0.1 mils (these are maximum deviations, not standard deviations). Assume that the equipment is set to such a thickness that essentially no material fails to meet the required 10.0-mil specification and that observed thickness of the material tends to follow the normal distribution.

   a. What potential material savings does the new equipment offer? Show as a percentage.

   b. The problem called for the assumption of the normal distribution. Could this assumption be relaxed without disturbing your response to Part a?

12-9. The following hypothetical projection of liability settlements of policyholders of an automobile insurance company is based on actual experience of recent years coupled with a projection of accident rate and settlement sum trends.

The probability that a "statistic" policyholder will, during a one-year interval, incur a liability which will be settled for a sum in the range specified	Range	Average sum for a liability settlement in the range specified
0.960,00	0	0
0.040,00	$ 1- $1,000	$ 200
0.003,00	1,000- 10,000	4,000
0.000,10	10,000- 100,000	50,000
0.000,01	100,000-1,000,000	300,000

1.003,11 ◄— Probability adds to more than 1.000,000 because some of the "statistic" policyholders are involved in more than one liability settlement in a given year.

Find the annual premium to be charged a policyholder if one-third of premium costs are needed to cover administrative, selling, income tax, and return requirements and if the liability limit selected by the policyholder is:

    a. $10,000.
    b. $100,000.
    c. $1,000,000.

12-10. Mr. Green and his family travel from New York City to Los Angeles on a four-lane highway. As they drive along, the family counts the number of Volkswagen cars going east and compares this with the number going west; all along their route the ratio of eastbound to westbound Volkswagens seems to be 8 to 1. Why?

12-11. Young Ben Blue lives in Manhattan near a subway express station. He has two girl friends, Betty in Brooklyn and Brooke in the Bronx. To visit Betty in Brooklyn he takes a train on the downtown side of the platform; to visit Brooke in the Bronx, he boards on the uptown side of the same platform. Since he likes both girls equally well, he simply takes the first train that comes along. In this way he lets chance determine whether he rides to the Bronx or Brooklyn. Ben reaches the platform at a random moment each Saturday afternoon. Brooklyn and Bronx trains arrive at the station equally often, every ten minutes. Yet for some obscure reason he finds himself spending most of his time with Betty in Brooklyn; in fact, on the average he goes there 9 times out of 10. Can you decide why the odds so heavily favor Brooklyn?

12-12. Compute variability of *PEX* in Example 12-3.

12-13. Under what circumstances would you expect an economy study decision to be very sensitive to the minimum attractive rate of return?

12-14. A parameter follows the frequency distribution:

Value	$60,000	$80,000	$100,000	$120,000
Probability	0.1	0.3	0.5	0.1

    a. Find the expected value by an analytic approach.
    b. Find the expected value by use of the random number table for 100 cycles.

12-15. Bruce Blue is considering a risky opportunity which requires an investment of $50,000; the two possible outcomes are (1) a uniform cash flow over each of the next three years sufficient to produce repayment and a 50% rate of return or (2) nothing. The outcomes are equally probable. Note that income taxes tend to reduce the variability of cash flows.

    a. If Blue's resources are sufficient and if he makes many similar investments in this and succeeding years, what is his estimated overall rate of return?

    b. Repeat your analysis for cash flow spread over ten years instead of three.

    c. It sometimes has been advocated that investment risk should be recognized by using a higher rate of return requirement on high-risk projects than on low-risk projects. Does such a proposal need to be applied with caution or qualification? Explain.

12-16. The following probabilistic after-tax cash flows have been forecast for a proposed project:

End of Year z	$P_{1z} = 0.75$	$P_{2z} = 0.25$
0	- $100	- $100
1	0	0
2	0	+$900

    a. Find $E(i)$.

    b. Find $E(PEX)$, assuming $MARR = 25\%$.

12-17. Refer to Figure 12-6. Using $MARR = 10\%$, find PEX and the probability of same for:

    a. The most unfavorable set of input parameters.

    b. The most favorable set of input parameters.

12-18. Refer to Figure 12-6. Simulate one estimate for each input parameter by use of a three-digit number randomly selected from Appendix D. Find PEX, using $MARR = 10\%$. If each member of the class makes one such simulation, the collective results can be compiled and compared with the analytic probabilities for PEX.

12-19. Figure 12-6 shows there is an 88% probability that the rate of return $\geqslant 0\%$. List the combinations of inputs which would produce a rate of return $< 0\%$. Show the probabilities for each combination and sum these as you attempt to verify the 88% figure.

12-20. Refer to Figure 12-6. Assume the three inputs of Step 2 are correlated over their pessimistic-optimistic range in such a way that they can be simulated by mechanically interlocking the three spinners so they will read identically—000, 111, or 222, and so forth. Without the aid of a computer, find the ten possible PEX outcomes and their associated probabilities. Then sketch these results into the graph of Step 5 of Figure 12-6. Notice that correlation > 0 increases the probability of achieving outcomes at either the pessimistic or optimistic extreme.

12-21. Alternatives A and B have identical lives and first costs; each has zero salvage. Several sets of outcome possibilities exist for the cash flow patterns of Alternatives A and B. You have analyzed each and produced the following comparison:

Outcome	PEX_A	p_A	PEX_B	p_B
Pessimistic	-$200,000	0.2	-$400,000	0.3
Most likely	200,000	0.7	300,000	0.4
Optimistic	600,000	0.1	700,000	0.3

   a. Find expected value of *PEX* for each alternative.

   b. State the circumstances under which you might choose Alternative A.

   c. Sketch a decision tree for this choice; show $E(PEX)$ for each alternative.

12-22. In 1730 Daniel Bernoulli presented an example which came to be known as the *St. Petersburg Paradox*. Suppose a fair coin (that is, the probability of heads = the probability of tails = 0.5) is tossed until a head appears. The gambler receives 2^n dollars, where n = number of the toss on which the first head appears ($2 if it first appears on the first toss, $4 if it first appears on the second toss, and so forth). What price should a gambler be willing to pay to play this game?

   a. Find expected value of the outcome of the game to determine the price the gambler should be willing to pay (if "house" wealth is unlimited).

   b. Find expected value of the outcome if the maximum possible reward (due to limited wealth of the "house") is $1024 = 2^{10}.

   c. Find expected value of the outcome if the maximum possible reward (due to limited wealth of the "house") is $1,048,576 = 2^{20}

13

Utility of Money
and the Cost of Variability

Implications Regarding Variability and Loss

SUPPOSE that a decision maker were given the alternatives:

A. Place a bet with a probability of 0.75 of winning $120,000 and a probability of 0.25 of losing $200,000.
B. Reject Alternative A (not bet).

Suppose further that the decision maker chooses to reject Alternative A. Now, according to expected value of the outcome for each,

$$E(x_A) = \$40,000$$
$$E(x_B) = \$0$$

it seems the bet should have been accepted. Furthermore, the decision maker tells us he would accept A if all the dollar values were scaled down by a ratio of 1000 to 1. In explaining this, he says that he can "afford" a $200 loss but that a $200,000 loss is too severe for a person of his means. Pursuing the matter, we ask him to provide us with some additional clues to his intuitive decision making by noting the certain result at which his choice between new alternatives, C and D, would be a matter of indifference (no preference, a break-even situation, or a standoff).

Point Number	Alternative C, a Certain Result	Alternative D, a Probabilistic Result	
		Gain $700,000 with probability p	Lose $400,000 with probability $(1 - p)$
1		0.00	1.00
2		0.20	0.80
3		0.40	0.60
4		0.60	0.40
5		0.65	0.35
6		0.80	0.20
7		1.00	0.00

The decision maker then notes the certain result which would cause him to feel indifferent toward the choice of C or D. Another method of developing the data is to supply the decision maker with the certain results for Alternative C, then ask him to note the p, Alternative D, which would cause him to feel indifferent toward the choice of C or D.

Point Number	Alternative C	Alternative D, a Probabilistic Result	
		Gain $700,000 with probability p	Lose $400,000 with probability $(1 - p)$
1	- $400,000	0.00	1.00
2	- 300,000	0.20	0.80
3	- 200,000	0.40	0.60
4	- 100,000	0.60	0.40
5	0	0.65	0.35
6	+ 300,000	0.80	0.20
7	+ 700,000	1.00	0.00

Where uncertainty exists (points 2, 3, 4, 5, and 6), our decision maker somewhat "discounted" the expected value of Alternative D. He thus displays conservatism.

To convert this data into a utility scale, we can note that the utility of the certain outcome of zero (see point 5) is equal to the expected utility of Alternative D for the probabilities noted, that is,

$$U(0) = 0.65 \, U(\$700,000) + 0.35 \, U(-\$400,000) = E(U_{Ds})$$

Then if we arbitrarily define a certain outcome of zero as having a utility of zero,

$$0 = U(0) = 0.65 \, U(\$700,000) + 0.35 \, U(-\$400,000)$$
$$U(-\$400,000) = -(0.65/0.35) \, U(\$700,000)$$

And if we let utility be measured in utiles, and arbitrarily let $U(\$700,000) = 700$ utiles, then

$$U(-\$400,000) = -(0.65/0.35)(700 \text{ utiles}) = -1300 \text{ utiles}$$

Because the selection of a zero point (origin) and definition of a utile (slope) was arbitrary, our utility scale is only a *relative* indicator, much like the temperature scale of a thermometer. Just as 60 C is not twice as warm as 30 C, an alternative with a utility of 400 utiles is not twice as desirable as one with a utility of 200 utiles. By repeating our computation for the other points:

Certain Outcome	Utility
-$400,000	-1300 utiles
-$300,000	- 900 utiles
-$200,000	- 500 utiles
-$100,000	- 100 utiles
0	0 utiles
+$300,000	300 utiles
+$700,000	700 utiles

Next, we construct a graph of his indifferences as Figure 13-1a. By plotting the slope (or taking the first derivative) of the total utility curve, we can show the marginal utility as Figure 13-1b. Several observations are appropriate to Figure 13-1a and situations where diminishing marginal utility of money is appropriate:

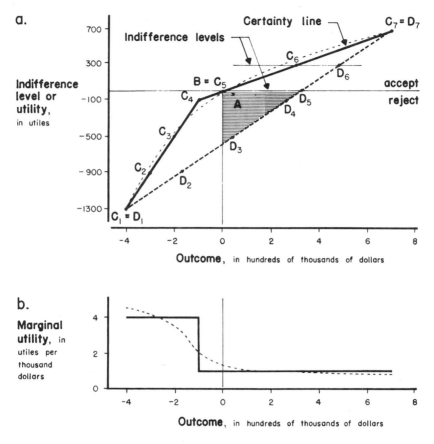

The continuous utility curves, shown as dotted lines, are probably a more realistic representation than the discrete utility curves, shown as solid lines.   For ease of explanation and computation we have chosen to use the discrete representation.

Fig. 13-1. A hypothetical "utility of money" function.

1. Indifference levels are horizontal lines and the preferred alternative is the one at the highest level. Preferences can be identified by computing the expected utility of each alternative, then choosing the alternative with the greatest expected utility.
2. If this decision maker limits himself to decisions with consequences within the range of the graph (maximum loss = $400,000 and maximum gain = $700,000), all the prob-

abilistic results must represent a weighing of results on the certainty line and hence no decision would exist outside the area bounded by the certainty line and the D line.

3. Alternative A (which our decision maker rejected) can be located by connecting the two certainty line outcomes, −$200,000 and +$120,000, and measuring the line thus constructed, placing A 0.75 of the length of the line from the −$200,000 point. A check or alternative solution is provided when a vertical line is drawn at the expected value of $40,000; it intersects the line connecting the −$200,000 and +$120,000 outcomes at A. Note that confirmation of the decision to reject is apparent in the figure.

4. The decision maker stated he would accept a scaled-down (by 1000 to 1) version of Alternative A. Inspection reveals that this very small dollar decision is much preferable to A even though it is simply a scaled-down version.

5. Alternatives whose probabilistic outcomes place them in the shaded area should be rejected even though their expected outcomes are positive. Expected outcome is not the sole criterion of economic acceptability.

6. In our example, Alternative B is identical to Alternative C_5. A horizontal line (indifference level) passed through this point provides us an accept-reject line (that is, accept those opportunities above the line, reject those below it).

7. The expected utility of Alternative A can be computed by noting that a certain loss of $200,000 has a utility of −500 utiles, and a certain gain of $120,000 has a utility of 120 utiles:

$$E(U_A) = -500(0.25) + 120(0.75) = -35 \text{ utiles}$$

Rejection of Alternative A is signaled.

8. Alternatives A_1, A_2, and A_3 have identical expected outcomes and identical expected utilities. They therefore share a common location on the utility map. They illustrate that, given declining utility for money and given alternatives with identical expected outcomes and different variances, we must *not* simply say choose the alternative with the smaller variance.[1] It *is* possible that an alternative with both smaller expected outcome and greater variance is preferable.

---

[1] See W. T. Morris, *The Analysis of Management Decisions*, rev. ed. (Homewood, Ill., Irwin, 1964), p. 275.

Alternative	Win	p	Lose	1 - p	Expected Outcome	Variance	Expected Utility
					thousands	*billions*	*utiles*
A_1	$120,000	0.75	$200,000	0.25	40	19	-35
A_2	230,000	0.50	150,000	0.50	40	36	-35
A_3	560,000	0.25	133,333	0.75	40	90	-35

Example 13-1.   Evaluating and Comparing Projects When Marginal Utility of Money Is Nonuniform and Projects Have Multioutcome Possibilities of Known Probability

Information: Assume the indifference data from which Figure 13-1a was constructed. Further assume that an alternative with zero expected outcome just meets the rate of return requirements. No projects comparable in size and independent in outcome from those below are anticipated. Consider the mutually exclusive alternatives below:

Alternative	Outcome #1	p_1	Outcome #2	p_2	$E(x)$
E	+$50,000	0.9	- $200,000	0.1	$25,000
F	+200,000	0.7	- 200,000	0.3	80,000
G	+300,000	0.6	- 200,000	0.4	100,000
H	+250,000	0.5	- 150,000	0.5	50,000

Objective: Rank the alternatives in order of preference by computing the expected utility for each.

Analysis:

$$E(U_E) = 0.9(50) + 0.1(-500) = -5 \text{ utiles}$$
$$E(U_F) = 0.7(200) + 0.3(-500) = -10 \text{ utiles}$$
$$E(U_G) = 0.6(300) + 0.4(-500) = -20 \text{ utiles}$$
$$E(U_H) = 0.5(250) + 0.5(-300) = -25 \text{ utiles}$$

And the ranking in descending order of desirability is E, F, G, and H. Because none meet the criterion that $E(U) \geqslant 0$, all should be rejected.

Utility, Prediction, and Variance

Distant cash flows generally are subject to greater variability than those of the near future; fortunately, time value of money considerations tend to offset this otherwise worrisome aspect of estimation.

Under conditions of decreasing marginal utility for money, positive increments of cash have less utility than negative ones; this suggests that single-valued estimates of receipts should be decreased and that single-valued estimates of disbursements should be increased to allow for the decreasing marginal utility. Such estimating practices are consistent with decreasing marginal utility concepts; both are essentially expressions of conservatism.

Implications Regarding Diversification

The concepts of nonuniform utility of money are applicable when expected value of an outcome is inappropriate because (1) the number of comparable projects having outcomes independent of those considered is small or (2) the number of comparable projects having outcomes dependent upon that illustrated is large. It should be apparent that a manager might reduce the risk of an unfavorable outcome via diversification (acceptance of a number of projects whose outcomes are independent of each other). Variability can be reduced and expected utility increased by acceptance of independent projects. Example 13-2 illustrates this point.

Example 13-2.   Diversification and Its Relation to Utility of Money

Information: Assume the indifference data from which Figure 13-1 was constructed. Many sets of mutually exclusive alternatives have been compared and the best of each has been selected for further consideration in the funding (capital budgeting) process. Alternatives have identical first costs; due to limited capital, no more than two can be selected. Alternative outcomes are independent of each other, even those having the same letter designation.

Objective: If each of the investment opportunities is considered in isolation from all others, it would seem that all should be rejected. If the opportunities are considered in pairs, their independent outcomes might make some groupings (diversification) more attractive; consider several sets to check this possibility.

Alternative Number	Outcome #1	p_1	Outcome #2	p_2	$E(x)$	$E(u)$
	thousands		*thousands*		*thousands*	*utiles*
1E	$ 50	0.9	- $200	0.1	$ 25	- 5
2E	50	0.9	- 200	0.1	25	- 5
3F	200	0.7	- 200	0.3	80	- 10
4F	200	0.7	- 200	0.3	80	- 10
5G	300	0.6	- 200	0.4	100	- 20
6G	300	0.6	- 200	0.4	100	- 20
7H	250	0.5	- 150	0.5	50	- 25
8H	250	0.5	- 150	0.5	50	- 25

Analysis: A decision tree is shown for the combined investment, 1E plus 2E.

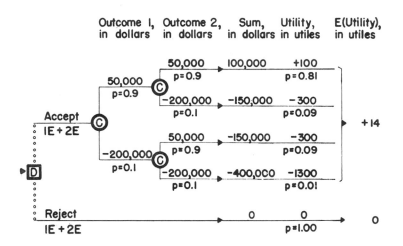

Expected outcome from the joint project

$$= E(x_{1E+2E}) = E(x_{1E}) + E(x_{2E}) = \$50,000$$

Or, by reference to outcomes and probabilities,

$$E(x_{1E+2E}) = \$100,000(0.9)(0.9) - \$150,000(0.9)$$
$$\cdot (0.1)(2) - \$400,000(0.1)(0.1) = \$50,000$$

$$E(U_{1E+2E}) = 100(0.9)(0.9) - 300(0.9)(0.1)(2)$$
$$- 1300(0.1)(0.1)$$

$$= +14 \text{ utiles}$$

Thus the combined investment opportunity is acceptable even though the individual investment opportunities had negative expected utility and were therefore not acceptable. In this instance diversification of investment over alternatives with independent outcomes has proven beneficial. That diversification is not always beneficial is illustrated by the joint project (7H + 8H):

$$E(x_{7H+8H}) = \$500,000(0.5)(0.5) + \$100,000(0.5)(0.5)(2)$$
$$- \$300,000(0.5)(0.5) = \$100,000$$

$$E(U_{7H+8H}) = 500(0.5)(0.5) + 100(0.5)(0.5)(2)$$
$$- 900(0.5)(0.5) = -50 \text{ utiles}$$

From this set of eight alternatives there are a total of 28 pairs which might be considered if we were to search for the pair which would maximize utility.

Our example illustrates that it is advantageous, if not mandatory, that the projects which comprise the capital budget asking be reviewed as a group rather than in isolation from each other. Our example suggests that the choice between mutually exclusive alternatives cannot always be made in isolation from the capital budgeting process. Later, in treating the capital budgeting topic, we will see additional circumstances under which the choice between mutually exclusive projects and the priority ranking of projects in the capital budgeting process are not always separable activities.

Example 13-3.   Evaluating and Comparing Projects When Marginal Utility of Money Is Nonuniform and Projects Have Multioutcome Possibilities of Unknown Probability

Information: Assume the indifference data from which Figure 13-1a was constructed. Further assume that zero expected outcome means the alternative just meets the rate of return requirement and hence just meets the threshold of acceptability. Consider Alternatives J and K:

Alter-native	Outcome #1	Probability	Outcome #2	Probability
J	$400,000	p_J	-$200,000	$(1 - p_J)$
K	200,000	p_K	- 100,000	$(1 - p_K)$

Objective: Find the minimum value of p_J and p_K for these alternatives to be acceptable; then compute several sets of p_J and p_K values where the alternatives would have identical expected utility.

Analysis: At break even,

$$E(U_J) = p_J(400) + (1 - p_J)(-500) = 0$$
$$p_J = 500/900 = 0.555$$
$$E(U_K) = p_K(200) + (1 - p_K)(-100) = 0$$
$$p_K = 100/300 = 0.333$$

The alternatives have equal utility when $E(U_J) = E(U_K)$, which is

$$p_J(400) + (1 - p_J)(-500) = p_K(200) + (1 - p_K)(-100)$$
$$900\,p_J - 500 = 300\,p_K - 100$$
$$p_J = (300\,p_K + 400)/900$$

So alternatives are equivalent when

p_J	p_K	
0.444	0.000	
0.555 ——	0.333 ——	break-even point
0.667	0.667	
0.777	1.000	

Ultimately, recognition of the nonuniform utility of money is simply a device for nonuniform weighting of the various dollar outcomes of an opportunity. Figure 13-1b shows that outcome dollars to the left of - $100,000 are four times as dear as those to the right of that point. Our example showed the weighting used by a decision maker with *declining marginal utility for money.* Note that our measurement was a subjective one, much more elusive (our hypothetical decision maker was kind enough to avoid contradictory choices and provided us a smooth utility curve) than might be apparent.

Other Models of the Utility of Money

When a decision maker exhibits diminishing marginal utility for money,[2] as he did in Figures 13-1 and 13-2, we can say that

---

[2] For an investigation into the utility functions of a number of managers, see R. O. Swalm, "Utility Theory—Insights into Risk Taking," *Harvard Business Review* 44 (Nov.-Dec., 1966), 123–36.

If all outcomes have been reduced
to present equivalent outcomes and
if a decision maker's utility function = $U(x)$,
e.g., $U(x) = 1800x - 800x^2$, where x =
outcome, in millions of dollars; and $-1 \leq x \leq 1$

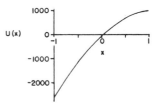

Marginal utility = $d(U(x))/dx = 1800 - 1600x$

And if outcome function is discrete, e.g.,
$x_1 = -0.5$, $p_1 = 0.3$; $x_2 = 0.5$, $p_2 = 0.7$;
where p_j is the probability of outcome x_j
and there are k possible outcomes.

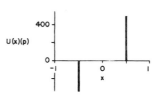

Expected outcome = $E(x) = \sum\limits_{j=1}^{k} x_j p_j$ =
$-0.5(0.3) + 0.5(0.7) = 0.2 = \$200,000$

Expected utility = $E(U) = \sum\limits_{j=1}^{k} U(x_j)(p_j)$ =
$-1100(0.3) + 700(0.7) = 160$ utiles

And if outcome function is continuous,
e.g., $p(x) = 0.75(-x^3 - x^2 + x + 1)$ where $p(x)$
is the probability density function of outcome x; $-1 \leq x \leq 1$; and $p(x) = 0$ elsewhere.

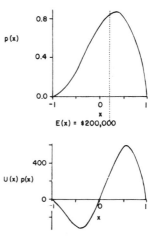

Expected outcome = $E(x) = \int x\, p(x)\, dx$ =
$\int_{-1}^{1} 0.75(-x^4 - x^3 + x^2 + x)\, dx = 0.2 = \$200,000$

Expected utility = $E(U) = \int U(x)\, p(x)\, dx$ =
$\int_{-1}^{1} (1800x - 800x^2)(0.75)(-x^3 - x^2 + x + 1)\, dx$ =
200 utiles

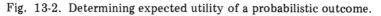

Fig. 13-2. Determining expected utility of a probabilistic outcome.

he is giving more weight to less desirable outcomes; that is, he makes his decisions from a *conservative* or *pessimistic* viewpoint.

Figure 13-3a shows a certainty line for a person whose decisions are based strictly on expected outcome; he weighs all dollars equally (uniform marginal utility), and for him single-valued estimates of expected outcome will give identical decisions to those produced when multivalued estimates are made.

Figure 13-3b shows a certainty line for a person who exhibits increasing marginal utility; he weighs the brightest possible outcome more heavily, and hence is willing to gamble even if the odds are against him and the stakes are high. His decisions are based on a view which is *speculative* or *optimistic.* One might wonder about the effect of our income tax laws upon a gambler's utility; tax laws require payment of tax on winnings, but gambling losses are not deductible except as offsets against income. This should at least reduce the tendency for a gambler to exhibit an increasing marginal utility for money.

Figure 13-3c shows a certainty line which combines some elements of both Figures 13-1a and 13-3b. As drawn, there are two points of substantial change (points of inflection) in the view of marginal utility for money. Several possible causes of a sudden change in marginal utility include:

1. There may be some loss which would cause bankruptcy or receivership; loss beyond this point would have meaning but less significance.
2. Losses in excess of a certain amount may be insured against.
3. The breakover point of no loss, no gain may be the basis on which a manager's ability is judged and thus represents a breakover from "wrong" to "right" decisions.
4. Income taxes.
   a. A taxpayer with carryover losses from earlier years might feel that the first x dollars of profit would be "all his" while additional profits would be taxable and therefore only "half his."
   b. The deductibility of losses and taxation of gains reduces the scale of losses and gains and thus reduces the impact which either, in the absence of taxes, would have.
   c. Capital gains taxes at half rate and losses deductible at full rate tend to "flatten" the curve of diminishing marginal utility.
   d. A change in the incremental income tax rate applicable

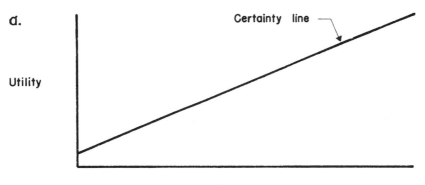

a.

Utility

Certainty line

Outcome

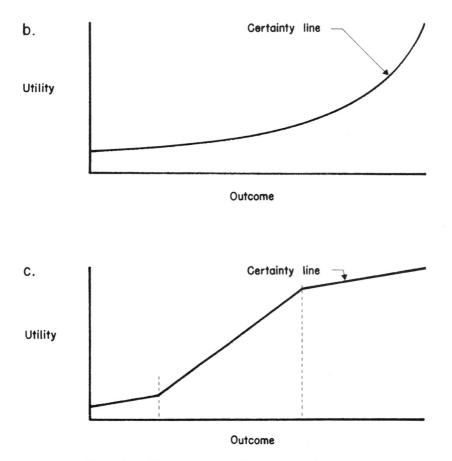

b.

Utility

Certainty line

Outcome

c.

Utility

Certainty line

Outcome

Fig. 13-3. Various models of the utility of money.

to taxpayer income could cause a sudden change in marginal utility.

Diminishing marginal utility for money by the firm and its owners is evidenced by:

1. Stockholder discontent with or marketplace devaluation of securities displaying excessive variability of earnings per share.
2. Management policies geared toward preservation of the firm rather than profit maximization.

Expected Utility, $E(U)$, As a Capital Expenditure Criterion

$$E(U) = E(U_{PEX}) = \sum_{j=1}^{k} p_j\, U(PEX_j)$$

where $U(PEX_j)$ = the utility of the PEX for the jth set of n end-of-year cash flows. There are k such sets.

p_j = the probability that PEX_j will occur

If all cash flows occur at the end of year zero, as in a bet, $z = n = 0$, the equation above simplifies to

$$E(U) = \sum_{j=1}^{k} p_j\, U(x_j)$$

For a group of mutually exclusive projects select the one with maximum $E(U)$, then accept it if either the project function is mandatory or $E(U) \geqslant 0$.

If, in the capital budgeting process, the cost of independent projects thus accepted exceeds the funds available, the rate of return required should be revised upward to reflect "opportunity cost." The final value for i should be the one which matches funds required with funds available.

Problems

13-1. Bruce Blue has a personal utility function identical to that pictured in Figure 13-1. Bruce is currently considering the purchase of an insurance policy to protect himself from the remote possibility of a loss of $200,-

000. The probability that the loss would occur once within any given calendar year is 0.01; the probability of more than one such occurrence in one year is nil.

   a. What is the maximum annual premium Mr. Blue is apparently willing to pay? Illustrate your analysis with a decision tree.

   b. What is the premium an insurance company would have to charge if administrative, selling, income tax, and other costs add 50% to the premium required to meet risk only?

13-2. Refer to Example 13-2 and compute (1) expected outcome and (2) expected utility for the following combined investments:

   a. 4F and 6G.

   b. 2E and 8H.

13-3. a. Refer to Example 13-2 and compute (1) expected outcome and (2) expected utility for the combined investment (1E + 2E) assuming that outcomes are *dependent;* that is, either both will be favorable or both will be unfavorable.

   b. What happens to expected outcome as a criterion of desirability when a project is considered in combination with many other projects with independent outcomes?

13-4. Consider a specific investment which logically might be made by either the federal government or a small, family-owned, manufacturing firm. Would the utility for money function be about the same for either potential buyer? Why?

13-5. Blue Corporation is involved in a patent suit which is thought to have three possible outcomes:

   A. If the case is settled out of court now, Blue will receive $80,000 (after taxes).

   B. If the case is continued and won, the present after-tax value to Blue is estimated as $500,000. The probability of winning is p_w .

   C. If the case is continued but lost, the present after-tax cost to Blue is estimated as $200,000. The probability of losing is $1 - p_w$.

   a. Find the minimum probability of winning necessary for Blue to pass up the settlement, assuming a uniform marginal utility for money over the relevant range of outcomes.

   b. Find the minimum probability of winning necessary for Blue to pass up the settlement, assuming the diminishing marginal utility for money indicated in Figure 13-1 and its accompanying data.

13-6. You have been invited to play a game in which you and your opponents must place one of two types of bets each round (the "stakes" are poker chips, no money is involved).

	Win	p_w	Lose	$1 - p_w$	$E(\ )$
Bet R	20	0.2	4	0.8	0.8
Bet S	2	0.5	1	0.5	0.5

The rules state that you cannot place multiple bets and you are dropped from the game if you run out of chips (no borrowing). The winner is the person with the most chips (or the survivor when all other players have been dropped from the game). Bets and outcomes for each player are known to all players.

a. What strategy (say first five bets for Player A) do you recommend if all players have just started the game and each has just four chips?

b. What strategy do you recommend for Player A if (1) the only surviving players are A and G, (2) A has 20 chips and G has 47 chips, and (3) the end of the game occurs after five more bets?

c. Refer to Part b; would you change your recommendation if the probabilities on Bet R were changed so p_w was now 0.15?

d. Refer to Part b; what counterstrategy do you recommend for Player G?

e. Comment on the marginal utility of chips to:

Player A in Part a.
Player A in Part b.
Player G in Part d.

13-7. Gene Green plans to purchase one of two securities. Security A costs $746.20 today and is regarded as very safe and stable with an essentially certain redemption value of $1000 six years from today, thus earning a return of 5.0%. Security B costs $746.20 today and is somewhat riskier; its probabilistic outcomes six years from now are estimated as:

Probability	Outcome Redemption Value	Utility, in Utiles
0.25	$ 400	- 48
0.00	1000	0
0.50	1200	4
0.25	1600	12

The above data results in an expected value of the outcome of $1100; this indicates a rate of return of about 6.7%.

Compute expected utility for each of the alternatives and state your conclusion. How does the security cost of $746.20 have a bearing?

13-8. Discuss the relationship between the criteria of *maximizing profit* and *maximizing utility*, noting the conditions under which the two tend to produce (1) identical preferences and (2) different preferences.

13-9. A football coach is considering two alternative plays with the estimated outcomes below:

	#1		#2		#3		Expected Value of Outcome
	Outcome	Probability	Outcome	Probability	Outcome	Probability	
Run	−3 yards	0.1	+2 yards	0.4	+5 yards	0.5	+3 yards
Pass	Lose ball*	0.05	0	0.65	+20 yards	0.3	+4 yards

*Treat as equivalent to a 40-yard loss.

a. Could he employ utility concepts in making his choice? Explain.

b. If his utility function could be determined explicitly, how could it be of benefit to the coach?

c. How could it be of benefit for the coach to keep his utility function from becoming known to his opposing teams and coaches?

d. Late in the first quarter, with his team on their own 48-yard line, third down coming up, and with three yards needed for a first and ten, the coach elected the pass play. Early in the fourth quarter, again with his team on their own 48-yard line, third down coming up, and with three yards needed for a first and ten, the coach elects the run play. What are some possible reasons for his choices, especially as they relate to utility concepts?

e. In what *one* way is the utility function of the football coach quite different from that of the corporate manager of long-term assets.

13-10. The outcome for Alternative M is estimated to follow a continuous uniform distribution with a lower limit of $- \$400,000$ and an upper limit of $+\$600,000$.

a. Find the expected outcome of Alternative M.

b. Find the expected utility of Alternative M, assuming the indifference data from which Figure 13-1a was constructed.

c. Find the expected utility of Alternative M, assuming the utility function given in Figure 13-2.

13-11. A decision maker's time-invariant utility function $= U = f(x)$, where x_z is the net cash flow for year z and $z = 0, 1, 2 \ldots n$. The required rate of return $= i$. Find the utility of an investment involving the cash flows x_0, x_1, x_2 (the cash flows of all other years are zero). Note that the introduction of probabilistic cash flows would not alter the nature of the question. Should you:

a. Find the utility of each years cash flow, *then* convert to present equivalent? That is,

$$U(\ ) = U(x_0) + (1 + i)^{-1} U(x_1) + (1 + i)^{-2} U(x_2)$$

b. Find the present equivalent of each cash flow, *then* convert to utility and sum? That is,

$$U(\ ) = U(x_0 + (1 + i)^{-1} x_1 + (1 + i)^{-2} x_2)$$

c. Other—please specify.

13-12. The following probabilistic after-tax cash flows have been forecast for a proposed project. $MARR = 10\%$.

Outcome	Probability	*ATCF* at the End of Year		
		0	1	2
Pessimistic	0.64	$- \$200,000$	0	0
Optimistic	0.36	$- \$200,000$	0	$\$800,000$

a. Find $E(i)$.

b. Find $E(PEX)$.

c. Find $E(U)$ assuming that the utility function of Figure 13-1 applies to *PEX* of each of the two possible outcomes.

13-13. Using the utility function given in Figure 13-1, find $E(U)$ for Alternatives A and B in Problem 12-21.

13-14. The following estimates apply to Project A; no other cash flows are anticipated.

First Cost	Probability	Prospective After-Tax Cash Flow at the End of Each of the Next Ten Years	Probability
$426,000	0.75	$100,000	0.80
$526,000	0.25	$ 40,000	0.20

    a. Find $E(PEX_A)$ if $MARR = 12\%$.
    b. Find $E(U_A)$ if the utility function of Figure 13-1 applies.

13-15. Mr. Portfolio has K dollars available for investment. His utility function = $U(x) = 200x - 10x^2$, where x = outcome in millions of dollars, and $-10 \leqslant x \leqslant 10$. The outcomes shown for Opportunities 1A and 2B are those which would occur if all K dollars were invested in that one opportunity. Let y = the portion of total funds invested in 1A and $(1 - y)$ the portion of total funds invested in 2B. Find the optimal "mix" of 1A and 2B (that is, find y such that $E(U)$ is maximized).

Opportunity	Outcome #1, in Millions	p_1	Outcome #2, in Millions	p_2	$E(x)$, in Millions	$E(U)$, in Utiles
1A	$2	0.8	-$4	0.2	$0.8	+96
2B	$6	0.5	-$4	0.5	$1.0	-60

13-16. For a continuous utility function to exhibit diminishing marginal utility, the first derivative must be _____ and the second derivative must be _____ .

14

Nonuniform (Dispersed) Service Lives

Group Properties

IT IS FREQUENTLY CONVENIENT to speak of or work with a set of similar assets called a *group property*. Examples of group properties include water meters, miles of pipe or pipeline, miles of electric cable, miles of sewer system, hydrants, power poles, transformers, 747's, diesel locomotives, miles of track, switching centers, centerless grinders, multiple-spindle drill presses, or miles of overhead conveyor track. Treating such things as property groups can be helpful by reducing the recording effort necessary in management control functions, including accounting.

Estimating Service Life and Dispersion

When we estimate service life of one item of property, our estimate is usually in the form of a single estimate. If our estimate applies to a group of similar properties, we may again form a single estimate of service life, which we consider to be an average of the individual service lives.

Suppose that both property sets below consist of two similar units having lives as shown:

Set A: 4 years, 4 years
Set B: 3 years, 5 years

If estimates are formed as to average life of the group property, the result for either Set A or Set B is an average service life of four years. Although the sets have identical average lives, the dispersion (or variability, variation, scattering) of Set B lives is greater than that of Set A.

Failure to include an estimate of dispersion infers that there is zero dispersion—that all units survive exactly the same number of years. This seems an unlikely expectation for most group properties. Conceivably, even one item of property would not be retired in its entirety at a given point in time. (For example, a building which is rather like a group property consisting of structural components, flooring, lighting fixtures, an air-conditioning system, elevator-escalator system, and other components, each part having a different life.) It seems reasonable that, even for an item of property whose retirement will occur all at once, estimated life could still include a "range" of service lives possible for that one item of property.

When the dispersion of service lives within a proposed group property is undeterminable (as for a newly developed type of facility), we suggest that the S_1 dispersion is a more logical estimate than the SQ "minimum cost" dispersion.[1]

Where some experience with similar units exists, average life and dispersion can be studied. Survivor experience could, for example, be plotted and compared to survivor curves such as those shown in Figures 14-1 through 14-6.[2] Using this as an estimate, we could then apply factors appropriate to such a dispersion. The problem of fitting survivor dispersions with such mathematical models as the Iowa type, orthogonal polynomial, or Gompertz-Makeham formula is deferred to other textbooks and courses.[3] When necessary, the problems of this book provide the survivor-dispersion type as data.

_____

[1] Normally labeled S_1, the curve types will also be shown as S1 because of their later presubscripted use as in the factor $_{S1}(a/p)_{50}^{10\%}$. The S_1 dispersion appears in Figure 14-5 and Figure H-2 of the Appendix.

[2] A convenient method of comparison by use of overlays of the Iowa type survivor curves permits graphical estimation of both dispersion type and average service life.

[3] See A. Marston, R. Winfrey, and J. C. Hempstead, *Engineering Valuation and Depreciation*, 2nd ed. (Ames, Iowa State University Press, 1953).

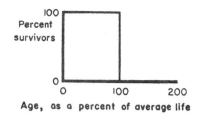

Rectangular or square (SQ) survivor
dispersion

Percent survivors = 100 for $z \leq n$
Percent survivors = 0   for $z \geq n$

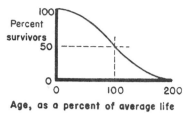

Iowa Type S_X (SX) survivor dispersion

$$\left(\begin{array}{c} \text{Percent} \\ \text{survivors} \end{array}\right) = 100 - \int_0^Z K_1 \left(\frac{2zn - z^2}{n^2}\right)^{K_2} dz$$

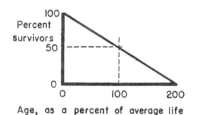

Iowa Type O_1 (O1) or straight-line
survivor dispersion

Percent survivors = $100 - \dfrac{100\,z}{2n}$

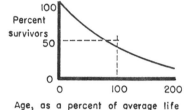

Negative exponential (NE) survivor
dispersion

Percent survivors = $100\,e^{-\frac{z}{n}}$

K = a constant
n = average life, in years
z = attained age, in years

$\left(\begin{array}{c} \text{Percent} \\ \text{survivors} \end{array}\right) = \dfrac{\text{units remaining in service at age } z}{\text{units originally placed in service}}$

Age, as a percent of average life = $\dfrac{z}{n}$

Fig. 14-1. Selected survivor dispersions.

What effect does dispersion of lives have upon the relative
economy of a group property? Does service life dispersion always have an adverse effect upon the economy of a project? A
reply to these questions is facilitated by Example 14-1.

Example 14-1.   Annual Cost of Group Properties

Information: Two sets, A and B, of group properties are to
be analyzed. Set A consists of two identical property units

drawn from or installed in a population of units for which experience indicates the lives are very uniform, all units having a service life of four years. Set B consists of two identical property units drawn from or installed in a population of units of which, experience indicates, about half will be retired at the end of three years, and the other half at the end of five years. Service need for the units is estimated to continue for a very long period of time and thus a series of replacements will be required. Each of the four units has a first cost of $1000 and salvage value at retirement for each is estimated to be zero. The minimum attractive rate of return is 8%. Note that average life for both Set A and Set B is four years; only the dispersion of service lives differs.

Objective: Find annual equivalent *ATCFR* for Set A and for Set B.

Analysis:

Set A: $ATCFR = \$2000\,(a/p)_4^{8\%} = \$603.84$.
Set B: What would happen if we apply the same $(a/p)_4^{8\%}$ factor to Set B?

	Pattern of Recovery				
	Year 1	Year 2	Year 3	Year 4	Year 5
3-year property	$301.92	$301.92	$301.92		
5-year property	301.92	301.92	301.92	$301.92	$301.92

Check the present worth of the recovery:

$$\text{Present worth} = \$301.92\,(p/a)_3^{8\%} + \$301.92\,(p/a)_5^{8\%}$$
$$= \$301.92\,(2.577 + 3.993)$$
$$= \$1983.61$$

Note that present worth of the recovery pattern shown falls short of the $2000 cost of Set B. The total dollars recovered in Set A equals the total dollars recovered in Set B, but the dispersion within Set B results in a one-year delay in the recovery of the last $301.92 amount. Had the lives in Set B been two years and six years, this delay would further increase the annual equivalent cost of recouping the investment. A survivor dispersion of zero as in Set A is an "ideal" or "minimum" cost situation; the greater the dispersion, the more the cost. Only if the time value of money is zero ($i = 0\%$) would dispersion be of no consequence. The object of a correct capital recovery factor is to recover in full at the stated rate of interest the first cost less salvage value; therefore

$$\begin{pmatrix} \text{Number} \\ \text{of units} \\ \text{placed} \end{pmatrix} \begin{pmatrix} \text{first} \\ \text{cost less} \\ \text{salvage} \\ \text{per unit} \end{pmatrix}$$

$$= \begin{pmatrix} \text{first} \\ \text{cost less} \\ \text{salvage} \\ \text{per unit} \end{pmatrix} \begin{pmatrix} \text{capital} \\ \text{recovery factor} \\ \text{for dispersion} \\ \text{specified} \end{pmatrix} \begin{pmatrix} \text{summation} \\ \text{of present} \\ \text{worth of} \\ \text{survivors} \end{pmatrix}$$

So

$$\begin{pmatrix} \text{Capital recovery} \\ \text{factor for dispersion} \\ \text{specified} \end{pmatrix} = \begin{pmatrix} \dfrac{\text{number of units placed}}{\text{summation of present}} \\ \text{worth of survivors} \end{pmatrix} \quad (14\text{-}1)$$

And in our example

$$_B(a/p)_4^{8\%} = 2/(2.577 + 3.993) = 0.30441$$

So with the derived capital recovery factor we may now compute annual $ATCFR$ for Set B as follows, using the notation $_B(a/p)_4^{8\%}$ for a dispersed capital recovery factor.

$$ATCFR = \$2000_B(a/p)_4^{8\%} = \$2000(0.30441) = \$608.82$$

Recognition of dispersion increased $ATCFR$ by less than 1%: much larger errors are possible; the circumstances under which more sizable errors arise will be considered shortly.

Example 14-2.   Present Worth of Costs for a Group Property

Information: Same as Example 14-1.

Objective: Compare the present equivalent $ATCFR$ for perpetual service from Set A and Set B.

Analysis: For each set we need only to multiply the annual *ATCFR* determined in Example 14-1 by the present worth factor, $(p/a)_\infty^{8\%}$:

$$PE(ATCFR_A) = (\$603.84)\,(12.5) = \$7548$$

and

$$PE(ATCFR_B) = (\$608.82)\,(12.5) = \$7610$$

Three Common Misinterpretations of the Present Equivalent Costs for a Group Property

In computing the present equivalent costs for a group property, we should avoid using average life as a "cycle time" over which first costs will be repeated during the period of study. In Example 14-2 we will be led to an erroneous result if we assume that the $2000 cost of Set B will be repeated each four years. Such an assumption gives results coinciding with those of the "no-dispersion" (Set A) case, thus underestimating actual costs when a dispersion exists.

$$PE(ATCFR) = \$2000\,(a/p)_4^{8\%}\,(p/a)_\infty^{8\%} = \$7548$$

A second misinterpretation occurs when it is assumed that the $_{xx}(a/p)_n^i$ factor for a dispersed group is simply an average factor $[\,(a/p)_{n_1}^i + (a/p)_{n_2}^i \cdots + (a/p)_{n_n}^i\,]/n$ or that short-lived items will be replaced by short-lived items and likewise for long-lived items. Thus in Example 14-1, over a 15-year period one might erroneously assume five sets of three-year life items and three sets of five-year life items. The error is apparent when we see that either assumption treats the group as if 62.5% of the units had a life of three years and 37.5% of the units had a life of five years. The distortion is also apparent, at least as it applies to Example 14-1, where we can see that the larger percentage of short-lived units would result in an overstatement of costs.

$$PE(ATCFR) = [\,\$1000\,(a/p)_3^{8\%} + \$1000\,(a/p)_5^{8\%}\,](p/a)_\infty^{8\%}$$
$$= \$7981$$

A third misinterpretation occurs when it is assumed that the long-lived item will be replaced by the short-lived item and vice-versa. This assumption applied to Example 14-1 would give replacement costs of $1000 in year 3, $1000 in year 5, $2000 in year 8, $1000 in year 11, and so forth. Again, however, there is a distortion in that the early replacement was filled by a long-lived unit; thus the method results in understatement of costs.

$$PE(ATCFR) = [\,\$2000 + \$1000(p/f)_3^{8\%} + \$1000(p/f)_5^{8\%}\,]$$
$$\cdot (a/p)_8^{8\%} (p/a)_\infty^{8\%} = \$7557$$

Calculation of the Capital Recovery Factor for a Given Survivor Dispersion

As noted in Example 14-1, the capital recovery factor for a given survivor dispersion will be denoted as $xx(a/p)_n^i$ and can be computed for various interest rates, survivor dispersions, and average service lives by the relationship

$$\begin{pmatrix} \text{Capital recovery} \\ \text{factor for dispersion} \\ \text{specified} \end{pmatrix} = \begin{pmatrix} \text{number of units placed} \\ \text{summation of present} \\ \text{worth of survivors} \end{pmatrix}$$

This procedure has been applied to various survivor dispersions to produce Appendix H.[4] Table 14-1 provides a few selected

TABLE 14-1. Selected Interest Factors for Group Properties

$$i = 7\%$$

Uniform series end-of-period amount a equivalent to present sum p (capital recovery factor) for a group property with survivors following the ____ dispersion.

$$xx(a/p)_n^i = xx(a/f)_n^i + i$$

n	Negative exponential NE	Straight-line or Iowa Type O_1	Iowa Type S_1	Square, rectangular, or no dispersion SQ	n
5	0.27692	0.25610	0.24923	0.24389	5
10	0.17346	0.15488	0.14770	0.14238	10
15	0.13897	0.12237	0.11501	0.10979	15
20	0.12173	0.10684	0.09942	0.09439	20
30	0.10449	0.09236	0.08507	0.08059	30
50	0.09069	0.08212	0.07565	0.07246	50

[4] See American Telephone and Telegraph, *Engineering Economy*, 2nd ed. (New York, 1963), pp. 224–35; Paul Jeynes, *An Abbreviated Course in Engineering Economics* (Newark, N.J., Public Service Electric and Gas Company, 1960), Vol. 1, pp. 58–60.

capital recovery factors for the dispersions given in Figure 14-1. The relationship between compound interest factors for single units also holds for the compound interest factors for dispersed units:

$$_{xx}(a/p)^i_n = {}_{xx}(a/f)^i_n + i$$

Thus the various available tables may be used readily whether the factors in the tables are sinking-fund factors or capital recovery factors for a given survivor dispersion.

Example 14-3. **Determination of Capital Recovery Factors for a Given Survivor Dispersion and Interest Rate**

Information: A group property has been analyzed for the purpose of estimating its survivor dispersion and average life. Survivor data are closely approximated by the Iowa Type O_1 (shown in Figure 14-2) with an average life of five years. An interest rate of 7% has been determined as appropriate. The O_1 survivor curve can be represented by

$$\text{Percent survivors} = 100 - (100z/2n)$$

where z = age attained, in years
n = average life, in years

Objective: Find the capital recovery factor for the given survivor dispersion and interest rate.

Analysis: The present worth of survivors can be approximated by (1) finding the present worth of a set of vertical strips representing the number surviving *at* a given age or by (2) finding the present worth of a set of horizontal strips representing the number surviving *to* a given age. In any case, ordinate values of the continuous function are approximated by treating average annual ordinate values as a discrete function. The summation may be represented by vertical strips as in Table 14-2 or by horizontal strips as in Table 14-3. In either table the present worth of survivors is 3.9008, so the approximate[5] capital recovery factor for the

---

[5] Approximate because the continuous survivor curve has been treated as a stair-step discrete function. Compare the factor in Table 14-1 (0.25610) with the approximation here (0.2564).

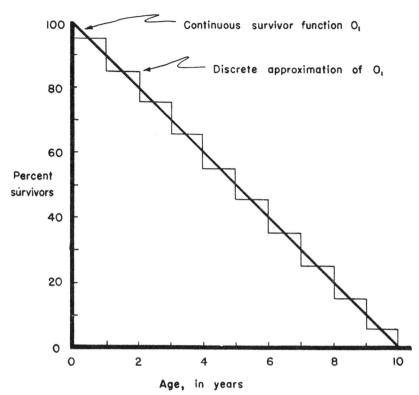

Fig. 14-2. Iowa Type O_1 survivor curve.

TABLE 14-2. Present Worth of Survivor Service Calculations Employing the (p/f) Factor

Age Attained	Average Percent Survivors during Year	Factor for Present Worth of a Future Sum, $(p/f)_n^{7\%}$	Present Worth of Survivors
years	*percent*		
1	95	0.9346	0.887,870
2	85	0.8734	0.742,390
3	75	0.8163	0.612,225
4	65	0.7629	0.495,885
5	55	0.7130	0.392,150
6	45	0.6663	0.299,835
7	35	0.6227	0.217,945
8	25	0.5820	0.145,500
9	15	0.5439	0.081,585
10	5	0.5083	0.025,415
			3.900,800

TABLE 14-3. Present Worth of Survivor Service Calculations Employing the (p/a) Factor

Years of Service	Percent of Units Giving Exactly ____ Years of Service	Factor for Present Worth of a Uniform Series $(p/a)_n^{7\%}$	Present Worth of Survivors
	percent		
0	5	0.000	0.0000
1	10	0.935	0.0935
2	10	1.808	0.1808
3	10	2.624	0.2624
4	10	3.387	0.3387
5	10	4.100	0.4100
6	10	4.767	0.4767
7	10	5.389	0.5389
8	10	5.971	0.5971
9	10	6.515	0.6515
10	5	7.024	0.3512
			3.9008

dispersed survivors $_{01}(a/p)_5^{7\%}$

$$= \frac{\text{number of units}}{\text{present worth of survivors}} \cong \frac{1.0000}{3.9008} = 0.2564$$

This factor is applicable to group properties maintained at original size by a policy of replacing all retired units promptly.

Circumstances under Which Dispersion of Property Lives Has Its Greatest Impact

If the dispersion of property lives is neglected, the capital re-recovery factor will be understated by

$$\text{Error} = [_{xx}(a/p)_n^i - (a/p)_n^i]/_{xx}(a/p)_n^i$$

The error is always one of understating costs or overstating rate of return. Table 14-4 provides a rough guide to the circumstances under which recognition of dispersion is most important in analysis. Note that error approaches a maximum when the product of average life and minimum attractive rate of return is about 200%-years.

While it would be of some help to provide capital recovery

TABLE 14-4. Error Introduced When the Effects of Survivor Dispersion Are Neglected

Average Life of Property	Minimum Attractive Rate of Return			
	3%	7%	12%	20%
years				
5	6	12	16	20
10	10	18	22	23
15	14	21	23	22
20	17	22	23	19
30	20	23	20	15
50	23	20	15	10

Note: Based on the negative exponential dispersion and where error = $[_{NE}(a/p)_n^i - (a/p)_n^i]/[_{NE}(a/p)_n^i]$ and is reported to the nearest whole percent.

factors for all the Iowa type survivor dispersions, we might question whether doing so is really essential and whether some selected dispersions might provide essentially the same information without obscuring the impact of dispersion by proliferation of data or tables. We observed earlier that the cost consequences of dispersion of lives is greatest at about $ni = 200\%$-years. Employing this and our computed data for the various survivor dispersions, we can produce Table 14-5 which lists the survivor dispersions in descending order of impact. (The order is only slightly altered for $ni \neq 200\%$-years.)

Several observations are helpful:

1. Where an error of 2% or less is tolerable, the impact of dispersion on equivalent cost is negligible for properties following the Iowa Type L_4, L_5, S_3, S_4, S_5, S_6, R_4, or R_5 dispersion.
2. Several logical groupings appear; it is apparent that survivor dispersions NE and O_3 exert about equal impact. It seems practical to select one dispersion from each group as representative of that set.

Note that while such selection is representative of the error outcome possibilities, it is *not* representative of the frequency with which such group properties will be encountered. Perhaps as many as 60% of the group properties considered will be of the L_4, L_5, S_3, S_4, S_5, S_6, R_4, R_5, or SQ type, while perhaps

TABLE 14-5. Impact of Survivor Dispersion under Selected Circumstances

Survivor Type	$_{xx}(a/p)_{33}^{6\%}$	$_{xx}(a/p)_{20}^{10\%}$	$_{xx}(a/p)_{7}^{30\%}$	Approximate Maximum Error Introduced by Neglecting Dispersion*
				percent
O_4	0.09731	0.16269	0.49460	30
NE	0.09120	0.15246	0.46335 ⎫	
O_3	0.09033	0.15101	0.45897 ⎭	23
O_2	0.08311	0.13891	0.42194	15
L_0	0.08054	0.13462	0.40895 ⎫	
O_1	0.08048	0.13450	0.40835 ⎭	13
L_1	0.07710	0.12887	0.39159 ⎫	
R_1	0.07686	0.12846	0.39010 ⎬	9
S_0	0.07648	0.12783	0.38831 ⎭	
L_2	0.07456	0.12463	0.37876 ⎫	
R_2	0.07424	0.12409	0.37693 ⎬	6
S_1	0.07420	0.12403	0.37683 ⎭	
L_3	0.07278	0.12166	0.36972 ⎫	
S_2	0.07266	0.12145	0.36902 ⎬	3.5
R_3	0.07242	0.12104	0.36775 ⎭	
S_3	0.07169	0.11983	0.36412 ⎫	
L_4	0.07152	0.11955	0.36327	
R_4	0.07139	0.11932	0.36255	
S_4	0.07097	0.11862	0.36043 ⎬	2
L_5	0.07088	0.11848	0.36000	
R_5	0.07072	0.11821	0.35916	
S_5	0.07059	0.11799	0.35849	
S_6	0.07040	0.11767	0.35753 ⎭	
SQ	0.07027	0.11746	0.35687	

*Error % = $[_{xx}(a/p)_n^i - (a/p)_n^i]/[_{xx}(a/p)_n^i]$ and is applicable to the non-salvage portion of first cost; the error is always one of underestimation of cost. This order is only slightly altered for $ni \neq 200$ percent-years.

only 0.1% will be of the O_2 type. Still more rare, perhaps, are dispersions of the O_4 type. Printing considerations tend to limit our selection to six; the mathematical simplicity of curve types NE, O_1, and SQ encourages us in their use. Based on these various considerations, the curve types for which capital recovery factors have been provided in this book are the NE, O_2, O_1, S_0, S_1, and SQ.

The following applications of these tables is suggested:

For survivor dispersion ___	Use $_{xx}(a/p)_n^i$ for the ___ dispersion
$L_4, L_5, R_4, R_5, S_3,$ $S_4, S_5, S_6,$ and SQ	SQ, error will not exceed 2%
$L_2, L_3, R_2, R_3, S_1,$ and S_2	S_1
$L_1, R_1,$ and S_0	S_0
$L_0,$ and O_1	O_1
O_2	O_2
NE and O_3	NE
O_4	very infrequently encountered; no dispersions of this type are included in this book

For the practitioner whose needs may be still more precise, a set of capital recovery factors for all Iowa type dispersions in a wide range of interest rates is planned.

While it might seem helpful to suggest dispersions for various types of property, we should be cautioned that identical properties such as telephone poles in Arizona and in Florida, may not be subject to identical conditions. Because of this, neither the average life nor dispersion of lives can be truly generalized. With this precaution in mind, a few hypothetical illustrations, which may be of some help in providing a feeling for the degree of dispersion, are suggested below.

Degree of Dispersion	Example	Hypothetical Iowa Type
Least dispersed	typewriters	S_3
↑	highway trucks	L_3
	passenger cars	L_2
	conveyor equipment	R_1
	electromechanical switching systems	L_0
↓	power poles	O_2
Most dispersed	telephone installations	O_3

A complete set of curves for the Iowa type dispersions is shown in Figures 14-3, 14-4, 14-5, and 14-6.

Applications

The approach suggested in Chapter 10 is little changed by the inclusion of dispersion effects.

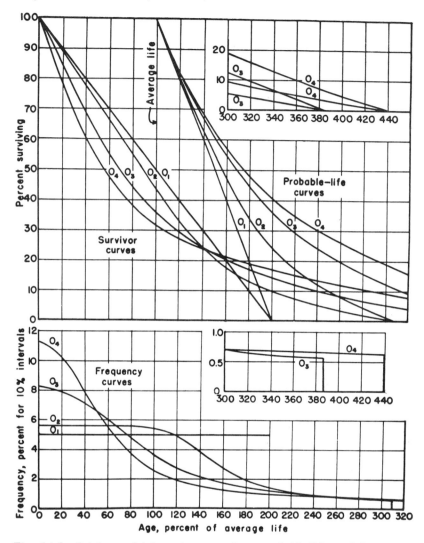

Fig. 14-3. Origin modal Iowa type survivor, probable-life, and frequency
curves. (Reproduced from an unpublished M.S. thesis by Frank
Couch, Iowa State University, Ames.)

In the more common of the dispersed group studies where
(1) average-life group procedure is used and (2) retired units are
immediately replaced:

$$BTCFR = [B_{xx}(a/p)_n^i - V_{xx}(a/f)_n^i - t(AED)]/(1-t)$$

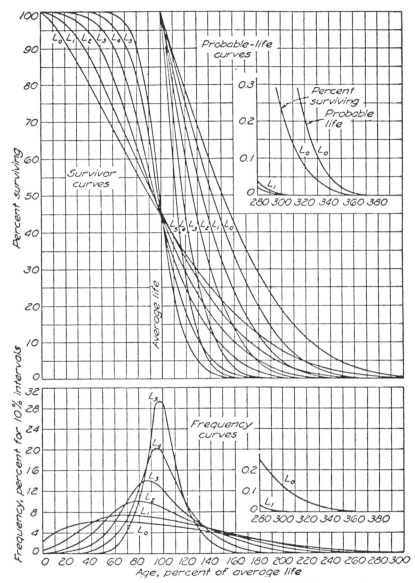

Fig. 14-4: Left modal Iowa type survivor, probable-life, and frequency curves. (Reproduced by permission from *Engineering Valuation and Depreciation* by Marston, Winfrey, and Hempstead, © 1953 by the Iowa State University Press, Ames, Iowa.)

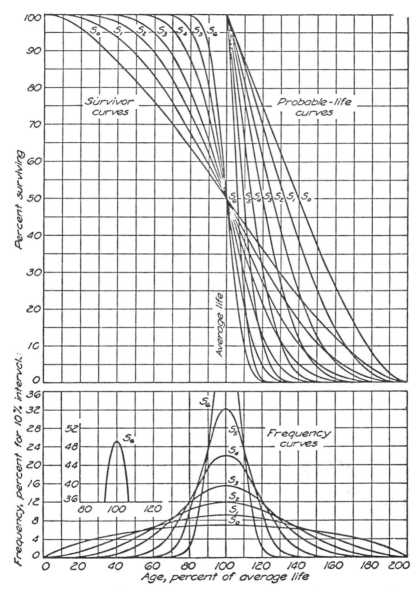

Fig. 14-5. Symmetrical Iowa type survivor, probable-life, and frequency curves. (Reproduced by permission from *Engineering Valuation and Depreciation* by Marston, Winfrey, and Hempstead, © 1953 by the Iowa State University Press, Ames, Iowa.)

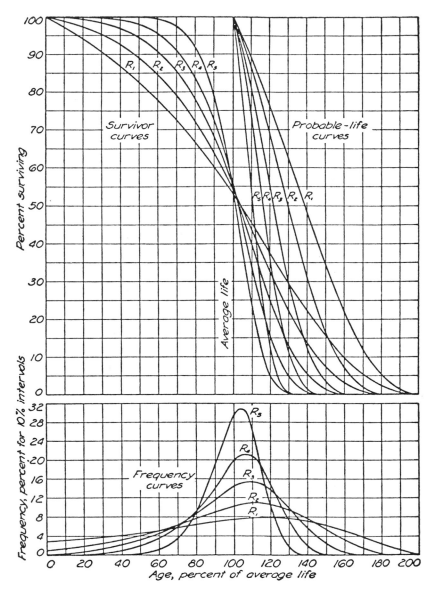

Fig. 14-6. Right modal Iowa type survivor, probable-life, and frequency curves. (Reproduced by permission from *Engineering Valuation and Depreciation* by Marston, Winfrey, and Hempstead, © 1953 by the Iowa State University Press, Ames, Iowa.)

and

$$AEC = AEM + [B_{xx}(a/p)_n^i - V_{xx}(a/f)_n^i - t(AED)]/(1 - t)$$
$$\text{(14-2)}$$

For present equivalent costs under those same circumstances it is recommended that the preceding calculations of AEC be made and then converted: $PEC = AEC(p/a)_n^i$.

In some instances the capitalized first cost of a unit in a group property is accompanied by certain *expensed* charges; a dispersed factor should be applied to the expensed items if both original *and* replacement units are accompanied by those expensed charges.

Example 14-4.   Application of Capital Recovery Factors for Dispersed Group Properties

Information: A proposed group property (such as a large quantity or number of pipe, cable, meters, poles, or transformers) consists of 1000 units. Each unit has an installed cost of $100, and in addition to this cost an expensed "cost of rearrangement" of $8 is incurred for each original or replacement unit. Each unit costs $6 per year for property taxes, insurance, maintenance, and other. Straight-line depreciation and average-life group procedure are used. Retired units are promptly replaced. $MARR = 10\%$, $t = 50\%$, average life of the units is 20 years. The property follows the Iowa type S_1 dispersion, and salvage is estimated as $10 per unit.

Objective: Find annual equivalent cost of the group property.

Analysis: Including dispersion (see the 10% table of Appendix H):

$BTCFR\ \#1 = [\$100,000\ _{S_1}(a/p)_{20}^{10\%} - \$10,000\ _{S_1}(a/f)_{20}^{10\%}$
$$- 0.5(\$4500)]/0.5$$

$= \$24,806 - \$481 - \$4500$		$= \$19,825$
$BTCFR\ \#2 = \$8000\ _{S_1}(a/p)_{20}^{10\%}$	=	992
Operating costs	=	6,000
AEC		$= \$26,817$

Neglecting dispersion:

$$BTCFR\ \#1 = [\$100,000\,(a/p)_{20}^{10\%} - \$10,000\,(a/f)_{20}^{10\%}$$
$$- 0.5\,(\$4500)]\,/\,0.5$$

$$= \$23,492 - \$349 - \$4500 \qquad\qquad = \$18,643$$

$$BTCFR\ \#2 = \$8000\,(a/p)_{20}^{10\%} \qquad\qquad\quad = \qquad 940$$

Operating costs $\qquad\qquad\qquad\qquad\qquad\qquad = \underline{\quad 6,000}$

AEC $\qquad\qquad\qquad\qquad\qquad\qquad\qquad\qquad = \$25,583$

Depreciation Calculations with Group Properties

Several options have a bearing upon the actual depreciation charges which will result for a dispersed group property. These are:

1. The choice between the widely used *average-life group procedure* and the *unit summation procedure,* the latter being also known as the *equal-life* or *reciprocal weighting procedure.*
2. The choice of classification methods; options include the *vintage* (consisting of items installed in the same year) and the *continuous* (replacement and additional units are also included with the original units).
3. The choice of period over which a property group is to be considered; whether average life, maximum life, or some other span of years.

In the problems and examples of this chapter we will treat group depreciation as if it produces results identical to those of item depreciation. This is correct for some combinations of the options noted and approximate for other combinations. Because the most common combinations of group depreciation options are correctly represented by item depreciation results, for the sake of brevity we will neglect the disparity which can arise. The common combination of options under which item depreciation is a correct representation is where average-life group procedure is employed and the group is maintained at original size with retired units being replaced promptly.

Relevance of Dispersion to the Treatment of Probabilistic Sets of Cash Flow Durations

Consideration of dispersion, as already discussed, is equally appropriate in the evaluation of a prospective cash flow stream which has a probabilistic set of durations. This point is illustrated in Examples 14-5 and 14-6.

Example 14-5.   A Probability-dispersed Set of Cash Flow Durations

Information: Forecasts of prospective sales and related costs have been employed in developing the following estimates of after-tax cash flows connected with a proposed new product line.

Annual After-Tax Cash Flows			
Amount	Probability	Duration	Probability
		years	
$10,000	0.3	5	0.3
20,000	0.7	15	0.7

The present investment required is $100,000 and salvage value is estimated to be negligible. Minimum attractive rate of return is 12%.

Objective: Find expected value of the annual after-tax cash flow amount and duration. Then consider the acceptability of the proposed project.

Analysis: The possible outcomes are

After-Tax Amount	Cash Flow Duration	Probability of Same	*PEX* = Present Equivalent of *ATCF* Less $100,000 Investment
	years		
$10,000	5	(0.3)(0.3) = 0.09	− $63,950
10,000	15	(0.3)(0.7) = 0.21	− 31,890
20,000	5	(0.7)(0.3) = 0.21	− 27,900
20,000	15	(0.7)(0.7) = 0.49	+ 36,220

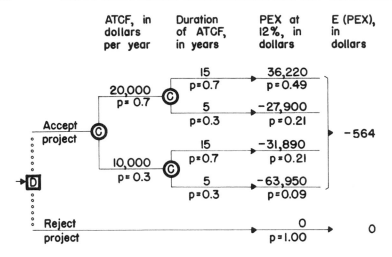

Note that the expected values of $17,000 and 12 years for *ATCF* amount and duration respectively imply acceptability since when

$$(a/p)^i_{12} = \$17,000/\$100,000, \quad i = 13.1\%$$

However,

$$E(PEX) = 0.09(-\$63,950) + 0.21(-\$31,890)$$
$$+ 0.21(-\$27,900) + 0.49(+\$36,220) = -\$564$$

and rejection is correctly signaled.

The use of an expected value of 12 years misled us. The penalties of a cash flow duration which falls short of 12 years are not fully compensated for by cash flow durations which exceed 12 years. (For example, receiving a cash flow of $1 in year 13 does not fully compensate us for having missed $1 of cash flow in year 11, if $i > 0\%$.)

Example 14-6.   Valuation of a Probability-dispersed Set of Cash Flow Durations

Information: An after-tax cash flow stream of $10,000 per year has been forecast for a proposed new product. The average duration of this cash flow stream is estimated as 12 years with a uniform distribution of probabilities over the range of 0 through 24 years.

Objective: Find the maximum present expenditure justifiable in the development of the proposed new product if the minimum attractive rate of return is 15%.

Analysis: The uniform distribution (see Figure 12-1) is identical to the frequency curve of Iowa Type O_1 (see Figure 14-3). Present worth of the after-tax cash flow stream

$$= {}_{01}(p/a)_{12}^{15\%}(\$10,000) = [1/{}_{01}(a/p)_{12}^{15\%}](\$10,000) = \$47,486$$

Had dispersion of the cash flow duration been neglected, the present worth would be overestimated as:

$$(p/a)_{12}^{15\%}(\$10,000) = 5.421(\$10,000) = \$54,210$$

Summary

Failure to allow for the dispersion of property lives will result in understatement of revenue requirements or overstatement of rate of return.[6] That error apparently approaches a maximum when the product of average life and minimum attractive rate of return is about 200%-years.

Even when survivor dispersion information regarding a proposed property group is lacking, estimation of, for example, the S_1 dispersion is more logical and reasonable than the SQ "ideal" or "minimum-cost" dispersion. Even when estimated survivor dispersions closely approximate the SQ type, the need for considerable precision in the result may still overrule the approximation involved when dispersion is neglected.

In many instances market forecasts for sales of new products involve a very uncertain duration of sales and hence of cash flow. In such cases it is appropriate to consider the dispersion of cash flow durations.

Problems

14-1. Find (1) average life and (2) capital recovery factor for a group property which exhibits the following dispersion. Use $MARR = 7\%$ and assume all retirements occur at the beginning of the year.

---

[6] Providing survivor dispersion not SQ, minimum attractive rate of return > 0%, and first cost > salvage.

End of Year	Survivors
	percent
1	100
2	90
3	70
4	40
5	0

14-2. Find the capital recovery factor for a group of four property units with the dispersion of lives shown. $MARR = 8\%$.

25% of the units are retired at the end of year 1
25% of the units are retired at the end of year 3
25% of the units are retired at the end of year 5
25% of the units are retired at the end of year 7

14-3. A certain group property has an average life of 20 years and follows the S_1 dispersion. First cost of the property is $100,000, and salvage value is expected to be negligible. Retired units are promptly replaced. Minimum attractive rate of return is 7% and income taxes are not applicable to the agency considering this group property.
   a. Find the annual equivalent cost of recouping the investment and interest on it ($ATCFR$).
   b. Find the percent error introduced into $ATCFR$ if dispersion is neglected.

14-4. Find the percent error introduced into $ATCFR$ when the SQ dispersion is used but the S_1 dispersion is appropriate; retired units are replaced promptly. Salvage as a percent of first cost is 0. Assume minimum attractive rate of return is 7% and solve for $n = 10$, $n = 30$, and $n = \infty$.

14-5. Given a group property with a negative exponential survivor dispersion and a minimum attractive rate of return of 7%, we might logically ask what no-dispersion, shorter-lived property group with prompt replacement of retired units would give comparable $ATCFR$ costs. Such a comparison can be made when the table below is completed. The capital recovery factors given are from Table 14-1 and are for the negative exponential dispersion. Use $MARR = 7\%$ to complete the comparison below.

A property with a negative exponential dispersion and average life of ___ years	Has a capital recovery factor of ___	About this same capital recovery factor is appropriate for a square dispersion and average life of ___ years
5	0.27692	
10	0.17346	
20	0.12173	
50	0.09069	

Interpolate and record your result to the nearest 0.1 year.

14-6. A certain group property follows the negative exponential survivor dispersion and has an average life of 30 years. The company analyzing the group property for possible acquisition has 100% equity ownership and an effective combined state and federal income tax rate of 55%. Straight-line depreciation and average-life group procedure are used for income tax purposes; retired units would be replaced promptly. First cost of the group property is $100,000 and estimated salvage is 10% of first cost. Minimum attractive rate of return for the company is 10%. Estimated operating costs are $5000 per year. Find annual equivalent cost for the group property.

14-7. A group property with an average life of 15 years follows the O_1 survivor dispersion. There is 100% equity ownership with a combined effective state and federal income tax rate of 50%. Straight-line depreciation and average-life group procedure are used for income tax purposes; retired units will be replaced promptly. There are 1000 units, each with a first cost of $100, estimated salvage value at retirement of $40; estimated cash operating costs are $8000 per year. The minimum attractive rate of return is 10%.

   a. Find annual equivalent cost for the group property.

   b. Compare your result with that obtained when the effect of dispersion is overlooked.

14-8. A group property with an average life of 30 years follows the S_1 survivor dispersion. There is 100% equity ownership with a combined effective state and federal income tax rate of 50%. Straight-line depreciation and average-life group procedure are used for income tax purposes; retired units are replaced promptly. First cost of the property is $120,000 and salvage value is an estimated 10%; both values will be incurred for replacement units. For tax purposes $100,000 of first cost is capitalized and $20,000 is expensed. Maintenance costs, operating costs, and property taxes total an estimated $9000 per year. Find annual equivalent cost using a minimum attractive rate of return of 7%.

14-9. a. Find the average life and capital recovery factor appropriate to that life for the following group property of four units, $MARR = 12\%$.

	Life
	years
Unit #1	4.4
Unit #2	6.0
Unit #3	17.0
Unit #4	17.0

b. Under what circumstances should the survivor dispersion be considered in the engineering economy study?

14-10. A proposed group property has an installed cost of $10 per unit. The group consists of 10,000 units with an average life of 20 years and following the Iowa Type O_1 dispersion; any retired units will be replaced promptly. Expensed costs of rearrangement amounting to $0.50 per unit occur at the time of the original installations, but not for the subsequent replacement units. Property tax, insurance, maintenance, and other costs of $1 per unit per year will be incurred. Straight-line depreciation and average life group procedure are used. $t = 50\%$, and $MARR = 10\%$.   Salvage of retired units is negligible. Find annual equivalent cost of the group property.

14-11. A study of Item J reveals that 25% of the units survive exactly six years, while 75% of the units survive exactly two years. Minimum attractive rate of return is 10%.

a. Find the average life of Item J.

b. Find the capital recovery factor for this group property and compare it to the factor which you would use if there were no dispersion.

c. Numerous random spot checks of property in service indicated that at any given time half the units in service tended to be the longer lived Item J's while the other half were the shorter lived Item J's. Reconcile these observations with the original data that 25% of the units survive six years and 75% survive two years.

d. A writer has suggested that the capital recovery factor for a dispersed group property is simply the *average* capital recovery factor for each unit; thus for a group of Item J:

Average capital recovery factor

$$= 0.25(a/p)_6^{10\%} + 0.75(a/p)_2^{10\%}$$
$$= 0.25(0.22961) + 0.75(0.57619)$$
$$= 0.48954$$

The above figure is, however, an overstatement of the correct value determined in Part b. Why?

e. Referring to Part c, we might compute an average capital recovery factor as follows:

Average capital recovery factor

$$= 0.50(a/p)_6^{10\%} + 0.50(a/p)_2^{10\%}$$
$$= 0.50(0.22961) + 0.50(0.57619)$$
$$= 0.40290$$

The above figure is, however, an understatement of the correct value determined in Part b. Why?

14-12. The following experience has been observed for a group property which originally consisted of 100 units.

Age, in Years	Units Surviving	Age, in Years	Units Surviving
0	100	6	32
1	99	7	17
2	94	8	6
3	83	9	1
4	68	10	0
5	50		

a. Find the average service life.
b. Which of the four dispersions shown in Figure 14-1 most closely fits this group?
c. Use $MARR = 10\%$ to find the capital recovery factor applicable to computation of the annual equivalent of first cost for units of this type.

14-13. It would be helpful to have some indicator of the impact of dispersion of service lives upon the economy study. One such indicator can be obtained by considering a project which shows a prospective rate of return of 15% in the absence of dispersion and computing what actual rate of return will be earned if the property is subject to the negative exponential dispersion. As an example, consider a property for which salvage = $0 and $n = 4$; then the after-tax cash flow required for a 15% return (zero dispersion) is

$$\text{(first cost)}(a/p)_4^{15\%} = 0.35027 \text{ (first cost)}$$

If the property actually follows the negative exponential dispersion, the rate of return would be

$$0.35027 \text{ (first cost)} = {}_{NE}(a/p)_4^i \text{ (first cost)}$$
$${}_{NE}(a/p)_4^{8\%} = 0.33987; {}_{NE}(a/p)_4^{10\%} = 0.36230$$

So by interpolation

$$i = 8\% + 2\% \, (1040/2243) = 8.9\%$$

a. Repeat the preceding analysis for $n = 1, 2, 8, 15, 30$, and 60. Comment on the disparity between rates of return when dispersion is recognized.

b. Reconcile your results with the rule of thumb that the error in *ATCFR* resulting from neglecting dispersion is a maximum at $ni = 200\%$-years.

14-14. A proposed group property consists of 1000 units to which the following estimates apply:

Installed cost (all capitalized) = $100 per unit
Salvage = $40 per unit; average life = 20 years
Dispersion = Iowa Type O_2
Cash operating costs = $8.33 per unit per year
Depreciation method is straight-line
$t = 50\%$; *MARR* = 10%

Find present equivalent cost of 20 years of service if retired units are promptly replaced to maintain group size at 1000 units.

14-15. An after-tax cash flow stream of $10,000 per year has been forecast for a proposed new product. The duration of the cash flow stream is estimated as ten years; the probabilistic duration is approximately that of the L_0 frequency curve of Figure 14-4. Use the dispersed annuity tables to find the maximum present expenditure justified in the development of the proposed new product. The minimum attractive rate of return is 10%; estimated salvage value is zero.

I5

Expansion and Economic-Package Concepts

Expansion of Capacity: Impact on Economic
Package/Economic Interval

GROWTH in demand for goods or services of a firm tends to be continuous while the growth in total facility capacity tends to be available only in discrete steps. As can be noted in Figure 15-1, growth in demand can be met by additions to capacity which are large and infrequent or small but frequent.

The problem here is one of determining the size of the addition (economic package) and/or the frequency of addition (economic interval) which maximizes profit within the other restrictions determined by the firm (such as quality of service, demands upon employees, and the like). It should be apparent that the extreme conditions are generally inefficient; large additions of capacity in excess of needs result in much idle plant ($\sim$ idle capital); frequent (and therefore relatively small) additions of capacity result in repetition of start-up costs, and the smaller units are generally less efficient and/or more costly per unit of output capacity. This latter concept is frequently referred to as "economies of scale." Consider, for example, an underground pipe; the major part of its installed cost is the

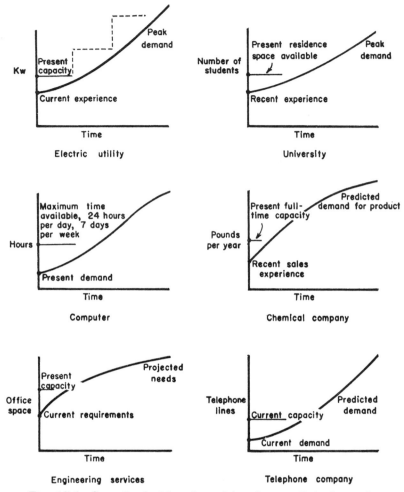

Fig. 15-1. Capacity decisions brought on by growth in demand.

start-up cost of removing paving, soil, and other obstacles. The size of pipe installed increases the cost of installation, but far less than proportionately. Alternatively, we might compare various sizes of air conditioners of a specific kind and note that the cost per Btu of capacity tends to be less for the large units than for small ones.

If the cost of installed extra capacity were directly proportional to size and if interest on money were more than zero, the economic package would be the smallest one available. If interest on money were zero and if the cost per unit of output capacity were less for larger packages, the economic package would be the largest one available. Since interest on money *is* a factor and since the cost per unit of output capacity generally *does* decrease for larger packages (due to economies of scale and start-up costs), our problem is one of balancing these considerations to determine an optimum size addition (economic package) or frequency of addition (economic interval).

In Example 15-1 we investigate the manner in which various parameters influence economic package or economic interval decisions. By the insights thus developed we hope to sharpen our intuitive understanding of the stimulus-response mechanisms involved.

Example 15-1. Economic Package / Economic Interval = Expansion Rate

a. *Information:* A long-range forecast for Green Corporation indicates that demand for a certain product or service will grow by 100,000 units per year for each of the next 12 years, then stabilize at 1,200,000 units per year thereafter for a very long time. Because no spare capacity is presently available, and because considerable time is required for the acquisition, installation, and construction related to providing the needed capacity, it is felt that a decision on and commitment to a specific addition size must be made within the next few months. Management feels that the addition size selected will not have an appreciable effect upon revenues, morale, quality of service, or other such factors, except that failure to add capacity will be catastrophic. It is estimated that no price increases will occur for capacity additions placed in future years and that no rapid or dramatic technological breakthroughs will occur.

Fractional addition packages or fractional addition intervals can be considered within the constraint of their availability. If, for example, construction is limited to the summer season, it may be that only integer year intervals can be considered. If the facility under consideration is a pipeline, we may be restricted (physically or economically) to standard pipe sizes available from manufacturers.

Note that "economies of scale" are present in both first cost and operating cost; that is, two packages of 200,000 units each will cost more to purchase and to operate than one package of 400,000 units. The life of any added facilities is estimated to be 25 years from date of installation; salvage at retirement is thought to

be negligible.   Cash operating cost (includes maintenance, property tax, and insurance) is estimated as 8.5%
of first cost annually.   The minimum attractive rate of
return = 12%; financing is 100% equity capital, straight-
line depreciation is employed, and the effective tax rate
on taxable income is 50%.   The costs of added capacity
have been estimated as

K Interval Between Additions, in Years	X Addition Package, in Units of Capacity	$100,000(K+1)$ First Cost of Each Package	$100,000(K+1)/X$ First Cost per Unit of Capacity	$8500(K+1)/X$ Operating Cost per Unit of Capacity
1	100,000	$200,000	$2.00	$0.170
2	200,000	300,000	1.50	0.128
3	300,000	400,000	1.33	0.113
4	400,000	500,000	1.25	0.106
6	600,000	700,000	1.17	0.099
12	1,200,000	1,300,000	1.08	0.092

Objective: Find the economic package size.

Analysis: For every dollar of first cost spent, the annual
equivalent cost flowing from that day forward is

$ATCFR = \$1\,(a/p)_{25}^{12\%}$ $= \$0.12750$
Income tax $= 1.00(\$0.12750 - \$0.04)$ $=\ \ 0.08750$
Operating cost $=\ \ \underline{0.08500}$
AEC per dollar of first cost $= AEC/B$ $= \$0.30000$

Let Z = the duration of linear growth in demand for the
product or service, in years.   In this example
$Z = 12$.
K = the interval between additions, in years.   In this
example it is mandatory that Z/K = an integer
greater than 0.
N = the duration of product or service need, in
years.   In this example $N \longrightarrow \infty$.
B = the first cost of each package.   Refer to the
estimates and note that $B = (K + 1)(\$100,000)$.
AEC/B = annual equivalent cost per dollar of first cost,
B.   Cost flows from the date of installation of
the original package to year N and is measured
in dollars per dollar of first cost.   In this example $N \longrightarrow \infty$, so the flow of costs is perpetual.
FEC_K = the future equivalent cost of adding capacity
each Kth year.
PEC_K = the present equivalent cost of adding capacity
each Kth year.

One of the difficulties which would be posed by a finite termination of need ($N < \infty$) is that units retained for less than a full life cycle (n years) may have greater annual equivalent cost than those retained for a full life cycle. There will be some instances in which this increase is trivial and still other instances in which the increased cost is so distant as to be negligible. These latter instances are reasonably common and have the advantage of being more readily formulated.

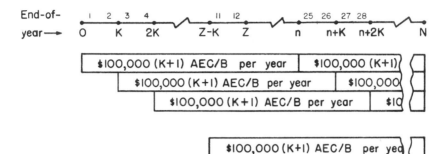

Providing that $Z \leqslant n$, the equivalent cost at year N of having placed an addition each Kth year is

$$FEC_K = \$100,000\,(K + 1)$$
$$\cdot AEC/B\,[\,(f/a)^i_N + (f/a)^i_{N-K} \cdots + (f/a)^i_{N-Z+K}\,]$$

So

$$PEC_K\,(1 + i)^N = \frac{\$100,000(K + 1)}{i}\,AEC/B\,[(1 + i)^N$$
$$- 1 + (1 + i)^{N-K} - 1 \cdots + (1 + i)^{N-Z+K} - 1]$$

Multiplying by $(1 + i)^{-N}$,

$$PEC_K = \frac{\$100,000(K + 1)}{i}\,AEC/B\left[1 + (1 + i)^{-K}\right.$$
$$\left. + (1 + i)^{-2K} + \cdots + (1 + i)^{-Z+K} - \frac{Z}{K}\,(1 + i)^{-N}\right]$$

Multiplying by $(1 + i)^{-K}$,

$$PEC_K\,(1 + i)^{-K} = \frac{\$100,000(K + 1)}{i}\,AEC/B\left[(1 + i)^{-K}\right.$$

$$+ (1 + i)^{-2K} \cdots + (1 + i)^{-Z+K}$$

$$\left. + (1 + i)^{-Z} - \frac{Z}{K}(1 + i)^{-N-K} \right]$$

Subtracting the latter equation from the former,

$$PEC_K = \frac{\$100,000(K+1)}{i\,[1 - (1+i)^{-K}]} AEC/B \left[1 - \frac{Z}{K}(1+i)^{-N} \right.$$

$$\left. - (1+i)^{-Z} + \frac{Z}{K}(1+i)^{-N-K} \right]$$

$$= \frac{\$100,000(K+1)}{i\,[1 - (1+i)^{-K}]} AEC/B \left\{ 1 - (1+i)^{-Z} \right.$$

$$\left. - \frac{Z}{K}(1+i)^{-N}[1 - (1+i)^{-K}] \right\}$$

$$= \frac{\$100,000(K+1)}{i} AEC/B \left[\frac{1 - (1+i)^{-Z}}{1 - (1+i)^{-K}} \right.$$

$$\left. - \frac{Z}{K(1+i)^N} \right] \qquad (15\text{-}1)^1$$

When $i > 0$ and $N \longrightarrow \infty$, $Z/[K(1+i)^N] = 0$, and

$$PEC_K = \$100,000(K+1)(AEC/B)(1/i)(p/a)^i_Z\,(a/p)^i_K$$

K Interval Between Additions, in Years	$15.485\,(a/p)^{12\%}_K$ Present Equivalent Cost per Dollar of First Cost of Each Package, $i = 12\%$	$\$100,000(K+1)$ First Cost of Each Package	PEC_K Present Equivalent Cost of Adding a Package Each Kth Year*
1	17.343	$ 200,000	$3,469,000
2	9.162	300,000	2,749,000
3	6.447	400,000	2,579,000
4	5.098	500,000	2,549,000
6	3.767	700,000	2,637,000
12	2.500	1,300,000	3,250,000

*Data rounded to the nearest thousand dollars.

---

[1] Equations 15-1 and 15-2 are applicable only where $Z \leqslant n$. Economic interval is only slightly shortened by relaxation of the restraints ($N \longrightarrow \infty$) ($Z < n$).

At $i = 12\%$:

$$PEC_K = \$100,000(K + 1)(2.5)(p/a)_{12}^{12\%}(a/p)_K^{12\%}$$
$$= \$100,000(K + 1)(15.485)(a/p)_K^{12\%} \quad (15\text{-}2)^1$$

We can now observe that economic interval is about 4 years, so economic package is about 400,000 units.

b. As an alternate solution we could differentiate Equation 15-1 with respect to K, set equal to zero, then solve for K. (The differentiation is left as an optional exercise for the student, Problem 15-7.) Under the restraints that $Z > 0 < K < \infty$, $N \longrightarrow \infty$ the results can be simplified as

$$0 = (1 + i)^K - 1 - (K + 1) \ln (1 + i)$$

When $i = 12\%$, optimal $K = 3.9$ years; at $K = 3.9$, $PEC_K = \$2,548,000$.

c. PEC_K values at $MARR = 12\%$ have been plotted in Figure 15-2. PEC_K values at other $MARR$ values can be determined from the equation

$$PEC_K = \$100,000(K + 1)[2(a/p)_{25}^i + 0.045]$$
$$\cdot (1/i)(p/a)_{12}^i (a/p)_K^i$$

Results can then be plotted. By connecting the minimum cost point of each PEC_K curve, we can determine a "marginal investment opportunity curve" which can be used in either of two ways:

1. Given $MARR = 8\%$, we can find that the economic package size is about 500,000 units.
2. Given the decision that a package of 300,000 units will be used, we can find that the after-tax rate of return foregone on the marginal investment in a larger package is about 20%.

Notice that these determinations do not require knowledge of the revenue level; they do require knowledge of the effect, if any, of package size on revenue. In this example, "addition size selected will not have an appreciable effect upon revenue . . . ."

We could also compare various available addition package alternatives to find the $MARR$ value at which they break even. When a 100,000 (K_1) and 100,000 (K_2) package break even, $PEC_{K1} - PEC_{K2} = 0$:

$$\$100{,}000(K_1 + 1)[2(a/p)^i_{25} + 0.045](1/i)(p/a)^i_{12}\,(a/p)^i_{K1}$$
$$- \$100{,}000(K_2 + 1)[2(a/p)^i_{25} + 0.045]$$
$$\cdot (1/i)(p/a)^i_{12}\,(a/p)^i_{K2} = 0$$

After canceling the many common terms,

$$(K_1 + 1)(a/p)^i_{K1} - (K_2 + 1)(a/p)^i_{K2} = 0$$

Next, substitute pairs of K_1 and K_2 values, then solve for i, using trial-and-error procedures. Compare the following tabulation of such results with the break-even *MARR* which could have been approximated by inspection of Figure 15-2.

Range of *MARR*	Economic Package Size	K = Economic Interval, in Years
Over 100.0%	100,000 units	1
30.3%–100.0%	200,000 units	2
15.1%– 30.3%	300,000 units	3
7.6%– 15.1%	400,000 units	4
2.6%– 7.6%	600,000 units	6
Under– 2.6%	1,200,000 units	12

d. "Economies of scale" or lack thereof play an important role in determining economic package or economic interval. Significant economies of scale in first cost can result (1) when installation of buried pipe or cable involves obstructions such as earth or concrete which must be removed each time there is an addition; (2) when disruption costs are large as with a highway rerouting or system outage; or (3) when planning, handling, and set-up costs are large relative to the "run" costs.

Economies of scale in the operating cost of a facility may result because an operator is required regardless of unit size or because small units are less efficient due to more resistance to flow, greater fuel consumption per unit of output, or similar reasons. When originally installed, the large unit presents capacity in excess of

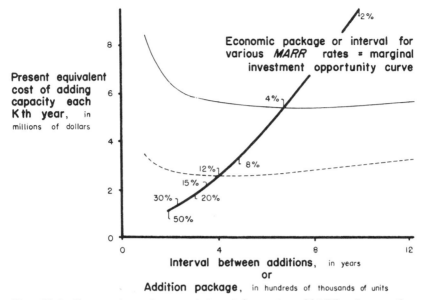

Fig. 15-2. Economic package or interval for various *MARR* values as the basis for a marginal investment opportunity curve.

needs, thus operating costs take some time to rise to their full potential.

The relationship of economies of scale to economic interval or economic package may be clarified by re-solving the original problem for first costs of $100,000 times $(K + 0)$ or $(K + 3)$ or $(K + 6)$ or $(K + 10)$ using

$$PEC_K = \$100,000(K + x)(15.485)(a/p)_K^{12\%}$$

By such calculations we can determine data from which Figure 15-3 can be plotted. Note that economic package size increases as economies of scale increase.

e. Economic package and economic interval are influenced by the estimated expansion rate. To trace this influence and to gain insights into the effect of unduly optimistic or pessimistic estimates of the expansion rate, we can repeat our computations for various expansion rates. Figures 15-4 and 15-5 are based on the equation

$$PEC_K = \left[\$1 \left(\begin{array}{c} \text{expansion rate, in} \\ \text{units per year} \end{array} \right) (K) + \$100,000 \right] \cdot (15.485)(a/p)_K^{12\%}$$

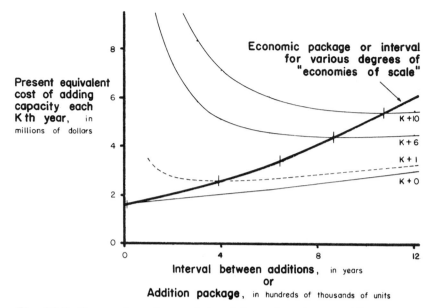

Fig. 15-3. Economic package or interval for various degrees of "economies of scale."

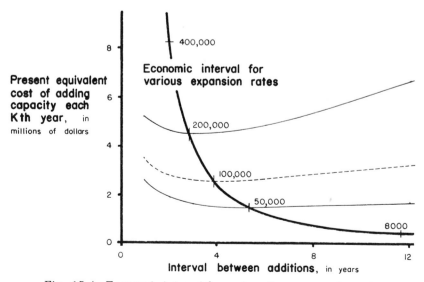

Fig. 15-4. Economic interval for various linear expansion rates.

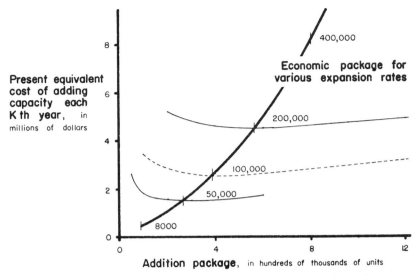

Fig. 15-5. Economic package for various linear expansion rates.

Some Cautions in Application of Economic Package Concepts to Expansion Needs

Certain features have been assumed in Example 15-1 which may be less than realistic. A review of these may help to avoid identical application of this approach to problems for which different conditions exist:

1. Growth in demand was assumed to be relatively uniform with time. It should be obvious that, for the opposite extreme where growth in demand is in large discrete "steps," *there may not be a problem of economic package;* the growth package in such cases may simply be the addition required to satisfy each large change (step) in the demand curve.
2. Where significant doubt (uncertainty) exists as to the likelihood of assumed future growth in demand, the analyst must also study the costs of potential "overbuilding" (idle capacity) and "underbuilding" (lost sales, decreased market share, and so forth) and make probabilistic estimates of future sales demand levels.
3. Sometimes certain intervals may become impractical, as when growth will soon force relocation of facilities. Frequently, when relocation is inevitable, adding any growth facilities at the present location is impractical.

4. Sometimes the timing of provision for growth may not be entirely discretionary. If a street is soon to be rebuilt, the timing of added sewer, electrical, water, gas, and other facilities will certainly be influenced. With prospective changes in municipal limitations on such factors as height, width, and depth of buildings, a contemplated addition may be relegated to a now-or-never status.

5. Technological change was assumed to be zero. The possibility of improved productive units would encourage smaller additions to capacity.

6. No change in price levels was anticipated; if the cost of added capacity increases while other factors remain constant, the firm should make relatively larger additions to capacity.

7. No allowance for the risk of product obsolescence has been included; this risk, if nonzero, would encourage smaller additions to capacity.

8. The action or probable action of competitors may be a factor.

9. In the example, product demand was assumed to continue "for an extended period." Where the period of product demand is shorter, it may be better to estimate a date at which the production will be terminated and the salvage values of productive units at that time.

10. Still other solutions may exist for the firm as it attempts to meet a growing demand. The electric utility not only has a choice of economic package but also may be able to purchase power for resale so that demand can actually exceed the company's generating capability. This is especially true with the interconnection of so many of our electric utilities.

Example 15-2.   Complete Provision of Future Needs versus a Two-Stage Program

a. *Information:* All future demands for a certain service can be met by Plan A which calls for installation of Unit A with a first cost of $300,000 and annual operating costs of $5000. Alternatively, this same demand can be met by Plan B which calls for installation of one Unit B now and one Unit B ten years from now, each having a first cost of $200,000 and annual operating costs of $3000. It is estimated that the original units will serve until abandonment 40 years from now when negligible salvage will be realized for either A or B. $MARR = 5\%$ and, because the organization is a municipality, income taxes do not apply (these services provided by the city are not commercially available and do not, therefore, compete with privately produced goods or services).

Objective: Determine which plan is the more economic. Compare the present worth of costs under either plan.

Analysis:

$PEC_A = \$300,000 + \$5,000(p/a)_{40}^{5\%} = \$385,795$

$PEC_B = \$200,000 + \$3,000(p/a)_{40}^{5\%} + \$200,000(p/f)_{10}^{5\%}$
$$+ \$3,000[(p/a)_{40}^{5\%} - (p/a)_{10}^{5\%}]$$
$$= \$200,000 + \$51,477 + \$122,780 + \$28,311$$
$$= \$402,568$$

b. *Information:* Same as Part a except assume a 25% increase in first cost and operating costs of the deferred Unit B has been forecast.

Objective: Find the effect an anticipated price increase has on a comparison of a present versus a deferred installation.

Analysis:

$PEC_A = \$385,795$

and

$PEC_B = \$200,000 + \$51,477 + \$250,000(p/f)_{10}^{5\%}$
$$+ \$3,750[(p/a)_{40}^{5\%} - (p/a)_{10}^{5\%}]$$
$$= \$440,341$$

Anticipated future price increases enhance the desirability of Alternative A.

c. *Information:* Same as Part a except assume that $MARR$ is unknown.

Objective: Find several combinations of $MARR$ and deferral periods for which the alternatives break even.

Analysis: Let x = period of deferral; then at some combinations of i and x values,

$$PEC_A - PEC_B = 0 = \$300,000 + \$5000(p/a)_{40}^i$$
$$- \$200,000 - \$3000(p/a)_{40}^i$$
$$- \$200,000(p/f)_x^i$$
$$- \$3000[(p/a)_{40}^i - (p/a)_x^i]$$
$$= \$100,000 + \$3000(p/a)_x^i$$
$$- \$200,000(p/f)_x^i - \$1000(p/a)_{40}^i$$

By substituting various values of i and solving for x, we can determine the marginal investment opportunity curve of Figure 15-6. Observe that an increase in *MARR* relatively inclines us toward Alternative B which involves the deferred investment.

d. *Information:* Same as Part a except the probability that the second B unit will be needed at the end of year ten is 0.8 and there is a 0.2 probability that the second B unit will never be needed.

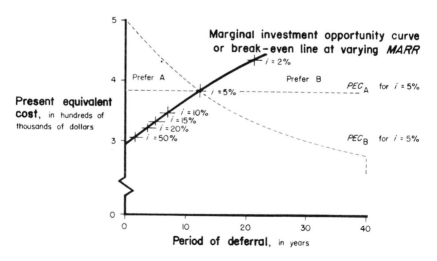

Given *MARR* = 5%, then the minimum deferral period to justify B is 12+ years

Given a deferral period of 10 years, then B is preferable to A if *MARR* ≤ 7+%

Given a deferral period of 5 years and that B has been selected, then the rate of return foregone on the incremental investment A–B ≅ 14%

Fig. 15-6. Break-even deferral period for various *MARR* values as the basis for a marginal investment opportunity curve in Example 15-2c.

Objective: Find the preferred alternative, assuming $MARR = 5\%$.

Analysis: In Part a it was determined that

$$PEC_A = \$385,795$$
$$PEC_{B1} = \$251,477$$
$$PEC_{B12} = \$402,568$$

where PEC_{B1} includes only the costs related to the first unit, while PEC_{B12} includes costs related to both the B units. Then

$$E(PEC_A) = (0.8 + 0.2)(\$385,795) = \$385,795$$
$$E(PEC_B) = 0.2(\$251,477) + 0.8(\$402,568) = \$372,350$$

So B is preferable to A. In Figure 15-7 a decision tree is used to emphasize the choice as a *sequential decision.*

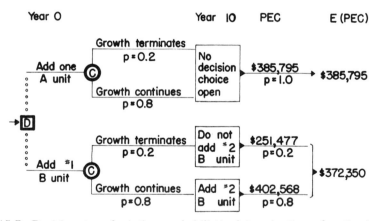

Fig. 15-7. Decision tree depicting probabilistic determination of optimal addition package size.

Problems

15-1. Two alternative plans are being considered. Plan C involves provision of a complete facility now, while Plan D provides for part of the facility now and the remaining part of it at a later date. Which alternative will tend to be favored by:

a. Inclusion of an anticipated increase in income tax rate beginning next year.

b. Use of a very low minimum attractive rate of return.

c. The decision maker's tendency to be conservative in "committing" the firm to long-term prospects.

d. Inclusion of anticipated price increases in the cost of the facility or its parts.

e. Inclusion of operating costs which are large relative to first cost and proportional to facility capacity rather than facility usage.

f. The supposition that rapid and dramatic technological breakthroughs will take place in the near future.

g. A limited supply of funds available for capital expenditures.

h. Economies of scale, as when one large unit costs less than two half-size units.

i. The possibility that actual growth in needed service will be considerably less than the estimated but uncertain growth rate.

j. High "disruption" costs, as when highway improvements require detours during the construction period or in the case of buried pipe or cable.

15-2. A certain service must be provided over each of the next 30 years; that need can be met by installing one full-size unit now (Alternative F) or by installing one half-size unit now and a second half-size unit several years hence (Alternative H). An H unit costs 70% as much as an F unit; that is, first cost, salvage, operating cost, and other cash flows for one H unit are 70% of what they are for an F unit. All replacement units have costs identical to those of the units they replace. Because of these special characteristics the ratio of (AEC for one H unit and its replacements) to (AEC for the F unit and its replacements) is 0.7 and is independent of $MARR$. Revenues from the two alternatives are likely to be identical.

a. Using $MARR$ = 5%, find the deferral period at which the two plans break even.

b. Using $MARR$ = 30%, find the deferral period at which the two plans break even.

15-3. a. Compare the first cost of four different sizes of window air-conditioning units. Make the units as comparable as possible in all respects except cost and capacity.

b. Repeat Part a for outboard motors.

c. Repeat Part a for earth-moving equipment such as a "cat-skinner."

15-4. D. Purple, Inc., notes that their growth in demand tends to surge upward every two or three years, then level out, then surge again. Does this observation have any bearing on economic interval problems for the company? Explain.

15-5. Refer to Example 15-1, then respond to the following questions.

    a. How sensitive is economic package to errors in the estimation of growth rate?

    b. The analysis indicated an economic interval of four years. Consider the possibility of error in estimation of the growth rate and the "commitment" of the firm to the plan for adding capacity. Is it a 12-year commitment? Why?

    c. Are errors of underestimating or overestimating the size of economic package of like impact when such data apply? Why?

15-6. A community's growing needs for a certain *perpetually* required service can be met by installing one full-size unit now (Alternative F) or by installing one half-size unit now and a second-half-size unit several years hence (Alternative H). An F unit costs 60% more than one H unit; that is, first cost is 60% more and so are salvage and all operating costs. All costs for replacement units will be identical to those for the units they replace. Because of these special characteristics the ratio of AEC for the F unit and its replacements to AEC for one H unit and its replacements is 1.6 and is independent of $MARR$. Revenues from the two alternatives are likely to be identical.

    a. Find the minimum number of years of deferral (years between now and when the second H unit is installed) for Alternative H to be equal or preferable to Alternative F if $MARR$ = 8%.

    b. Find the rate of return foregone on the marginal investment opportunity if an H unit is installed now even though the second H unit will be needed just two years from now. (That is, find the $MARR$ value at which the alternatives break even.)

    c. Suppose the probability that the second unit will be needed at the end of year six is 0.8 and there is a 0.2 probability that the second H unit will never be needed. If the second H unit proves unnecessary, the original F unit will be replaced by a series of H units at the end of its 25-year life. Find the preferred alternative, assuming $MARR$ = 8%.

15-7. Refer to Equation 15-1, differentiate with respect to K to find optimal economic interval and PEC_K for that interval.

15-8. See Example 15-2c. Assume that installation of the second B unit can be deferred five years.

a. Draw a cash flow diagram for incremental investment A - B.

b. Find the rate of return on the incremental investment A - B. Relate your result to Figure 15-6.

15-9. Refer to Example 15-1. Assume $MARR = 8\%$. Use the equation given in Part b to find, to the nearest year, the interval between additions which would minimize PEC. Check your result against Figure 15-2.

15-10. Refer to Example 15-1. Assume $MARR = 8\%$. Use the equation given in Part c to find PEC_5. Check your result against Figure 15-2.

16

Commitment and
Investment Irreversibility

Commitment: The Risk of Taking Action

IF WE COMMIT ourselves to the action of purchasing a new car, we may discover that the apparent value of the car decreases by $500 the moment it is driven from the dealer's showroom. Since we expect to retain the car for quite some time (perhaps six years), recognize that this same thing will happen to all others who purchase new cars, and consider that this economic phenomenon, though interesting, is something over which we have little control, we might be inclined to dismiss consideration of the economic consequences of actions involving commitments. Yet if something should happen (risk) to our source of income, cash required to meet expenses, family size, drivers license, or ability to drive, we might be forced to change our plans and to give up this new car, and to do so at some sacrifice (cost) compared to our original intention. Commitment thus involves *risk* and that risk may have an identifiable cost. The differential between value prior to ownership and after acquisition causes us to act in a manner which might be likened to the concept of inertia; when a portion of initial cost (as with cost of installation) cannot be recouped even through immediate resale, we have, to an extent, become captively committed to the property and tend to be something less than a free agent in the market. The degree of commitment varies with the type of property but in any case commitment involves risk; it is specifically treated later in this chapter.

Examples involving small commitment include automobiles (especially the more popular models) and general purpose equipment (such as an electric drill); while somewhat greater commitment generally prevails for highly specialized equipment, perhaps custom equipment unsuited to the needs of others. Even life insurance or mutual fund investments involve some commitment; one should note *in advance of purchase* that cancellation costs make eventual withdrawal from the plan a risk which has a cost. In much this same manner a realtor may advise a prospective landlord not to furnish a rental property; the completely furnished apartment presents much less of a barrier (inertia) to moving than does the unfurnished apartment where the rational occupant considers the cost, trouble, time, and possible damage to furniture in moving from one location to another. Even the government "E" series savings bond involves a small degree of commitment; if resold prior to maturity date, a slightly lower rate of return is realized.

Examples of substantial investment commitments whose reversal was later considered or acted upon include:

1. The development of a supersonic transport (SST) plane. After more than $800 million of the anticipated developmental costs were spent, Congress refused to authorize the remaining funds. Environmental considerations were important factors in this reversal of commitment.
2. The Florida barge canal. The 107-mile canal was authorized by Congress in 1962 with the hope of reducing the cost of shipping by barge. In the following nine years about 26 miles of the canal was dug, and about $50 million was spent on or committed to various phases of construction. Total cost of the completed project was estimated to be $180 million. On January 19, 1971, President Nixon halted construction to prevent potentially serious environmental damages. He noted that in the earlier studies "the destruction of natural ecological values was not counted as a cost, nor was a credit allowed for actions preserving the environment. The step I have taken today will prevent a past mistake from causing permanent damage."

3. The Pruitt-Igoe federal public housing project in St. Louis. Built in 1955 at a cost of $36 million, it was to house 2800 families. Termed by some a "high-rise hell," its soaring vacancy rates, maintenance costs, and social failures led to proposals to demolish it at a cost of more than $3 million.
4. In January 1971, United Air Lines cancelled eight orders and 15 options for new jumbo McDonnell-Douglas DC-10 airbuses and postponed delivery of four Boeing 747's by 20 months. Neither McDonnell nor United would comment on the amount of the penalty payment involved.

The cost of "feasibility" or "evaluation" studies is often small, but not always; the evaluation study for the pipeline which would bring crude oil from Prudhoe Bay in northern Alaska to the all-weather port of Valdez cost more than $3 million. So did a similar study of a project to bring natural gas by pipeline from Prudhoe Bay to Manitoba, Canada. In each case the evaluation costs are dwarfed by the project costs which will probably exceed $5 billion.

When cost and risk elements of commitment are being estimated, the ease with which the property can be converted to cash (liquidity) should be considered. Some examples of varying liquidity include:

Liquidity	Example
Most liquid	cash
↑	bonds, stocks
	inventory
	vehicles
	land
↓	buildings
Least liquid	special purpose assets

Ease of conversion may be indicated both by the time required to dispose of the property and by the number of prospective buyers.

Finally then, the reader is asked to recognize commitment as the *prospect* of facing a sunk-cost situation later. Prospective sunk cost (commitment) involves risk, and probabilistic estimates of possible cost consequences are advised.

Multistage Commitment to an Imperfectly Reversible Investment

The time lag between commitment to buy and delivery of new plant and equipment tends to be greater for buildings than for equipment and tends to be increased when rapid expansion of capital expenditure levels builds up an unusual backlog of orders. Some indicators of the time lag are noted by MAPI:[1]

1. The Treasury estimated that, of the equipment eligible for investment credit, 40% has an order-to-delivery period of less than six months, 40% between six months and a year, and 20% over a year (the average for the last group being about two years). Buildings and structures are not included in the preceding.
2. Surveys by the Department of Commerce indicate an average interval between commitment and payment of eight months for manufacturing and thirteen months for public utilities.
3. National Industrial Conference Board (NICB) surveys show an average of nine to ten months between appropriations and expenditures in manufacturing.

The recognition that decisions deal with future courses of action and therefore the cost consequences of funds already committed can no longer be avoided is generally known as the sunk-cost concept. Commitment to an investment may occur in several stages. Consider the following *sequential decisions* which result in an increasing degree of commitment:

Date or Year*	Stage of Commitment to Equipment	Actual Expenditure	Cumulative Expenditure	Salvage at That Date
−2.00	search and evaluate	$ 1,650	$ 1,650	$ 0
−1.75	guarantee order	5,080†	6,730	0
−1.00	transport to site	3,640	10,370	0
−0.75	inspect and accept	65,170	75,540	40,960
−0.50	install	9,530	85,070	34,320
−0.25	debug and train	7,830	92,900	39,060

*These dates are referenced to a time, zero, which is arbitrarily defined here as the date when use of the equipment begins.

†This sum is "earnest money" which accompanies order. It is forfeited if the would-be purchaser cancels the order; it is thus irretrievable.

---

[1] See *Capital Goods Review,* Machinery and Allied Products Institute, 1200 Eighteenth Street N.W., Washington, D.C., No. 68, Dec. 1966, p. 4.

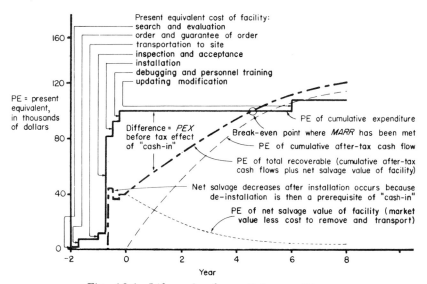

Fig. 16-1. Life cycle of a capital expenditure.

The increasing commitment and other data are illustrated in Figure 16-1. Figures 16-1, 16-2, 16-3, 16-5, and 16-6 are based on data which appear in Table 16-1, which precedes the problem section at the end of this chapter; data are deferred in the hope that readers will gain concepts and insights without being deterred by the detail from which such arise.

In Figure 16-1 the "present equivalent of cumulative expenditure" exceeds the "present equivalent of total recoverable" during the early years of equipment life. This illustrates both the *imperfect reversibility* of the investment and the investor's *exposure to risk.* Imperfect reversibility suggests that once committed to this equipment, an owner can escape from ownership during the early years only at some sacrifice. The investment illustrated is only moderately irreversible; much greater irreversibility exists for such investments as buried pipe or poured concrete.

Commitment: Its Impact upon B, the Basis for Computing After-Tax Cash Flow Required

Commencing with the initial investment commitment, an owner generally experiences some sunk cost. The then-relevant basis, B, for computing the after-tax cash flow requirement in a replacement study is the sum of (1) the remaining investment required (for acquisition, transportation, installation, debugging,

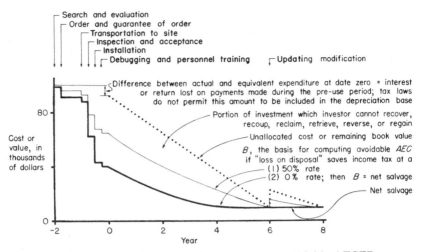

Fig. 16-2. Basis of property for computing avoidable ATCFR.

and so forth) to make the equipment operational; (2) the net salvage (from gross proceeds less cost of removal, restoration, transportation, and so forth) foregone by not relinquishing ownership now; and (3) the income tax adjustment for gain or loss on disposal foregone by not relinquishing ownership now. Prior to any investment commitment (such as the earnest money of an order guarantee) items (2) and (3) are zero; this was the type of problem treated in preceding chapters. After the equipment is in operation, item (1) is generally zero. Figure

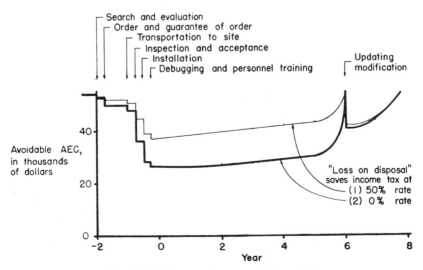

Fig. 16-3. Time-variant avoidable *AEC.*

16-2 displays the time-variant nature of B, the basis for computing the after-tax cash flow requirement.

Sharp decreases in B lead to sharp decreases in ATCFR and emphasize that with each new stage of commitment the manager has relinquished a degree of control over annual costs related to the equipment. *Avoidable AEC* is thus time variant as portrayed in Figure 16-3. The decrease in avoidable *AEC* means not only decreased control but also means that in the early years of equipment use only drastically improved equipment can "prove in" and displace the present equipment. This effect is similar to that of inertia; once committed to an equipment, its displacement is not a frictionless process.

Example 16-1. **Inertia-Like Aspects of Commitment: An Imperfectly Reversible Investment**

a. *Information:* At year - 2.00 the Brianwade Corporation decides to consider equipment alternatives for a certain operation. No commitment has yet been made. $MARR = 10\%$, $t = 50\%$, and $r = 0\%$. Revenues will be the same for the various equipment alternatives. Data appropriate to Equipment A and related to that given earlier and in Table 16-1 are:

Present (year zero) equivalent investment required = $100,000 (the actual cash outlay = $92,900)
Estimated salvage = $10,000 at end of year six
Depreciation is by the SL method; tax-allowable depreciation is based on actual cash outlay, so $D = (\$92,900 - \$10,000)/6 = \$13,817$ per year
Operating costs are $10,000 during year one and increase by $6000 per year; this flow is continuous rather than end-of-year

Objective: Find avoidable *AEC* at year - 2.00 for Equipment A.

Analysis:

$ATCFR = \$100,000(a/p)_6^{10\%} - \$10,000(a/f)_6^{10\%}$
$$= \$21,665$$
Income tax $= 1.00(\$21,665 - \$13,817)$ $\quad = \quad 7,848$
Operating costs $= \$10,000(a/a')^{10\%}$
$\qquad\qquad + \$6000(g/g')^{10\%}(a/g)_6^{10\%} = \underline{\quad 24,487}$
Avoidable *AEC* at year - 2.00 $\qquad\qquad = \$54,000$

b. *Information:* It is now year - 0.75; Equipment A has just been received, inspected, accepted, and paid for. Equipment B with AEC of $49,000 is discovered. Other data appropriate to Equipment A and related to that given earlier and in Table 16-1 are:

Present (year zero) equivalent of additional investment still required to pay for installation, debugging, and training

$$= \$9530(1.10)^{0.50} + \$7830(1.10)^{0.25} = \$18,000$$

Present (year zero) equivalent of net salvage from relinquishing ownership today

$$= \$40,960(1.10)^{0.75} = \$44,000$$

Present (year zero) equivalent of income tax impact from loss on disposal, assuming a one-year lag

$$= \left[\binom{\text{actual expend-}}{\text{iture to date}} - \binom{\text{actual}}{\text{proceeds}} \right] (50\%)(1.10)^{-0.25}$$

$$= (\$75,540 - \$40,960)(50\%)(0.97646) = \$16,880$$

Objective: Determine whether or not Equipment A should be replaced by Equipment B.

Analysis: Speaking in present (year zero) equivalent sums, $44,000 of the $82,000 invested to date can be recouped through immediate resale, leaving a $38,000 difference. The lagged income tax effect from loss on disposal saves almost half of that difference; only $38,000 - $16,880 = $21,120 is beyond managerial control. The portion over which control can still be exercised is $100,000 - $21,120 = $78,880. At year - 0.75 management still retains control over (1) whether the remaining investment sums for installation, debugging, and training ($18,000) will be spent; (2) whether the present salvage ($44,000) will be foregone via continued ownership; and (3) whether the tax impact ($16,880) from loss on disposal will be foregone via continued ownership. These three elements are added together to form the "cash alternative" basis of $78,880. Then

$$ATCFR = \$78,880(a/p)_6^{10\%} - \$10,000(a/f)_6^{10\%} = \$16,816$$

Income tax $= 1.00(\$16,816 - \$13,817)$	$= \quad 2,999$
Operating costs $=$ same as in Part a	$= \quad 24,487$
Avoidable AEC_A at year - 0.75	$= \$44,302$

So Equipment B should be rejected. See Figure 16-4.

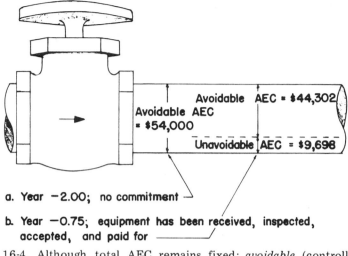

a. Year −2.00; no commitment

b. Year −0.75; equipment has been received, inspected, accepted, and paid for

Fig. 16-4. Although total AEC remains fixed; *avoidable* (controllable) AEC shrinks with each new step of commitment.

Comments Related to Example 16-1.

Example 16-1 emphasizes the inertia-like aspects of commitment. So long as we are uncommitted, equipment with $AEC <$ \$54,000 is preferable to A. After committing ourselves to ordering, transporting, receiving, inspecting, accepting, and paying for A, we find that the challenging equipment must be still more attractive ($AEC <$ \$44,302) to displace our original choice. Note however, that Equipment A must still produce annual equivalent revenue (as in a product pricing decision) of \$54,000 if it is to produce the stipulated $MARR$.

Our example also raises another issue; the basis, B, used for cash flow purposes does not always coincide with the basis used for income tax depreciation purposes. For a new or newly acquired property the basis, B, tends to be the same for both $ATCFR$ and AED computations.[2] For a property already in service it is likely that present net salvage measured by *market* value B_v is not identical to that measured by remaining *book* value B_t. For a property already in service, subject to straight-line depreciation and item accounting, and with a remaining life of n years,

---

[2] Exceptions exist. The Internal Revenue Service does not allow return foregone on payments made during the preuse period (interest lost during construction) to be expensed or capitalized by inclusion in the property basis used for computing depreciation expense.

$$BTCFR = \left[\left(\begin{array}{c}\text{present net (market)}\\\text{salvage plus income tax}\\\text{effect from disposal}\end{array}\right)(a/p)_n^i - \left(\begin{array}{c}\text{future}\\\text{net}\\\text{salvage}\end{array}\right)(a/f)_n^i\right.$$

$$\left.- t\left(\begin{array}{cc}\text{present} & \text{future}\\\text{net (book)} - & \text{net}\\\text{salvage} & \text{salvage}\end{array}\right)\Big/n\right]\Big/(1-t)$$

$$= [B_v(a/p)_n^i - V(a/f)_n^i - (B_t - V)(t/n)]/(1-t)$$

Risk of Probabilistic Investment "Cash-in"

The difference between the "present equivalent of total recoverable" and the "present equivalent cumulative expenditure" of Figure 16-1 is equal to *PEX*. Figure 16-5 displays *PEX* as a time-variant function and emphasizes the imperfect reversibility of the investment. It also shows us the sunk cost concept from a fresh point of view as we peer *forward* to that risk rather than *backward on that irretrievable loss.*

We tend to be "locked into" past investments; they may be impossible to fully reclaim, recoup, recover, regain, retrieve, or reverse. The risk of premature "cash-in" should be evaluated

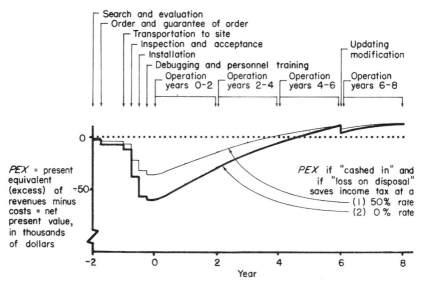

Fig. 16-5. Multivalued representation of *PEX*.

before we have committed ourselves to a potential investment. We can evaluate the potential investment by estimating the time-variant probability of "cash-in," then multiplying by the time-variant *PEX* to produce an expected value of *PEX*, which could be formulated for the continuous case as:

$$E(PEX) = \int_{\substack{\text{earliest} \\ \text{commitment time}}}^{\substack{\text{latest} \\ \text{"cash-in" time}}} \left(\begin{array}{c} PEX \text{ as a function} \\ \text{of time} \end{array} \right) \cdot \left(\begin{array}{c} \text{probability of "cash-in"} \\ \text{as a function of time} \end{array} \right) (\text{d time})$$

This is shown in Figure 16-6 which includes a *decision tree* illustrating the *sequential decision* nature of commitment.

In lieu of a decision based on expected *PEX*, we could employ the utility of money concept discussed in Chapter 13 by replacing each *PEX* value with the associated utility index, then multiplying by the probability of each outcome. One would then select the alternative with maximum expected utility.

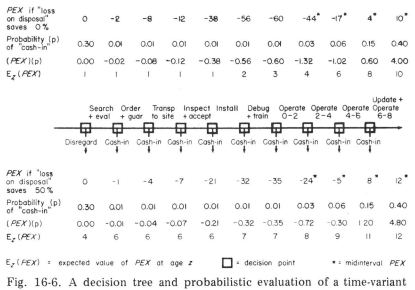

PEX if "loss on disposal" saves 0%	0	-2	-8	-12	-38	-56	-60	-44*	-17*	4*	10*
Probability (p) of "cash-in"	0.30	0.01	0.01	0.01	0.01	0.01	0.01	0.03	0.06	0.15	0.40
(*PEX*)(p)	0.00	-0.02	-0.08	-0.12	-0.38	-0.56	-0.60	-1.32	-1.02	0.60	4.00
E_z(*PEX*)	1	1	1	1	1	2	3	4	6	8	10

| | Search + eval | Order + guar | Transp to site | Inspect + accept | Install | Debug + train | Operate 0-2 | Operate 2-4 | Operate 4-6 | Update + Operate 6-8 |

PEX if "loss on disposal" saves 50%	0	-1	-4	-7	-21	-32	-35	-24*	-5*	8*	12*
Probability (p) of "cash-in"	0.30	0.01	0.01	0.01	0.01	0.01	0.01	0.03	0.06	0.15	0.40
(*PEX*)(p)	0.00	-0.01	-0.04	-0.07	-0.21	-0.32	-0.35	-0.72	-0.30	1.20	4.80
E_z(*PEX*)	4	6	6	6	6	7	7	8	9	11	12

E_z(*PEX*) = expected value of *PEX* at age *z*   ☐ = decision point   * = midinterval *PEX*

Fig. 16-6. A decision tree and probabilistic evaluation of a time-variant *PEX*.

TABLE 16-1. Data Supporting Figures 16-1 through 16-6

		Date	1-1-88	4-1-88	1-1-89	4-1-89	7-1-89
z is given		Date Label	-2.00	-1.75	-1.00	-0.75	-0.50
$A = 1.10^{-z}$		$(p/f)_z^{10\%}$	1.210	1.181	1.100	1.074	1.049

		Stage of commitment to investment	Search and evaluate	Guarantee order	Transport to site	Inspect and accept	Install
C is given $D = C(A)$ if $z \leqslant 0$ $D = C(A)(f/f')^{10\%}$ if $z > 0$	Expenditure (-) or ATCF (+)	Actual	-1.65	-5.08	-3.64	-65.17	-9.53
		PE_0	-2	-6	-4	-70	-10
$E_j = \Sigma_{-2}^{z} D_j$		Cumul PE_0	-2	-8	-12	-82	-92
F is given	Salvage	Actual	0	0	0	40.96	34.32
$G = F(A)$		PE_0	0	0	0	44	36
$H = E + G$	"Cash-in" PEX_0; no tax impact		-2	-8	-12	-38	-56
$I = F$ if $z > 0$ $J = 100 + H$ if $z \leqslant 0$	B_v on date shown; no tax impact	Actual					
		PE_0	98	92	88	62	44
$K = \Sigma_{-2}^{z} C_j$ if $z \leqslant 0$ $K = 92.90 - 13.82\,z$ if $0 < z \leqslant 6$ $K = 22.4 - 6.2\,(z - 6)$ if $6 < z \leqslant 8$	Unallocated cost		1.65	6.73	10.37	75.54	85.07
$L = (K - F)/2$	Tax impact of "loss on disposal"	Loss/2	0.82	3.36	5.18	17.29	25.37
$M = L\,(p/f)_{z+1}$		PE_0, 1 yr lag	0.90	3.61	5.18	16.88	24.19
$N = H + M$	"Cash-in" PEX_0 with tax impact		-1.10	-4.39	-6.82	-21.12	-31.81

$O = 1 + L/1.1$
   if $z > 0$
$P = J + M$
   if $z \leqslant 0$
	B_v on date
	shown with
	tax impact
		Actual					

		PE_0	98.90	95.61	93.18	78.88	68.19
Q	Avoidable AEC using B_v from lines _____	I or J	53.08	50.33	48.49	36.55	28.28
R		O or P	53.50	51.99	50.87	44.30	39.39

* = ATCF receipts flow continuously over the year ending at the date shown.
PE_0 = present equivalent sum at date zero.
Data in columns C through R is in thousands of dollars. $i = 10\%$, $t = 50\%$, $r = 0\%$
B_v is the investment base as used in Example 16-1 and elsewhere in the text.
Update is required to extend facility life through years 7 and 8.

TABLE 16-1. (Continued)

10-1-89	1-1-90	1-1-91	1-1-92	1-1-93	1-1-94	1-1-95	1-1-96	1-1-96	1-1-97	1-1-98
-0.25	0	1	2	3	4	5	6	6	7	8
1.024	1.0000	0.9091	0.8264	0.7513	0.6830	0.6209	0.5645	0.5645	0.5132	0.4665
Debug and train	Operate	Operate	Operate	Operate	Operate	Operate	Update	Operate	Operate	
-7.83	30*	27*	24*	21*	18*	15*	-12.40		10*	7*
-8	28.61	23.41	18.92	15.05	11.72	8.88	7		5.38	3.43
-100	-71.39	-47.98	-29.06	-14.01	-2.29	+6.59	-0.41		+4.97	+8.40
39.06	30	22	16	12	10	10	10		10	10
40	27.27	18.18	12.02	8.20	6.21	5.64	5.64		5.13	4.66
-60	-44.12	-29.80	-17.04	-5.81	+3.92	+12.23	+5.73		+10.10	+13.06
	30	22	16	12	10	10	10		10	10
40										
92.90	79.08	65.26	51.44	37.62	23.81	10.00	22.40		16.20	10.00
26.92	24.54	21.63	17.72	12.81	6.90	0	6.20		3.10	0
25.06	20.28	16.25	12.10	7.95	3.90	0	3.18		1.45	0
-34.94	-23.84	-13.55	-4.94	+2.14	+7.82	+12.23	+8.91		+11.55	+13.06
	52.31	41.67	32.11	23.65	16.27	10.00	15.64		12.82	10.00
65.06										
26.45	26.91	27.52	27.78	29.15	30.14	55.35	41.06		44.20	
37.96	38.68	39.94	40.73	42.57	43.94	55.35	42.18		44.76	

Expenditures, salvage, and other flows are discrete.

All expenditures are capitalized; SL depreciation method is used.

Loss on disposal impact is not diminished by long-term capital gains elsewhere in the business, nor is it deferred by a transaction which causes the loss to be added to the capitalized cost of the replacement equipment.

Problems

16-1. A comparative cost study between proposed (challenger) and presently owned equipment (defender) is under way. Facts and estimates pertaining to the defender:

Original cost five years ago	= $13,000
Replacement cost new today	= $16,000
Remaining book value today	= $ 8,000
Gross resale value today	= $ 5,000
Todays cost of repair, removal, transportation to selling site, and selling commission	= $ 1,000
Combined effective state and federal income tax rate	= 50%

Depreciation is computed on an item (not group) basis

Find the present value, B, to be used in computing $ATCFR$ of the defender.

16-2. A dealer in musical instruments agrees that certain prospective customers exhibit considerable reluctance in the purchase of new musical instruments, especially for children who are beginners.

    a. What are some ways in which the dealer can reduce the barriers of "commitment?"

    b. Could similar policies be followed by the manufacturers of very new or very expensive equipment? Note any necessary limitations.

16-3. Distinguish between the concept of *commitment* in an investment and *liquidity* of an investment.

16-4. How could you attempt to pinpoint the exact time of commitment, the point of no return, in acquisition of such an asset as a large electric generator for a municipal utility?

16-5. A "wait-and-see" investment decision is sometimes made with the hope of "buying time and information." Does this type of decision bear any resemblance to an insurance investment? Explain. What are some of the dangers of a wait-and-see policy?

16-6. Write the equation for computing avoidable AEC at any date prior to use of a facility. Use the data of Example 16-1 and Table 16-1 for operating costs, tax rate, and so forth. Simplify your equation so that the only term left as a variable is B_v.

16-7. a. Figures 16-5 and 16-6 contain "PEX with 50% tax saving" and "PEX with 0% tax saving." Find the lines of Table 16-1 which provide those data.

    b. Figure 16-3 contains "avoidable AEC with 50% tax saving" and "avoidable AEC with 0% tax saving" plots. Find the lines of Table 16-1 which provide the data for those plots.

c. Figure 16-2 contains three plots: "Unallocated cost," "B with 50% tax saving," and "B with 0% tax saving." Find the lines of Table 16-1 which provide the data for those plots. Specify the range of time, z, over which your observation applies.

d. Figure 16-1 contains four plots: "PE of cumulative expenditures," "PE of cumulative $ATCF$," "PE of net salvage," and "PE of total recoverable." Find the lines of Table 16-1 which provide the data for those plots. Specify the range of time, z, over which your observation applies.

16-8. Refer to Figure 16-6; see the PEX values where "loss on disposal" saves 0%. Sketch a decision tree and compute $E(PEX)$ prior to "search and evaluation" for the following revised values of probability, p, of "cash-in":

$$0.0, 0.0, 0.0, 0.0, 0.0, 0.0, 0.0, 0.1, 0.2, 0.3, 0.4$$

16-9. Five months ago Green Corporation studied some equipment alternatives; Equipment A was selected. An order was placed and filled; the equipment was received and installed but has not yet been run. Today Equipment B with AEC of $25,000 is discovered and the question is whether A should be replaced. Other data pertaining to A are:

Equipment cost	= $90,000
Freight cost on equipment	= $ 3,000
Installation cost of equipment	= $ 7,000
Gross salvage available today	= $64,000
Gross salvage available ten years hence	= $14,000
Removal (de-installation) cost	= $ 4,000
Annual cash operating cost	= $ 7,000

$MARR = 10\%; t = 50\%; r = 0$

Depreciation is on an item (not group) basis by the SL method; as yet, none has been charged.

a. Find avoidable AEC_A as of today to see whether replacement is appropriate.

b. Find avoidable AEC_A as of five months ago, prior to any investment commitment.

16-10. A proposed addition to the Green office building calls for use of the land where some of Green's present parking facilities are now located. You are asked to make an engineering economy study of the proposed addition.

a. Should any land costs be assigned to first cost of the addition? Defend your response.

b. Should first cost of the addition include the cost of tearing out the parking facilities? Defend your response.

c. Should first cost of the addition include the cost of replacing the torn-out parking facilities? Defend your response.

d. Should first cost of the addition include the cost of added parking facilities needed only if the addition is constructed? Defend your response.

16-11. Yesterday the Green Corporation purchased a new widget assembler for $100,000; its salvage value today is only $50,000. The widget assembler will involve annual costs of $42,000 for operation, maintenance, property taxes, and insurance; its estimated salvage value ten years from today is $0. $r = 0$; $t = 50\%$; SL depreciation is used; no capital gains are expected in the current year. A new super widget assembler is now available for $200,000; it also has an estimated salvage value of $0 ten years from today; it will reduce the annual cost of operation, maintenance, property taxes, and insurance by $30,000.

a. Which is the pertinent first-cost figure, B, for use in computing avoidable annual equivalent cost for the one-day-old widget assembler?

b. Find the $MARR$ value at which the alternatives break even. Then comment on how recognition of the sunk cost concept influences one in the decision to keep or replace an equipment.

16-12. Green Corporation is considering the replacement of a present equipment (defender = D). New equipment (challenger = C) having an estimated annual equivalent cost of $20,000 is available. For the defender you have the following data:

Original cost (four years ago) = $100,000, including installation and so forth
Depreciation method employed = SL
Estimated gross salvage six years from now = $25,000
Estimated cost of removal, selling, and transportation = $5000
Estimated gross resale value today = $37,000
Estimated cash operating costs per year = $8000
$MARR = 10\%$; $t = 50\%$; $r = 0\%$

    a. Find avoidable AEC_D as of today to see whether replacement is appropriate.

    b. Find avoidable AEC_D as of four years ago, prior to any investment commitment.

16-13. Many years ago Mr. Green purchased some corporation bonds for $600 each. Each bond pays interest of $15 per six months. Mr. Green's next interest receipt will be six months from now; if he holds the bonds to maturity, his last interest receipt will occur at the date of maturity, six and one-half years from now. At maturity the bonds will be redeemed by the corporation for $1000 each. Mr. Green's incremental income is subject to a combined federal and state income tax rate of 60%. His personal MARR is 2% per six months (= 4.04% per year). You are asked to neglect brokerage costs in making the computations requested. Whenever the bonds are sold, Mr. Green will pay capital gains tax on the increased value (x-$600) at half the rate applicable to ordinary income.

    a. Find todays maximum price at which Mr. Green should be willing to purchase additional bonds of this same type. Assume that he would hold such bonds to maturity.

    b. Find todays minimum price at which Mr. Green should be willing to sell the bonds he acquired many years ago.

    c. Explain why your answers to Parts a and b are different. Summarize the decision rules developed there.

17

Replacement, Retirement, and Obsolescence

"Life" of Properties As Used in Economy Studies

IN PREVIOUS CHAPTERS the life of a property has generally been given as part of the data. In this chapter we will consider the various meanings of property life illustrated in Figure 17-1, as well as the data and procedures helpful in determining the optimal or economic life of a property.

1. *Accounting life* is the period of time over which property cost will be allocated by the accountant. His options sometimes include the possibility of immediate allocation (expensing) of cost or of allocating cost over the five-year period permitted under special conditions. Since the specific allocation has a bearing upon income taxes in the economy study, it is of concern here.

2. *Economic life = service life* is the period of time extending from date of installation to date of retirement (by demotion or disposal) from the *intended* service. Retirement is signaled by a future economy study when annual equivalent cost of a new property is less than annual equivalent cost of

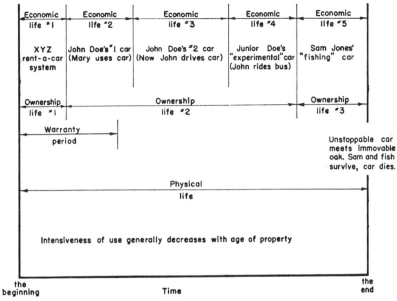

Fig. 17-1. Various "lives" of a property.

keeping the present property one or two more years. An estimate of retirement date or economic life, as signaled in an economy study now, *may* be based on the period which minimizes annual equivalent cost of the proposed property.

3. *Ownership life* is the period between date of acquisition and date of disposal of the property by a specific owner. A given item of property may have several service lives for a given owner; an electric generator might serve for 20 years in producing power (primary service), and then be demoted to serve as a standby source of power (secondary service) for 10 more years during its ownership life of 30 years. When there is just one service level to be performed (no demotion possible), service life and ownership life are identical.

4. *Physical life* is the period between original acquisition and final disposal of a property over its succession of owners. Even when not usefully employed the property may be physically sound. Ordinarily, this is longer than the lives described in (1), (2), and (3) above.

5. *Warranted, guaranteed, or assured life* is the minimum period of service as guaranteed by the manufacturer of an item. The warranty, guaranty, or assurance all can be forms of insurance for the buyer; their bearing on the lives described in (1), (2), (3), and (4) above is remote, however.

Estimation of Economic Life

Reviewing the causes of retirement listed in Chapter 9 reveals that a number of causes, such as inadequacy or obsolescence, can lead to retirement of property before it is worn out. The estimate that we make should be based upon consideration of all the factors which can lead to retirement and should be an estimate of *economic* life; a special effort should be made to *avoid* use of accounting, ownership, warranty, or physical life. The disparity between physical life and economic life is not necessarily a small one; although a good many Model A Fords still survive, the economic life of automobiles for a given firm or organization will generally be less than six years, and in the case of rent-a-car systems, economic life may be less than two.

As a presently owned equipment grows older, there is a tendency for (1) maintenance and other operating costs to increase and (2) quality of the good or service produced to deteriorate. The first tendency leads to reduced usage, which in turn raises the cost per unit of output. The second tendency leads to decreased revenue per unit of output. Meanwhile, the new models (challengers) are being developed. These tend to have still smaller operating costs, greater output quality, and greater output capacity. These factors can lead to an economic life which is significantly less than the physical life.

In Table 10-3 and Example 12-2d it can be seen that income tax considerations somewhat reduce the impact of errors in the estimation of economic life, especially when the life involved is more than 14 years. Nonetheless, a continuing effort should be made to reduce (1) the discrepancy between estimated and actual service life and (2) any consistent bias of estimates.

Optimal Retirement Age – Optimal Acquisition Age = Optimal Service Interval = Economic Life

The span of time starting at the optimal acquisition age of an equipment and ending at its optimal retirement age is called *economic life.* If the optimal acquisition age is zero, new rather than used equipment should be purchased; and in that special case the optimal retirement age coincides with the optimal service interval called economic life.

The usual criterion for determining optimal service interval, retirement age, or acquisition age is the minimization of *AEC.* Under unusual circumstances it may be appropriate to use the criterion of maximizing *PEX*; the reader is cautioned that maximization of *PEX* is not necessarily identical to minimizing *AEC* (as when only one life cycle is to be considered). The *AEC* computation has the advantage of not requiring an estimate of the magnitude of annual revenues. It is probably easiest to apply to the case where (1) the costs (in constant dollars) of present and replacement units are identical, (2) the annual revenues are uniform, and (3) the duration of service need is long. The first two characteristics may result when future technological breakthroughs just offset increased cost per unit of certain inputs and when future sales volume increases just offset decreased margin per unit.

In Example 17-1 we gradually increase the information input; this permits increased depth of analysis and improved output, but it also means increased complexity. Still greater complexity is involved as one adds such complications as use of an after-tax approach coupled with questions of gain or loss on disposal, depreciation method employed, and the debt ratio to be assumed. The example suggests that at least some engineering economy problems are nearly infinitely variable in depth.

Example 17-1.   **Determination of Optimal Acquisition and Retirement Ages**

a. *Information:* Green Corporation is considering the purchase of a new equipment for which

First cost = \$10,000 = B
Salvage = \$400 at the end of year four = V_4
Before-tax rate of return required = 20% = $BTRR$
Operating costs for maintenance, property tax, insurance, and so forth = \$2000 at the end of year one, and increase by \$1000 each year.

Objective: Find the annual equivalent cost of owning the equipment for four years. Use the before-tax approach.

Analysis:

$BTCFR = \$10,000\,(a/p)_4^{20\%} - \$400\,(a/f)_4^{20\%} = \$3788$
Operating cost $= \$2000 + \$1000\,(a/g)_4^{20\%}$ $= \underline{3274}$
AEC_{0-4} $\$7062$

b. *Information:* Same as Part a plus: estimated future net

salvage values for the equipment are

Age, in years	0	1	2	3	4	5	6	7	8
Net salvage	$6400	$3200	$1600	$800	$400	$200	$100	$50	$25

Objective: Find the integer equipment life which would minimize *AEC* if the equipment is purchased new. Use the before-tax approach.

Analysis: In Part a we found that AEC_{0-4} = $7062.

Try $n = 5$:

$$BTCFR = \$10,000\,(a/p)_5^{20\%} - \$200\,(a/f)_5^{20\%} = \$3317$$
$$\text{Operating cost} = \$2000 + \$1000\,(a/g)_5^{20\%} \quad = \underline{3641}$$
$$AEC_{0-5} \qquad\qquad\qquad\qquad\qquad\qquad = \$6958$$

Try $n = 6$:

$$BTCFR = \$10,000\,(a/p)_6^{20\%} - \$100\,(a/f)_6^{20\%} = \$2997$$
$$\text{Operating cost} = \$2000 + \$1000\,(a/g)_6^{20\%} \quad = \underline{3979}$$
$$AEC_{0-6} \qquad\qquad\qquad\qquad\qquad\qquad = \$6976$$

So $n = 5$ minimizes *AEC*.

Alternate Analyses: An alternate method for determining n is to write the general equation for *AEC*, then differentiate with respect to n, set equal to zero, and solve.

Still another alternate involves use of year-by-year costs, C_z. Using the after-tax approach and based on Equation 10-3,

$$C_z = M_z + [B_{z-1}\,(1 + i_a) - V_z - t(D_z)]/(1 - t)$$

And if the tax depreciation allocation, D_z, were equal to the decline in value, $B_{z-1} - V_z$, then

$$C_z = M_z + [B_{z-1}\,(1 + i_a - t) - V_z\,(1 - t)]/(1 - t)$$
$$= M_z + B_{z-1}\,[1 + i_a/(1 - t)] - V_z$$
$$= M_z + B_{z-1}\,(1 + i_b) - V_z$$

which is the equation produced by the before-tax approach and could be stated as:

$$C_z = \begin{pmatrix} \text{decline in} \\ \text{value during} \\ \text{year } z \end{pmatrix} + \begin{pmatrix} \text{pretax return required on} \\ \text{investment remaining at} \\ \text{beginning of year } z \end{pmatrix}$$

$$+ \begin{pmatrix} \text{operating} \\ \text{costs during} \\ \text{year } z \end{pmatrix}$$

$C_5 = \$200 + \$80 + \$6000 = \$6280$

Since $AEC_{0-4} = \$7062$, retention through year five will reduce AEC. Compute next,

$$C_6 = \$100 + \$40 + \$7000 = \$7140$$

Observe that retention through year six would increase AEC; therefore $n = 5$ is optimal.

By computations similar to the preceding, we could develop and graph as Figure 17-2a the following:

z	C_z	AEC_{0-z}
1	$10,800	$10,800
2	5,240	8,273
3	5,120	7,405
4	5,560	7,062
5	6,280	6,958
6	7,140	6,976
7	8,070	7,060
8	9,035	7,181

c. *Information:*   Same as Part b plus: Estimated revenues are a uniform $8260 per year. Duration of service need is one life cycle; no replacement equipment units are to be considered. Equipment is purchased new.

Objective: Find the integer equipment life which would maximize *PEX*. Use the before-tax approach.

Analysis: The equipment should be retained so long as marginal cost is less than marginal revenue (and thus as long as C_z is less than $8260). From the table of Part b,

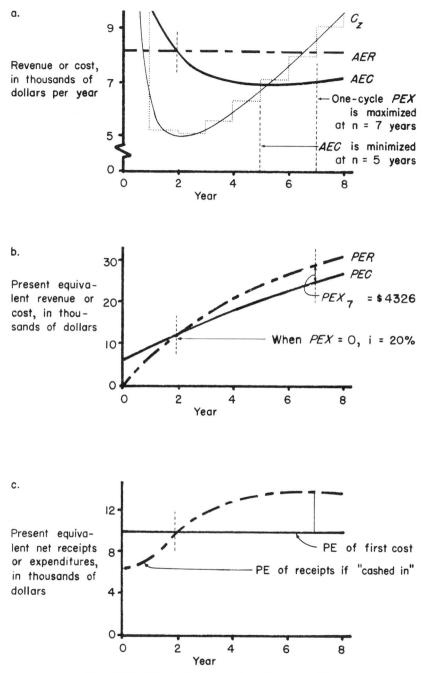

Fig. 17-2. Illustrations based on Example 17-1.

observe $C_7 = \$8070$ and $C_8 = \$9035$; therefore, $n = 7$ maximizes *PEX*.

We could verify this result through the tabulations below; they also appear in Figure 17-2.

$$PER_n = \$8260(p/a)_n^{20\%}$$
$$PEC_n = \$10,000 - \text{salvage } (p/f)_n^{20\%} + \$2000\,(p/a)_n^{20\%}$$
$$+ \$1000\,(p/g)_n^{20\%}$$

n	PER	PEC	PEX
1	$ 6,881	$ 8,999	-$2118
2	12,811	12,639	172
3	17,396	15,601	1795
4	21,385	18,284	3101
5	24,706	20,808	3898
6	27,473	23,200	4273
7	29,777	25,451	4326
8	31,694	27,551	4143

We might also have used equations such as

$$PEX_7 = (\$8260 - \$7060)\,(p/a)_7^{20\%} = \$4326$$

The unusual constraint of "one life cycle" ruled out replacement units. Under the much commoner circumstance that the service period is long enough to involve a series of replacement units, the minimization of *AEC*, maximization of *AEX*, or maximization of *PEX* are identical criteria. Recall that minimization of *AEC* does not require estimation of revenues.

d. *Information:* Same as Part b plus: The cost to purchase (acquire, transport, install, and debug) used equipment of the same type has been estimated as:

Age, in years	1	2	3	4	5	6	7	8
Cost to purchase	$6800	$5200	$4400	$4000	$3800	$3700	$3650	$3625

Objective: Find the integer acquisition and retirement ages that would minimize *AEC*.

Analysis:

Interval	Decline in Value*	Pretax Return on Beginning-of-Year Investment*	Operating Costs	Year-by-Year Cost if: Owned when year began	Year-by-Year Cost if: Purchased when year began†
0–1	$3200	$1280	$2000	—	$10,800
1–2	1600	640	3000	$5240	9,560
2–3	800	320	4000	5120	9,440
3–4	400	160	5000	5560	9,880
4–5	200	80	6000	6280	10,600
5–6	100	40	7000	7140	11,460
6–7	50	20	8000	8070	12,390
7–8	25	10	9000	9035	13,355

*If owned when year began.
†In this example the "cost to purchase" exceeds "net salvage" by $3600 regardless of age. Year-by-year cost if "purchased when year began" is therefore $3600 $(1 + i) = \$4320$ more than if "owned when year began."

1. Compute year-by-year costs as shown above. Verify that C_z and *AEC* are ∪-shaped cost functions (by inspection, by graph, or by showing that the second derivative is positive).
2. Select the one-year ownership interval which minimizes *AEC* (interval 2–3).
3. Seek the optimal interval by observing that interval 1–3 would be better, that interval 1–4 would be still better, and that interval 1–5 is optimal.

Interval	Year-by-Year Costs Incurred	*AEC*
2–3	$9440	$9440
1–3	$9560; $5120	7542
1–4	$9560; $5120; $5560	6997
1–5	$9560; $5120; $5560; $6280	6864
0–5	$10,800; $5240; $5120; $5560; $6280	6958
2–5	$9440; $5560; $6280	7293

This procedure could be continued to seek optimal acquisition and retirement dates to the nearest 0.1 year. This unsophisticated, intuitive, optimal-seeking method offers some consolations; although calculus procedures can be attempted and partial differentials computed, the resulting expressions can only be solved by trial-and-error computations (hand or computer assisted) which must be applied to expressions considerably more complex than those involved above. Howard provides still another alternative method of solution with a dynamic programming procedure.[1]

Alternate Analysis: Compute and tabulate *AEC* for all possible combinations of asset acquisition age, n_a, and asset retirement age, n_v. Note that life of asset $= n_v - n_a = n$.

$$AEC = \left(\begin{array}{c}\text{cost to}\\\text{purchase}\end{array}\right)(a/p)_n^{20\%} - \left(\begin{array}{c}\text{net}\\\text{salvage}\end{array}\right)(a/f)_n^{20\%}$$

$$+ \$2000 + \$1000\,(n_a) + \$1000\,(a/g)_n^{20\%}$$

Asset age, at date of retirement*

		1	2	3	4	5	6	7	8
Asset	0	10800	8273	7407	7063	6957	6976	7061	7180
age,	1		9560	7542	6997	6864	6901	7019	7175
at	2			9440	7676	7293	7264	7373	7540
date	3				9880	8244	7940	7965	8108
of	4					10600	9027	8764	8815
acqui-	5						11460	9919	9676
sition	6							12390	10865
	7								13355

*AEC values above result from a computer-assisted analysis and sometimes vary by $1 or $2 from results based on 4- or 5-place interest tables.

e. *Information:* Same as Part d plus: Assume that one-year-old equipment was purchased, then at age five when replacement was supposed to take place the shortage of capital caused the company to postpone replacement for one year. Consider only integer-year values.

Objective: Measure the urgency of replacement.

Analysis: $C_6 = \$7140$, while $AEC_{1-5} = \$6864$. This suggests that the cost of postponing replacement by one year is $276. Still another year of deferral with $C_7 = \$8070$ would result in extra cost of $1206, so the urgency of replacement grows rapidly.

---

[1] See R. A. Howard, *Dynamic Programming and Markov Processes* (Cambridge, M.I.T. Press, 1960), pp. 54–59; A. Kaufmann and R. Cruon, *Dynamic Programming* (New York, Academic Press, 1967).

The company is not able to finance all projects which provide a pretax return of 20% or better. The actual cutoff rate of return is apparently somewhat greater; we can solve for the $BTRR$ at which we would be indifferent between ownership interval 1-5 and 1-6; that is, find i where

$$AEC_{1-5} - AEC_{1-6} = 0$$

$$\$6800\,(a/p)_4^i - \$200\,(a/f)_4^i + \$3000 + \$1000\,(a/g)_4^i$$
$$- \$6800\,(a/p)_5^i + \$100\,(a/f)_5^i - \$3000 - \$1000\,(a/g)_5^i = 0$$

$$\$6800\,[(a/p)_4^i - (a/p)_5^i] - \$200\,(a/f)_4^i + \$100\,(a/f)_5^i$$
$$+ \$1000\,[(a/g)_4^i - (a/g)_5^i] = 0$$

At $i = 25\%$; the difference is $-\$10$
At $i = 30\%$; the difference is $+\$14$
so $i = 25\% + 5\%\,(10/24) = 27.1\%$

At $BTRR$ 27.1% we are indifferent between replacement at the end of the fifth or sixth year. If the equipment is then retained for still another year, our decision is consistent with a cutoff rate of 49.1%, showing the rapidly increasing urgency of replacement. Optimal acquisition and retirement ages are responsive to $BTRR$; by calculations similar to the preceding we can find:

$BTRR$	Optimal Acquisition Age	Optimal Retirement Age
6.1%	0	4 or 5
14.5%	0 or 1	5
27.1%	1	5 or 6
49.1%	1	6 or 7
56.9%	1 or 2	7

The tabulation shows how $BTRR$ influences the optimal acquisition and retirement ages. It can also be used conversely; by observing the policy followed or decisions reached, we can find an inferred capital budget cutoff rate. It is crucial that these capital budget "inferences" be consistent.

Figure 17-3 helps to emphasize that the timing of acquisition and retirement can be viewed as a marginal investment opportunity problem.

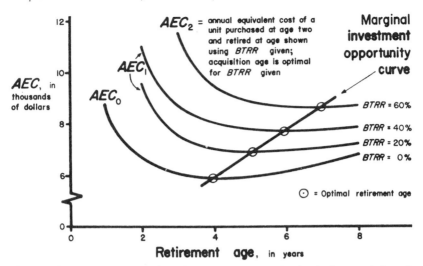

Fig. 17-3. Acquisition and retirement ages as the basis for an inferred marginal investment opportunity rate of return or vice-versa.

f. *Information:* Same as Part d plus: Estimated revenues are a uniform $7000 per year. Due to the risk of possible obsolescence caused by cessation of market demand or the development of technologically improved models, the duration of equipment use is probabilistic. There is a 0.1 probability that use will persist for exactly one year; 0.2 that use persists for exactly two years; 0.3 that use persists for exactly three years; and 0.4 that use persists for exactly four years.

Objective: Assume that a one-year-old unit is purchased; then find *PEX*.

Analysis:

Ownership Interval	$AEX =$ $AER - AEC$	$(p/a)_n^{20\%}$	Probability	$AEX(p/a)_n^{20\%}$ (Probability)
1-2	-$2560	0.833	0.1	-$213
1-3	- 542	1.528	0.2	- 166
1-4	3	2.106	0.3	2
1-5	136	2.589	0.4	141
Expected value of *PEX* =				-$236

Note that recognition of the risk of possible obsolescence leads us to rejection of this investment opportunity.

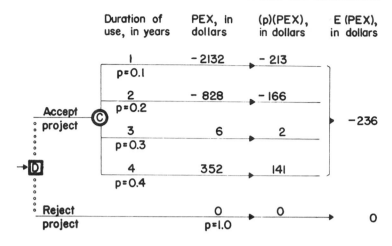

Duration of use, in years	PEX, in dollars	(p)(PEX), in dollars	E (PEX), in dollars

Timing of Retirement and Replacement

Property retirement (the withdrawal of an asset from its primary service function) can be accomplished by disposal (selling, junking, or abandoning) or demotion (relegating to secondary service, for example, as a standby unit). Factors leading to the retirement of property were discussed in Chapter 9. A review of these may suggest that the managerial decision to retire a property is usually based on noneconomic considerations; one might, however, argue that questions such as, Are the property's functional characteristics satisfactory? and Has the need terminated? can best be answered by an economy study. In such instances and in the case of retirement based specifically on economic considerations, there is a need for enumeration and evaluation of these factors which have a bearing on the retirement decision.

Many considerations apparently external to an economy study of retirement can have a bearing on the final decision. Among these "apparently external" considerations are the growth and financing ability of the firm. Because a firm tends first to meet its basic production or service demands before considering modernization programs, the firm with rapid growth in demand and limited financial resources may be able to provide *growth* units but unable to provide also for *replacement* units; thus property retirement is retarded. Conversely, when production or service demands decrease, there is an excess of equipment, and retirement is accelerated. Changing demands can thus be something of an overriding consideration in the retirement decision.

The determination of economic life, illustrated in this chapter, suggests (1) the concept of a series or chain of replacements of similar economic life and (2) a timing of retirement which can be a useful guide. Because the decision to retire and replace equipment is made at a time *subsequent* to installation, the accumulation of such further information as better estimates and knowledge regarding improved equipment, tax measures, or sales requirements is possible. Then, as the presently owned equipment (the defender) is compared against a proposed equipment (the challenger), the problem may be one of comparing the cost of keeping the defender one more year versus the annual equivalent cost of the challenger.[2]   As the year-by-year costs of the defender increase with its age, there comes a time when the defender costs exceed those of the challenger. If the defender is retained beyond the break-even point, its costs continue to grow and replacement becomes more urgent, as illustrated in Figure 17-4.

In Figure 17-5 some challengers of the future are shown; if the improved challenger of 7-1-19x1 does become available, a new replacement study may well take place earlier than the computed break-even point. Finally, if the further improved

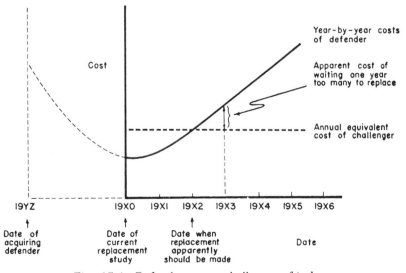

Fig. 17-4. Defender versus challenger of today.

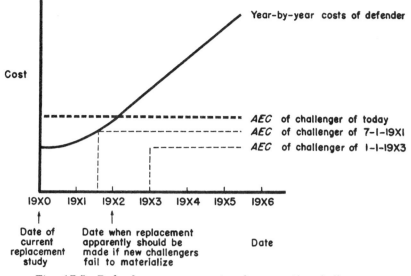

Fig. 17-5. Defender versus current and prospective challengers.

challenger of 1-1-19x3 does seem likely, it may be still better to postpone replacement until that challenger becomes available. In any case, retention of the defender beyond the break-even point has a cost which grows with time, and thus replacement becomes more urgent. This cost of waiting can in some instances be worthwhile if it permits purchase of improved equipment having economies that more than pay for the cost of waiting. It is possible that a decision to postpone heavy commitment to any course of action may "buy time and information." Obviously too, a "wait and see" policy is not without its pitfalls.

When replacement is not signaled by the economy study, more information may become available before the next challenge to the defender and hence the next comparison should include that additional information. In the case of replacement decisions, *postponement* means a postponement of the decision "when to replace," not the decision to postpone replacement until a specified future date. The economy study which signaled postponement is also helpful in other ways:

1. As part of a system of regular periodic investment "checkups" to provide managerial control of capital equipment.
2. As a source of information for *estimates* of future replacement dates, thus facilitating long-range planning of funds and expenditures.

Finally, the exact timing of replacement may be more important than is inferred by Figure 17-5 for at least two reasons: (1) Because technological change tends to be sudden and dramatic rather than uniform and gradual, new challengers with significantly improved features can arise sporadically and can change replacement plans substantially and (2) there is a very clear long-time trend of automation/mechanization followed by still greater automation/mechanization of productive facilities. The replacement of labor, an input factor with relatively constant annual cost, by machines that have annual equivalent costs very dependent upon period of use certainly infers that the timing of replacement is becoming more critical, as in Figure 17-6.

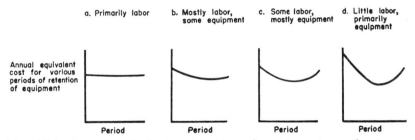

Fig. 17-6. Annual equivalent cost patterns for various stages of automation/mechanization showing increasing sensitivity to replacement policies.

Obsolescence: Its Effect upon Annual Equivalent Cost and Economic Life

The data developed in Example 17-1d indicate an economic life of four years and an *AEC* of $6864 under the assumption of zero technological improvement (obsolescence). If technologically improved equipment or competing products become available, the *AEC* of $6864 may prove overly optimistic. Allowance for the hazard of obsolescence might be accomplished by several methods:

1. By the application of formulas or charts such as those of MAPI which incorporate an implicit allowance for obsolescence.[3] Results similar to those of MAPI will be produced if estimated life is shortened to one-half that which would be employed if technological obsolescence were not in pros-

---

[3] See G. Terborgh, *Business Investment Management*, (Washington, D.C., Machinery and Allied Products Institute, 1967).

pect.[4]   The principal disadvantages of the MAPI procedure
are that the obsolescence allowance implicitly included does
not permit flexibility to recognize the risk of obsolescence
being greater for special purpose equipment than for general
purpose equipment and the degree to which decisions
are sensitive (or insensitive) to obsolescence allowances is
obscured.

2. By the explicit shortening of estimated life to some value less
   than that which would be employed if technological obsolescence were not in prospect.

3. By making estimates of the probability of advanced replacement due to obsolescence, then applying these probabilities
   as illustrated in Example 17-1f.

4. By making estimates of the annual "cost of obsolescence" as
   paid via reduced profits by the firm which retains its present
   equipment and thus does not avail itself of the higher speeds,
   greater live load capacities, increased uniformity of output,
   or other improvements available to the firm (and its
   competitors).

A final note of caution; technological improvements do not
always cause a shorter economic life. Replacement of an existing equipment may be postponed beyond normal retirement so
that the new unit can include improvements soon to become
available.

Comparing Alternatives with Unequal Lives and Service Needs

In many cases comparisons involve alternatives which have
different lives. Up to this point, generally we have treated only
problems with life data such as:

Alternative A has an economic life of 6 years.
Alternative B has an economic life of 9 years.
Service need is for 18 years.

Our comparisons could be conducted on a rate of return, annual equivalent cost, or present equivalent cost basis without
any special difficulties. As we attempt to include more of the
complexities of reality, we are likely to find problems with life
data such as:

---

[4] Shown in an unpublished thesis (1958) by the author.

Alternative C has an economic life of 5 years.
Alternative D has an economic life of 13 years.
Service need is for 8 years.

If the estimates above should materialize, cessation of use (and probably retirement) of either C or D comes at a time dictated by service needs. It is evident that additional data should be helpful, specifically:

Salvage value of the second C unit after just three years of use.
Salvage value of the original D unit after eight years of use.
Perhaps the alternative of using two C units for four years each
   should be explored.

In this type of problem, the comparison of costs requires considerable caution; a comparison which assumes a definite service need of 11, 12, or 13 years tends to favor D, while a comparison which assumes a definite service need of 14 or 15 years tends to favor C.

We might also observe that for certain types of equipment annual equivalent cost is quite sensitive to variation in the life over which it is retained. In such instances the estimation of the service-need period is an especially critical requirement of a proper comparison.

Some Cautions in Replacement Studies

Trade-in Value. When one equipment is replaced by another, the seller may accept trade-in of the old equipment as partial payment for its replacement. If a dealer offers an allowance of $1000 on an older car when traded for a newer model listed at $3000, an exchange price ($2000) has been bid, but salvage value of the older car is not yet clear. Only by knowing "cash price" of the newer model, say $2400, can we have a true estimate ($400) of salvage value of the older one. The distinction is important. Even though the difference in either case is $2000, the erroneous overstatement of challenger and defender costs by $600 does not generally result in compensating errors; the $600 overstatement of costs of the new challenger can be recouped over a longer period than is likely for the aging defender. Misuse of trade-in value data can thus introduce a bias favorable to the challenger.

Exchange Price. Again considering the data above, we might be tempted to look upon the $2000 exchange price as a firm bid representing an incremental investment. If this reasoning is pursued, our economy study would show zero incremental investment for defender and $2000 for challenger, thus understating the value of each by $400. The error does not fall equally on a new challenger compared to an aging defender whose value must be annualized over a shorter remaining life. Such misuse of exchange value data can thus introduce a bias favorable to the defender.

Inflation and Deflation. The prospect of inflation either can *accelerate* or can *retard* replacement. If price increases were to affect first cost more than other costs, replacement could be retarded. On the other hand, prospective increases in wage levels can make the replacement equipment which requires fewer operating and maintenance hours a more attractive opportunity and hence accelerate replacement.

Problems

17-1. A new automobile with a list price of $5000 is being considered as a replacement for a used automobile presently owned by the company. The new automobile can be acquired by trading in the old one plus $3000 cash. Alternatively, the company can sell outright the present automobile for $1000 and purchase the new car outright for $4000 cash. Current remaining book value (unallocated cost) of the present automobile is $1500.

    a. If annual equivalent cost of the present automobile is based on its remaining book value, the resulting *AEC* tends to be (overstated, understated, correctly stated).

    b. If annual equivalent cost of the new automobile is based on the $5000 list price and if annual equivalent cost of the used automobile is based on the trade-in allowance of $2000, the study tends to be biased in favor of (proposed, present, neither) automobile.

    c. If annual equivalent cost of the new automobile is based on the $3000 cash paid and if annual equivalent cost of the used automobile is based on the fact that because "already owned" it requires zero outlay and therefore has zero present basis, the study tends to be biased in favor of (proposed, present, neither) automobile.

17-2. Find AEC_4, AEC_5, and AEC_6 in Example 17-lb assuming $t = 50\%$, $MARR = 10\%$, and that the other data of the example apply. Straight-line depreciation will be based on the salvage values given; if $n = 4$, $D = \$2400$; if $n = 5$, $D = \$1960$; and so forth.

17-3. The following estimates apply to a certain equipment for which the firm's before-tax composite rate of return required is 20%. Revenues are stated as end-of-year equivalents:

Year	Net Salvage Value at Beginning of Year	Decline in Value at End of Year	Equivalent End-of-Year Operating Costs
1	$10,000	$5200	$2000
2	4,800	700	3000
3	4,100	600	4000
4	3,500	500	5000
5	3,000	400	6000
6	2,600	300	7000
7	2,300	200	8000
8	2,100	100	9000

a. Assume (1) revenues are $5250 per year and (2) equipment is purchased new and retained for four full years. Write an equation which would permit you to find the before-tax rate of return.

b. Assume equipment is purchased new. Find AEC for various ownership periods; indicate the n value which minimizes AEC.

c. Assume (1) equipment is already owned, (2) revenues are $6000 per year, and (3) the service provided by the equipment is not a mandatory one. At the end of which year should equipment be retired?

d. Assume (1) equipment is purchased new, (2) no replacements are planned, and (3) revenues are $8000 per year. At the end of which year should equipment be retired? That is, find the integer ownership period which would maximize PEX.

e. Assume (1) equipment is purchased new, (2) an infinitely long chain of replacements identical in cost and other aspects is forecast, (3) revenues are time invariant (uniform), and (4) either $AEX \geqslant 0$ or service is mandatory. Find the integer ownership period which will maximize PEX.

17-4. Assume the data of Problem 17-3 and that $t = 50\%$, $MARR = 10\%$. Straight-line depreciation will be based on the salvage values given; if $n = 4$, $D = \$1750$; if $n = 5$, $D = \$1480$; and so forth.
  a. Same as Problem 17-3a, except that you are to write an equation which would permit you to find the after-tax rate of return.
  b. Same as Problem 17-3b.
  c. Same as Problem 17-3c.
  d. Same as Problem 17-3d.
  e. Same as Problem 17-3e.

17-5. A new equipment model is being considered for possible replacement of an existing equipment. Straight-line depreciation will be used over the ten-year life of the new equipment; it is expected to have negligible salvage at that time. The existing equipment is fully depreciated; its present net market value and present net book value equal $1000. Estimated future net market value of the existing equipment is also $1000.

  The annual end-of-year cost of operation, maintenance, property taxes, and insurance for each equipment is estimated as:

	Existing	New
At 100% of capacity	$5555	$3500
At 50% of capacity	$2851	$2600

  Output capacity of the two alternatives is identical. Full capacity is defined as 40 hours per week, 50 weeks per year. $MARR = 10\%$, $t = 50\%$.
  a. Find the maximum sum which this prospective purchaser can justify paying for the new equipment, assuming operation at 100% of capacity.
  b. Repeat Part a, assuming operation at 50% of capacity.
  c. Explain why intensity of use affects value of an equipment to the prospective owner.

17-6. Refer to Example 17-1d where annual equivalent cost for the age interval $1-5 = AEC_{1-5} = \$6864$. It can be shown that $AEC_{1-6} = \$6901$ and $AEC_{1-7} = \$7019$. Is this not proof that the example equipment is very *insensitive* to the period of retention? Explain.

17-7. A certain equipment costs $6000 today and has operation, property taxes, and insurance that cost $1000 the first year, $1200 the second, and grow by $200 per year. Salvage value is estimated to be negligible by the end of year six. Use a before-tax rate of return requirement of 15% to find the year-by-year salvage values necessary for annual equivalent cost of the equipment to be independent of the period of retention.

17-8. Two alternative means of performing a certain manufacturing function are being considered. The duration of product life (due to possible obsolescence of the product) is rather uncertain. $MARR = 12\%$.

Product Life, in Years	Estimated Probability of That Life	AEC_A for Life Shown	AEC_B for Life Shown
1	0.1	$25,000	$12,000
2	0.1	14,000	11,000
3	0.2	10,000	10,300
4	0.4	9,500	10,000
5	0.2	9,200	10,200

a. Which alternative has costs more sensitive to product life?

b. The median of estimated produce life is 4.0 years, the mode is 4.0 years, and the mean is 3.5 years. On this basis which alternative would seem to be preferred (before making any calculation)?

c. What is the minimum product life required for Alternative A to be preferable to Alternative B?

d. Compare $E(PEC)$ values to find the preferred alternative.

e. Suppose that further study reveals a new set of probabilities; find the expected PEC for Alternative A if the following revised probabilities are appropriate:

Product Life, in Years	Estimated Probability of That Life	AEC_A for Life Shown
1	0.0	$25,000
2	0.1	14,000
3	0.4	10,000
4	0.4	9,500
5	0.1	9,200

f. Find the value to the firm of the information developed in the "further study" of Part e.

17-9. A trade-in allowance of $9000 has been offered for Defender X if exchanged for Challenger Y listed at $20,000.

a. Why is it likely that use of $B_{challenger}$ = $20,000 and $B_{defender}$ = $9000 will lead to a biased comparison of alternatives favoring the challenger?

b. How would you attempt to better measure $B_{challenger}$ and $B_{defender}$?

17-10. Blue Corporation has decided to replace their present equipment, X, with a new one, Y. Remaining book value of X is $7000. The two courses of action now open seem to be:

A. Trade in X for Y; an allowance of $8000 has been offered for X on the purchase of Y which has a list price of $14,000. For income tax purposes, add the remaining book value of X to the $6000 cash "boot" to give Y a depreciation base of $13,000.

B. Sell X outright to a company which has offered $3000 for it; in this case the best negotiated no-trade-in price for Y seems to be $10,000. For income tax purposes, Blue can then immediately write off the $4000 difference between book and realized value of X as a $4000 "loss on disposal." The depreciation base would be $10,000.

Other facts which may be relevant:

Estimated life of Y = ten years with zero salvage at the end of its life

Depreciation is by the straight-line method

Blue Corporation's marginal income tax rate = 50%

MARR = 10%

No capital gains will be realized in the current tax year

Maintenance and other cash operating costs = $1000 per year

Find *AEC* for Alternatives A and B.

17-11. You have gathered data for a comparison of an existing equipment (defender) and a proposed equipment (challenger). One important aspect of this comparison is that the function to be performed by either challenger or defender is expected to be "phased out" four years from now. Cost estimates:

Defender: Year-by-year costs of ownership plus operation, maintenance, property taxes, insurance, and the like are $5500 at the end of year one, $6000 at the end of year two, and growing each year by $500.

Challenger: First cost = $10,000, salvage = $0 regardless of date of retirement; year-by-year costs of operation, maintenance, property taxes, insurance, and the like are uniform at $2000 per year.

Using *MARR* = 10%, *t* = 50%, you have produced the following comparison:

Replacement Made___ Years from Now	Years of Use for Equipment	Annual Equivalent Cost		Year-by-Year Costs of Defender
		Chal-lenger	De-fender	
0	4	$5,809	$6190	$5500
1	3	6,709	6468	6000
2	2	8,524	6738	6500
3	1	13,000	7000	7000

Sample calculations:

$$AEC_{challenger} = \$2000 +$$
$$[\$10,000(a/p)_4^{10\%} - 0.5(\$2500)]/0.5 = \$5809$$
$$AEC_{defender} = \$5500 + \$500(a/g)_4^{25\%} = \$6190$$

a. Based on the preceding information should the defender be replaced now? Explain the basis for your decision.
b. Find annual equivalent costs for retaining the defender this year then replacing it with the challenger for the three remaining years.
c. In most cases the urgency of replacement grows with time; why is that not true in this situation?

17-12. PRL Corporation is comparing renting (Plan R) versus purchasing (Plan P) a certain equipment. The rental agreement would permit cancellation without penalty at any time by PRL. Ownership costs have been estimated on a year-by-year basis and have been converted to annual equivalent costs, *AEC*, based on PRL's *MARR* of 10%. Because the investment decision is quite irreversible, considerable "commitment" is involved; this is reflected by the fact that initial year-by-year costs are much greater than subsequent ones.

	Year 1	Year 2	Year 3	Year 4	Year 5
Year-by-year cost	$4416	$1000	$1000	$1000	$1000
AEC if termination is at end of year shown	$4416	$2789	$2249	$1980	$1819
Probability that need will terminate at end of year shown	0.1	0.1	0.1	0.1	0.6
Probability that need will continue to end of year shown	1.0	0.9	0.8	0.7	0.6

Find the annual rental cost at which Plan R would break even with Plan P; that is, $E(PEC_R) = E(PEC_P)$.

17-13. Refer to Example 17-1. Assume $MARR = 0\%$, $t = 0\%$, and the equipment is purchased new. Find AEC for intervals 0–1, 0–2, 0–3, 0–4, and 0–5.

17-14. Given the following data, including $BTRR = 25\%$, find the integer equipment acquisition and retirement ages which would minimize AEC. All dollar values are in thousands.

	Decline in Value of Equip-ment*	Pretax Return on Beginning-of-Year Investment*	Operating Costs of Equipment	Year-by-Year Cost if:	
				Owned when year began	Purchased when year began
Year					
1	$64	$32	$24	$120	$140
2	32	16	32	80	100
3	16	8	40	64	84
4	8	4	48	60	80
5	4	2	58	64	84
6	4	1	63	68	88
7	0	0	72	72	92
8	0	0	80	80	100

*If owned when year began.

17-15. Given the following data, including $BTRR = 20\%$, find the integer equipment acquisition and retirement ages which would minimize AEC. All dollar values are in thousands.

Age Interval, in Years	Purchase Price of Equipment at Beginning of Interval	Salvage (resale) Value of Equipment at Beginning of Interval	Operating Costs of Equipment
0-1	$135	$100	$38
1-2	50	35	48
2-3	20	10	58
3-4	15	10	68
4-5	15	10	78
5-6	15	10	88
6-7	15	10	98

17-16. A passenger automobile with certain equipment is thought to be subject to the purchase and resale schedule that follows. It is felt that this schedule applies equally well to the comparable model car of both past and future years. Data is "best negotiated price," not "list" or "book value."

Year	Cost if Purchased at End of Year	Resale Value if Sold at End of Year
0	$4000	$3500
1	3200	2780
2	2500	2150
3	1900	1610
4	1400	1160
5	1000	800
6	700	530
7	500	350
8	400	260
after 8	400	260

a. The State Highway Department is considering purchase of the vehicle noted. It is felt that all the data except the $4000 "cost if purchased new" applies. That figure would be revised to $3700 because the State Highway Department obtains significant price reductions from fleet purchasing and state sales tax will not apply.

Marginal opportunity cost of funds = 10%. Annual end-of-year cost of insurance, license, gasoline, oil, lubrication, and repair for 20,000 miles per year of travel = $950 in year one and increase by $150 per year.

Annual end-of-year indirect costs of lost time due to breakdown, delay reduced pride of ownership, and reduced availability, reliability, or use = $50 in year one and increase by $100 per year.

Income taxes, inflation, and technological consideration do not apply. .

Find the optimal timing of purchase and retirement of the vehicle for these particular circumstances. Consider only integer years of ownership.

b. Gene Green is considering purchase of the vehicle noted. $MARR$ = 10%, t = 0%. Annual end-of-year costs of insurance, license, gasoline, oil, lubrication, and repair for 20,000 miles per year or travel = $950 in year one and increase by $150 per year. Annual end-of-year indirect costs of lost time due to breakdown, delay, and reduced pride of ownership and reduced availability, reliability, or use = $50 in year one and increase by $100 per year. Inflation and technological considerations do not apply.

Find the optimal timing of purchase and retirement of the vehicle for these particular circumstances. Consider only integer years of ownership.

c. Ben Blue is considering purchase of the vehicle noted. $MARR$ = 10%, t = 0%. Annual end-of-year cost of insurance, license, gasoline, oil, lubrication, and repair for 5000 miles per year of travel = $275 in year one and increase by $75 per year. Annual end-of-year indirect costs of lost time due to breakdown, delay, and reduced pride of ownership and reduced availability, reliability, or use = 0. The estimation of indirect costs as negligible may arise because the car is used strictly for personal rather than business trips, it is a "second" car, or alternate transportation is readily available. Inflation and technological considerations do not apply.

Find the optimal timing of purchase and retirement of the vehicle for these particular circumstances. Consider only integer periods of ownership.

d. Using calculus, solve Part a for $MARR$ = 10%, then repeat for $MARR$ = 20%.

18

Nonmonetized Factors

Nonmonetary Considerations

TWO VITAL QUESTIONS not previously treated are considered in this chapter:

1. Must we always evaluate in monetary terms all the factors which influence our decisions?
2. Must we always require that an investment alternative meet the threshold of monetary acceptability?

The general pattern of this book has been to start with very restricted mathematical models, such as the bond model, and then gradually to relax these restrictions to permit greater realism and applicability. As restrictions are relaxed, our model grows more precise but also more unwieldy. We find that at some level of restriction further realism is unwarranted. At that level we find some factors not yet treated still remain which can influence our decisions. These factors, not yet evaluated in monetary terms, are really the leftover, remaining, or residual factors. The literature generally refers to these remaining influences as nonmonetary, irreducible, unevaluable, noneconomic, or intangible factors;[1] we prefer to use the terms *nonmonetized, nonreduced,* or *unevaluated* factors. The former terms imply that an identifiable group of factors, such as safety, exists and defies monetary evaluation. The latter terms emphasize that it is the decision maker who chooses whether or not to estimate the monetary value of a given factor such as safety. Thus the monetary value of safety is estimated in one instance and neglected in another; the choice is dependent upon whether the cost in time and effort is warranted by the circumstances. Circumstances which suggest the cost in time and effort is not warranted include:

1. Where it is practical to solve for the minimum monetary value of unevaluated factors required for project justification; the next step would be evaluation of whether or not the nonmonetized factors were worth such an amount. This approach is consistent with our practice of leaving the most uncertain of factors as the unknown.
2. Where it is obvious that unevaluated factors will not change the decision indicated in their absence, as when (a) the alternative favored by unevaluated factors is already justified on the basis of monetary factors considered or (b) the unevaluated factors are too trivial in consequence.
3. Where future estimates will not be enhanced by current estimating experience because (a) the unevaluated factor is infrequently experienced or (b) feedback information which would aid future studies is unavailable, too costly, or too delayed.

Because unevaluated factors are remaining considerations in our problems, we should again note that a factor such as flexibility might be monetized in one instance and not in another. Any list of possible nonmonetized factors should be so viewed; such examples include esthetic, political, scheduling, legal or security considerations, conservation aspects, social consequences, ethical implications, national goals, safety, flexibility, control, uncertain future price levels, and degree of commitment.

In earlier chapters we have suggested that, under some

---

[1] The term intangible seems inappropriate because many intangibles such as patents, copyrights, trademarks, mineral rights, trade secrets, contracts (including easements, leases, franchises, and options to buy) have a clearly identifiable value and are frequently reduced to monetary terms.

circumstances, an attempt to maximize utility rather than profit is appropriate. We have also suggested that some decisions only indirectly relate to monetary considerations; these include (1) the maximization of military effectiveness; (2) the stabilization of business cycles; (3) the maximization of social good; and (4) the ameliorization of national pride, concern, and awareness. When operating under such criteria, the broader term, *unevaluated,* better emphasizes that the *as yet* unevaluated influences must nonetheless be reckoned with in some manner, although not necessarily a monetary one.

Implication of Nonmonetized Factors upon the Stated Goal of Profit Maximization

Suppose we solve for the minimum monetary value of unevaluated factors required for project justification and find that monetary cost apparently exceeds nonmonetary value. Must we then reject the alternative?

To illustrate the issues involved ask yourself whether corporate responsibility should include:

1. Retraining an older, technologically displaced worker when a trained younger worker is available.
2. Updating and upgrading an employee with many years of service when a young, better trained but less experienced employee is available.
3. Hiring handicapped workers.
4. Providing a higher level of pollution control, working conditions, safety, or product responsibility than that required by law.
5. "Selling" the public on new procedures, methods, plans, or philosophies of the corporation.
6. A conscious effort to make the lives of its employees more fruitful.
7. Hiring high school students for summer work on the grounds that they need the wages and experience, even though their employment cannot be economically justified.
8. Providing free plant tours.
9. Producing a product whose quality and price make it truly superior to its competitors.
10. Providing research and development effort for the sake of the future.
11. Attempting to achieve employee participation in rather than acceptance of the changing production environment.

Proposal 2

Relating to Limiting Charitable, Educational
and Similar Contributions by the Company

Mrs. Evelyn Y. Davis, P.O. Box 2329, Grand Central
Station, New York, N.Y. 10017, who is the holder of
record of 10 shares of Common Stock of the Company,
has informed the Company that she intends to present
the following proposal at the meeting:

"**RESOLVED:** That the Corporation's Certificate
of Incorporation be amended by adding thereto the
following provisions: 'No corporate funds of this
corporation shall be given to any charitable,
educational or other similar organization, except
for purposes in direct furtherance of the business
interests of this corporation and subject to the
further provision that the aggregate amount of

such contributions shall be reported to the
shareholders not later than the date of the annual
meeting.' "

The statement submitted by this stockholder in
support of her proposal is as follows:

"**REASONS:** Your Company gives away millions
of your dollars to charity each year, money which
belongs to you.

"The cost of new safety equipment makes such a
limit especially desirable at Ford.

"If you AGREE, please mark your proxy FOR this
resolution, otherwise it is automatically cast
against."

Adoption of this proposal requires approval by a
majority of the votes that could be cast by stockholders
who are present in person or by proxy at the meeting,
computed in the case of each share as described in
the second paragraph on page 2 of this Proxy
Statement.

**The Board of Directors recommends a vote
"against" Proposal 2.**

The proposal requires an amendment to the Company's
Certificate of Incorporation. Under Delaware law,
such amendments are initiated by the Board of
Directors, and the Board is required to make a
declaration as to the advisability of the amendment.
The Board of Directors believes that the Company's
support of educational and charitable organizations
has been reasonable in amount and that the proposed
restriction would not be in the best interests of the
Company or its stockholders.

The Board of Directors believes that the Company
has an obligation, as a responsible corporate citizen,
to contribute to the support of educational and
charitable organizations. This is especially true of local
organizations serving the communities in which the
Company maintains facilities and in which Ford people
live and work. The contributions benefit the Company
by maintaining its favorable reputation and improving
the economic and social climate so vital to the
Company's success.

The Company contributed $11.4 million during 1968
for educational and charitable purposes, including
donations of automotive equipment to 848 schools,
colleges and universities. Of the total, $10 million
went to Ford Motor Company Fund, a non-profit
organization supported primarily by contributions

from the Company. The Fund makes substantial
contributions for educational purposes. It also
contributes to community funds, hospitals and other
social welfare organizations, both on a national level
and in areas where Company operations are located.

A significant part of the donations made by the
Company and the Fund have been used in efforts to
meet the crisis in our cities. Business and industry
have an obligation to serve the nation in times of
crisis, whether the danger is internal or external. It is
clear, moreover, that whatever seriously threatens the
stability and progress of the country and its cities also
threatens the growth of the economy and the Company.
Prudent and constructive Company efforts to help
overcome the urban crisis are demanded not only by
the Company's responsibility as a corporate citizen,
but by management's duty to safeguard the
stockholders' investment.

Corporate support of charitable and educational
organizations is encouraged by both corporate and
tax laws. The Delaware Corporation Law gives each
corporation created under its provisions the power to
"make donations for the public welfare or for
charitable, scientific or educational purposes." The
proposed resolution, which would prohibit all such
contributions unless made "in direct furtherance of
the business", would impose an ambiguous test and
unduly restrict management in discharging this
important responsibility.

Spaces are provided in the accompanying form of
proxy for specifying approval or disapproval of this
proposal, which is identified as Proposal 2.

**The Board of Directors recommends a vote
"against" Proposal 2.**

Fig. 18-1. The corporate citizen.

12. An active participation in the affairs of the community and its problems.

Can a corporation answer such questions? Can the corporation avoid responding to such questions?

Corporate responsibility includes a responsibility to employees, consumers, the community, stockholders, and the nation. It is not without conflict; consider the attractive structure with pleasing lines and landscaping versus the stark but adequate structure whose lack of landscaping saves caretaker costs. Let that choice be made more difficult by the fact that the corporation we have in mind is one of very modest means whose stockholders are primarily elderly widows depending upon their dividend income for support and one whose "customers" are volunteer agencies which provide aid to handicapped and underprivileged children. Consider too that the size of corporate contributions to the local United Fund may have to be weighed against the best interests of employees, consumers, and stockholders. We might even wonder whether the corporation should "use" its contributions in efforts to promote its public image. Should the corporation respond to the wishes of its employees, stockholders, and consumers in finding the degree to which it should act for social good? Or should the corporation actively seek approval for what appear to be appropriate corporate goals? Who should set goals, and who should provide approval?

The whole point of this discussion and the qualification upon our usual criterion of maximization of profit is this: *Resources allocated on a purely profit-maximizing basis are likely to be overallocated to commercial demands and underallocated to social needs.*

The Corporation As Citizen

Figure 18-1 reminds us of the delicate balances in the corporation's relationship to its stockholders, employees, and national community. While the quantitative evaluation of the worthiness of social goals and needs may be elusive, our financial response is clearly quantitative.

Example 18-1.   Evaluating the Nonmonetized Factors of a Make-or-Buy Decision

Information: Green Corporation is considering the "make" versus "buy" alternatives for Part Number 0022S. Blue

Corporation has quoted a firm price of $0.10 per unit on annual orders of 10,000 or more. Green expects to use these at the rate of 20,000 per year; after adjustment for all appropriate considerations regarding overhead, their own best estimate of cost is $0.12 per unit. Green Corporation feels the decision is still difficult to make because the following unevaluated factors favor the "make" alternative:

1. 0022S is new; it is likely to involve some design changes during the next year and implementation, coordination, speed, and accuracy of such changes are felt to be significantly better when Green is the producer.
2. Past company experience has frequently been that production costs decrease as experience is gained (with time) during the production of an item. If Blue is the producer, their production savings are not expected to be passed on; and Green or any other potential producer would be in a still less competitive position a year from now than they would be today.
3. Although 0022S is not of great consequence in itself, it is an essential component of a high-cost system. Supply failures due to production difficulties or design changes are therefore felt to be quite crucial.

One unevaluated factor favors the "buy" alternative: Blue Corporation specializes in the manufacture of mechanical components similar to 0022S, while Green Corporation, an electronics firm, is only moderately interested in acquiring technical know-how and equipment of this type.

Problem: Determine whether Green should make or buy Part 0022S.

Analysis: The extra cost of making the part is, at most, ($0.02) (20,000) = $400 the first year. Apparently, if purchased, the part could be expected to continue to cost $0.10 per unit, while if manufactured by Green, the cost of $0.12 per unit would gradually decrease. The reduced risk of supply failure is quite important in this instance; because of this, it is decided that Green should manufacture 0022S.

"Environmental Quality" Considerations in Capital Expenditure Decisions

We all share certain common property resources which yield service in their natural state. These resources include the air mantle, atmosphere, various bodies of water, complex ecological systems, the electromagnetic spectrum, and perhaps the absence of sound. Traditionally, these resources have been free to use or abuse. Such underappraisal of the value of environmental quality has lead to decisions which endanger man himself. Environmental resources can be repriced through regulation, financial incentives, education programs, and/or technical assistance. To be effective, regulation must be reasonable, efficient, equitable, enforceable, widely understood, reasonably popular, and not too costly. Enforcement efforts may involve fines and/or special levies on polluters. Financial incentives to better protect environmental quality include (1) the five-year writeoff allowed on antipollution equipment; (2) federal sharing of municipal and state expenditures for pollution control facilities; and (3) special loans, grants, subsidies, or other aids. Voluntary compliance of the majority will be an essential requirement regardless of the means of repricing.

Environmental quality questions involve complex tradeoffs. If Blue pollutes Green's environment, should Blue pay Green? Or should Green pay Blue to control pollution? Will pollution regulations shift costs from downstream to upstream users? Will pollution regulations cause firms which generate large volumes of oxygen-demanding waste water to move to areas with heavy rainfall and fast-moving streams? Will the outlawing of throwaway containers retard or prevent development of self-destructing containers? Will the "system" effect of water pollution regulation be such that polluters simply move to ocean-shore locations?

Income lost and medical bills paid by people who suffer from respiratory and other ills caused by air pollution are enormous. Other direct costs of pollution include cleaning bills, deterioration of buildings and equipment, damage to crops and other vegetation, loss of fish and other forms of aquatic life, and costs of treating water for home or industrial use. The problem of protecting and improving environmental quality resembles a massive case of deferred maintenance which may cost billions of dollars to correct.

Problems

18-1. Ten men have been trapped by an underground mine explosion. Rescue teams are brought in and great effort is made to remove debris from the area of supposed entrapment. Communication efforts are made but fail. Several days pass without encouraging prospects, then after several more days the rescue efforts are withdrawn. Explain how the rescue effort and withdrawal actions are consistent with each other and with a set of values enumerated in monetary units.

18-2. Green Corporation's product development chemists have just produced a new liquid effective in the removal of paint. It is superior to existing paint removers and appears quite promising. The paint remover is very toxic, however. If a child should accidentally drink some of this liquid, it could prove fatal within a period of 30 minutes. Continuous inhalation with inadequate ventilation, even outdoors in "pocket" areas or on windless days, could be dangerous to adults. Company chemists feel that about 1 person in 100 will have skin reactions (burning, itching) from use of the chemical. What is your advice at this point?

18-3. Suppose that the Green Corporation has given you the task of determining the proper level of annual contributions to various forms of charity. Let us call the possible recipients of Green's charity #1, #2, and so forth.

a. List some of the pertinent factors to consider in arriving at a "proper" overall level of charitable contributions.

b. Draw up a set of guidelines by which Green Corporation can determine whether charity #13 should receive any funds at all.

c. Suggest some guides by which Green Corporation can determine how total planned contributions can be divided among the recipients.

d. What can Green Corporation do about the possibility that some charitable requests not presently known but nonetheless likely will occur?

18-4. The Green Corporation is adding capacity. Alternative A calls for a full-size installation now, while Alternative B calls for a half-size installation now and another half-size unit four years from now. Either alternative gives the same eventual capacity, and eight years from now still greater capacity probably will be required. Assume equivalent costs are identical for the alternatives. List some of the unevaluated factors which should be considered.

18-5. It has been said that overattention to economic considerations subordinates designing for safety to designing for economic gain. Do you agree? Explain.

18-6. The damage to a local restaurant from a recent fire is extensive; the restaurant has been closed for cleanup work. The decision now faced by Mr. Green, the owner-manager, is whether (1) to spend the funds necessary to refurnish, remodel, and rebuild or (2) to sell the burned-out shell and the few undamaged items at whatever the market will bring. Mr. Green has agreed to the settlement sum offered by the insurance company. The settlement plus considerable additional funds supplied by Mr. Green would be required to reopen the remodeled restaurant, but the funds are available. Mr. Green states: "Before the fire, I never really considered whether I should quit business. During my twenty years of ownership, I have never had a single serious offer to buy. I have never publicly offered the restaurant for sale, but I really doubt that many people are willing to invest the money *and* a work week of sixty hours with few holidays and very short vacations. I suppose the big problem is that one simply cannot hire a dependable "fill-in" manager, and the business cannot support both a full-time manager and an owner. What has been a very profitable business for me is not attractive to someone who wishes to be simply an owner."

What additional information bearing on the owner-manager's decision would you like to have? Explain how this added information might influence the decision.

18-7. City Council members are discussing the pros and cons of a policy which would require all city contracts and purchases to be awarded to the low bidder. One council member states that there must be some evaluation of the bidders themselves with a preference given to those who have provided good work in the past and who have "stood behind" their work and products. Another member states that any departure from the objective low-bidder policy introduces politics into the selection and ultimately wastes public money. What policy do you recommend and why?

18-8. Green Corporation is currently considering two mutually exclusive alternatives. Alternative X involves purchasing a certain parcel of land for $100,000 now, while Alternative W calls for purchase of an equivalent parcel of land five years from now at a cost of $400,000. Because Green Corporation is a regulated public utility, the revenues from the two alternatives will not necessarily be the same. Regulatory policy is not known, but five different treatments of Alternative X appear possible.

Policy A: Disallow the "land held for future use" as part of the rate base and disallow related taxes during the first five years on the grounds that the land is not used in the utility system during that period. During the next 35 years permit inclusion of the land in the rate base at original cost of $100,000.

Policy B: Allow $100,000 cost of land as part of the rate base during the first five years, and during the next 35 years permit its inclusion in the rate base at original cost of $100,000.

Policy C: Allow $100,000 cost of the land as part of the rate base during the first five years, then permit its inclusion in the rate base at a "fair value" (from weighing original cost, replacement cost, current market value, and other evidences of value) of $250,000 during the next 35 years.

Policy D: Disallow the "land held for future use" as part of the rate base and disallow related taxes during the first five years, then permit inclusion of land in the rate base at replacement cost or market value of $400,000 during the next 35 years.

Policy E: Allow $100,000 cost of the land as part of the rate base during the first five years, and during the next 35 years permit its inclusion in the rate base at replacement cost or market value of $400,000.

Assume that inflation is not involved in the problem. (The dramatic increase in land value may be due to development of the nearby land area.) Property taxes are estimated as $2000 per year for the first five years for Alternative X and then will be $10,000 per year for the next 35 years under either alternative. $t = 50\%$. The rate of return allowed by the regulatory body is 10%; for simplicity rather than realism, assume financing is entirely by equity capital.

a. Compute the present equivalent revenue, *PER*, under each of the policies.

b. Assume present equivalent costs of: $PEC_x = \$270,000$; $PEC_w = \$540,000$. Then comment on which regulatory treatment seems the most appropriate and why.

VI

Capital Budgeting Considerations

*An engineer is one who can do with a dollar
what any bungler can do with two.*

ARTHUR WELLINGTON in *The Economic Theory of
Railway Location*, 1887.

19

Expenditure Opportunities and Sources of Funds

Capital Expenditures: Definition and Illustration

A CAPITAL EXPENDITURE is a cash outlay recouped slowly (for a period in excess of one year and with the exception of small outlays which are more conveniently treated as current expenses). Although coinciding closely with the accountant's concept of items which are to be capitalized rather than expensed, this capital expenditure definition includes such costs as advertising, training, and research, which may be expensed by the accountant. The capital expenditure embraced by a computer alternative includes not only the cost of the computer itself but also such items as:

Freight in.
Installation costs.
Sales tax.
Preproduction programming.
Initial training of personnel.
Costs of changeover from Brand X programs.
Added working capital for spare parts, tapes, storage files.

Housing costs where added space, specifically for the computer,
  is purchased.[1]
Interest foregone on funds during preuse—postpayment period.

In the case of a highway project, the capital expenditure em-
braces not only the cost of construction itself but also such
items as:

Right-of-way.
Design.
Survey.
An allocation of highway department overall service costs, in-
  cluding such items as inspection, contract letting, and blue-
  printing.

Types of Capital Expenditure Alternatives

Five classifications of alternatives are suggested as
distinguishable.

1. *Perfect substitutes.* Mutually exclusive courses of action for
   which the acceptance of one course rules out the acceptance
   of other courses of action in the same set. **Example:**
   (1) Overhead cable versus underground cable. (2) Various
   proposed highway routes between two adjacent cities.
2. *Imperfect substitutes.* Courses of action for which the acceptance of one course discourages the acceptance of other
   courses of action in the same set. **Example:** Bridge versus
   ferry service.
3. *Independent.* Courses of action for which the acceptance of
   one course bears no influence on the acceptance of other
   courses of action in that set. (Note that "perfect" independence also requires such restrictions as (1) capital not to
   be rationed, (2) manpower required for planning and implementing not to be overlapping, and (3) projects not to be
   interrelated by requirements for balanced spending by
   category, region, or other factor.) **Example:** A cost reduction project in the milling machine department and a new
   equipment item in the electroplating department.
4. *Imperfect complement.* Courses of action for which the acceptance of one course encourages the acceptance of other

---

[1] Frequently the "housing" or "space" costs are included in annual
costs as a "rental" cost.

courses of action in the same set. **Example:** Possession of an automobile encourages possession of a garage and vice-versa, but neither is a prerequisite of the other.

5. *Perfect complement.* Courses of action for which the acceptance of one course requires (is contingent upon) acceptance of another course of action in the same set. (These relationships are generally not symmetric; acceptance of A is contingent upon acceptance of B, but acceptance of B is not contingent upon acceptance of A. Note that B is a *prerequisite* of A.) **Example:** (1) TV set as a prerequisite for TV antenna. (2) Computer as prerequisite for its satellite equipment.

Note that the "perfect" terminology of (1), (3), and (5) must be in a relative degree rather than absolute.

Listing Capital Expenditure Opportunities

There are four basic phases to the task of producing a list of independent capital expenditure opportunities: (1) the search for possible alternatives, (2) analysis of the various sets and (generally) elimination of all but one of each set of mutually exclusive alternatives (this is the phase to which the major part of our attention has been focused in preceding chapters), (3) the collection of "winners" from each set of competing alternatives, and (4) sometimes the adjustment of computed rate of return (or analogous measure of acceptability) to reflect differences in risk associated with each of the independent alternatives. Such a listing of capital expenditure opportunities can help management to:

1. Plan the long-term future of the organization.
2. Avoid conflicting or overlapping projects.
3. Avoid overcommitment.
4. Stabilize the long-term capital expenditure pattern.

Ranking Capital Expenditure Opportunities

Once a list of opportunities has been produced, the next problem encountered is that of determining some means by which projects can be ranked according to their desirability. The problem is partly answered by the practice of listing as a group the "levelized" opportunities with the ensuing side-by-side comparison providing some help in the determination of priority. The remaining problem is the objective measurement of "desirability." Some possibilities are:

1. Rate of return on investment.
2. Ratio of present worth of receipts less present worth of expenditures to present worth of investment.
3. Present worth of receipts less present worth of expenditures.
4. Postponability, as measured by the dollar consequence upon the company of one year deferral of the project.
5. Urgency.
6. Present value sacrificed if the program were cut.
7. Effect of cut in the program on specified fish and game populations.
8. Impairment of logistic capability if the program were cut.
9. Change in number of rescues and reported violations if the program were cut.
10. Physical accomplishment sacrificed if the program were cut.
11. Present value of the loss of man-hours, the increase in future operating costs, or the increase in future outlays for medical care that would be incurred if the program were cut.[2]

Obviously the measurement of "desirability" is a problem; the revenue attributable to a specific transformer in an electric utility's distribution system may defy meaningful measurement. Such difficulties certainly hamper application of the first three of the listed measures of desirability. Example 20-4 shows, however, that even in such circumstances equivalent cost and rate of return techniques remain operational. In either case, our theoretical model is based on the assumption that projects are equivalent in all respects, including risk and excepting cash flows and their timing. In practice this required homogeneity is a serious operational difficulty. Only by ranking projects can the cutoff rate of return be conveniently determined. (The inconvenience of the trial-and-error approach theoretically necessary for the equivalent cost methods is considerable.) Ranking by present-worth techniques requires knowledge of the cutoff rate of return, although frequently the cost of external capital is

---

[2] For 6-11 and many additional possible criteria see R. N. McKean, *Efficiency in Government through Systems Analysis,* 2nd ed. (New York, John Wiley and Sons, 1964).

incorrectly applied. It will be shown subsequently that cutoff rate of return and cost of external capital are seldom identical, especially under the usual condition of a rationed supply of funds.

The ranked capital expenditure opportunities presented in graphic form can be considered an "investment opportunity" or "demand for funds" curve as in Figure 19-1. The figure suggests

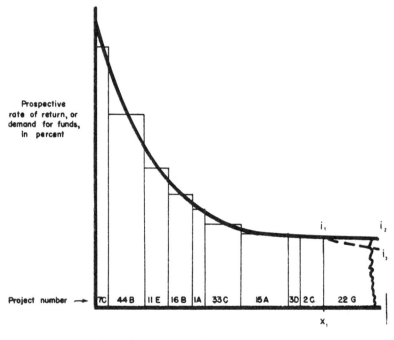

Fig. 19-1. The "investment opportunities" or "demand for funds" function.

that beyond some investment level, X_1, the capacity of the firm to handle adequately and economically a still larger expenditure level for the period is reduced. This point may be reached because the supply of such resources as manpower, management, or money is semilimited; that is, further quantities can be obtained only through the payment of substantial economic penalties. Occasionally such penalties may be acceptable to meet the unusual demands of national defense, emergency aid, response to catastrophe, and the like. To reflect this effect, the prospective rate of return, i_1 to i_2, may be adjusted downward as for i_1 to i_3.

Once the demand curve has been determined, the next step is the determination of the supply of funds and perhaps the supply curve.

Sources of Funds

A funds flow statement like that given in Figure 2-3 is very helpful in determining either the external funds required for a known capital expenditure level or the capital expenditure level permissible for a given supply of external funds.

Funds for the acquisition of new plant and equipment can come from both internal and external sources. Funds internally derived include those representing (1) cash receipts less *all* cash disbursements (= depreciation expense plus net income less dividends = after-tax cash flow less interest on debt and dividends on equity capital) (see also Fig. 10-2) and (2) decreased working capital. Cash derived from decreasing working capital is generally a "one-time" flow rather than a sum which might be generated over each of a number of years. Large changes in working capital can be the result of such policy changes as: (1) inventory reductions via production shutdowns, modernized purchasing procedures or pruning of slow-moving items from the stock or (2) accounts receivable reduction via a changeover from credit to cash only policy or by requiring prepayment of accounts by customers (as for some utility companies) or selling on a cash rather than consignment basis.

Externally derived funds are generally classified as debt (bonds, notes, equipment certificates) or equity (common stock or preferred stock). The supply curve of available funds may be arranged in ascending order of cost of capital.

Approximating the Cutoff Rate of Return

In the remainder of this chapter considerable space is devoted to discussion and illustration of means by which the cost of capital may be estimated. The relevance of this material is in its helpfulness to management in their task of determining a cutoff point in the organization's capital expenditure budget (as illustrated in Fig. 20-1). It is significant that the treatment of means by which the cutoff rate of return can be approximated may be more urgent than treatment of means of measuring the cost of capital when: (1) funds are rationed and (2) rate of return methods rather than equivalent cost methods are applied in evaluation procedures. Where these conditions prevail, it may be preferable to reverse the sequence of Chapter 20 and the remainder of this chapter.

Measures of the Cost of Funds

Determination of a supply curve of funds requires measurement of the cost of capital. Several measures of this cost are suggested.

Internal Funds. When the firm has funds in excess of capital expenditure needs, its alternative uses of that excess of internally generated funds include:

1. Purchase of securities (perhaps common stock) of another firm.
2. "Calling" or repurchase of the company's own outstanding debt capital.
3. Purchase of the company's own outstanding equity capital (then held as "treasury stock").
4. Increased dividend payments.[3]

In purchasing the common stock of another firm, income tax considerations may encourage the search for stocks whose prospective return tends to be primarily in price appreciation rather than in substantial dividend payments. Because of income taxes, brokerage fees, and the like, such investments might yield a net return of 4% to 8%. The assumption of rates as high or higher than external cost of capital would be questionable if not dangerous.

"Calling" or repurchase of the company's own outstanding debt capital could produce apparent savings as follows. Assume the following data pertaining to an outstanding bond issue:

	Per Bond
Call price of bond	$1050
Annual interest payment	40
Annual cost of handling interest payments, and so forth	2
Price if redeemed at maturity, including handling	1020
Handling costs of reacquisition (calling) of bond	20
Years to maturity = 6 years	

---

[3] Limited by the current level of earned surplus and sometimes limited by provisions of a preferred stock or bond issue.

Compute the interest rate saved by calling:

$$PEX = \$0 = -\$1050 - \$20 + \$42(p/a)_6^i + \$1020(p/f)_6^i$$
At $i = 4\%$:$PEX = -\$44$
At $i = 3\%$:$PEX = +\$12$
so $i = 3\% + 1\% (12/56) = 3.2\%.$

The real savings may be even greater since the debt capital still outstanding is made more secure (as indicated by such ratios as "times charges earned"[4] or "debt to fixed assets") and leverage on outstanding equity capital is reduced. Indeed, if the theory that total capital cost tends to be constant independent of the capital structure of the firm is applicable, then real savings may approach the cost of external capital.[5] Again we can suggest that savings produced by calling of corporate bonds are 3.2% to perhaps 8%.

The equity stock repurchased from the market by the firm is called treasury stock. Since the dividends on this are then retained by the company, that portion certainly represents savings. Growth in earnings per share may lead to an increased market price for the stock and potential gains may thus arise. If our example which follows is realistic, then the cost of internal funds as indicated by this alternative use is somewhere between 5% (dividends only considered) and 10% (dividends and price appreciation considered). For equity stock assume:

	Per Share
Current market price of stock	$100.00
Current rate of annual earnings anticipated and required to maintain market price of stock	4.80
Annual cost of handling, recording, and so forth, of dividends/ownership	0.20
Growth of annual earnings anticipated and required to maintain market price of stock (neglecting any growth due to plowback of earnings)	0.50
Brokerage fees, administrative, and handling costs	= 2% of transaction

---

[4] Times charges earned = $TCE = \dfrac{\text{taxable income} + \text{interest on debt capital}}{\text{interest on debt capital}}$

[5] See treatment by F. Modigliani and M. Miller in: E. Solomon, ed., *The Management of Corporate Capital* (New York, Macmillan, 1964), pp. 150-80.

Compute cost of this equity capital:

$$PEX = \$0 = -\$100.00 - \$2.00 + (\$4.80 + \$0.20)(p/a)^i_\infty$$
$$+ \$0.50\,(p/g)^i_\infty$$
$$= -\$102.00 + \$5.00(1/i) + \$0.50(1/i^2)$$

By the quadratic formula, factoring, or trial-and-error solution with interpolation, $i = 9.9\%$.

Compute composite cost of "called" bonds and treasury stock assuming that existing proportions of outstanding debt and equity capital are maintained:

$$
\begin{aligned}
&\text{Debt } 20\% @ 3.2\% &= 0.64\% \\
&\text{Equity } 80\% @ 9.9\% &= 7.92\% \\
&\text{Composite return} &= 8.56\%
\end{aligned}
$$

The calculations above show an apparent cost of internally generated funds of about $8\frac{1}{2}\%$.

Increased dividend payments are a possible use for excess funds and can furnish some guidelines to the question of cost of internal funds. Changes in dividend payments are of concern to the stockholder; he might rationally object to increased dividends on the basis that it produces extra cash for him which is taxed as income, and if the residue is reinvested brokerage fees may arise. From the stockholder's point of view it might be much more efficient and less costly for the firm to reinvest its own funds. The prospect of future increased earnings resulting from reinvested present earnings may make the firm's stock more attractive and benefit both its market price and the apparent cost of future equity funds. Reduced dividend payments may meet resistance from those persons whose cash planning is based on a historic level or pattern of dividends and hence may reduce market price of the stock, thus increasing the apparent cost of future equity funds. Because of stockholders' widely varying income tax brackets and widely varying dependence upon dividend income, changes in dividend payments are difficult to use as a basis for inference of the cost of internally generated funds.

On the basis of all the preceding alternative uses for excesses of internally generated funds, we might estimate that the cost is somewhere between 5% and 9%.

External Funds: Debt, Equity, and Composite. When capital expenditure needs exceed the internal funds available, the organization should question the availability and cost of external funds. Several options are open under such circumstances:

1. Financing by means of debt capital.
2. Financing by means of equity capital.
3. Leasing facilities.
4. Limiting the acceptance of capital expenditures so that available funds will not be exceeded (rationing capital).

The cost of debt capital might be estimated as follows. Assume the following data pertaining to a prospective bond issue:

	Per Bond
Estimated issue price of bond	$1000
Underpricing, legal, administrative, and underwriting costs in sale of bond issue	50
Current interest yield on comparable issues in dollars per year	45
Annual cost of handling interest payments, and so forth	2
Price if redeemed at maturity, including handling	1020
Planned term of issue = 20 years	

Compute cost of this debt capital:

$$PEX = \$0 = +\$1000 - \$50 - (\$45 + \$2)(p/a)_{20}^i - \$1020(p/f)_{20}^i$$
$$= \$950 - \$47(p/a)_{20}^i - \$1020(p/f)_{20}^i$$

At i = 4%: $950 - $47(13.590) - $1020(0.4564) = -$150

At i = 5%: $950 - $47(12.462) - $1020(0.3769) = +$20

so i = 4% + 1% (150/170) = 4.9%.

The cost of equity capital might be estimated as follows. Assume the following data pertaining to a prospective issue of additional common stock:

	Per Share
Current market price of stock	$100.00
Underpricing, legal, administrative, and underwriting costs in sale of stock issue	10.00
Current rate of annual earnings anticipated and required to maintain market price of stock	4.80
Annual cost of handling, recording, and so forth, of dividends and ownership	0.20
Growth of annual earnings anticipated and required to maintain market price of stock (neglecting any growth due to plowback of earnings)	0.50

Compute cost of equity capital:

$$PEX = \$0 = +\$100.00 - \$10.00 - (\$4.80 + \$0.20)(p/a)^i_\infty$$
$$- \$0.50(p/g)^i_\infty$$
$$= +\$90.00 - \$5.00(1/i) - \$0.50(1/i^2)$$

By the quadratic formula, factoring, or trial-and-error solution with interpolation, $i = 10.7\%$.

Compute composite cost of the prospective bonds and stock, assuming the same proportions of outstanding debt and equity capital are maintained:

$$\begin{array}{ll} \text{Debt } 20\% @ 4.9\% & = 0.98\% \\ \text{Equity } 80\% @ 10.7\% & = \underline{8.56\%} \\ \text{Composite cost} & = 9.54\% \end{array}$$

The apparent cost of externally generated funds is about $9\frac{1}{2}\%$.

A number of terms helpful in the communication of certain aspects of financing are given in Table 19-1; the illustration employs some data from Chapter 2.

Return on or Cost of Equity Capital

Equity capital cost or return can be approached from the viewpoint of (1) a prospective purchaser of common stock who wishes to estimate its prospective rate of return or (2) a firm which wishes to estimate its cost of equity capital, expressed as an interest rate. One method of determination can serve either viewpoint.

TABLE 19-1. **Definition and Illustration of Some Terminology of Financing**

Terms	Utility	Industrial
Number of shares of common stock*	10 million	39.2 million
Book value of common stock (see Fig. 2-1) $= \dfrac{\text{par value + capital surplus + earned surplus}}{\text{number of shares of common stock}}$	\$ 62.00	\$ 20.13
Earnings per share of common stock (see Fig. 2-2) $= EPS = \dfrac{\text{operating return on equity}}{\text{number of shares of common stock}}$	\$ 5.40	\$ 4.28
Dividends per share of common stock (see Fig. 2-3) $= DPS = \dfrac{\text{dividends on equity capital}}{\text{number of shares of common stock}}$	\$ 3.60	\$ 2.14
Payout $\% = \dfrac{\text{dividends on equity capital}}{\text{operating return on equity}}$	$66\frac{2}{3}\%$	50%
Plowback \$ = operating return on equity- dividends	\$18 million	\$84 million
Plowback $\% = \dfrac{\left(\begin{array}{c}\text{operating return}\\ \text{on equity}\end{array}\right) - \text{dividends}}{\text{operating return on equity}}$	$33\frac{1}{3}\%$	50%
Market price per share*	\$100.00	\$100.00
Price-earnings ratio $= \dfrac{\text{market price per share}}{\text{earnings per share}}$	18.5	23.4
Estimated growth (arithmetic) in *EPS* per year before effect of plowback is considered*	\$ 0.405	\$ 0.50

*Arbitrarily specified for this illustration.

Let DPS_z = dividends per share for year z (treated as end-of-year)

EPS_z = earnings per share for year z (treated as end-of-year)

PPS_z = price per share at end of year z

i = annual effective rate of return on, or cost of, equity capital

D/E = dividend payout ratio = DPS_z/EPS_z

Then the present value per share of the future cash flow stream is

$$PPS_0 = \sum_{z=1}^{\infty} \frac{DPS_z}{(1+i)^z} = \sum_{z=1}^{n} \frac{DPS_z}{(1+i)^z} + \frac{PPS_n}{(1+i)^n} \qquad (19\text{-}1)$$

If reinvested funds earn the same rate of return as present capital earns, the dividend payout ratio, D/E, does not influence the prospective rate of return or cost of capital, i. This assumption would permit us to develop an alternate expression for *PPS*. First, note that reported year-by-year earnings per share are defined as $EPS_1, EPS_2, \ldots, EPS_n$. Next, remove the effect of retained earnings so that the revised earning per share stream can be evaluated as if all earnings had been paid out as dividends. Revised earnings would be:

$$EPS_1 \; ; EPS_2 \; - \; i(1 - D/E)\, EPS_1 \; ; EPS_3$$
$$- \; i(1 - D/E)(EPS_1 + EPS_2) \cdots$$

The present value of that future cash flow stream is

$$PPS_0 = \frac{EPS_1}{(1 + i)} + \frac{EPS_2 - i(1 - D/E)\, EPS_1}{(1 + i)^2}$$
$$+ \frac{EPS_3 - i(1 - D/E)(EPS_1 + EPS_2)}{(1 + i)^3} \cdots$$

$$(19\text{-}2)$$

Equation 19-2 is recommended for cases where it is simpler to model *EPS* than *DPS*, and where $D/E = 0$.   Where $D/E = 1$, Equations 19-1 and 19-2 are identical.

   To illustrate application of Equation 19-2 we might consider *EPS* data for any of the companies of Table 19-2. For companies B_1 or B_2, with $i = 10\%$:

$$PPS_0 = (\$6.00/1.1) + (\$6.40/1.1^2) + (\$6.80/1.1^3) \cdots$$
$$= \$6.00(p/a)_\infty^{10\%} + \$0.40(p/g)_\infty^{10\%} = \$100.00$$

   If dividends per share are growing (geometrically) by a constant *rate* per year forever and if the growth rate, r, is less than the valuation rate, i; then from Equation 4-11:

$$PPS_0 = DPS_1 \; {}^r(p/c)_n^i = DPS_1 \, \frac{1}{1 + r}(p/a)_n^x$$

When $r < i$,

$$x = \frac{1 + i}{1 + r} - 1 = \frac{i - r}{1 + r}$$

and when n is infinite,

TABLE 19-2. Financial Data for Several Hypothetical Firms All Having an Equity Capital Cost of 10%

			End of Year			
		0	1	2	3	4
Company A_1	Market price	$100.00	$100.00	$100.00	$100.00	$100.00
No growth in EPS,	EPS		10.00	10.00	10.00	10.00
100% payout of	DPS		10.00	10.00	10.00	10.00
EPS	PPE ratio		10	10	10	10
Company A_2	Market price	$100.00	$105.00	$110.25	$115.77	$121.56
No growth in EPS	EPS		10.00	10.50	11.02	11.58
except plowback	DPS		5.00	5.25	5.51	5.79
earning 10%, 50% payout of EPS	PPE ratio		10.5	10.5	10.5	10.5
Company A_3	Market price	$100.00	$110.00	$121.00	$133.10	$146.41
No growth in EPS	EPS		10.00	11.00	12.10	13.31
except plowback	DPS		0	0	0	0
earning 10%, 0% payout of EPS	PPE ratio		11	11	11	11
Company B_1	Market price	$100.00	$104.00	$108.00	$112.00	$116.00
Arithmetic growth	EPS		6.00	6.40	6.80	7.20
in EPS of $0.40 per	DPS		6.00	6.40	6.80	7.20
year, 100% payout of EPS	PPE ratio*		17.3	16.9	16.5	16.1
Company B_2	Market price	$100.00	$106.00	$112.20	$118.61	$125.23
Arithmetic growth	EPS		6.00	6.60	7.22	7.86
in EPS of $0.40 per	DPS		4.00	4.40	4.81	5.24
year augmented by plowback earning 10%, $66^2/_3$% payout of EPS	PPE ratio		17.7	17.0	16.4	15.9
Company C_1	Market price	$100.00	$105.00	$110.00	$115.00	$120.00
Arithmetic growth	EPS		5.00	5.50	6.00	6.50
in EPS of $0.50 per	DPS		5.00	5.50	6.00	6.50
year, 100% payout of EPS	PPE ratio*		21.0	20.0	19.2	18.5
Company C_2	Market price	$100.00	$107.50	$115.37	$123.64	$132.32
Arithmetic growth	EPS		5.00	5.75	6.54	7.36
in EPS of $0.50 per	DPS		2.50	2.88	3.27	3.68
year augmented by plowback earning 10%, 50% payout of EPS	PPE ratio		21.5	20.1	18.9	18.0
Company C_3	Market price	$100.00	$110.00	$121.00	$133.10	$146.41
Arithmetic growth	EPS		5.00	6.00	7.10	8.31
in EPS of $0.50 per	DPS		0	0	0	0
year augmented by plowback earning 10%, 0% payout of EPS	PPE ratio		22.0	20.2	18.7	17.6
Company D_1	Market price	$100.00	$110.00	$120.00	$130.00	$140.00
Arithmetic growth	EPS		0	1.00	2.00	3.00
in EPS of $1.00 per	DPS		0	1.00	2.00	3.00
year, 100% payout of EPS	PPE ratio		∞	120.0	65.0	46.7
Company D_2	Market price	$100.00	$110.00	$120.50	$131.53	$143.10
Arithmetic growth	EPS		0	1.00	2.05	3.15
in EPS of $1.00 per	DPS		0	0.50	1.02	1.58
year augmented by plowback earning 10%, 50% payout of EPS	PPE ratio		∞	120.5	64.2	45.4

TABLE 19-2. (Continued)

		End of Year				
		0	1	2	3	4
Company D₃	Market price	$100.00	$110.00	$121.00	$133.10	$146.41
Arithmetic growth	*EPS*		0	1.00	2.10	3.31
in *EPS* of $1.00 per	*DPS*		0	0	0	0
year augmented by	*PPE* ratio		∞	121.0	63.4	44.2
plowback earning						
10%, 0% payout						
of *EPS*						
Company E₁	Market price	$100.00	$104.00	$108.16	$112.49	$116.99
Geometric growth	*EPS*		6.00	6.24	6.49	6.75
in *EPS* of 4% per	*DPS*		6.00	6.24	6.49	6.75
year forever, 100%	*PPE* ratio		17.3	17.3	17.3	17.3
payout of *EPS*						
Company E₂	Market price	$100.00	$107.00	$114.43	$122.31	$130.67
Geometric growth	*EPS*		6.00	6.54	7.12	7.73
in *EPS* of 4% per	*DPS*		3.00	3.27	3.56	3.87
year forever aug-	*PPE* ratio		17.8	17.5	17.2	16.9
mented by plow-						
back earning 10%,						
50% payout of						
EPS						
Company E₃	Market price	$100.00	$110.00	$121.00	$133.10	$146.41
Geometric growth	*EPS*		6.00	6.84	7.77	8.81
in *EPS* of 4% per	*DPS*		0	0	0	0
year forever aug-	*PPE* ratio		18.3	17.7	17.1	16.6
mented by plow-						
back earning 10%,						
0% payout of *EPS*						
Company F₁	Market price	$100.00	$102.75	$104.33	$104.40	$104.40
Geometric growth	*EPS*		7.25	8.70	10.44	10.44
in *EPS* of 20% per	*DPS*		7.25	8.70	10.44	10.44
year for the next 3	*PPE* ratio		14.2	12.0	10.0	10.0
years, then growth						
ceases; † 100%						
payout of *EPS*						
Company F₂	Market price	$100.00	$106.38	$112.44	$118.10	$124.05
Geometric growth	*EPS*		7.25	9.06	11.26	11.82
in *EPS* of 20% per	*DPS*		3.62	4.54	5.63	5.91
year for the next 3	*PPE* ratio		14.7	12.4	10.6	10.6
years, then growth						
ceases;† augmented						
by plowback earn-						
ing 10%, 50%						
payout of *EPS*						
Company F₃	Market price	$100.00	$110.00	$121.00	$133.10	$146.41
Geometric growth	*EPS*		7.25	9.42	12.10	13.31
in *EPS* of 20% per	*DPS*		0	0	0	0
year for the next 3	*PPE* ratio		15.2	12.8	11	11
years, then growth						
ceases;† augmented						
by plowback earn-						
ing 10%, 0% pay-						
out of *EPS*						
Company G₁	Market price	$100.00	$107.19	$112.29	$112.29	$112.29
Geometric growth	*EPS*		2.81	5.61	11.23	11.23
in *EPS* of 100%	*DPS*		2.81	5.61	11.23	11.23
per year for the	*PPE* ratio		38.1	20.0	10.0	10.0
next 3 years, then						
growth ceases;†						
100% payout of						
EPS						

*Approaches 10.0 as n increases without limit.

†**Caution.** The assumption of geometric growth results in an explosive model (present worth of earnings is infinite) if n is infinite and the rate of growth assumed is equal to or more than the interest rate.

$$\frac{1}{1+r}(p/a)_n^x = \frac{1+r}{(1+r)(i-r)} = \frac{1}{i-r}$$

So where $r < i$ and n is infinite,

$$PPS_0 = DPS_1/(i-r) \qquad\qquad (19\text{-}1a)$$

or

$$i = DPS_1/PPS_0 + r \qquad\qquad (19\text{-}1b)$$

If "earnings per share less plowback" are growing by a constant *rate* per year forever and if the growth rate, r, is less than the valuation rate, i, equations can be derived as they were for *DPS*:

$$PPS_0 = EPS_1/(i-r) \qquad\qquad (19\text{-}2a)$$

or

$$i = EPS_1/PPS_0 + r \qquad\qquad (19\text{-}2b)$$

as in Table 19-2 for Company E.

Where payout is less than 100%, we need only to reduce the observed rate of growth in earnings to allow for that due to plowback. Note that the period of *ownership* need not be infinite.

The preceding equations and the company data in Table 19-2 indicate that where reinvested funds earn at the same rate as present capital, the dividend payout policy has no bearing on the cost of capital. This does not mean that management is unrestricted in its dividend payout policy; experience indicates that variation from an established payout percent range may meet with stockholder opposition. Payout percent is neither good nor bad; high payout ratios tend to attract buyers who wish liberal dividend income, while low payout ratios tend to attract buyers who, perhaps for tax reasons, may prefer to take their income via capital appreciation. We might further expect that in times of greater economic uncertainty or for the conservative or pessimistic prospective stockholder, there would be a preference for stocks with high payout ratios; a preference for low payout, high growth can be expected where the circumstances are reversed.

Given current earnings, the most important determinants of

the price-earnings ratio are the valuation rate and the prospective growth of those earnings. When the valuation rate = 10% and payout is 100%,

Arithmetic Growth As a Percent of Base Year *EPS*	Theoretical Price-to- Earnings Ratio	Geometric Growth As a Percent of Last Year *EPS*	Theoretical Price-to- Earnings Ratio
0	10.0	0	10.0
5	15.5	2	12.7
10	21.0	4	17.3
20	32.0	6	26.5
40	54.0	8	54.0

If there were no costs associated with the sale, purchase, handling, or planning of stock issues, the rate of return appropriate to the tabled data, $i = 10\%$, could be interpreted both as a prospective cost of equity capital to the company and as a prospective rate of return to the buyer of such stock. The costs of sale, purchase, handling, and planning are analogous to friction; because the acquisition of funds is not a frictionless process, the firm may have a cost of equity capital exceeding 10%, while the stockholder earns a rate of return less than 10%.

Cautions in Estimating Cost of Capital. Finally, let us note several cautions in the estimation of cost of capital as it relates to the source of funds.

1. $\begin{pmatrix} \text{Cost of capital for} \\ \text{governmental projects} \end{pmatrix} \geqslant \begin{pmatrix} \text{taxpayers' marginal} \\ \text{opportunity cost of capital} \end{pmatrix}$
Cost of capital from tax funds $> 0\%$. When agencies of government obtain funds from general governmental revenues, such as via appropriations, there is a temptation to classify these as interest-free funds. Here our concept of alternative uses of funds is especially helpful in providing insight into the problem; if the money is not taken from the taxpayer via taxes, what alternative uses does the taxpayer have? Advance payments on a house mortgage, cash rather than time purchase of furniture, advance payments on an auto, or investments in various securities are among the possibilities. The widespread applicability of the first opportunity certainly supports an "opportunity" interest rate

of about 6%. On this basis, we caution that it both is inefficient and is a misallocation of funds for a government to invest funds at 3% when the taxpayer could invest those same
funds at a higher rate of return. It is unfortunate that use of
extremely low interest rates for governmental projects seems
to be the rule rather than the exception.

2. Note that when external funds are raised via debt capital one
   year and equity capital the next, a question is raised as to
   how the cost of external funds on a year-by-year basis should
   influence economy studies. We recommend the use of a
   single rate of return requirement in engineering economy
   studies and base this on (1) studies which indicate that debt-
   equity proportions have little influence on overall cost of
   capital,[6] (2) the practical need for an accept-reject device
   which is relatively stable on a year-by-year basis, and (3) the
   impracticality of associating specific funds with specific
   projects.

3. Prospective cost of capital $\neq$ regulatory cost of capital.
   Regulated public utilities are, under certain circumstances,
   permitted to earn a specified rate of return which is called
   the cost of capital. We have several objections to using such
   a figure as prospective cost of capital for engineering economy purposes. The regulatory cost of capital is generally a
   historic one. Debt capital obtained many years ago at 3% is
   permitted to influence regulatory cost of capital; with current market rates at much higher levels, such historic costs
   are irrelevant to the cost of new funds to finance new projects. The proportions of debt and equity used in regulatory
   cost of capital are the historic ones reflected by book values
   rather than the current ones reflected by market values. The
   regulatory cost of capital is generally applied to a rate-base
   figure based on both book and current market values, and
   hence its application is not even comparable to that of the
   prospective cost of capital. Prospective cost of capital is
   determined in the marketplace, while regulatory cost of
   capital is a "finding" of the regulatory or judicial body.

4. Prospective cost of capital $\neq$ minimum attractive rate of return $= MARR$. As pictured in Figure 20-1 and as discussed in
   that section, the prospective cost of capital coincides with

---

[6]See A. Barges, *The Effect of Capital Structure on the Cost of Capital*
(Englewood Cliffs, N.J., Prentice-Hall, Inc., 1963); F. Modigliani and M.
Miller in: E. Solomon, ed., *The Management of Corporate Capital* (New
York, Macmillan Co., 1964), pp. 150–80.

the minimum attractive rate of return only when both (1) internally generated funds are not sufficient to meet investment demand needs and (2) the firm obtains *all* the necessary external funds to provide for projects which can be justified at a *MARR* identical to the cost of capital; that is, there is no capital rationing. Fortune magazine reports a statistic which suggests that capital rationing applies to most U.S. industrial firms;[7] surveys and experience indicate that insufficient capital (cutoff rate of return requirement greater than the cost of external funds) is experienced by the great majority of firms.

The Buy versus Lease versus Status Quo Decision

The preceding section dealing with external funds included leasing as a means of financing equipment. The buy versus lease question generally is regarded as a *financing* decision, just as the buy versus status quo question generally is regarded as an *equipment* decision.

In general it is recommended that separable decisions be made separately, not mixed; not all choices can be so nicely partitioned, however. Consider the comparisons in Figure 19-2 and their possible outcomes:

	Case 1	Case 2	Case 3	Case 4
Equipment decision favors (status quo versus buy)	buy	buy	status quo	status quo
Financing decision favors (buy versus lease)	buy	lease	buy	lease
Conclusion	buy	lease	status quo	?

In Case 3, for example, if we have found that "status quo" is better than "buy," and have also found that "buy" is better than "lease," we conclude that status quo is "best" of the three alternatives. Case 4 shows, however, that we cannot always separate the *equipment* and *financing* decisions. In Case 4 we

---

[7]See *Fortune*, 72 (August, 1965): 96, where it is reported that 65 of the 500 largest U.S. industrials ended the year 1964 with more than 20% of assets in cash items.

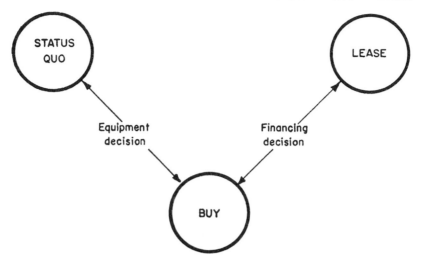

Fig. 19-2. Apparent separability of the "status quo versus buy" and the "buy versus lease" decisions.

must compare status quo with lease to reach a conclusion, and this by definition will be a *mixed* decision involving both the equipment and financing decisions. The point here is a necessary one since many treatments of leasing have correctly compared *buy* versus *lease* but have neglected to provide a means of comparing *status quo* versus *lease*.

Leasing

Leasing is a method of financing an asset via rental payments over the period of use or period of the lease agreement. Leasing is a popular method of financing; in many instances a building is constructed, then sold, then leased back. The modern supermarket, office building, railroad car, airliner, automobile, office copier, or business machine are examples of the wide array of facilities which may be leased.

Lease specifications are generally detailed in a formal written contract. The contract may contain such specifications as:

1. The amount, timing, and termination date of rental payments.
2. "Escape" provisions, if any. Is the contract cancelable for some fixed penalty fee? Can the lessee sublet the facilities if his own need for the equipment terminates or subsides?
3. Purchase privileges. Are rental payments applicable in part

or in total to subsequent purchase of the item? While such a privilege may be valuable, tax laws will not permit deduction of full rental payment where the purchase option gives the lease the character of a *conditional sales contract.*

4. Renewal privileges. Is the lease renewable? Is the renewal price specified?

5. Other features. Can lessee modify the item? Does the 99-year lease of an existing building permit the lessee subsequently to modify the structure? Who pays for maintenance, property taxes, and insurance? Who is liable in case of accidents?

The specific type of lease which is of special concern is one which (1) tends to be a fixed obligation with little "escape" possible under adverse conditions and (2) is in effect for relatively long periods, say three years or more. Such leases are frequently referred to as financial leases. The long-term nature of the lease may be a result of either the contract itself or subsequent operating conditions which somehow preclude termination of the lease.

The financial lease is of special concern because it carries with it certain subtle but important disadvantages. Its impact is similar to that of added debt capital; both will:

1. Reduce the organization's ability to attract further debt capital. The increased fixed charges (for debt repayment and return or lease payment) decrease the "times fixed charges earned" ratio. Conservative practice or provisions of presently outstanding securities may prohibit the sale of further debt capital when that ratio drops to 2.0 or less.

2. Increase variability (leverage) in prospective earnings on equity capital.

Financial leasing thus impairs the position of both debt and equity capital. The financial effects of borrowing equipment are the same as those of borrowing funds. To allow for these effects, we next develop an approach in which we hypothetically replace enough debt capital with composite capital so that the financial effects of leasing are fully offset (no impairment takes place).

If the debt-equity relationship is already at an optimum, additional debt capital or leasing simply increases the risk on and hence cost of existing and future debt and equity capital.

If the debt-equity relationship is *not* already at an optimum, the question of future financial structure and the savings from its improvement should be kept separate from the question of acceptability of any proposed equipment. If leasing would actually result in an improved financial structure, identical results could be produced through increments of debt financing.

An Equation for Evaluating the Impact of a Financial Lease

Let ALR = a lease "inescapability" rating (from 0.0 for the easily escaped costs of an equipment lease, to 1.0 for an inescapable financial lease). The amount of debt capital which would produce the same annual obligation as a lease calling for payment of ALP dollars at the end of each of the next x years is:

$$ALP(ALR)(p/a)^{i_d}_x$$

If that amount of debt capital were replaced with an identical amount of composite capital, the before-tax cash flow requirement, BTCFR, in year x would increase by:

$$ALP(ALR)(p/a)^{i_d}_x \, [i_a/(1-t) - i_d]$$

The present equivalent of such increased *BTCFR* for an n year lease is:

$$ALP(ALR) \, [i_a/(1-t) - i_d] \, K$$

Where

$$K = (p/a)^{i_d}_n (p/f)^{i_a}_1 + (p/a)^{i_d}_{n-1} (p/f)^{i_a}_2 + \cdots + (p/a)^{i_d}_1 (p/f)^{i_a}_n$$

$$= \frac{1 - (1 + i_d)^{-n}}{i_d(1 + i_a)} + \frac{1 - (1 + i_d)^{-n+1}}{i_d(1 + i_a)^2} + \cdots + \frac{1 - (1 + i_d)}{i_d(1 + i_a)^n}$$

$$= 1/i_d \, \{ (p/f)^{i_a}_1 + (p/f)^{i_a}_2 + \cdots + (p/f)^{i_a}_n - [(p/f)^{i_a}_1 (p/f)^{i_d}_n$$
$$+ (p/f)^{i_a}_2 (p/f)^{i_d}_{n-1} + \cdots + (p/f)^{i_a}_n (p/f)^{i_d}_1] \}$$

$$= 1/i_d \, \{ (p/a)^{i_a}_n - [(p/f)^{i_a}_1 (p/f)^{i_d}_n + (p/f)^{i_a}_2 (p/f)^{i_d}_{n-1} + \cdots$$
$$+ (p/f)^{i_a}_n (p/f)^{i_d}_1] \}$$

Multiply by $(1 + i_d)/(1 + i_a)$:

$$K \left(\frac{1 + i_d}{1 + i_a} \right) = \frac{1}{i_d} \left\{ \frac{1 + i_d}{1 + i_a} (p/a)_n^{i_a} \right.$$

$$\left. - \left[(p/f)_2^{i_a} (p/f)_{n-1}^{i_d} + \cdots + (p/f)_{n+1}^{i_a} (p/f)_0^{i_d} \right] \right\}$$

Subtract the last equation from the one which precedes it:

$$K \left(\frac{1 + i_a - 1 - i_d}{1 + i_a} \right) = \frac{1}{i_d} \left\{ \frac{i_a - i_d}{1 + i_a} (p/a)_n^{i_a} \right.$$

$$\left. - \left[(p/f)_1^{i_a} (p/f)_n^{i_d} - (p/f)_{n+1}^{i_a} \right] \right\}$$

So

$$K = \frac{1}{i_d} \left\{ (p/a)_n^{i_a} - \frac{1}{i_a - i_d} \left[(p/f)_n^{i_d} - (p/f)_n^{i_a} \right] \right\}$$

Let PLI = the present equivalent "financial impairment effect" from leasing. Then the present equivalent of increased $BTCFR$ equals

$$PLI = ALP(ALR) \left[\frac{i_a}{i_d (1 - t)} - 1 \right] \left\{ (p/a)_n^{i_a} - \frac{1}{i_a - i_d} \right.$$

$$\left. \cdot \left[(p/f)_n^{i_d} - (p/f)_n^{i_a} \right] \right\} \qquad (19\text{-}3)$$

Let ALI = the annual equivalent "financial impairment effect" from leasing. Then the annual equivalent of increased $BTCFR$ equals

$$ALI = ALP(ALR) \left[\frac{i_a}{i_d(1 - t)} - 1 \right] \left\{ 1 - \frac{(a/p)_n^{i_a}}{i_a - i_d} \right.$$

$$\left. \cdot \left[(p/f)_n^{i_d} - (p/f)_n^{i_a} \right] \right\} \qquad (19\text{-}4)$$

Example 19-1.   Evaluating the Impact of a Financial Lease

Information: Green Corporation is considering the use of certain equipment to replace the present (status quo) method which requires no equipment and has an annual equivalent cost of $22,000 per year. The proposed method calls for equipment that can be obtained by a financial lease, $ALR = 1.0$, calling for end-of-year payments of $20,000 for each of the next three years. The need for the service provided by either method terminates in three years.

The Corporation is subject to a combined effective state and federal income tax rate of 50%. Half the present financing consists of debt capital costing 8%; the remaining capital is equity with an opportunity cost of 16%. Note that

$$i_c = 0.5(0.08) + 0.5(0.16) = 0.12 = 12\%$$
$$i_a = i_c - tr_d i_d = 0.12 - 0.5(0.5)(0.08) = 0.10 = 10\%$$

Objective: Use an annual equivalent cost approach to compare the status quo and leasing alternatives.

Analysis: From Equation 19-4,

$$ALI = ALP(ALR)\left[\frac{i_a}{i_d(1-t)} - 1\right]$$

$$\cdot \left\{1 - \frac{(a/p)_n^{i_a}}{i_a - i_d}\left[(p/f)_n^{i_d} - (p/f)_n^{i_a}\right]\right\}$$

$$= \$20,000(1.0)\left[\frac{0.10}{0.08(0.5)} - 1\right]$$

$$\cdot \left\{1 - \frac{0.40211}{0.10 - 0.08}\left[0.7938 - 0.7513\right]\right\}$$

$$= \$4355$$

So

$$AEC_L = \$20,000 + \$4,355 = \$24,355$$

Since the "full" costs of leasing exceed the $22,000 cost of status quo, the latter is the preferred alternative.

Our results from the hypothetical restructuring of capital are verified below. Note the annual revenue required is:

$$AEC_L = \$26,186 - [\$1908\,(f/p)_1^{10\%} + \$3963]\,(a/f)_3^{10\%}$$
$$= \$24,355$$

End of Year	1	2	3
Debt capital replaced by composite capital $= ALP(ALR)(p/a)_x^{i_d}$	$51,540	$35,660	$18,520
Increased $BTCFR$ $= ALP(ALR)(p/a)_x^{i_d}$ $\cdot \left(\dfrac{i_a}{1-t} - i_d \right)$	6,186	4,278	2,223
Lease payments $= ALP$	20,000	20,000	20,000
Revenue required	26,186	24,278	22,223

TABLE 19-3. Hypothetical Data for Selected Debt Ratios of Company K

Given Data			
Debt ratio	0.0	0.2	0.4
Total capital	$1,000,000	$1,000,000	$1,000,000
Debt capital	0	$200,000 @ 5%	$400,000 @ 6.5%
Shares of common stock (market = book = $100/share)	10,000	8000	6000
Pretax earnings before deduction of interest on debt*	max $230,000 min -$10,000	$230,000 -$10,000	$230,000 -$10,000
Interest on debt	0	$10,000	$26,000
Computed Data			
Taxable income*	max $230,000 min -$10,000	$220,000 -$20,000	$204,000 -$36,000
Pretax earnings per share = pretax EPS* max min	$23.00 -$1.00	$27.50 -$2.50	$34.00 -$6.00
Probability that "pretax earnings before deduction of interest on debt" will be less than "interest on debt"	0.04	0.08	0.15
Increase in max pretax EPS		$4.50	$6.50
Decrease in min pretax EPS		$1.50	$3.50
$= \dfrac{\text{beneficial increment}}{\text{detrimental increment}}$		3.00	1.86

Note: See also—G. W. Smith, "Decreasing Utility for Money and Optimal Corporate Debt Ratio." *The Engineering Economist* (Spring 1968): 165–71.

*Probability distribution is uniform and continuous.

Factors Which Influence the Optimal Debt Ratio

Increasing the ratio of debt capital to total capital increases *leverage* or *sensitivity* of pretax earnings per share to changes in "pretax earnings before deduction of interest on debt." If total capital needs and probabilistic "pretax earnings before deduction of interest on debt" are held fixed while the debt ratio is increased, the result is an increased variability of equity capital "pretax earnings per share." The "nonuniform utility of money" concept (investor conservatism) is helpful in revealing this growing variability as an investor barrier to still greater debt ratios.

Table 19-3 contains data pertaining to a hypothetical Company K under various debt ratios. Figure 19-3 presents the

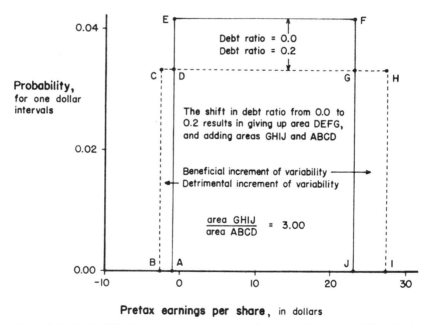

Fig. 19-3. Probabilistic pretax earnings per share as a function of the debt ratio of Company K.

probabilistic pretax *EPS* for debt ratios of 0.0 and 0.2 for the company. Acceptability of the higher debt ratio would depend upon:

1. How prospective debt capital investors react to the increased probability (from 0.04 to 0.08) that "pretax earnings before

deduction of interest on debt" will be less than "interest on debt."

2. How prospective equity capital investors react to the increased variability of pretax *EPS*. Because the beneficial increment is three times the detrimental increment, only the very conservative prospective investor would find the 0.2 debt ratio unattractive.

By varying the given input data, we can further explore the factors which influence optimal debt ratio. From such study we will discover that optimal debt ratio is decreased by:

1. A decreased expected value of "pretax earnings before deduction of interest on debt." This point is illustrated by Company L of Table 19-4.

TABLE 19-4. Hypothetical Data for Selected Debt Ratios of Company L

		Given Data		
Debt ratio		0.0	0.2	0.4
Total capital		$1,000,000	$1,000,000	$1,000,000
Debt capital		0	$200,000 @ 5%	$400,000 @ 6.5%
Shares of common stock (market = book = $100/share)		10,000	8000	6000
Pretax earnings before deduction of interest on debt*	max $170,000 min –$70,000		$170,000 –$70,000	$170,000 –$70,000
Interest on debt		0	$10,000	$26,000

		Computed Data		
Taxable income*	max $170,000 min –$20,000		$160,000 –$80,000	$144,000 –$96,000
Pretax earnings per share = pretax *EPS* *	max min	$17.00 –$7.00	$20.00 –$10.00	$24.00 –$16.00
Probability that "pretax earnings before deduction of interest on debt" will be less than "interest on debt"		0.29	0.33	0.40
Increase in max pretax *EPS*			$3.00	$4.00
Decrease in min pretax *EPS*			$3.00	$6.00
= $\dfrac{\text{beneficial increment}}{\text{detrimental increment}}$			1.00	0.67

*Probability distribution is uniform and continuous.

2. An increased variability of "pretax earnings before deduction of interest on debt."
3. An increased risk aversion (conservatism) of potential debt and equity investors. Such risk aversion is portrayable with the "utility of money" concepts and graphs of Chapter 13.
4. An increased positive responsiveness of cost of debt capital to debt ratio. (In our hypothetical data debt capital will cost 5% if the debt ratio is 0.2 but will increase to 6.5% if a debt ratio of 0.4 is used).
5. A negative correlation between debt ratio and "pretax earnings before deduction of interest on debt." This could occur for a regulated public utility if an increased debt ratio causes a decreased allowed return and thus a decreased allowed "pretax earnings before deduction of interest on debt."
6. An inability to carry back or carry forward losses when taxable income is negative.

Problems

19-1. A certain firm plans to obtain 10% of its future funds from debt, 40% from equity, and 50% from retained funds. It is estimated that the debt capital will yield its owner 5%; after considering all relevant costs such as those for legal fees, recording, printing, administration, discounting, flotation, and handling, the estimated net cost to the company is 6%. Equity capital is similarly estimated to provide owners a return of 9% while costing the company 10%. Last year the company earned only 4% on invested capital.
a. Find the composite cost of external funds.
b. Is the result in Part a identical to the cutoff rate of return? Explain.

19-2. a. Why should a nonprofit organization whose funds are obtained strictly by donations, use a *MARR* greater than 0%?
b. Suggest a means by which such an organization could estimate *MARR*.

19-3. Compare and contrast (1) regulatory cost of capital, (2) prospective cost of capital, and (3) minimum attractive rate of return.

19-4. Find the cost of debt capital for a firm if:

Gross proceeds = 2000 bonds @ $980 = $1,960,000
Costs of legal fees, printing, flotation fees, and administration = $30,000
Future annual cost of interest payment on 2000 bonds with $1000 maturity value and 5% coupon rate = $100,000/yr

Future annual cost of record keeping and handling of above = $20,000/yr

Future redemption cost at maturity 20 years hence = $2,000,000

Treat annual costs as if they were end-of-year costs.

19-5. A prospective investor states that his personal *MARR* is 10% and he is considering purchasing some shares of the GRN Company. He estimates that the end-of-year earnings will be $2 per share this year and will increase by $0.50 per share each year in the future; this growth in earnings will be augmented still further by growth due to retained earnings. Dividend payout is 0%. Find the maximum price which he theoretically should be willing to pay for a share of this stock. Neglect broker commission and handling costs.

19-6. a. Find the prospective rate of return on equity capital if the present price per share is $100 and dividends per share are forecast as a constant $8 per year forever.

b. Find the prospective cost of equity capital which is sold today to yield the company net proceeds of $90 per share. Estimated dividends plus record-keeping and handling charges will amount to an estimated $9 per share per year forever.

19-7. Using a rate of return requirement of 25% find the maximum present price you would be willing to pay for a share of stock in a company which retains all earnings and for which the price per share two years hence is an estimated $100.

19-8. Find the prospective rate of return on equity capital if the present price per share is $150 and forecast dividends per share are $5 at the end of year one, $6 at the end of year two, and increasing by $1 per year forever.

19-9. Using a rate of return requirement of 10% find the maximum present price you would be willing to pay for a share of stock in a company which has a 50% dividend payout and for which forecast earnings per share are $10 at the end of year one, $11 at the end of year two, and increasing by $1 per year forever.

19-10. Dividend payout of the Green Corporation is 0%. A prospective investor estimates that both (1) return on funds reinvested by the corporation and (2) his own rate of return requirement equal 10%. Estimated year-by-year earnings per share are: $5.00, $5.50, $6.05, $6.66, $7.33, ... Use Equation 19-2 to estimate the present price per share.

19-11. An investor states he will buy 100 shares of GRN Company stock if it has a prospective rate of return of 12% or more. He feels the following estimates are appropriate:

Present PE ratio = 24.0
Past and prospective dividend policy = 100% payout
Present price per share = $120.00
Estimated price per share one year from now = $127.50
Estimated growth in earnings = $0.70 per share per year

    a. Find the prospective rate of return if he buys the stock now, then sells it a year later.
    b. Find the prospective rate of return if he buys the stock and his family and heirs retain it forever.

19-12. Find the current price to earnings ratio you would expect for the stock of a company to which the following estimates apply:

Present earnings = $1 per share per year
Growth in earnings = increasing by 8% each year forever
Investor $MARR$ = 10%
Dividend payout = 100%
Neglect effects of brokerage fees, and so forth

19-13. Find the cost of new equity capital for a company if

Present market price of common stock = $50 per share
Estimated net proceeds per share if 100,000 shares are sold = $4,400,000
Estimated earnings per share rate necessary to support the present tendency to consider this a stable growth stock = $2.20 next year and grow by 6% each year
Dividend payout = 100%

19-14. a. Projected dividends per share of the Green Corporation are $8 at the end of year one and then grow at the rate of 2% per year forever. Neglect transaction costs and find the present market price per share which would result in a 10% return to the investor.
    b. Find the prospective cost of equity capital which is sold today to yield the company net proceeds of $100 per share. Estimated per share dividends plus record-keeping and handling charges are $7 at the end of year one, then increase at the rate of 5% per year forever.

19-15. BBKLPW Enterprises is financed 50% by debt capital costing 4%, and 50% by equity capital costing 10%.

They are considering the acquisition of an asset costing $100,000 and having negligible salvage value 20 years from today. BBKLPW notes that $t = 0$, $i_c = 7\%$, and that:

$$a_1 = \$100,000(a/p)_{20}^{7\%} = \$9439$$

Yet with $50,000 of 4% money and $50,000 of 10% money:

$$a_2 = \$50,000(a/p)_{20}^{4\%} + \$50,000(a/p)_{20}^{10\%} = \$9552$$

Perform the calculations necessary to show why a_1 and a_2 differ.

19-16. Use Equation 19-4 in finding AEC for a lease calling for end-of-year payments of $20,000 for each of the next four years. Other data:

$$ALR = 0.5 \qquad i_a = 0.10$$
$$t = 0.375 \qquad i_d = 0.08$$

19-17. The following data apply to Green Corporation:

$$i_a = 0.10 \qquad i_d = 0.08 \qquad r_d = 0.50$$
$$i_c = 0.12 \qquad i_e = 0.16 \qquad t = 0.50$$

Depreciation method used is straight-line. Find AEC for each of the following alternatives.

Alternative L. Accept a purely financial lease ($ALR = 1.0$) calling for payments of $14,903 at the end of each of the next ten years. The leased equipment is expected to incur operating costs of $9000 per year.

Alternative B. Buy an equipment which has a first cost of $100,000; its salvage ten years hence is thought to be negligible. Estimated operating costs are $9000 per year.

Alternative S. Continue the present system of operation which incurs operating costs of $30,000 per year.

19-18. PRL Corporation is considering use of certain equipment. Plan L involves a financial lease which calls for five end-of-year lease payments of $1516 each. Under this plan the company is not permitted to sublet, convert, trade, or return the equipment, nor to otherwise escape the required annual payments ($ALR = 1.0$).

Plan R is a nonfinancial lease which calls for end-of-year rental payments. Under this plan the company can cancel the rental agreement on very short notice. Removal of the equipment would be rapid and without cost to PRL. Property taxes, insurance, maintenance, and all other cash operating costs will be identical for the two alternatives. Other data:

$$t = 50\% \qquad r_d = 25\% \qquad i_d = 8\% \qquad i_e = 12\%$$

Find the annual rental cost at which the two plans would break even.

20

Matching the Sources and Applications of Funds

Factors Influencing the Overall Level and Specific Makeup of Capital Expenditures

IN ADDITION to the factor of supply, which includes both the amount and the cost of capital, there are other factors of importance to the determination of the total amount and specific makeup of the capital expenditures of a period:

1. *Balance.* The oil company must have a program of expenditures which will provide a "balanced" growth of exploration, refining, and marketing activities. Many firms likewise are concerned with the balance of "defensive" expenditures—those producing cost reduction—and "aggressive" expenditures—for facilities to produce new products.
2. *Forecasted sales demand.* Industry growth of product sales frequently means choosing between loss of sales to a competitor or acquisition of additional facilities to fill the increased demand. Studies show companies consider maintenance or growth of their "share of the market" important.[1]

---

[1] See R. Eisner, *Determinants of Capital Expenditures* (Urbana, Univ. Ill., 1956).

3. *Coordination of physical and financial plans.* Employee training requirements may put a ceiling on the growth rate which a firm can undergo. There is also a balancing of work loads to consider, for example, when installation or construction is to be done by company personnel. Heller concludes that postwar expenditures of 1946–1948 were held down by the scarcity of manpower and brainpower.[2]

4. *Financial policy.* Policy, especially in the governmental agency, can have an impact on capital spending. Will excess funds be withdrawn if not spent in the current period? Will failure to spend a budgeted amount affect the budget for the next period?

5. *Technological change.* Such new techniques as those for processing data, cutting metal, or wiring circuits can lead to obsolescence of and tendency toward retirement of existing equipment.

6. *Governmental measures.* From time to time the federal government changes tax rates, changes depreciation life and rate regulations, changes tax treatment of capital gains or losses, or provides some other measure such as the investment credit, at least partly with the hope of influencing the rate of capital spending for the firms affected. For the firm with an excess of cash, federal antitrust laws and interpretations can be a decisive factor in decisions of expansion or acquisition. Capital expenditures are sometimes necessary for compliance with such legal requirements as building codes, safety regulations, city zoning, or waste disposal and treatment rulings.

Determination of the Cutoff Point

If we could quantify all the preceding factors in our "desirability" scale, we might produce a priority ranking of alternatives which would be quite generally useful.

In lieu of this, unavailable approach projects could be listed in descending order of a rate of return based on the monetized factors as shown in Figure 20-1. There it can be seen that cutoff rate of return is *influenced* by the supply of and demand for funds. Ultimately, however, the cutoff rate of return is *determined* by management.

When projects are arranged in descending order of pro-

---

[2] See W. Heller, "The Anatomy of Investment Decisions," *Harvard Business Review* 29 (March 1951): 102.

spective rate of return and funds are arranged in ascending order of prospective cost of capital, these demand and supply curves can meet at a dollar level in excess of, just equal to, or less than the internally generated funds of the period.

Let x_x = externally supplied funds

$\quad\;\; x_i$ = internally generated funds

$\quad\;\; k$ = a constant

$\quad\;\; i_i$ = cost of internally generated funds expressed as an effective rate of interest per year

$\quad\;\; i_r$ = minimum effective rate of return per year available on projects financed from internal funds only

$\quad\;\; i_x$ = cost of external funds expressed as an effective rate of interest per year

Assume that demand and supply curves are the result of adequate search efforts.

If demand and supply curves intersect at a dollar level in excess of internally generated funds (see Fig. 20-1a), the minimum investment level, x_i and maximum $x_i + k$ indicate that externally supplied funds, x_x, should be in the range 0 to k; consequently, the cutoff rate of return under this frequently encountered relationship of demand and supply would be in the range i_x to i_r. The externally supplied funds, x_x, are frequently less than k; this apparent suboptimization is generally attributed to conservatism, allowance for uncertainties (analogous to a safety factor), or management policies such as restricting expenditures to those which can be financed internally.

If demand and supply curves intersect at a dollar level just equal to the internally generated funds (see Fig. 20-1b), the investment level, x_i, indicates that no external funds should be sought; the cutoff rate of return is in the range i_i to i_x.

If demand and supply curves intersect at a dollar level less than internally generated funds (see Fig. 20-1c), the investment level, x_i, indicates that no external funds should be sought; the cutoff rate of return is less than i_i. While this situation may be somewhat unusual, the oversupply of cash by some firms is suggested in *Fortune* where it is reported that 16 of the 500 largest U.S. industrials ended the year 1968 with more than 25% of assets in cash items.[3]

It is both important and significant to note that the cutoff

---

[3] See *Fortune* 79 (May 1969): 84.

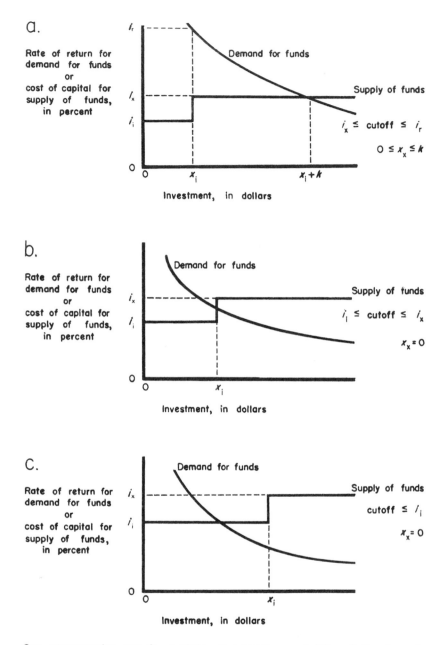

See accompanying text for symbols. Investment opportunities of the demand curve are equivalent in all respects (including risk) except for size and prospective rate of return.

Fig. 20-1. The cutoff rate of return as influenced by the supply of and demand for funds.

rate of return can and does vary widely from the cost of capital. Only after careful consideration of the capital budgeting process are we ready to suggest *MARR* rates which reflect the actual capital rationing (or oversupply) under which the firm operates. The interaction and inseparability of the two processes, (1) choosing between mutually exclusive alternatives and (2) approval and funding of independent alternatives, should be apparent. Ranking alternatives or making accept-reject decisions by a present worth procedure based on the cost of externally supplied capital can be erroneous; the cutoff rate of return is not independent of funds available nor is it independent of projects available. With present worth methods, ranking should be deferred until the cutoff rate of return is known.

Reinvestment Rate of Return, Cutoff Rate of Return, and Minimum Attractive Rate of Return

At least three terms are used widely in the literature to describe the interest rate which represents the present or future threshold of project acceptability. These terms are:

1. Minimum attractive rate of return.
2. Cutoff rate of return.
3. Reinvestment rate of return.

In a detailed treatment of capital budgeting problems it is important to characterize their differences. It is quite possible (1) that economy studies could be performed using a *MARR* of 20%; (2) that the capital budgeting process reveals the firm's available funds are so rationed that the effective cutoff rate is 25%; and (3) that management feels the young firm, currently operating on a shoestring, will find funds more available as the years pass and its own future becomes more secure, and on this basis predicts future reinvestment rates of return of 15%.

Our treatment here will be limited to those situations where the three terms are approximately equal; with this limitation in mind we have used the terms interchangeably. Because it is likely that capital is rationed and investment opportunities are not risk-free, it is very *unlikely* that cost of external capital is identical at any time to any of the three terms. The reader is cautioned that this fourth term is definitely not to be used interchangeably with the other three terms.

Some Considerations in the Year-by-Year Decisions with
Regard to External Financing

A number of problems arise in securing external funds; to respond adequately to these problems requires some consideration of alternative courses of action.

One problem is that the *exact* amount of external financing required is uncertain. While the projected data in a funds flow statement is very helpful, it is inexact in terms of both actual project costs and funds required. Some means of alleviating this problem are suggested:

1. Temporarily finance with short-term notes; when the project is complete and more precise financing requirements are apparent, permanent financing can be undertaken.
2. Using "callable" bonds or preferred stocks permits an organization to apply excess funds (present or future) to the recall of a matching amount of outstanding bonds or shares of preferred stock.[4]

A second problem is that small security issues are uneconomic. Brokerage fees, handling costs, legal and printing expenditures, and commission schedules tend to make issuing of small amounts of debt or equity capital through ordinary investment channels excessively costly. Small-scale financing might be:

1. Accomplished by permitting present stockholders to purchase additional shares (the number of these being based on present holdings). By pricing new shares below current market value, a company can increase attractiveness of the issue and its likelihood of success. A discount can also be offered by issuance of "rights" or "warrants" which permit the conversion to stock or bonds. By use of an expiration

---

[4] The callability, if present, is specified in the terms of the issue. Details such as the sequence of calling and the schedule of year and call price are likewise specified. Callability has a second virtue which permits the organization to take advantage of lower future interest costs by refunding of the outstanding issue with a new lower cost issue. The callability privilege is similar to the "acceleration" privilege of some mortgage contracts which permit more rapid (accelerated) repayment of the mortgage loan without penalty. Since the buyer of a callable bond should consider the likelihood of recall and since recall tends to place a ceiling on the market value of such securities, the callability is generally not a cost-free feature of a security.

date on the rights, the firm can control the timing of conversion and the subsequent realization of funds.

2. Deferred several months or years until the accumulating needs for funds have grown to a size which *is* economic.

A third problem is that the timing of receipt of funds from external financing and the timing of payments required from such funding are not certain. This is especially notable as the lead time for planning and execution of funding may be six months or even a year in advance of payment requirements. Funds received in advance of needs are generally invested in temporary securities, then converted to cash as needed. Because the temporary funds generally earn (say 4%) less than the corporate cost of capital (say 9%), the corporation treasurer is correctly concerned with the timing of funds. An interesting reversal of the problem occurs, however, as we consider the municipal or state treasurer whose external funds obtained from tax-exempt securities costing, say, 3½% can be temporarily invested in short-term U.S. obligations paying 4%. While the temporary excess of funds may be a costly and not entirely avoidable problem for the corporation, it is likely to be a windfall for the municipality.

A fourth problem is the selection of the specific type of issue to be used in the current financing, the choice between common stock, preferred stock, bonds, and other methods such as notes and equipment trust certificates. Even when such a choice is made, further details may still require specification; whether the bond should be subordinate to an existing issue, whether it should be cumulative, callable, convertible, serial, debenture or mortgage, and so on. The choice is not an unrestricted one; the maximum amount of general obligation bonds outstanding is generally limited by the community's assessment base; revenue bonds may require a certain level of uncommitted cash flow. Previous security issues may also restrict subsequent issues, especially with regard to prior claims on earnings or assets. In general, debt capital is easier to obtain than equity capital; however, continual tapping of just one source of funds will alter the ratio of debt to total capital (debt ratio). Altering the debt ratio may cause an increase in the cost of capital from its minimum; and if the debt ratio becomes too large, debt financing may become limited as a future source of funds. Flexibility of choice between debt and equity forms of financing is important so that the firm can take advantage of a good market or avoid a bad market for a given type of financing.

The optimum debt ratio, if one exists, is a matter of considerable study in itself.[5]

Some Capital Budgeting Problems

While the rate of return and present worth techniques of displaying descending priority of investments are very helpful, there are too many instances in which revenue, and hence rate of return, is not readily determined. (For example, how much of the total revenue of an electric utility is attributable to a specific transformer?) The investor-owned public utility may even introduce the additional question of whether or not all of their investment proposals tend to have identical rates of return due to regulatory determination of the "fair rate of return." Example 20-4 illustrates that even under the foregoing circumstances, the rate of return technique is operational.

While it might seem that the firm should seek to obtain sufficient funds to reach equilibrium in the demand-supply curves of investment opportunities and funds available, respectively, studies show this is generally not the practice.[6] Furthermore it is the contention of Hart that, as a result of uncertainty, "capital rationing is an essential trait of the capitalist economy."[7] The analyst must be able to optimize capital expenditure choices and do so even if funds are rationed and regardless of whether such rationing is the result of uncertainty, conservatism, or nonoptimality of expenditure level.

Advocates of present worth techniques of capital budgeting have frequently criticized rate of return techniques because of the nonunique solution type of problem (a rare but possible situation treated earlier) and because the internal rate of return of a project is not always sufficient data for a decision. This latter problem is reconciled in Example 20-2 where rate of return techniques can be employed to display a much broader view of the sensitivity (or insensitivity) of ranking to the cutoff rate of return. Unfortunately, the advocates of present worth techniques have given scant attention to determina-

---

[5] For an excellent search into the relationship of debt ratio and cost of capital see A. Barges, *The Effect of Capital Structure on the Cost of Capital* (Englewood Cliffs, N.J., Prentice-Hall, 1963).

[6] See for example, G. Donaldson, *Corporate Debt Capacity*, Graduate School of Business Administration, Harvard Univ., 1961.

[7] A. G. Hart, *Anticipations, Uncertainty and Dynamic Planning*, Studies in Business Administration, Vol. 11, No. 1 (Chicago, Univ. Chicago Press, 1940).

tion of the "proper" interest rate to be applied.[8] Worse yet, some of the advocates state or infer that the cost of external capital is the appropriate rate. Present worth methods require knowledge of reinvestment rates (generally the same as current cutoff rate of return and generally significantly greater than cost of external capital). Since cutoff rate of return is dependent upon the total expenditure level, our purpose of ranking as an aid to the determination of expenditure level and cutoff rate of return seems to be inverted by the present worth techniques. Properly applied, the rate of return and present worth methods yield identical accept-reject decisions if the fluctuation in cutoff rate of return is negligible. The ranking will not necessarily coincide because one method gives first priority to the alternative which provides the greatest margin of safety as measured by the excess of profit *rate* over that at the cutoff rate, while the other method gives first priority to the alternative which provides the greatest margin of safety as measured by the excess of profit *amount* over that at the cutoff rate.

In the end we must recognize that a process of ranking projects in descending order of rate of return in the funding process can only be operational if carefully qualified. The qualifications include the need to recognize somehow the following considerations:

1. Independent projects frequently are not "perfectly" independent. Independence is reduced by expenditure plans which call for *balanced* spending (such as over time, geographic areas, or exploration-production-marketing categories).
2. Mandatory projects sometimes show a very small or even negative rate of return; funding of such projects seems inconsistent with our scheme unless we treat the project as one which preserves the entire system and therefore produces a very high rate of return. This latter view is also dangerous since the real effect of such projects may be to reduce overall profitability of the firm.
3. The preceding difficulty can also be experienced in a

---

[8] See, for example, H. M. Weingartner, *Mathematical Programming and the Analysis of Capital Budgeting Problems* (Englewood Cliffs, N.J., Prentice-Hall, 1963), p. 8: "The choice of the appropriate rate by which future flows are to be discounted is not our primary concern. As mentioned, Lorie and Savage take this to be the cost of capital, although how this concept is to be measured is not discussed by them, and will not be discussed by us."

reversed fashion, as when a project with a very high rate of return is proposed as an addition to an existing facility with a substandard rate of return. While the new project with a high rate of return seems urgent, we must first be certain that the facility itself should be retained.

4. Prerequisite projects; consider, (a) Project 27B has a prospective rate of return of 10%, (b) Project 44D has a prospective rate of return of 20%, and (c) Project 27B is a prerequisite of Project 44D. Because 27B is a prerequisite of 44D, funds must be allocated to it first, even though it has the lower rate of return. If the prerequisite project has the higher rate of return, the difficulty vanishes because the rate of return scheme allocates funds in the proper sequence— prerequisite, then the contingent project. When the prerequisite project has the lower rate of return (as it does in the illustration), we should consider rate of return of the entire project (27B + 44D) in the ranking process. The remaining possible difficulty occurs when rationing of funds permits us to fund currently just one of a set of prerequisite projects (27B); in this instance it may seem we have been inconsistent with our scheme of ranking project priority by rate of return.

Still other complications emphasize the complexity of the capital budgeting task. Many projects are indivisible; in large projects this characteristic causes spasmodic rather than smooth continuous demands for funds and people.

Although our suggestion here is that of a single long-term future cutoff rate of return based on present and past company experience and future expectations, the relative rationing (and hence the theoretical cutoff) is likely to vary from time to time due to variability of profits, the impact of a large project, or fluctuations in product demand. This difficulty is clearly described by Weingartner who notes, "Because the choices and constraints of one period almost certainly affect the choices and constraints of other periods, our present practices of planning for expenditures on a one-period basis will likely fail to optimize the allocation of funds."[9]

And finally, note that when the cutoff rate is not precisely known, the choice of alternatives from a set of mutually exclusive proposals may not be entirely separable from the capi-

[9] H. M. Weingartner, *Mathematical Programming and the Analysis of Capital Budgeting Problems* (Englewood Cliffs, N.J., Prentice-Hall, 1963).

tal budgeting decisions. Such a situation is illustrated in Example 20-1.

Example 20-1.   **Decisions on Mutually Exclusive Alternatives and Decisions on Independent Alternatives Are Sometimes Inseparable**

Information: A number of cost-reduction proposals have been made for the Orange Manufacturing Company. Each proposed change requires a present investment which is expected to produce a uniform annual after-tax reduction in disbursements over each of the next five years. No salvage value is expected for any of the alternatives. Alternatives with the same prefix number are mutually exclusive with respect to one another. Those with different prefix numbers are independent.

Proposal	Investment	Uniform Annual After-Tax Reduction in Disbursements
1A	$10,000	$ 2,505
1B	20,000	5,548
1C	30,000	7,904
2A	10,000	3,344
2B	20,000	6,399
2C	40,000	11,700
3A	10,000	3,494
3B	20,000	6,266
3C	30,000	9,321
4A	10,000	3,200

All other investment opportunities of the company show a prospective rate of return of 10%, and there is a large backlog of such opportunities. Minimum attractive rate of return has not been specified. Investment opportunities are equivalent in all respects (including risk) except size and prospective rate of return.

Objectives:
a. Compute rate of return on each investment and increment of investment.
b. Assign a priority ranking to the various investments and increments of investment using the rate of return approach.

c. Use your results above to select the acceptable projects and show the cutoff rate of return when funds available are:

(1) $130,000
(2) $ 80,000
(3) $ 60,000
(4) $ 50,000
(5) $ 10,000

d. Assume that internally generated funds are expected to amount to $32,000, that the company has no aversion to external financing, and that the cost of external capital is 13%. State your recommendations.

Analysis:
a. Since each of the investments is recouped through savings occurring over the next five years, a list of $(a/p)_5^i$ factors will be of help in finding the rate of return:

i	$(a/p)_5^i$	i	$(a/p)_5^i$
5	0.23097	15	0.29832
6	0.23740	16*	0.30553
7	0.24389	18*	0.31996
8	0.25046	20	0.33438
10	0.26380	22*	0.34937
12	0.27741	25	0.37185
14*	0.29135		

*By interpolation of the interest tables.

For 1A,

$$(a/p)_5^i = \$2505/\$10,000 = 0.25050$$

Since $(a/p)_5^{8\%} = 0.25046$ and $(a/p)_5^{10\%} = 0.26380$, the rate of return on 1A equals

$$i = 8\% + 2\% \,(4/1334) = 8.0\%$$

And similarly for each of the other investments. The letter S used for 1S, 2S, 3S, and 4S refers to the status quo or do-nothing alternative.

Proposal	Investment	Uniform Annual After-Tax Reduction in Disbursements	Rate of Return on Extra Investment over Alternative		
			S	A	B
			percent		
1A	$10,000	$ 2,505	8.0		
1B	20,000	5,548	12.0	15.8	
1C	30,000	7,904	10.0	10.9	5.7
2A	10,000	3,344	20.0		
2B	20,000	6,399	18.0	16.0	
2C	40,000	11,700	14.2	12.2	10.2
3A	10,000	3,494	22.0		
3B	20,000	6,266	17.1	12.0	
3C	30,000	9,321	16.7	14.0	16.0
4A	10,000	3,200	18.0		

b. Use network diagrams:

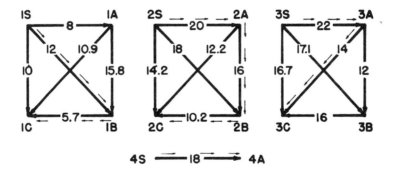

Then summarize in order of prospective rate of return:[10]

3A	$10,000 invested at 22.0%
2A	10,000 invested at 20.0%
4A	10,000 invested at 18.0%
2B-2A	10,000 invested at 16.0%
3C-3A	20,000 invested at 14.0%
1B	20,000 invested at 12.0%
2C-2B	20,000 invested at 10.2%

[10] We are neglecting the possibility that available funds will not fully fund the marginal project, thus neglecting the possibility of *indivisibility* of projects. This possibility is considered later.

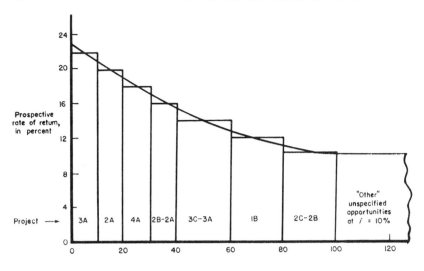

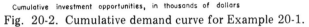

Cumulative investment opportunities, in thousands of dollars

Fig. 20-2. Cumulative demand curve for Example 20-1.

c. Referring to Figure 20-2, we can see that with funds of
$130,000 the projects will be 3A, 2A, 4A, 2B-2A, 3C-3A,
1B, 2C-2B, and $30,000 in "other" projects (that is, al-
ternatives 1B, 2C, 3C, 4A, and $30,000 in "other"
projects).

Funds Available	Projects	Apparent Cutoff Rate of Return
		percent
$80,000	1B, 2B, 3C, 4A	10.2–12
60,000	2B, 3C, 4A	12–14
50,000*	2B, 3A, 4A and $10,000 in "other" or held over to the next period or	14
	2A, 3C, 4A, or 2B, 3C, 4A with projects stretched out over several budgeting periods.	
.10,000	3A	20–22

*We earlier chose to neglect indivisibility, and in so doing,
eliminated 3B. While 2B, 3B, 4A seem attractive, their selection
seems to place undue emphasis on indivisibility. The problem of
indivisibility tends to disappear when sources of funds and expenditure opportunities are viewed as (1) a continual rather than
once per year flow and (2) consisting of thousands of items rather
than the few shown in Figure 20-2.

Note that a present worth study which fails to recognize that cutoff rate of return and cost of external capital are not necessarily equal can lead us to erroneous conclusions.

Perhaps still more important is the observation that the problems of capital budgeting are not necessarily separable from the problems of selecting the "best" from a set of mutually exclusive alternatives. They may be separable when the long-run *MARR* is relatively stable and predictable.

d. All projects earning 13% or more should be accepted. This means we should accept 2B, 3C, and 4A, a total investment of $60,000 requiring $28,000 of external funds.

Example 20-2.   **Relevance of Reinvestment Rate of Return in the Ranking of Independent Investment Alternatives**

Information: A number of independent investment opportunities are being considered by the Blue Corporation. Each has a set of estimated after-tax cash flows associated with it:

| | Cash Flow at End of Year | | | |
Proposal	0	1	2	3
1	- $1000	+$1500		
2	- 1000		+$1960	
3	- 1000			+$2197

Investment opportunities are equivalent in all respects (including risk) except size, duration, and prospective rate of return.

Objectives:
a. Compute rate of return on each investment and increment of investment.
b. Assign a priority ranking to the various investments, using the rate of return approach.
c. Assume that the long-run cutoff rate of return is identical to the reinvestment opportunity rate of return and is expected to fluctuate around 20%. Use this information to produce another means of comparing the alternatives.

Analysis:

a. For Proposal 1, $1000 (1 + i) = $1500, so $i = 50\%$; for Proposal 2, $1000 (1 + i)^2 = $1960, so $i = 40\%$; and so forth. The rate of return on the incremental investment, (2 - 1), is based on the difference in cash flows, a pattern of $0, -$1500, +$1960. With $1500 (1 + i) = $1960, $i = 30.7\%$; similar calculations for the other increments yield:

Proposal	Cash Flow at End of Year				Internal Rate of Return†	Rate of Return on Extra Investments over Alternative	
	0*	1	2	3		1	2
						percent	
1	−$1000	+$1500			50.0		
2	− 1000		+$1960		40.0	30.7	
3	− 1000			+$2197	30.0	21.0	12.1

*In cases where year 0 commitment varies between alternatives, it is suggested that the cash flow stream of one be multiplied by a factor making year 0 cash flows identical.

†Internal rate of return on a project is computed without regard to reinvestment of cash flows generated by it.

b. As in Example 20-1, we need only to list the investments in descending order of rate of return:

Funds Available	Alternatives Chosen	Cutoff Rate of Return
$ 0	None	50% or more
1000	1	40%–50%
2000	1,2	30%–40%
3000	1,2,3	30% or less

An apparent contradiction between *PEX* and *ROR* as the ranking criterion arises as follows. Assume that the cost of capital is 10% and that only $1000 is available. By calculation (or by observation from Fig. 20-3) it can be shown that Project 3 will maximize *PEX* at $i = 10\%$. Critics of the *ROR* approach attempt to use this as an illustration that *PEX* and *ROR* lead to different conclusions.[11] *But how can Project 1, earning 50%, be declined?* The *opportunity* cost of capital is obviously much higher than 10%. The difficulty is simply the result of over-specification. If the *cutoff* is given as 10%, then

---

[11] E. Solomon, ed., *Management of Corporate Capital* (New York, Macmillan, 1964), pp. 56–57; R. N. McKean, *Efficiency in Government through Systems Analysis* (New York, Wiley, 1964), pp. 89–90.

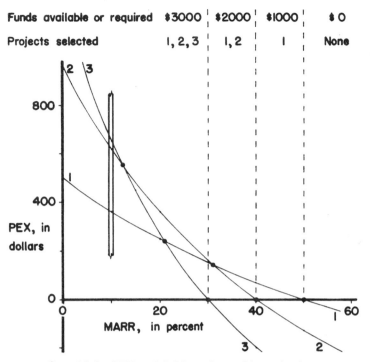

Fig. 20-3. *PEX* and *ROR* as the ranking criterion.

$3000 in funds must be available; if only $1000 is available, then the cutoff is between 40% and 50%, and Project 1 is the proper choice.

c. By assuming receipts are reinvested as they arise and yield a 20% return, we can produce an equivalent worth at the end of year three for each of the alternatives:

Pro-posal	Cash Flow at End of Year				Rate of Return on Combined Internal and External (to project) Investment
	0	1*	2*	3	
					percent
1	-$1000	-$1500 +1500	+$1800 - 1800	+$2160	29.3
2	- 1000		+ 1960 - 1960	+ 2352	33.0
3	- 1000			+ 2197	30.0

*Reinvestment of receipts leaves a net cash flow of zero for the end of years one and two for each of the proposals.

The ranking of projects based on a descending order of combined rate of return is 2, 3, and 1. That same ranking will result if *FEX* or *PEX* is the criterion. The reinvestment assumption is helpful because it is obvious we do wish to maximize future worth of the firm; this is consistent with maximizing present worth *providing* the latter is based on the reinvestment rate of return. Example 20-3 demonstrates that this approach is operational even if the reinvestment rate of return varies from period to period.

Example 20-3.   **Short Payoff Period versus High Rate of Return**

Information: Two independent investment opportunities are being considered by the Aqua Corporation (formed when the Green Company and Blue, Inc., merged). Each has a set of estimated cash flows associated with it:

Proposal	End-of-Year Cash Flow	Internal Rate of Return	Payoff Period
		percent	*years*
5	- $100 now, +$23 for each of the next 9 years	17.7	4.3
6	- $100 now, +$35 for each of the next 4 years	15.0	2.9

The two investment opportunities are equivalent in all respects (including risk) except size, duration, and prospective rate of return.

Funds are presently relatively plentiful (a low cutoff rate of return) but estimates indicate that a sizable shortage of funds will be encountered in the near future. This trend is due to several factors: (1) Plant capacity, which has been able to meet demands of the past, is no longer sufficient; (2) depressed earnings per share are anticipated over the next nine years while the firm experiences both growing pains and high initial costs of the planned automation of facilities; and (3) the growing reluctance of management toward any outside financing. This latter factor is related to some recent unfavorable experiences coupled with a more uncertain future. Predicted future cutoff rates for capital budgeting purposes are:

Year	0-1	Years 1-9	After 9 Years
Reinvestment rate = cutoff rate =	14%	25%	?

Objective: Determine whether either of the proposals is acceptable.

Analysis: If the year-by-year receipts are reinvested at 25% (the predicted future cutoff rate), the future worth of the present investment will have grown to a worth at the end of year nine of:

Proposal	Worth of Invested and Reinvested Funds at End of Year 9	
5	$23(f/a)_9^{25\%} =$	$593
6	$35(f/a)_4^{25\%}(f/p)_5^{25\%} = \$35(3.052)(5.766) =$	616

And it would appear that with rising cutoff rates the payoff period criteria might be producing a satisfactory comparison. Note, however, that an equal or better alternative exists if a short-term (for the period 0-1) investment return of at least 3.4% is available. Then this and the reinvested funds would accumulate at the end of year nine to

$$\$100(f/p)_1^{3.4\%}(f/p)_8^{25\%} = \$100(1.034)(5.960) = \$616$$

One frequent defense of payoff-period technique is that it recognizes the quick cash turnover better than does the rate of return comparison.[12] We have seen that the lower rate of return alternative can be preferable only if its quick cash turnover permits reinvestment at rates which are subsequently higher. Regardless of whether we are considering mutually exclusive or independent alternatives, the actual constraints are severe enough to raise the question of whether, in fact, such circumstances could possibly arise with a frequency worthy of a general method. In our example, Proposal 6 must:

1. Earn enough more on the reinvested funds (25% was given) that rate of return on the combined invested and reinvested funds would be greater than the rate of return

---

[12] See for example, N. N. Barish, *Economic Analysis* (New York, McGraw-Hill, 1962), p. 180.

on the combined invested and reinvested funds of Proposal 5 (unknown at this point but greater than 17.7%).
2. Pass the present cutoff rate so cutoff rate $\leqslant$ 15%.
3. Be superior to the alternative of temporary investment of funds at a low rate, say 4%, with subsequent reinvestment at 25%.

Example 20-4.   Determinability of Rate of Return

Information: Energetic Electric Company, a medium-sized investor-owned electric utility, has asked that we review their capital budgeting practices. One of the first characteristics which their management points out is the indeterminate nature of revenues for each investment[13] and hence rate of return or present value of the excess of revenues over disbursements; secondly, they note that at least in a sense, the rate of return is uniform on all investments and is essentially the same as "fair rate of return" as allowed by the regulatory commission.

The combination of these two characteristics raises a question as to whether either the rate of return or present value ranking methods are operational for capital budgeting purposes when revenues are not determinable.

As an example, the company has asked that we consider the following set of independent alternatives, each of which has already passed the *MARR* test or is mandatory. The company feels that a ranking of some kind is desirable so that efforts to "balance" many factors such as those listed below can also take into account the factor of project urgency. Typical factors are:

1. Availability of cash.
2. Assignment of personnel connected with construction, installation, and inspection.
3. Timing of such related projects of the city as street programs, area development, and so forth.
4. Construction and delivery schedules and variances due to weather, and so forth.
5. Emergency efforts and funds required by storm, flood, lightning, and other accidental causes.

---

[13] Although "system" revenues may be known, an allocation of these to component parts of the system may, if determinable at all, contain so many subjective judgments that they render the allocated component revenue meaningless.

From our conversation it has also become apparent that the alternatives have varying degrees of postponability. In an effort to quantify "postponability," we have listed for each project a mutually exclusive alternative, XP, the costs incurred if Project X is postponed. The after-tax cash flow pattern for Projects X and XP is also given.

Project 1. *Provision of an employee cafeteria.* If postponed, costs themselves are not expected to increase, but the cost of employee dissatisfaction, inconvenience, and the like has been estimated as $5000 for a one-year postponement.

Project 2. *Vehicle replacement.* If replacement is postponed one year, the annual equivalent costs of new vehicles are expected to remain unchanged; but costs such as those for maintenance and operation are expected to be $2000 more this year than the annual equivalent costs incurred if replacement takes place now. Estimated first cost of the new vehicles is $20,000.

Project 3. *Replacement of small turbine engine.* The replacement unit costs an estimated $20,000. If replacement is postponed one year, extensive repairs will need to be made now, although some of these costs can be recouped through next years higher salvage value (as a result of the reconditioning).

Project 4. *Acquire land.* A parcel of land well suited to company needs is being offered for $20,000. Although the land is not needed for several years, the location and price are very favorable. It is estimated that an equivalent parcel purchased two years from now will cost $30,000. Property taxes can be assumed as $800 per year in either case.

Project 5. *Office building.* A new building costing $150,000 is planned. Present rented facilities cost $15,000 per year and are felt to have various "handicaps" valued as equivalent to an extra annual cost of $8000 per year. The planned building would have property taxes, insurance, and other costs estimated as $10,000 per year.

Project 6. *Substation facility.* A new facility with costs of $40,000 now and $2000 per year has been selected as most economic of a set of alternatives. With a shortage of cash the firm is willing to consider an alternative facility costing $20,000 now

and $5000 per year. Both alternatives have lives
of 15 years.

Project 7. *Generating unit.* A new generating unit costing
$200,000 has received preliminary approval. A
unit with half the capacity of the $200,000 unit
can be purchased for $105,000 and will take care
of needs for the next year. The larger unit will
meet demand requirements of the next two
years.   Either unit has an estimated life of 20
years.

Investment opportunities are equivalent in all respects (including risk) except size, duration, and possibly, the prospective rate of return.

Cash flows for the projects are summarized below:

| | [Equivalent] After-Tax Cash Flow at End of Year | | | |
	0	1	2	3
Project 1	-$20,000	-$ 2,200	-$2,200	-$2,200
1P		- 25,000	- 2,200	- 2,200
Project 2 - $[K_2]$	- 20,000	-$[K_2]$	-$[K_2]$	-$[K_2]$
2P - $[K_2]$	- 2,000	-$[K_2]$ - 20,000	-$[K_2]$	-$[K_2]$
Project 3 - $[K_3]$	- 20,000	-$[K_3]$	-$[K_3]$	-$[K_3]$
3P - $[K_3]$	- 1,000	+ 1,600 - 20,000		
	- 3,000	-$[K_3]$	-$[K_3]$	-$[K_3]$
Project 4	- 20,000	- 800	- 800	- 800
4P			- 30,000	- 800
Project 5	-150,000	- 10,000	- 10,000	- 10,000
5P	- 23,000	-150,000	- 10,000	- 10,000
Project 6	- 40,000	- 2,000	- 2,000	- 2,000 ···
6P	- 20,000	- 5,000	- 5,000	- 5,000
Project 7	-200,000	-$[K_7]$	-$[K_7]$	-$[K_7]$
7P		-$[K_7]$ - 1,000		···
	-105,000	-105,000	-$[K_7]$ - 1,000	-$[K_7]$ - 1,000

Objective: Assume that revenues will be unaffected whether
we choose the immediate or postponed alternative. Analyze
the alternatives to determine "postponability" as measured
by rate of return on the increment of investment required
to acquire the proposed item now rather than one year from
now.

Analysis:
Compute rate of return on incremental investment
(1 - 1P):

$$0 = -\$20,000(1 + i) + \$22,800 \quad \text{so } i = 14.0\%$$

Compute rate of return on incremental investment
(2 - 2P):

$$0 = -\$18,000(1 + i) + \$20,000 \quad \text{so } i = 11.1\%$$

Compute rate of return on incremental investment (3 - 3P):

$$0 = -\$16,000(1 + i) + \$18,400 \quad \text{so } i = 15.0\%$$

Compute rate of return on incremental investment (4 - 4P):

$$0 = -\$20,000(1 + i)^2 - \$800(1 + i) + \$29,200$$

by use of the quadratic formula we find $i = 18.8\%$.
Compute rate of return on incremental investment (5 - 5P):

$$0 = -\$127,000(1 + i) + \$140,000 \quad \text{so } i = 10.2\%$$

Compute rate of return on incremental investment (6 - 6P):

$$0 = -\$20,000 + \$3,000(p/a)_{15}^{i}$$

by interpolation we find $i = 12.4\%$.
Compute rate of return on incremental investment (7 - 7P):

$$0 = -\$95,000 + \$105,000(p/f)_{1}^{i} + \$1,000(p/a)_{20}^{i}$$

by trial and error we find $i = 17.5\%$.
Then, ranking by rate of return on the increment of investment which provides for immediate investment, we would show in decreasing order of desirability (increasing order of postponability):

Project Number and Description	Rate of Return on Increment of Investment in Nondefer Project	Funds Required
	percent	
4—Land	18.8	$ 20,000
7—Generator	17.5	200,000
3—Turbine	15.0	20,000
1—Cafeteria	14.0	20,000
6—Substation	12.4	40,000
2—Vehicles	11.1	20,000
5—Office building	10.2	150,000
		$470,000

Problems

20-1. Alternatives 7B and 9E are independent of each other. The after-tax cash flows for 7B are (-$100, +$130) and for 9E are (-$100, $0, +$156.25). Let the cutoff (= reinvestment rate of return) vary from 0% to 50% in a table showing the projects which are acceptable and their rank. The alternatives are equivalent in all respects, including risk, except for size, duration, and prospective rate of return.

20-2. In the management game of Problem 20-5 each of the investments has a uniformly distributed probabilistic set of cash flow outcomes. We would like to examine the effect of investing funds in a group of several whose outcomes are independent of each other. To do this we can use Appendix K to obtain 10 two-digit random numbers, omitting any two-digit number ending in zero. Then refer to the Table of Outcomes at the end of this chapter to find the sum of the cash flows for one each of Investment D, F, and H according to the first random number. Repeat for each of the remaining two-digit random numbers. Then compare the distribution of your sums with those below:

D	- 1	9	19	29	39	49
F	- 3	5	13	21	29	37
H	- 11	- 1	9	19	29	39
Sum	- 15	13	41	69	97	125

Times your sum between those limits	□	□	□	□	□

Explain why the distribution of sums does not seem to follow the uniform probability distribution. What implications does this have for possible game strategies?

20-3. Suppose that a firm obtains additional debt capital at a net cost of 6%. If the firm adopts a financing policy where all future capital will come from debt capital only, can the firm then use 6% as the composite cost of capital? Explain.

20-4. GRN, Inc., is considering a number of investment opportunities; each involves a present investment which will produce a uniform annual increase in after-tax cash flow over each of the next six years; the opportunities are equivalent in all respects, including risk, except for size and prospective rate of return. No salvage value is expected for any of the alternatives. The XS alternatives are the "status quo" alternatives. Alternatives with the same prefix number (for example, 1S, 1A, 1B, and 1C) are mutually exclusive with regard to one another; those with different prefix numbers are independent of each other (for example, 1B, 2S, 3A). GRN finds that many investment opportunities with a prospective rate of return of 9% are available. Calculation has provided us with the following data:

Proposal	Investment	Uniform Annual Increase in After-Tax Cash Flow	ROR on Extra Investment over That of Alternative		
			S	A	B
			percent		
1A	$20,000	$ 6,014	20.0		
1B	30,000	8,146	16.0	7.5	
1C	40,000	10,570	15.0	9.7	11.9
2A	10,000	2,432	12.0		
2B	30,000	7,093	11.0	10.5	
2C	40,000	9,184	10.0	9.3	6.9
3A	10,000	2,398	11.5		
3B	20,000	5,144	14.0	16.4	
3C	30,000	7,402	12.5	13.0	9.4

a. Assign a priority ranking to the various investments and increments of investment using the rate of return approach.
b. Use your results above to select the acceptable projects and show the cutoff rate of return when funds available are:

(1) $130,000   (3) $50,000
(2) $ 80,000   (4) $10,000

c. Assume that internally generated funds are expected to amount to $27,000, that the company has no aversion to external financing, and that the prospective cost of external capital is 11%. State your recommendations.
d. What do you conclude if you are told that (1) the maximum total expenditure will be $40,000 and (2) $MARR = 10\%$?

20-5. *A noncomputer game dealing with capital budgeting.*

Management games have been used as teaching devices by which students may employ strategies and study their consequences within the harmless framework of a game in which the loss of thousands of lives or millions of dollars is only a "paper" loss; only the ego can be damaged. Some experience, insight, or feeling for decision factors may be gained through games; the interaction of factors and "balancing" requirement for the numerous variables can become more real. Games, which are necessarily simplifications of reality, can never substitute for experience; experience is not always a practical teacher, however, for her lessons are sometimes fatal or disastrous and her pace can be painfully slow.

With these qualified hopes for games as a possible aid to the student and his understanding of society (business, political, military, or other), the following is offered.

To start: Each player (or team) starts the game with 400 million dollars which will be abbreviated M$ (megabucks). Each player can choose from eight different investments labeled A, B, C, D, E, F, G, and H. These may be selected any number of times and in any combination; since G involves investment of 50 M$, a player could immediately choose eight of Investment G. He could likewise choose 1A (400 M$), or 1D (200 M$) plus 2E (100 M$ each), or 1F (100 M$) plus 1G (50 M$) plus 5H (50 M$ each), and so forth.

Penalties: The player is probably well advised to leave some cash on hand to cover periods of low income and "penalty" costs (up to 20 M$) which sometimes occur. Penalties for all players arise when the outcome drawn specifies a penalty; a penalty of 10 M$ is also charged to any player whose cash flow from all investments this year falls short of last years cash flow by 50 M$ or more. If a player runs short of cash, he must sell enough of his investments to cover the shortage; investments which are sold in this manner are bought by the bank, not by other players, and the price is 5 M$ per remaining period of investment life. Since all investments cost 10 M$ per period of investment life, the penalty is a substantial one which should be avoided if possible.

Reinvestment: It is impossible to borrow money or to obtain additional external capital; the only added capital is that from internal generation via cash flow (return on and repayment of capital).

Termination: The game may be terminated in either of two ways. (1) Continue play until all players but one have failed to meet the increasing "dividend" demand of stockholders which begins modestly at 8 M$ the first year and increases 4 M$ per year thereafter. Such termination is likely, but especially with a large number of players the game might go on for many periods (years). It should be noted that termination in this manner is not a matter of "going broke," but of failing to meet a rather strong growth in dividend demand by stockholders who then remove the player from office. (2) Continue play for some number of periods (years) agreed upon in advance. During a player's first exposure to the game six or seven years might be used; some players may be forced out, even in this short interval. When play is terminated in this manner, players value their investments at 10 M$ per remaining life period and add cash on hand; the player with the greatest worth wins.

Before play commences, players should be familiar with the "investment alternatives": Eight investments are described below; the statistical distribution of possible cash flows is *uniform* (rather than the normal distribution with a cluster near the median).

Invest-ment	Invest-ment	Life, in Periods	Cash Flow per Period		Approximate Rate of Return on Average Cash Flow
			Mini-mum	Maxi-mum	
	M$		*M$*		*percent*
A	400	40	20	40	7.0
B	400	40	0	80	9.8
C	200	20	11	26	6.7
D	200	20	- 1	49	10.3
E	100	10	9	19	6.6
F	100	10	- 3	37	11.0
G	50	5	11	13	6.4
H	50	5	-11	39	12.4

Play: After players have committed their investments, the years outcome (cash flow) for each investment is determined and posted and the penalty announcement, if any, is made. Each player should compare his cash flow this year with that of last year; if there is a decrease of 50 M$ or more, the player has an added penalty of 10 M$. Commitment of player investment can be secret or open as agreed upon in advance. In any case care should be exercised to "retire" investments when their indicated lives have passed. After cash flows are announced for Period 1,

TABLE OF OUTCOMES

Dice #1	Dice #2	Random Number Digit #1	Digit #2	Card Number	A	B	C	D	E	F	G	H	Penalty
1 or 2	1	even	1	1	36	32	14	49	15	21	11	9	0
	2		2	2	32	80	17	39	19	5	13	19	0
	3		3	3	28	0	20	29	11	37	13	−11	20
	4		4	4	40	16	14	9	15	5	11	39	0
	5		5	5	40	16	20	9	19	21	11	39	0
	6		6	6	36	16	23	29	9	13	12	− 1	20
3 or 4	1		7	7	20	48	23	39	11	13	11	29	0
	2		8	8	20	64	23	39	13	− 3	13	19	0
	3		9	9	32	0	11	19	17	− 3	12	29	20
	4	odd	1	10	28	48	11	−1	17	21	12	9	0
	5		2	11	20	64	26	49	11	37	13	9	0
	6		3	12	24	0	11	−1	19	29	11	−11	20
5 or 6	1		4	13	36	80	17	29	15	37	12	29	0
	2		5	14	24	80	26	−1	13	29	13	39	0
	3		6	15	40	32	26	19	9	− 3	12	19	20
	4		7	16	28	64	14	19	17	5	12	− 1	0
	5		8	17	24	48	17	9	9	29	11	−11	0
	6		9	18	32	32	20	49	13	13	13	− 1	20

Method of Determination		
even	0	Go on to the next random number
odd	0	

Cash Flow Record

Date _____   Player or Team _____

Year of Decade

Investment	1	2	3	4	5	6	7	8	9	10
Cash flow										
plus: Carryover cash	→	→	→	→	→	→	→	→	→	→
plus: Proceeds from sale of investments										
less: Dividends	8	12	16	20	24	28	32	36	40	44
less: Penalties										
equals: Available*										
less: New investments										
equals: Carryover cash										

* Must be ≥ 0; if not, sell enough investments at 5 M$ per remaining year of investment life to satisfy requirement.

players may complete their "cash flow record" for the period; if available capital is negative, enough property must be sold at 5 M$ per remaining year of investment life to make available capital zero or greater. If sufficient capital is available, additional investment(s) may be made; any excess is carried into the following period. At this point players are ready for Period 2 and play proceeds as before. Subsequent periods are played in a like manner.

Outcome: The outcomes of each round are listed in the "table of outcomes." Several alternative methods of determining outcomes are suggested:

1. Give one die to each of two people calling one person #1, the other #2; each round of outcomes is the result of each cast of the dice. Alternatively, a pair of dice, distinguishable as #1 and #2 (by color or size), can be cast by one person.
2. A random number table can be used; choose a starting point in some random fashion, then proceed in an ordered fashion (across and up the page for example), reading a pair of digits for each round.
3. Construct 18 cards representing each horizontal line of the table of outcomes; draw randomly for each round, reinserting the drawn card and reshuffling for each round.

21

Implementation, Control, and Follow-up of Capital Expenditures

Economy and the Search for Better Alternatives

IT IS SELF-EVIDENT that searching out alternative courses of action is a prerequisite to comparison of alternatives. Because comparisons can only reveal preferences among the alternatives identified, it might seem that the search effort should be conducted without limitation. However, we also note that the search effort costs both time and money. From this we can begin to conceive of an amount of search which has an economic limit. Example 21-1 illustrates this point.

Example 21-1.   Economy in the Search for a Better Alternative

Information: A present rearrangement of plant facilities will cost $4000 now and will save an estimated $2000 per year in operating costs for each of the next nine years. It is felt that an additional year of search will reveal an alternative which will reduce operating costs by $2200 per year and would still cost only $4000 at the time of the rearrangement. Because of the one-year delay the savings in operating costs are estimated to occur only over the eight years following the deferred rearrangement. Income taxes are 50%, and financing is by equity capital. Minimum attractive rate of return is 12%. Because rearrangements can be expensed, the after-tax cash flows for the alternatives are half the given pretax amount.

Objective: The rate of return on the prospective current rearrangement (48.6%) is less than that for the distant rearrangement (53.2%) which requires further search. Determine whether further search is worthwhile.

Analysis:

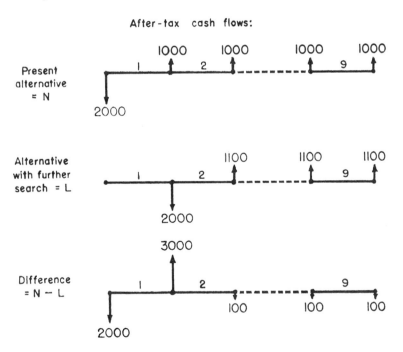

After-tax cash flows:

and for that difference: At $i = 12\%$,

$$PEX = -\$2000 + \$3000(p/f)_1^{12\%} - \$100(p/a)_8^{12\%}(p/f)_1^{12\%}$$
$$= -\$2000 + \$2679 - \$444$$
$$= +\$235$$

Since *PEX* is positive, the rate of return on the advanced (extra) investment in N earns more than the required 12%;

Alternative N should be adopted and search efforts should be discontinued now even though there may be some presently undiscovered Alternative L.

Procedures for the Implementation and Control of Capital Expenditures

After investment opportunities have been perceived and described, their attractiveness is evaluated; a sizable portion of this book has been directed to such efforts. Now we wish to go further in describing the characteristics of a system of review, control, and administration of the continuing flow of investment opportunities.[1] To do this, we will list the attributes we feel are necessary to such a system:

1. Screening or approval required should be related to project size. Size should be measured by the *rate* of expenditure rather than amount; that is, large numbers of inexpensive items can be just as significant as a single large project. This simple point apparently eludes a few firms which permit large numbers of small expenditures to be relatively uncontrolled while large one-time expenditures are subjected to a much more rigorous screening. By controlling expenditures according to their annual rate, the firm can also alleviate the inclination to split projects into component parts simply to reduce the pressures and delays of review.
2. Divisional and departmental allocations should be determined with great care; serious difficulties are invited when allocations are based on one factor only, such as projected rate of return, past allocations, or current request levels.
3. Proposed expenditures should be classified as being justified on economic grounds or on some other basis. It is important that the proportion of projects in each class be carefully controlled within predetermined bounds at the approval stage.
4. Projects should be classified according to use; this provides an opportunity to allow for differences in risk; facili

---

[1] For a thoughtful probing of the organizational arrangements by which a firm reaches capital investment decisions, see William T. Morris, *Capital Budgeting Systems for Large Organizations*, Fourth Summer Symposium Papers of EED-ASEE, available through *The Engineering Economist*, Stevens Institute of Technology, Hoboken, N.J.

tates the planning of orderly growth; and permits response to changes in long-range goals, current economic conditions, or other stimuli.

5. Special channels should be provided for screening and consideration of emergency requests or requests which must be acted upon with a minimum of delay, such as special investment opportunities which present a "now or never" type of choice. Provisions should include a means by which such projects, if approved, can be financed.

6. Careful coordination and full communication should be provided between the various interested divisions, including personnel, finance, and others which are directly or indirectly affected. Representation through committee membership and informational routing of plans is generally essential. Formal procedures, including request forms containing data such as project originator, cash flows, and description are suggested.

7. The means by which approved projects are to be implemented should be specified. Are bids required? Are performance bonds required? Is the method of choosing from among the bidders specified? Are special purchase or requisition forms required? Certain phases here obviously involve the need for legal consultation.

8. Feedback as to system effectiveness as well as project progress is essential.

9. Long-run goals and direction should be made clear. Preliminary planning periods of five to ten years are not unreasonable, especially for those producers whose demands are somewhat predictable and whose facility order-to-installation cycle is long (as for electric utilities). The choice among diverse investment opportunities loses meaning in the absence of such direction.

10. A course of action should be specified for those circumstances where appropriation required exceeds that requested by a substantial amount.

11. Actual cash flows should be monitored and compared with the planned cash flows. Corrective actions should be employed where necessary and this experience should be incorporated in future estimating practices.

12. A governing body, probably a budget committee, should be empowered to change expenditure plans in response to changes in product markets, national outlook, stock market, or other influential indicators.

13. The use of a projected funds flow statement coupled

with a statement of current conditions is suggested. The funding and control further imply the need for an influence upon if not control of dividend policy, probably by the budget committee.

14. Pertinent data such as sales forecasts, production budgets, departmental budgets, and the like should play an integral role in planning efforts.

15. Employee educational programs designed to produce vigorous search efforts for investment opportunities and consistent evaluation analyses of proposed actions are essential to long-term vitality of the firm.

Operational Aspects of Capital Expenditure Procedures and Systems

When any systematic procedure is implemented, we find that certain influences which have been neglected in designing the procedure now play a part in its actual operation.[2] Common among these are the conscious and unconscious traits, biases, and other characteristics of the people who operate such systems. Our purpose here is to note a few cautions in the implementation of any system which proposes to provide control of capital expenditures.

Successful discovery of expenditure opportunities is the key to organizational vigor and longevity. Expenditure opportunities are subjected to many screenings, beginning with those of its originator. The procedural screening process must be carefully controlled not only to do a proper job of screening but also to perform this function in such a way that search or reporting efforts are not discouraged. An organization can significantly influence both the conscious and unconscious screening of its prospective idea people; yesterdays rejection of a dissimilar project probably influences todays would-be innovator.

Bower reports a study of one company in which forecast performance was compared with actual performance for 50 investment projects:[3]

---

[2] For a helpful basic guide to forms, check lists, and procedures in a system-related view, see N. E. Pflomm, *Managing Capital Expenditures,* a research report from the National Industrial Conference Board, Studies in Business Policy, No. 107, NICB, New York, 1963.

[3] J. L. Bower, *Managing the Resource Allocation Process* (Boston, Harvard Univ., 1970).

Type of Project	(PEX Actual)/(PEX Forecast)
Cost reduction	1.1
Sales expansion	0.6
New products	0.1

The report is alarming; an "ideal" ratio of 1.0 for a group of projects only assures us of the absence of a consistent bias. The findings confirm our expectation that operating costs and construction costs tend to be more accurately estimated and controllable than sales volume and market price. Consider a cost-reduction project, for example, where the estimated reduction in manpower requirements may be implemented and thus become a self-fulfilling prophecy.

In attempting to screen projects by size, we stated that size should be measured as an expenditure *rate*, rather than amount. Bower reports an instance in which an entire plant was built and equipped by a series of expense orders, each of which was less than $50,000. Only when the division requested the climaxing indivisible chimney project costing more than $50,000 did top management discover the plant and facilities it would serve.

Benefit-cost ratios take on a new meaning when a manager weighs the "benefits of being right" against the "costs of being wrong" in project sponsorship. There is a danger that managers will evaluate projects against their perception of the organization's criteria for evaluating and rewarding managers and then place their bets. Is excess capacity evidence of poor judgment? Is maintaining a market position more important than maintaining profitability? Will a manager be shifted to new positions every two or three years and thereby escape performance measurement based on accountability for the projects he has sponsored? Is a manager's relationship to his peers more enduring and more significant in his future than his successful sponsorship of projects? The answers to such questions affect the allocation of an organization's resources.

Each successful project sponsorship increases a manager's ability to sponsor new projects by adding to the confidence which is placed on his judgment. That confidence must be protected, not jeopardized. But would not a manager's "perfect" record of successful projects indicate overly conservative sponsorship?

Still another system caution is that the innovators are not equal in their aggressiveness in seeking change, nor in their ability to communicate new ideas; nonetheless these factors bear upon the ultimate decision to accept or reject a project.

Acceptance of an expenditure opportunity is often thought to involve risk; risk is, however, only more subtle in the rejection decision where one accepts the risk and consequences of waiting too long to act, of failing to act decisively, or of being too conservative.

The organization which succeeds in generating many good and acceptable opportunities will find that decisions employing imperfect techniques and made under circumstances of incomplete information must be made within a severely limited time interval.

Problems

21-1. Review the capital expenditure procedures of an organization with which you are familiar, then check it against the list of attributes given in this chapter. Comment on control features unique either to the suggested system or to the actual system.

21-2. Write a brief report explaining the present capital expenditure procedures of an organization with which you are familiar; include forms employed and diagram the sequence of required approvals. Comment upon changes which might improve present procedures.

21-3. Two weeks ago the board of directors of Green Corporation gave final approval to next years capital expenditure budget of $20 million. Drastic changes on both a national and international level now make it appear that the planned budget should be cut in half. It would be a fairly simple matter to cut each divisional, departmental, and other appropriation in half to accomplish this needed cutback; would this be a good idea? Give your recommendations for any departure from this.

VII

Appendices

A

Notation

THE FOLLOWING LIST contains the symbol name of a variable, its verbal description or definition, and where applicable, an equation which provides a mathematical description or definition of that variable. The reader will find the list helpful either as a glossary or as a convenient access point for equations relating the variables of engineering economy analyses. A similar list including suggested FORTRAN symbols is provided in Appendix B for those involved with computer-programmed analyses. **Caution:** Equations below apply to the after-tax approach. To use the before-tax approach, substitute i_b for i_a, then use $t = 0$ except in equations for i_a, i_b, and i_c.

Symbol   Definition and Equation

a = the annual EOY discrete-flow sum in a uniform series of sums

a' = the annual continuous-flow sum in a uniform series of sums

a'/a = a continuous-flow sum equivalent to EOY sum
   $= [\ln (1 + i_a)] / i_a$

a/a' = EOY sum equivalent to a continuous-flow sum
   $= i_a / \ln (1 + i_a)$

$a/c = AE$ of geometric series $= (a/f)(f/c)$

$a/f = AE$ of future sum $= i/[(1 + i)^n - 1]$

$a/g = AE$ of gradient series $= \dfrac{1}{i} - \dfrac{n}{(1 + i)^n - 1}$

$a/p = AE$ of present sum $= i/[1 - (1 + i)^{-n}]$

$AEC =$ annual equivalent cost $= AE$ revenue required

$= AEM + [B_z (a/p)_n^i - V(a/f)_n^i - t(AED)]/(1 - t)$

$AEC_L = AEC$ of leasing $= AEM + ALP + ALI$

$AEC_V = AEC$ avoided if "defender" relinquished

$= AEM + [B_{vz}(a/p)_n^i - V(a/f)_n^i - t(AED)]/(1 - t)$

$AED_{XXX} = AE$ tax depreciation allocations by XXX method

$= (a/p) PED_{XXX}$

$AEM = AE$ cash operating (M) costs $= (a/p) PEM$

$AER = AE$ revenues received $= (a/p) PER$

$AEX = AE$ excess of revenues minus costs $= (a/p) PEX$

$= (AER - AEC)(1 - t)$

$= (AER - AEM)(1 - t) - B_z(a/p) + V(a/f) + t(AED)$

$ALI =$ annual equivalent "financial impairment effect" from leasing

$$= ALP(ALR)\left[\frac{i_a}{i_d(1 - t)} - 1\right]\left\{1 - \frac{(a/p)_n^{i_a}}{i_a - i_d}\right.$$
$$\left. \cdot [(p/f)_n^{i_d} - (p/f)_n^{i_a}]\right\}$$

$ALP =$ annual leasing payment

$ALR =$ a lease "escapability" rating (from 0.0 for easily escaped costs of equipment lease, to 1.0 for an inescapable financial lease)

$ATCFP_z =$ after-tax cash flow (prospective) at EOY z

$= R_z - M_z - t(R_z - M_z - D_z - I_z)$

$ATCFR_z =$ after-tax cash flow (required) at EOY z

$= (1 + i_c)(B_{v(z-1)} - V_z)$

$B_{rz} =$ regulatory depreciation (rate) base unallocated at age z

$B_{tz} =$ tax depreciation base unallocated at EOY z. Also called the remaining book value or unallocated cost. If property is purchased new, $B_{tz} = B_{tO} - DC_z$. Even if $z = 0$, B_{tz} may differ from B_z or B_{vz} due to ILDC, investment credit, a market value different from book value, and so forth

$B_{tz\,DDB} = B_t (1 - 2/n)^z$ until switch or $DC_z = B_t - V$

$$B_{tz\,SF} = V + (B_t - V)\left[\frac{(1 + i)^n - (1 + i)^x}{(1 + i)^n - 1}\right]$$

$B_{tz\,SL} = V + (B_t - V)(1 - z/n)$

$$B_{tz\,SYD} = V + (B_t - V)\left(\frac{n^2 + n - 2nz - z + z^2}{n^2 + n}\right)$$

B_{vz} = investment base avoided if defender relinquished.
= cash foregone by retaining defender = after-tax "make-operative" costs + salvage foregone at EOY z

Before commitments are made: $B_{vz} = B_z$
Between commitment and operation: $B_z > B_{vz} > V_z$
After facility is operable: $B_{vz} = V_z$

B_z = investment base for "acquire" decision. After-tax cost of purchase + transport + install + working capital + *ILDC* - discounts - income tax savings via investment credit, all at age z. In before-tax approach, ignore income taxes

BCR_C = conventional benefit/cost ratio
= (user *PER*)/(supplier *PEC*)

BCR_M = modified benefit/cost ratio
$$= \frac{(\text{user } PER) - (\text{supplier } PEM)}{\text{supplier } PEBTCR}$$

$BTCFP_z$ = before-tax cash flow (prospective) at EOY z
= $R_z - M_z$

$BTCFR_z$ = before-tax cash flow (required) at EOY z
= $[B_{v,z-1}(1 + i_a) - V_z - tD_z]/(1 - t)$

$BTRR$ = before-tax rate of return required; see i_b

c = first cash flow in geometric series of discrete cash flows. Flow during year $z = c (1 + r)^{z-1}$

c' = first cash flow in geometric series of continuous cash flows. Flow during year $z = c' (1 + r)^{z-1}$

C_z = cost or revenue required at EOY z
= $[M_z + B_{v,z-1}(1 + i_a) - V_z - tD_z]/(1 - t)$

D_q = percentage depletion rate, such as 0.22, 0.15

D_r = regulatory depreciation allocation allowed at EOY z

D_z = tax depreciation allocation at EOY z

$D_{z\,DDB}$ = $B_t(2/n)(1 - 2/n)^{z-1}$ until switch or $DC_z = B_t - V$. Law permits switch to SLN, SYD, ASY

$D_{z\,SF}$ = $(B_t - V)(i_a)(1 + i_a)^{z-1}/[(1 + i_a)^n - 1]$

$D_{z\,SL}$ = $(B_t - V)/n$

$D_{z\,SYD}$ = $(B_t - V)(2)(n + 1 - z)/(n^2 + n)$

$D_{z\,1AS}$ = Let $X = 2B_t(1 - 2/n)/(n^2 - n)$; $D_1 = 2 B_t/n$; $D_2 = X(n - 1); D_3 = X(n - 2); \ldots$ until $DC_z = B_t - V$

$D_{z\,1SY}$ = Let $X = 2(B_t - V)(1 - 2/n)/(n^2 - n)$; $D_1 = 2B_t/n$; $D_2 = X(n - 1); D_3 = X(n - 2); \ldots$

$D_{z\,125}$ = $B_t(1.25/n)(1 - 1.25/n)^{z-1}$ until $DC_z = B_t - V$ or switch to SL when $D_{z+1} < (B_{tz} - V)/(n - z)$

DC_z = tax depreciation accrued at EOY $z = D_1 + D_2 + \cdots + D_z$

$DC_{z\,DDB}$ = $B_t[1 - (1 - 2/n)^z]$ until $DC_z = B_t - V$ or switch

$DC_{z\,SF}$ = $(B_t - V)\left[\dfrac{(1 + i_a)^z - 1}{(1 + i_a)^n - 1}\right]$

$DC_{z\ SL} = (B_t - V)\,z/n$

$DC_{z\ SYD} = (B_t - V)(2zn - z^2 + z)/(n^2 + n)$

$DC_{z\ 1AS} = B_t\,[2/n + (n - 2)(2n - z)(z - 1)/(n - 1)(n^2)]$ until
$\qquad\qquad DC_z = B_t - V$

DPR = dividend payout ratio = DPS_z/EPS_z

DPS_z = dividends per share in year z; treat as EOY

$e = 2.7182818\ldots$ the base of natural logarithms

EPS_z = earnings per share in year z; treat as EOY

$E(i_a)$ = expected $i_a = i_a$ such that $E(PEX) = 0$

$E(PEC)$ = expected PEC

$E(PEX)$ = expected $PEX = \Sigma_{z=0}^{n}\ E(x_z)/(1 + i_a)^z$
$\qquad\qquad = \Sigma_{J=1}^{\text{last}}\ p_{PEX}\ (PEX_J)$

$E(U)$ = expected utility = $\Sigma_{z=0}^{\text{last}}\ p_{PEXJ}\,(U_{PEXJ})$

$E(x_z)$ = expected value of the cash flow x at EOY z
$\qquad\qquad = \Sigma_{J=1}^{\text{last}}\ p_{xzJ}\,(x_{zJ})$

f = a future discrete-flow sum at EOY n

f' = a future continuous-flow sum during year n

$$f/c = FE \text{ of geometric series} = \frac{(1 + i_a)^n - (1 + r)^n}{i_a - r}$$

$$f/g = FE \text{ of gradient series} = \frac{1}{i_a}\left[\frac{(1 + i_a)^n - 1}{i_a} - n\right]$$

$f/p = FE$ of a present sum $= (1 + i_a)^n$

FEC = future equivalent cost = FE revenue required
$\qquad\quad = FEM + [B_z\,(f/p) - V - t(FED)]/(1 - t)$

FED_{XXX} = FE tax depreciation allocations by XXX method
$\qquad\qquad = \Sigma_{z=1}^{n}\ D_z(1 + i)^{n-z}$

FEM = FE cash operating (M) costs = $\Sigma_{z=1}^{n}\ M_z\,(1 + i)^{n-z}$

FER = FE revenues received = $\Sigma_{z=1}^{n}\ R_z\,(1 + i)^{n-z}$

FEX = FE excess of revenues minus costs
$\qquad\quad = (FER - FEC)\,(1 - t)$

g = gradient, the annual increase in an arithmetic series
of discrete cash flows. Flow at EOY z is $(z - 1)\,g$

g' = gradient, the annual increase in an arithmetic series
of continuous cash flows. Flow during year z is
$(z - 1)\,g'$

i_a = effective annual after-tax ROR, required $(MARR)$ or
prospective = i_a such that $\Sigma_{z=0}^{n}\ ATCFP_z/(1 + i_a)^z = 0$

$$i_a = i_c - tr_d\,i_d = (1 - t)\,r_d\,i_d + (1 - r_d)\,i_e$$
$$i_a = (1 + i_n)^m - 1; \quad \text{if } m \longrightarrow \infty,\ i_a = e^{i_n m} - 1$$

i_b = effective annual before-tax ROR, required $(BTRR)$
or prospective, $i_b \approx i_a/(1 - t) = MARR/(1 - t)$

i_c = effective annual after-tax ROR, required or prospective, on composite capital
$\qquad = r_d\,i_d + (1 - r_d)\,i_e = i_a + tr_d\,i_d$

i_d = effective annual after-tax ROR, required or prospective, on debt capital

i_e = effective annual after-tax ROR, required or prospective, on equity capital

i_i = cost of internally generated funds

$i_n m$ = nominal annual ROR. If $m \longrightarrow \infty$, $i_n m = \ln(1 + i_a)$

i_n = ROR, required or prospective, per interest-compounding period = $(1 + i_a)^{1/m} - 1$

i_r = minimum effective annual ROR prospective on projects financed from internal funds only

i_x = cost of external funds

I_z = interest sum due on debt capital at EOY z
    = $r_d i_d$ (unrecouped B_{z-1})

$ILDC$ = interest lost during construction. Let x_z = payment made to contractor z years prior to facility completion. Then $ILDC$ = sum of $x_z(z)(i_a)$ from z = first to z = last payment

m = number of interest-compounding periods per year

M_z = cash operating costs (maintenance + operation + insurance + property tax + material + labor) at EOY z

$MARR$ = minimum attractive rate of return. The effective annual after-tax (hurdle) ROR required to meet the investor's threshold of acceptability = "opportunity cost" of capital; see i_a

n_a = age of asset at date of acquisition

$n_c = n$ = life of asset, in years = $n_v - n_a$

n_t = tax life, in years. The shortest tax life permitted is: if $n < 5$, $n_t = n$; if $5 < n < 6.25$, $n_t = 5$; if $6.25 < n$, $n_t = 0.8 n$

n_v = age of asset at date of retirement

N = number of years that demand continues

NP = prospective payoff period = number of years such that $\Sigma_{z=0}^{NP} BTCFP_z = 0$

NR = required (maximum acceptable) payoff period

p = a present discrete-flow sum at EOY zero

p' = a present continuous-flow sum during year zero

$p/a = PE$ of annual series = $[1 - (1 + i_a)^{-n}]/i_a$

$p/c = PE$ of geometric series = $\left[1 - \left(\dfrac{1+r}{1+i_a}\right)^n\right] \Big/ (i_a - r)$

$p/f = PE$ of future sum = $(1 + i_a)^{-n}$

$p/g = PE$ of gradient series = $\dfrac{1}{i}\left[\dfrac{(1+i)^n - 1}{i(1+i)^n} - \dfrac{n}{(1+i)^n}\right]$

p_{x3} = probability that event x_3 occurs

$PEBTCP = PE$ of $BTCFP = PER - PEM$
    = $\Sigma_{z=1}^{n} (R_z - M_z)/(1 + i_a)^z$

$PEBTCR = PE$ of $BTCFR = PEC - PEM$
    = $[B_z - V(p/f) - t(PED)]/(1 - t)$

PEC = present equivalent cost = PE revenue required

$\quad = PEM + [B_z - V(p/f) - t(PED)]/(1 - t)$

PED_{XXX} = PE tax depreciation allocations by XXX method

$\quad = \Sigma_{z=1}^{n} D_z/(1 + i_a)^z$

$PED_{FYW} = (B_t - V)(1/5)(p/a)^{(i/5)}$

PED_{max} = use SYD if $V/B_t < -1/n_t$; 1SY if $-1/n_t < V/B_t < 0$; 1AS if $0 < V$

$PED_{SF} = (B_t - V)(a/f)_{n_t}^{i}(p/f)_1^i \, n_t$

$PED_{SL} = (p/a)_{n_t}^{i}(B_t - V)/n_t$

$PED_{SYD} = (B_t - V)[2/(n_t^2 + n_t)] \, [n_t(p/a)_{n_t}^{i} - (p/g)_{n_t}^{i}]$

$PED_{XWZ} = (B_t - V)(p/f)_z^i$

PEM = PE cash operating (M) costs = $\Sigma_{z=1}^{n} M_z/(1 + i_a)^z$

PER = PE revenues received = PE benefits realized

$\quad = \Sigma_{z=1}^{n} R_z/(1 + i_a)^z$

PEX = PE excess of revenues minus costs

$\quad = (PER - PEC)(1 - t)$

$\quad = (PER - PEM)(1 - t) - B_z + V(p/f) + t(PED)$

$\quad = \Sigma_{z=0}^{n} ATCF_z/(1 + i_a)^z$

PLI = present equivalent "financial impairment effect" from leasing

$$= ALP(ALR)\left[\frac{i_a}{i_d(1 - t)} - 1\right]\left\{(p/a)_n^{i_a} - \frac{1}{i_a - i_d}\right.$$
$$\left. \cdot \, [(p/f)_n^{i_d} - (p/f)_n^{i_a}]\right\}$$

PPE_z = (market price)/(current EPS) ratio at EOY z

PPS_z = price per share at EOY z

$PPS_0 = \Sigma_{z=1}^{n} DPS_z/(1 + i_a)^z + PPS_n/(1 + i_a)^n$

r = geometric series growth rate (as in 1, 2, 4, 8, . . . where $r = 1.0$)

r_d = debt ratio = debt/(debt + equity)

R_z = revenues prospective (or realized) at EOY z. In public projects use "price public would be willing to pay" minus "actual costs paid by public" at EOY z

t = effective combined federal and state income tax rate on incremental income

$\quad = 1 - [1 - TSR - TFR + TSR(TFR)]/$
$$\qquad\qquad\qquad [1 - TSR(TFR)(TDD)]$$

TCE_z = (before-tax) times interest charges earned for year z

$\quad = (R_z - M_z - D_z)/I_z$

TDD = deductibility of federal income tax in computing state income tax; $0 \leqslant TDD \leqslant 1$

TFR = federal tax rate on incremental income

TSR = state tax rate on incremental income

$U(PEX)$ = utility of a certain PEX outcome, in utiles

$U(X)$ = utility of a certain outcome x, in utiles

V = after-tax net salvage value at EOY n

V_z = after-tax net salvage value at EOY z = gross salvage minus costs to remove and sell plus income tax effect of capital gain or loss at EOY z. In before-tax approach, ignore income taxes

z = attained age of a property in years. Used as a subscript for $B, C, D, \ldots V$

Z = number of years that growth persists

σ = standard deviation = (variance)$^{0.5}$

σ^2 = variance = (standard deviation)^2
$$= \Sigma_{z=1}^{last} p_{xz}[x_z - E(x_z)]^2$$

Symbol	Abbreviation
AASHO	= American Association of State Highway Officials
AE	= annual equivalent
ASY	= SYD applied to B_t, not $(B_t - V)$; accrual limit
	= $(B_t - V)$; allowed since 1971
BPR	= Bureau of Public Roads
DDB	= double-declining-balance depreciation
EOY	= end-of-year
FE	= future equivalent
FICA	= Federal Insurance Compensation Act
FYW	= five-year writeoff form of depreciation
HRB	= Highway Research Board
ln	= natural logarithm
L2	= Iowa-type survivor dispersion L_2
MAPI	= Machinery and Allied Products Institute
max	= maximum
min	= minimum
NE	= negative exponential survivor dispersion
PE	= present equivalent
O2	= Iowa-type survivor dispersion O_2
ROR	= rate of return; see i_a
R2	= Iowa-type survivor dispersion R_2
S2	= Iowa-type survivor dispersion S_2
SF	= sinking-fund method of depreciation
SL	= straight-line method of depreciation
SYD	= sum-of-the-years-digits depreciation
XW	= expensed writeoff: impact delayed to EOY z
∞	= infinity
Σ	= sigma = sum of
1AS	= use DDB in year 1, ASY thereafter. Use only if $V > 0$. Accrual limit = $B_t - V$. Maximizes PED if $V > 0$, and if t and i_a are both positive and time invariant
1SY	= use DDB in year 1, SYD thereafter

125 = 125% declining-balance depreciation
150 = 150% declining-balance depreciation
 3B = an investment alternative. Mutually exclusive with
    any other investment (such as 3C) having the same
    prefix number.   Independent of any other invest-
    ment (such as 5C) having a different prefix number

Unless otherwise specified:

After-tax approach is to be used
Base for taxes, acquire/retain, or regulation are identical.
  $B_{tz} = B_z = B_{vz} = B_{rz}$
Cash flows are end of year
Compounding frequency is annual; $m = 1$, so $i_a = i_n m$
Debt ratio is zero; $r = 0$, so $i_a = i_c = i_e$; $I_z = 0$
i_a is positive and time invariant
Investment credit percentage = 0
Item (not group) accounting is used, so gain or loss on disposal
  can occur
Life for tax depreciation, acquire/retain, or regulation are
  identical $n_t = n$
Life given (n) is not subject to dispersion; when dispersion ap-
  plies, see Chapter 14
Optimal acquisition age (n_a) = 0; therefore, optimal retirement
  age (n_v) is identical to economic life (n)

B

Computer-aided Analysis for Capital Expenditure Decisions

THE USE OF COMPUTERS has had great impact on all fields of study; engineering economy is no exception. Compound interest tables such as Appendix I were once a closely guarded possession of the publisher. Today with a relatively simple computer program, the student can produce his own tables for any combination of parameters he chooses. He can produce in a matter of seconds a printed output that once took thousands of man-hours of calculation, checking, typesetting, proofreading, and correction. Today we can make multivalued estimates for each parameter, then with trivial computer time find the multivalued output on which a decision is to be based. Without the computer the unwieldiness of such calculations ruled out or at least severely restricted such analyses.

Computer-aided analyses are especially advantageous for problems requiring iterative solutions or for recurring types of problems which involve considerable detail. The programs suggested here provide a sample of those which might be developed.

1. Refer to Chapter 4, then write a flow chart and a computer program for the computation of compound interest factors. Neglect spacing and format considerations except to start a new line for each change of n. $n = 5, 10, 15, \ldots 60, i > 0$.

2. Refer to Chapter 7, then write a flow chart and a computer program for the computation of rate of return when given year-by-year cash flows. Make sure the routine will warn NEGATIVE for negative rates of return and NONUNIQUE where that possibility arises.

3. Refer to Chapter 9, then write a flow chart and a computer program for the computation and listing of year-by-year depreciation charges as a percent of first cost, B, if DDB is used the first year, and ASY thereafter. Make sure the routine meets the various legal restrictions and results in the maximum permissible charges. $V/B = -0.4, 0.0, 0.4, 0.8; n = 10$.

4. Refer to Chapter 9, then write a flow chart and a computer program for the computation and printing of present equivalent depreciation expense under the SYD depreciation method. $i = 10\%; V/B = -0.4, -0.2; 0.0; n = 5, 10, 15, \ldots 60$.

5. See your computer center's "library" of computer programs. List one of the programs applicable to engineering economy studies and briefly describe its function. Show input data requirements and output data capability.

6. See the graduate theses of your department. Note one which involved development of a computer program for an engineering economy type of problem. Describe its function, input data requirements, and output data capability.

7. See Table 10-3. Write a flow chart and a computer program to determine a similar table for any values of i_a and t. Exclude investment credit from your computations. Demonstrate using $i_a = 15\%, t = 40\%$.

8. See Example 17-1, Part d. If i is not given, the costs would be:

	Year-by-Year Cost if:	
Interval	Owned when year began	Purchased when year began
0–1	. . .	$8,800 + 10,000i$
1–2	$4600 + 3200i$	$8,200 + 6,800i$
2–3	$4800 + 1600i$	$8,400 + 5,200i$
3–4	$5400 + 800i$	$9,000 + 4,400i$
4–5	$6200 + 400i$	$9,800 + 4,000i$
5–6	$7100 + 200i$	$10,700 + 3,800i$
6–7	$8050 + 100i$	$11,650 + 3,700i$
7–8	$9025 + 50i$	$12,625 + 3,650i$

Write a flow chart and a computer program to efficiently determine optional acquisition and retirement dates for any given values of i. Solve for $i = 10\%, 50\%, 100\%$.

9. Perform a Monte Carlo simulation of outcomes as follows. Refer to Figure 12-6, then write a flow chart and a computer program, using the input data given. Compute *PEX* for each of 100 sets of input estimates. Use SIMPLOTTER to graph the multivalued output.
10. Write a flow chart and a computer program involving the following steps. Assume the continuous outcome function and utility function of Figure 13-2. Divide the x-axis into intervals of width 0.2. Then cross-multiply and sum the midinterval $U(x)$ and $p(x)$ values, to determine $E(U)$.

THE FOLLOWING LISTS PROVIDE A CROSS-REFERENCE BETWEEN THE SYMBOLS USED IN THIS BOOK, A SET OF SUGGESTED FORTRAN SYMBOLS, AND THE MEANINGS OF THOSE SYMBOLS. WE HAVE ATTEMPTED TO MAKE THE SYMBOLS AS COMPATIBLE AS POSSIBLE, RECOGNIZING THAT IN FORTRAN WE CANNOT USE LOWER CASE LETTERS, SPECIAL CHARACTERS OTHER THAN THE $ SYMBOL, OR SUBSCRIPTS. BY DEVELOPING A CAREFULLY-TESTED SYMBOL SYSTEM WE HOPE TO MINIMIZE THE READER'S EFFORT INVOLVED IN "TRANSLATING" BOOK OR COMPUTER PROGRAM MATERIAL. THE READER WILL ALSO FIND THIS A CONVENIENT ACCESS POINT FOR EQUATIONS RELATING VARIABLES OF ENGINEERING ECONOMY ANALYSES.

CAUTION: EQUATIONS BELOW APPLY TO THE AFTER-TAX APPROACH. TO USE BEFORE-TAX APPROACH, SUBSTITUTE IB FOR IA, THEN USE TT=0 EXCEPT IN EQUATIONS IA,IB,IC.

CAUTION: FORTRAN VARIABLE NAMES WHICH BEGIN WITH I,J,K,L, M,OR N ARE PREDEFINED AS INTEGER. OVERRIDE THIS CONVENTION BY STARTING YOUR PROGRAM WITH THE STATEMENT: REAL IA,IB, IC,ID,IE,II,IM,IN,IR,IX,NT

ALPHABETIZED AND DEFINED ACCORDING TO FORTRAN SYMBOLS:

BOOK	FORTRAN	DEFINITIONS AND EQUATIONS
a	A	= THE ANNUAL EOY DISCRETE-FLOW SUM IN UNIFORM SERIES OF SUMS. A=F*AEF=G*AEG=P*AEP=A$*AEA$
a/a'	AEA$	= EOY SUM EQUIVALENT TO A CONTINUOUS-FLOW SUM = IA/ALOG(1+IA)
AEC	AEC	= ANNUAL EQUIVALENT COST =AE REVENUE REQUIRED = AEM+(BVNA*AEP-VNV*AEF-TT*AED)/(1-TT)
AEC_L	AECL	= AEC OF LEASING =AEM+ALP+ALI
AEC_V	AECV	= AEC AVOIDED IF "DEFENDER" RELINQUISHED = AEM+(BVNA*AEP-VNV*AEF-TT*AED)/(1-TT)
AED_{xxx}	AEDXXX	= AE TAX DEPRECIATION ALLOCATIONS BY "XXX" = AEP*PEDXXX
a/f	AEF	= AE OF FUTURE SUM =IA/((1+IA)**NN-1)
a/g	AEG	= AE OF GRADIENT SERIES = 1/IA-NN/((1+IA)**NN-1)
a/c	AEH	= AE OF GEOMETRIC SERIES =AEF*FEH=IA/((1 +IA)**NN-1)*((1+IA)**NN-(1+RH)**NN)/(IA-RH)

BOOK   FORTRAN   DEFINITIONS AND EQUATIONS

```
AEM     AEM     = AE CASH OPERATING (M) COSTS          =AEP*PEM
a/p     AEP     = AE OF PRESENT SUM          =IA/(1-(1+IA)**(-NN))
AER     AER     = AE REVENUES RECEIVED                 =AEP*PER
AEX     AEX     = AE EXCESS OF REVENUES MINUS COSTS    =AEP*PEX
                = (AER-AEC)*(1-TT)
                = (AER-AEM)*(1-TT)-BVNA*AEP+VNV*AEF+TT*AED

ALI     ALI     = ANNUAL EQUIVALENT       "FINANCIAL IMPAIRMENT
                  EFFECT" FROM LEASING
                = ALP*ALR*(IA/(ID*(1-TT))-1)*(1-IA/(1-(1+IA)*
                  *(-NN)*(IA-ID))*((1+ID)**(-NN)-(1+IA)**(-NN
                  )))
ALP     ALP     = ANNUAL LEASING PAYMENT
ALR     ALR     = LEASE "INESCAPABILITY" RATING (FROM 0.0 FOR
                  EASILY ESCAPED COSTS OF EQUIPMENT LEASE, TO
                  1.0 FOR AN INESCAPABLE FINANCIAL LEASE).
ATCFPz  ATCFPZ  = AFTER-TAX CASH FLOW PROSPECTIVE    AT AGE Z
                = RZ-MZ-TT*(RZ-MZ-DZ-IZ)
ATCFRz  ATCFRZ  = AFTER-TAX CASH FLOW REQUIRED        AT AGE Z
                = (1+IC)*BV(Z-1)-VZ
a'      A$      = THE ANNUAL CONTINUOUS-FLOW SUM IN A UNIFORM
                  SERIES OF SUMS                      =A*A$EA
a'/a    A$EA    = A CONTINUOUS-FLOW SUM EQUIVALENT TO EOY SUM
                = ALOG(1+IA)/IA

BCRc    BCRC    = BENEFIT/COST RATIO (CONVENTIONAL)
                = (USER PER)/(SUPPLIER PEC(=PEBTCR+PEM))
BCRm    BCRM    = BENEFIT/COST RATIO (MODIFIED)
                = ((USER PER)-(SUPPLIER PEM))/SUPPLIER PEBTCR
                = PE TAX DEPRECIATION ALLOCATIONS
Bna     BNA     = BZ WHEN Z=NA, THE AGE AT ACQUISITION
Brz     BRZ     = REGULATORY DEPRECIATION (RATE) BASE UNALLO-
                  CATED AT AGE Z
BTCFPz  BTCFPZ  = BEFORE-TAX CASH FLOW PROSPECTIVE    AT AGE Z
                = RZ-MZ
BTCFRz  BTCFRZ  = BEFORE-TAX CASH FLOW REQUIRED       AT AGE Z
                = (BV(Z-1)*(1+IA)-VZ-TT*DZ)/(1-TT)
Btna    BTNA    = TAX DEPRECIATION BASE IF ACQUIRED AT AGE NA
BTRR            = BEFORE-TAX RATE OF RETURN REQUIRED     SEE IB

Btz     BTZ     = TAX DEPRECIATION BASE UNALLOCATED AT AGE Z.
                  ALSO CALLED REMAINING BOOK VALUE OR UNALLO-
                  CATED COST. IF PROPERTY PURCHASED NEW, BTZ
                  =BTO-DCZ. EVEN IF Z=0, BTZ MAY DIFFER FROM
                  BZ OR BVZ DUE TO ILDC, INVESTMENT CREDIT, A
                  MARKET VALUE DIFFERENT FROM BOOK VALUE,ETC.
BtzDDB  BTZDDB  = BTNA*(1-DBDR/NT)**(Z-NA)
                  DBDR=2.00      UNTIL SWITCH OR DCZ=BTNA-VTNV
BtzSFU  BTZSFU  = VTNV+(BTNA-VTNV)*((1+IA)**NT-(1+IA)**(Z-NA)
                  )/((1+IA)**NT-1)
BtzSLN  BTZSLN  = VTNV+(BTNA-VTNV)*(NT-Z+NA)/NT
BtzSYD  BTZSYD  = VTNV+(BTNA-VTNV)*(NT*NT+NT-2*NT*(Z-NA)-Z+NA
                  +(Z-NA)*(Z-NA))/(NT*NT*NT)
Buz     BUZ     = UNRECOUPED INVESTMENT AT AGE=Z
Bvna    BVNA    = BVZ WHEN Z=NA, THE AGE AT ACQUISITION
Bvz     BVZ     = INVESTMENT BASE AVOIDED IF DEFENDER RELIN-
                  QUISHED. EQUALS CASH FOREGONE BY RETAINING
                  DEFENDER = AFTER-TAX "MAKE-OPERATIVE" COSTS
                  + SALVAGE FOREGONE AT AGE Z
```

BOOK	FORTRAN		DEFINITIONS AND EQUATIONS

			BEFORE COMMITMENTS ARE MADE: BVZ=BZ
			BETWEEN COMMITMENT AND OPERATION: BZ>BVZ>VZ
			AFTER FACILITY IS OPERABLE: BVZ=VZ
B_z	BZ	=	INVESTMENT BASE FOR "ACQUIRE" DECISION. AF-
		=	TER-TAX COST OF PURCHASE+TRANSPORT+INSTALL+
			WORKING CAPITAL+ILDC-DISCOUNTS - INCOME TAX
			SAVINGS VIA INVESTMENT CREDIT,ALL AT AGE Z.
			IN BEFORE-TAX APPROACH,IGNORE INCOME TAXES.

C_z	CZ	=	COST OR REVENUE REQUIRED AT AGE Z
		=	MZ+(BV(Z-1)*(1+IA)-VZ-TT*DZ)/(1-TT)
DC_z	DCZ	=	TAX DEPRECIATION ACCRUED AT AGE Z
		=	D1+D2+...+DZ
DC_{zDDB}	DCZDDB	=	BTNA*(1-(1-DBDR/NT)**(Z-NA))
			DBDR=2.00 UNTIL SWITCH OR DCZ=BTNA-VTNV
DC_{zSFU}	DCZSFU	=	(BTNA-VTNV)*((1+IA)**(Z-NA)-1)/((1+IA)**NT-1)
DC_{zSLN}	DCZSLN	=	(BTNA-VTNV)*(Z-NA)/NT
DC_{zSYD}	DCZSYD	=	(BTNA-VTNV)*(2*NT*(Z-NA)-(Z-NA)*(Z-NA)+(Z-NA)/(NT*NT+NT)
DC_{zIAS}	DCZIAS	=	BTNA*(2/NT+(NT-2)*(2*NT-Z+NA)*(Z-1-NA)/((NT-1)*NT*NT)) UNTIL DCZ=BTNA-VTNV
DPR	DPR	=	DIVIDEND PAYOUT RATIO =DPSZ/EPSZ
DPS_z	DPSZ	=	DIVIDENDS PER SHARE IN YEAR Z; TREAT AS EOY
D_q	DQ	=	PERCENTAGE DEPLETION RATE 0.22,0.15,ETC
D_{rz}	DRZ	=	REGULATORY DEPRECIATION ALLOCATION ALLOWED AT AGE Z
D_z	DZ	=	TAX DEPRECIATION ALLOCATION AT AGE Z
D_{zDDB}	DZDDB	=	BTNA*DBDR/NT*(1-DBDR/NT)**(Z-1-NA) TIL DCZ=BTNA-VTNV OR SWITCH TO SLN,SYD,ASY. DBDR=2.
D_{zFYW}	DZFYW	=	(BTNA-VTNV)/5 FOR Z=(NA+1) THROUGH (NA+5)
D_{zSFU}	DZSFU	=	(BTNA-VTNV)*IA*(1+IA)**(Z-1-NA)/((1+IA)**NT-1)
D_{zSLN}	DZSLN	=	(BTNA-VTNV)/NT
D_{zSYD}	DZSYD	=	(BTNA-VTNV)*2*(NT+1-Z+NA)/(NT*NT+NT)
D_{zIAS}	DZIAS	=	LET X1AS=BTNV*2*(1-2/NT)/(NT*NT-NT); D(NA+1)=BTNA*2/NT;D(NA+2)=X1AS*(NT-1);D(NA+3)=X1AS*(NT-2);... UNTIL DCZ=BTNA-VTNV
D_{zISY}	DZ1SY	=	LET X1SY=(BTNV-VTNA)*2*(1-2/NT)/(NT*NT-NT); D(NA+1)=BTNA*2/NT;D(NA+2)=X1SY*(NT-1);D(NA+3)=X1SY*(NT-2);...D(NV)=X1SY
D_{z125}	DZ125	=	BTNA*DBDR/NT*(1-DBDR/NT)**(Z-1-NA)=BTZ*DBDR/NT; DBDR=1.25
			APPLIES UNTIL DCZ=BTNA-VTNV OR SWITCH TO SL WHEN D(Z+1)<(BTZ-VTNV)/(NT-Z+NA)
D_{z150}	DZ150	=	SAME AS DZ125 EXCEPT DBDR=1.50

$E(i_a)$	EIA	=	EXPECTED IA IA SUCH THAT EPEX=0
E(PEC)	EPEC	=	EXPECTED PEC
E(PEX)	EPEX	=	EXPECTED PEX =SUM EXZ/(1+IA)**(Z-NA) FROM NA TO NV =SUM PPEXJ*PEXJ FROM J=1 TO LAST
EPS_z	EPSZ	=	EARNINGS PER SHARE IN YEAR Z; TREAT AS EOY
E(U)	EU	=	EXPECTED UTILITY = SUM PPEXJ*UPEXJ FROM J=1 TO LAST
e	EXP	=	2.7182818... THE BASE OF NATURAL LOGARITHMS
$E(x_z)$	EXZ	=	EXPECTED VALUE OF THE CASH FLOW X, AT AGE Z
		=	SUM XZJ*PXZJ FROM J=1 TO LAST
		=	XZ1*PXZ1+XZ2*PXZ2+...+XZLAST*PXZLAST
f	F	=	A FUTURE DISCRETE-FLOW SUM AT EOY NN

BOOK	FORTRAN		DEFINITIONS AND EQUATIONS

f/a FEA   = FE OF ANNUAL SERIES $((1+IA)**NN-1)/IA$

FEC   FEC   = FUTURE EQUIVALENT COST =FE REVENUE REQUIRED
                = $FEM+(BVNA*FEP-VNV-TT*FED)/(1-TT)$

FED_{xxx} FEDXXX   = FE TAX DEPRECIATION ALLOCATIONS BY "XXX"
                = $FEP*PEDXXX$

f/g FEG   = FE OF GRADIENT SERIES
                = $(((1+IA)**NN-1)/IA-NN)/IA$

f/c FEH   = FE OF GEOMETRIC SERIES
                = $((1+IA)**NN-(1+RH)**NN)/(IA-RH)$

FEM   FEM   = FE CASH OPERATING (M) COSTS   =$FEP*PEM$

f/p FEP   = FE OF A PRESENT SUM   =$(1+IA)**NN$

FER   FER   = FE REVENUES RECEIVED   =$FEP*PER$

FEX   FEX   = FE EXCESS OF REVENUES MINUS COSTS   =$FEP*PEX$
                = $(FER-FEC)*(1-TT)$

f' F$ = A FUTURE CONTINUOUS-FLOW SUM DURING YEAR NN

g G   = GRADIENT = THE ANNUAL INCREASE IN AN ARITH-
                METIC SERIES OF DISCRETE CASH FLOWS.   FLOW
                AT EOY NN IS $(NN-1)*G$

g' G$ = GRADIENT = THE ANNUAL INCREASE IN AN ARITH-
                METIC SERIES OF CONTINUOUS CASH FLOWS. FLOW
                DURING YEAR NN IS $(NN-1)*G$

c H   = FIRST CASH FLOW IN GEOMETRIC SERIES OF DIS-
                CRETE CASH FLOWS
                FLOW AT EOY NN IS $H*(1+RH)**(NN-1)$

c' H$ = FIRST CASH FLOW IN GEOMETRIC SERIES OF CON-
                TINUOUS CASH FLOWS
                FLOW DURING YEAR NN IS H*(1+RH)**(NN-1)$

i_a IA   = EFFECTIVE ANNUAL AFTER-TAX ROR,   REQUIRED
                (MARR) OR PROSPECTIVE = IA SUCH THAT $0 =$SUM
                $ATCFP/((1+IA)**(Z-NA))$ FOR $Z=NA,NA+1,...NV$
                = IA SUCH THAT $0=-BVNA+VNV/(1+IA)**NN+SUM$ (BT
                $CFPZ*(1-TT)+DZ*TT/(1+IA)**(Z-NA)$ FOR $Z=NA+$
                $1,NA+2,...NA+NN$ NOTE THAT IA CAN BE EX-
                PRESSED AS A FUNCTION INDEPENDENT OF ID, RD
                = $IC-TT*RD*ID$ =$(1-TT)*RD*ID+(1-RD)*IE$
                = $(1+IN)**NM-1$.   IF NM INFINITE, IA=EXP**IM-1
                COMPUTER NOT LIMITED TO TABLED VALUES OF IA

i_{az} IAZ   = $(BVZ+(1-TT)*(RZ-MZ)+TT*DZ)/(BV(Z-1))-1$

i_b IB   = EFFECTIVE ANNUAL BEFORE-TAX ROR,   REQUIRED
                (BTRR) OR PROSPECTIVE
                APPR = $IA/(1-TT)$ = MARR/(1-TT)$

i_c IC   = EFFECTIVE ANNUAL AFTER-TAX ROR, REQUIRED OR
                PROSPECTIVE, ON COMPOSITE CAPITAL
                = $RD*ID+(1-RD)*IE = IA+TT*RD*ID$

i_d ID   = EFFECTIVE ANNUAL AFTER-TAX ROR, REQUIRED OR
                PROSPECTIVE, ON DEBT CAPITAL

i_e IE   = EFFECTIVE ANNUAL AFTER-TAX ROR, REQUIRED OR
                PROSPECTIVE, ON EQUITY CAPITAL
                = $(IC-RD*ID)/(1-RD)= (IA-(1-TT)*RD*ID)/(1-RD)$

i_i II   = COST OF INTERNALLY GENERATED FUNDS   0.07?

ILDC   ILDC   = INTEREST LOST DURING CONSTRUCTION. LET XZ=
                PAYMENT MADE TO CONTRACTOR ZX (0.75?) YEARS
                PRIOR TO FACILITY COMPLETION.   THEN ILDC =
                SUM $XZ*ZX*IA$ FROM FIRST TO LAST PAYMENT.

i_m IM   = NOMINAL ANNUAL ROR   IN*NM
                IF NM INFINITE,   IM=ALOG(1+IA)

i_n IN   = ROR, REQUIRED OR PROSPECTIVE, PER INTEREST-
                COMPOUNDING PERIOD. $(1+IA)**(1/NM)-1$

BOOK   FORTRAN   DEFINITIONS AND EQUATIONS

i_r IR   = MINIMUM EFFECTIVE ANNUAL ROR PROSPECTIVE ON
                    PROJECTS FINANCED FROM INTERNAL FUNDS ONLY
i_x IX   = COST OF EXTERNAL FUNDS   0.11?
I_z IZ   = INTEREST SUM DUE ON DEBT CAPITAL AT AGE Z
                  = RD*ID*(BU(Z-1))

MARR   = MINIMUM ATTRACTIVE RATE OF RETURN. THE EF-
                    FECTIVE ANNUAL AFTER-TAX (HURDLE) REQUIRED
                    ROR TO MEET THE INVESTOR'S THRESHOLD OF AC-
                    CEPTABILITY = "OPPORTUNITY COST" OF CAPITAL
                    SEE IA
M_z MZ   = CASH OPERATING COSTS (MAINTENANCE+OPERATION
                    +INSURANCE+PROPERTY TAX+MATERIAL+LABOR+ETC)
                    AT AGE Z
n_a NA   = AGE OF ASSET AT DATE OF ACQUISITION
N   NC   = NUMBER OF YEARS THAT DEMAND CONTINUES
Z   NH   = NUMBER OF YEARS THAT GROWTH PERSISTS
m   NM   = NUMBER OF INTEREST-COMPOUNDING PERIODS/YEAR
n_n NN   = ASSET LIFE IN STATED USE, IN YEARS   =NV-NA
NP   NP   = PROSPECTIVE PAYOFF PERIOD = NUMBER OF YEARS
                    NP, SUCH THAT 0=SUM BTCFPZ FOR Z=NA,NA+1,...
                    NA+NP
NR   NR   = REQUIRED (MAXIMUM ACCEPTABLE) PAYOFF PERIOD
n_t NT   = TAX LIFE, IN YEARS.   THE SHORTEST TAX LIFE
                    PERMITTED IS: IF NN<5, NT=NN; IF 5<NN<6.25,
                    NT=5; IF 6.25<NN, NT=0.8*NN
n_v NV   = AGE OF ASSET AT DATE OF RETIREMENT   =NA+NN

p   P   = A PRESENT DISCRETE-FLOW SUM AT EOY ZERO
p/a   PEA   = PE OF ANNUAL SERIES   =(1-(1+IA)**(-NN))/IA
PEBTCP   PEBTCP = SUM (RZ-MZ)/(1+IA)**(Z-NA) FOR Z=NA+1,NA+2,
                    ...NA+NN
PEBTCR   PEBTCR = PE OF BTCFR   = PEC-PEM
                  = (BVNA-VNV*PEF-TT*PED)/(1-TT)
                  = (BZ-V*PEF-TT*PED)/(1-TT)
PEC   PEC   = PRESENT EQUIVALENT COST=PE REVENUE REQUIRED
                  = PEM+(BVNA-VNV*PEF-TT*PED)/(1-TT)
PED_{xyx} PEDXXX = PE TAX DEPRECIATION ALLOCATIONS BY "XXX"
                  = SUM DZ/(1+IA)**(Z-NA) FOR Z=NA+1,NA+2,...NA
                    +NT
PED_{FYW} PEDFYW = (BTNA-VTNV)*(1-(1+IA)**(-5))/(5*IA)
PED_{MAX} PEDMAX = USE SYD IF VTNV/BTNA<-1/NT;   1SY IF -1/NT<
                    VTNV/BTNA<0;   1AS IF 0<VTNV
PED_{SFU} PEDSFU = (BTNA-VTNV)*NT*IA/((1+IA)*((1+IA)**NT-1))
PED_{SLN} PEDSLN = (BTNA-VTNV)*(1-(1+IA)**(-NT))/(IA*NT)
PED_{SYD} PEDSYD = (BTNA-VTNV)*2*(NT*IA-1+(1+IA)**(-NT))/((NT*
                    NT+NT)*IA*IA)
PED_{XYZ} PEDXWZ = (BTNA-VTNV)/((1+IA)**ZW
PED_{IAS} PED1AS = BTNA*2/(NT*NT-NT)*(1/(1+IA)+(1-2/NT)/IA*(NT
                    -1/IA+(1+IA)**(-K1AS)*(1/IA+1-(NT-K1AS+1)*(
                    2+IA*K1AS-IA*NT)/2)))-VTNV/(1+IA)**K1AS
                    WHERE INTEGER K1AS=NT+1.5-(0.25+VTNV/BTNA*(
                    NT-1)/(NT-2)*NT*NT)**0.5
PED_{ISY} PED1SY = BTNA*2/(NT*IA*(1+IA))*(IA+(1-2/NT-VTNV/BTNA
)*(1-(1+IA)**(1-NT))/(IA*(NT-1))))
p/f   PEF   = PE OF FUTURE SUM   =(1+IA)**(-NN)
p/g   PFG   = PE OF GRADIENT SERIES
                  = ((1-(1+IA)**(-NN))/IA-NN/((1+IA)**NN))/IA

| BOOK | FORTRAN | DEFINITIONS AND EQUATIONS |

p/c PEH   = PE OF GEOMETRIC SERIES
              = $(1-((1+RH)/(1+IA))^{**}NN)/(IA-RH)$

PEM   PEM   = PE CASH OPERATING (M) COSTS
              = SUM $MZ/(1+IA)^{**}Z$ FOR $Z=1,2,...NN$

PER   PER   = PE REVENUES RECEIVED = PE BENEFITS REALIZED
              = SUM $RZ/(1+IA)^{**}(Z-NA)$ FOR $Z=NA+1,NA+2,...NV$

PEX   PEX   = PE EXCESS OF REVENUES MINUS COSTS
              = $(PER-PEC)^*(1-TT)$
              = $(PER-PEM)^*(1-TT)-BVNA+VNV^*PEF+TT^*PED$ =SUM
                $ATCFPZ/(1+IA)^{**}(Z-NA)$ FOR $Z=NA+1,NA+2,...NV$

PLI   PLI   = PRESENT EQUIVALENT   "FINANCIAL IMPAIRMENT
                EFFECT" FROM LEASING
              = $ALP^*ALR^*(IA/(ID^*(1-TT))-1)^*((1-(1+IA)^{**}(-NN$
                $))/IA)-(1/(IA-ID))^*((1+ID)^{**}(-NN)-(1+IA)^{**}($
                $-NN)))$

PPE_z PPEZ = (MARKET PRICE)/(CURRENT EPS) RATIO AT EOY Z
                IF DPR=1, AND IF EPS GROWTH IS ARITHMETIC:
                THEORETICAL PPE $=(G/EPSO^*(1+IA)+IA)/(IA^*IA)$
                IF DPR=1,IA=RH,AND EPS GROWTH IS GEOMETRIC:
                THEORETICAL PPE $=(1+RH)/(IA-RH)$

PPS_z PPSZ = PRICE PER SHARE AT EOY Z. PPSO=(SUM DPSZ/(
                $1+IA)^{**}Z$ FOR $Z=1,2,...NN)+PPSNN/(1+IA)^{**}NN$

$p(x_3)$ PX3 = PROBABILITY THAT X3 OCCURS

p' P$ = A PRESENT CONTINUOUS-FLOW SUM DURING YEAR 0

r_d RD   = DEBT RATIO   = DEBT/(DEBT+EQUITY)

r RH   = GEOMETRIC SERIES GROWTH RATE (AS IN 1,2,4,8
                ...WHERE RH=1.0)

R_z RZ   = REVENUES PROSPECTIVE (OR REALIZED)AT AGE Z.
                IN PUBLIC PROJECTS, USE "PRICE PUBLIC WOULD
                BE WILLING TO PAY" MINUS "ACTUAL COSTS PAID
                BY PUBLIC" AT AGE Z.

σ STD   = STANDARD DEVIATION   = $VAR^{**}0.5$

TCE_{Bz} TCEBZ = (BEFORE-TAX) TIMES INTEREST CHARGES EARNED
                FOR YEAR Z   = $(RZ-MZ-DZ)/IZ$

TDD   TDD   = DEDUCTIBILITY OF FEDERAL INCOME TAX IN COM-
                PUTING STATE INCOME TAX. $0<=TDD<=1$

TFR   TFR   = FEDERAL TAX RATE ON INCREMENTAL INCOME

TSR   TSR   = STATE INCOME TAX RATE ON INCREMENTAL INCOME

t TT   = EFFECTIVE COMBINED FEDERAL AND STATE INCOME
                TAX RATE ON INCREMENTAL INCOME
              = $1-(1-TSR-TFR+TSR^*TFR)/(1-TSR^*TFR^*TDD)$

U(PEX) UPEX = UTILITY OF A CERTAIN PEX OUTCOME, IN UTILES

U(X)   UX   = UTILITY OF A CERTAIN OUTCOME X, MEASURED IN
                UTILES

σ^2 VAR = VARIANCE   = $STD^{**}2$
                SUM $PXZ^*(XZ-EXZ)^{**}2$ FOR $Z=1,2,...LAST$

V_{nv} VNV = AFTER-TAX NET SALVAGE VALUE AT AGE Z =GROSS

V_{tnv} VTNV = BEFORE-TAX NET SALVAGE VALUE AT AGE NV

V_{tz} VTZ = BEFORE-TAX NET SALVAGE VALUE AT AGE Z=GROSS
                SALVAGE - COSTS TO REMOVE AND SELL AT AGE Z
                VTZ DIFFERS FROM VZ IF: (1)PROPERTY IS SOLD
                AT A PRICE OTHER THAN BTZ, DEPRECIATED BOOK
                VALUE, AND (2)ITEM DEPRECIATION IS USED, OR
                GROUP DEPRECIATION IS USED AND "GENERAL IM-
                PACT ON DEPRECIATION CHARGES" IS EVALUATED.

V_z VZ   = AFTER-TAX NET SALVAGE VALUE AT EOY Z =GROSS
                SALVAGE - COSTS TO REMOVE AND SELL + INCOME

BOOK   FORTRAN   DEFINITIONS AND EQUATIONS

		TAX EFFECT OF CAPITAL GAIN OR LOSS AT EOY Z IN BEFORE-TAX APPROACH, IGNORE INCOME TAXES
	X1AS	= SEE DZ1AS
	X1SY	= SEE DZ1SY
z	Z	= ATTAINED AGE OF A PROPERTY IN YEARS. USED AS A SUBSCRIPT FOR B,C,D,...V.
	ZW	= YEARS OF DELAY IN TAX IMPACT FROM XWZ

BOOK   FORTRAN   ABBREVIATIONS

Book	Fortran	
AASHO	AASHO	= AMERICAN ASSOCTN OF STATE HIGHWAY OFFICIALS
AE	AE	= ANNUAL EQUIVALENT
ln	ALOG	= NATURAL LOGARITHM
ASY	ASY	= SYD APPLIED TO BTNA, NOT (BTNA-VTNV);ACCRU-AL LIMIT=(BTNA-VTNV); ALLOWED SINCE 1971.
BPR	BPR	= BUREAU OF PUBLIC ROADS
DDB	DDB	= DOUBLE DECLINING BALANCE DEPRECIATION WITH OPTIMAL SWITCH TO SLN
/	E	OFTEN USED HERE FOR "EQUIVALENT" AS IN A/P.
EOY	EOY	= END-OF-YEAR
FE	FE	= FUTURE EQUIVALENT
FICA	FICA	= FEDERAL INSURANCE COMPENSATION ACT
FYW	FYW	= FIVE-YEAR WRITEOFF FORM OF DEPRECIATION
HRB	HRB	= HIGHWAY RESEARCH BOARD
∞	INF	= INFINITY
L2	KL2	= IOWA-TYPE SURVIVOR DISPERSION L2
NE	KNE	= NEGATIVE EXPONENTIAL SURVIVOR DISPERSION
O2	KO2	= IOWA-TYPE SURVIVOR DISPERSION O2
R2	KR2	= IOWA-TYPE SURVIVOR DISPERSION R2
S2	KS2	= IOWA-TYPE SURVIVOR DISPERSION S2
MAPI	MAPI	= MACHINERY AND ALLIED PRODUCTS INSTITUTE
MAX	MAX	= MAXIMUM
ROR	ROR	= RATE OF RETURN
SF	SFU	= SINKING-FUND METHOD OF DEPRECIATION
SL	SLN	= STRAIGHT-LINE METHOD OF DEPRECIATION
Σ	SUM	= SIGMA = SUM OF
SYD	SYD	= SUM-OF-THE-YEARS-DIGITS DEPRECIATION
XW_z	XWZ	= EXPENSED WRITEOFF: IMPACT DELAY OF ZW YEARS
1AS	1AS	= USE DDB IN YEAR 1, ASY THEREAFTER. METHOD MAXIMIZES PED IF VTNV>0, AND IF BOTH TT AND IA ARE POSITIVE AND TIME-INVARIANT. ACCRUAL LIMIT=(BTNA-VTNV).
1SY	1SY	USE DDB IN YEAR 1, SYD THEREAFTER.
125	125	= 125% DECLINING BALANCE DEPRECIATION WITH OPTIMAL SWITCH TO SLN
150	150	= 150% DECLINING BALANCE DEPRECIATION WITH OPTIMAL SWITCH TO SLN
3B	3B	AN INVESTMENT ALTERNATIVE. MUTUALLY EXCLU-SIVE WITH ANY OTHER INVESTMENT (SUCH AS 3C) HAVING THE SAME PREFIX NUMBER. INDEPENDENT OF ANY OTHER INVESTMENT (SUCH AS 5C) HAVING A DIFFERENT PREFIX NUMBER
$\approx$	APPR=	= IS APPROXIMATELY EQUAL TO

BOOK ⁄ FORTRAN   ABBREVIATIONS

>	>	= IS GREATER THAN
≧	=>	= IS GREATER THAN OR EQUAL TO
<	<	= IS LESS THAN
≦	<=	= IS LESS THAN OR EQUAL TO
•	$	OFTEN USED HERE FOR "CONTINUOUS CASH FLOW".

UNLESS OTHERWISE SPECIFIED:

- AFTER-TAX APPROACH IS TO BE USED
- BASE FOR TAXES, ACQUIRE/RETAIN, OR REGULAT-
  ION ARE IDENTICAL.   BTZ=BZ=BVZ=BRZ
- CASH FLOWS ARE END-OF-YEAR
- COMPOUNDING FREQUENCY IS ANNUAL. NM=1;IA=IN
- DEBT RATIO IS ZERO.   RD=0,SO IA=IC=IE;IZ=0
- IA IS POSITIVE AND TIME-INVARIANT
- INVESTMENT CREDIT PERCENTAGE = 0
- ITEM, NOT GROUP, ACCOUNTING IS USED,SO GAIN
  OR LOSS ON DISPOSAL CAN OCCUR
- LIFE FOR TAX DEPRECIATION,   ACQUIRE/RETAIN,
  OR REGULATION ARE IDENTICAL   NN=NT
- LIFE GIVEN NN, IS NOT SUBJECT TO DISPERSION
  WHEN DISPERSION APPLIES, SEE CHAPTER 14.
- OPTIMAL ACQUISITION AGE (NA) = 0; THEREFORE
  OPTIMAL RETIREMENT AGE (NV) IS IDENTICAL TO
  ECONOMIC LIFE (NN)

C

Listing Alternatives in Ascending Order of Investment Size

THROUGHOUT THIS BOOK we have considered alternatives for which investment size ranking could be accomplished by inspection. When we consider investments having unequal lives or unequal salvage ratios, our task may be more involved.

Consider the three mutually exclusive investment alternatives, F, G, and H, to which the following estimates apply.

Alterna-tive	First Cost	Salvage	Life, in Years	Annual After-Tax Cash Flow
F	$ 5,000	$13,848	10	$ 0
G	11,152	0	10	2000
H	20,062	0	5	5782

We note that if investment size were measured by the dollar-years of unrecouped investment at $i = 0\%$,

for F: $ 5,000 + $ 5,000 + · · · + $5000 = $50,000
for G: $11,152 + $ 9,152 + · · · + $1152 = $36,912
for H: $20,062 + $14,280 + · · · + $2716 = $45,556

This at least raises the question of which is the correct ascending order of investment size. An increment analysis of these alternatives may prove helpful:

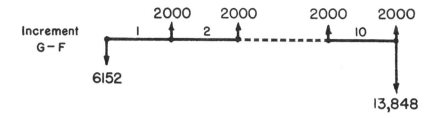

Before attempting to solve for rate of return, we should check the year-by-year accumulated cash flow. This check indicates two changes in sign and thus signals the possibility of a dual solution. Because the final value of the year-by-year accumulated cash flow was zero, we have, without further calculation, one solution, 0%. By trial and error we can find a second solution, 20%. When this process is repeated for increments (H - G) and (H - F) we again find dual solutions of 0% and 20%. Equivalence of the three alternatives at both 0% and 20% becomes apparent after the following calculations:

At $i = 0\%$:
$$\begin{cases} PEX_F = PER - PEC = \$13,848 - \$5000 = \$8848 \\ PEX_G = \$2000(10) - \$11,152 = \$8848 \\ PEX_H = \$5782(5) - \$20,062 = \$8848 \end{cases}$$

At $i = 20\%$:
$$\begin{cases} PEX_F = \$13,848(p/f)_{10}^{20\%} - \$5000 = -2764 \\ PEX_G = \$2000(p/a)_{10}^{20\%} - \$11,152 = -2764 \\ PEX_H = \$5782(p/a)_{5}^{20\%} - \$20,062 = -2764 \end{cases}$$

Rate of return on the overall investment can be computed for F,

$$(f/p)_{10}^{i} = \$13,848/\$5000 = 2.770$$

Since

$$(f/p)_{10}^{10\%} = 2.594; (f/p)_{10}^{12\%} = 3.106$$

$$ROR_F = 10\% + 2\% (176/512) = 10.7\%$$

and by similar calculations,

$$ROR_G = 12.4\%$$
$$ROR_H = 13.6\%$$

The information developed thus far still does not enable us to distinguish an ascending order of investment size. Although

Investment F @ 10.7% + Investment (G - F) @ 20% could equal Investment G @ 12.4%

it is also possible that

Investment G @ 12.4% + Investment (G - F) @ 0% could equal Investment F @ 10.7%

and similarly for other pairings.

Analysis of reinvestment rate of return possibilities might be of some help in determining relative investment size. If a reinvestment rate of return of, say, 25% can be realized and if an original sum of $20,062 is available for investment, the accumulated worth of the original sum plus reinvested funds at the end of year ten is

for F: $15,062$(f/p)_{10}^{25\%}$ + $13,848 = $154,120
for G: $ 8,910$(f/p)_{10}^{25\%}$ + $ 2,000$(f/a)_{10}^{25\%}$ = $149,485
for H: $ 5,782$(f/a)_{5}^{25\%}$ $(f/p)_{5}^{25\%}$ = $144,817

In a similar manner the following analysis can be developed:

Alternative	Reinvestment ROR					
	0%	5%	10%	15%	20%	25%
	thousands of dollars					
F	28.9	38.4	52.9	74.7	107.0	154.1
G	28.9	39.6	54.9	76.6	107.0	149.5
H	28.9	40.8	56.9	78.5	107.0	144.8

If investment is mandatory, then

When reinvestment rate of return < 20%, choose H
When reinvestment rate of return > 20%, choose F

Next, recall that when reinvestment rate of return is very low we are more inclined to accept large-scale investments (and vice versa for high reinvestment rates of return). Therefore F < G < H.

D
Random Digits

```
79171  75518  19897  26332  86029  23466  57107  21482  99622  49782
54293  69434  41154  06490  30699  96339  71281  08915  37021  53047
64183  96318  68338  75183  88753  29002  02409  56329  28538  18627
45101  72488  01863  43492  66560  25189  95716  61287  90920  78911
78922  23564  09548  51091  62610  01405  89061  74153  73398  64361

12734  49748  58181  23023  54726  08383  89593  15168  63382  45847
90670  99251  06009  78989  57543  30068  09153  68987  31028  47418
39741  09373  76333  22109  47940  58488  06704  74535  40979  53767
45771  49974  45496  83092  85376  68606  99707  31250  25319  42866
28542  07997  63931  13535  92530  68214  88955  47553  58440  03839

46343  45648  17931  66626  24358  10817  98306  08497  40265  57068
81378  85357  07226  55918  19811  18193  72130  38904  68767  49045
49344  62590  87075  25692  80717  31386  74498  81259  36930  98764
53706  20161  76294  81791  29294  67942  89694  91599  25516  03192
42194  24073  70964  08279  77719  42039  27917  31320  27319  79889

98965  80918  77026  71825  34832  57970  07942  05358  59301  70776
05023  76817  85669  03117  05261  65187  60372  04596  19386  41951
91813  95165  53259  48374  52232  55265  75327  02385  57505  81637
20827  87330  77974  61803  43689  18946  11845  98389  64252  78418
59032  00279  43409  34159  18152  33702  04826  04647  07280  16858

39392  17944  24784  04050  71202  40432  92609  34857  57072  28575
29907  76454  68303  16985  52142  67306  57343  86527  18053  82374
42948  78320  49303  11688  78528  04500  43526  42501  15597  58160
97788  39778  70220  76311  32536  50975  62939  04484  12444  38325
40109  80922  74984  22045  98471  51451  84191  11931  89208  46057

24767  07925  18139  40193  57477  36157  20866  20189  24027  04074
82652  55060  89940  36591  07199  82454  07593  03294  70325  17589
65559  80499  65269  90084  94759  27851  53665  52365  75175  81017
14364  02898  96951  79641  74790  62077  86664  70844  17126  11701
16630  02975  64331  59577  52427  77307  93693  84921  61644  49124

41663  90123  85739  75892  70146  75720  24463  39753  03796  98149
66621  42460  52317  36989  89112  92304  92172  20836  00896  88234
61120  77490  78335  99063  27719  73779  28223  86164  61554  55314
02670  35654  12619  36173  33614  62791  16901  85244  99424  82383
39339  72483  94828  29370  39622  93633  81688  82683  15808  65749

66685  82767  31996  30041  61864  26274  08986  38404  77127  60326
09906  55684  60262  94917  94702  83738  76255  87460  12893  32155
23332  85280  23361  26417  68030  51958  05153  02919  96286  36325
74569  42473  86408  75910  32407  90613  67337  79770  36574  03273
09596  64376  91344  78137  98561  08765  17027  43725  62352  52640
```

E

Continuous and Annual Cash Flow Conversion Factors

If interest is compounded continuously, an effective annual interest rate of __	Will result from a nominal annual interest rate of __	And an end-of-year amount of \$1 is equivalent to \$__ flowing uniformly and continuously throughout the year	And an end-of-year amount of \$__ is equivalent to \$1 flowing uniformly and continuously throughout the year
$= i_a$	$= ln(1 + i_a)$	$a'/a = c'/c = f'/f$ $= g'/g = p'/p$ $= \dfrac{ln(1 + i_a)}{i_a}$	$a/a' = c/c' = f/f'$ $= g/g' = p/p'$ $= \dfrac{i_a}{ln(1 + i_a)}$
0.0025	0.002 496	0.9988	1.001
0.005	0.004 987	0.9975	1.002
0.01	0.009 950	0.9950	1.005
0.02	0.019 802	0.9901	1.010
0.03	0.029 558	0.9853	1.015
0.04	0.039 220	0.9805	1.020
0.05	0.048 790	0.9758	1.025
0.06	0.058 268	0.9711	1.030
0.07	0.067 658	0.9665	1.035
0.08	0.076 961	0.9620	1.039
0.10	0.095 310	0.9531	1.049
0.12	0.113 328	0.9444	1.059
0.15	0.139 761	0.9317	1.073
0.20	0.182 321	0.9116	1.097
0.25	0.223 143	0.8926	1.120
0.30	0.262 364	0.8745	1.143
0.40	0.336 472	0.8412	1.189
0.50	0.405 465	0.8109	1.233

F

A Caution in Separating Composite Capital into Its Components

IN COMPUTING the sum required to provide repayment of and interest on an investment, a paradox occurs as follows. Let

Financing = 50% debt capital costing 4% and 50% equity capital costing 10%

Project first cost = $100,000 with a negligible salvage value 20 years from today

Using composite cost of capital,

$$a_1 = \text{repayment and interest} = \$100,000 \, (a/p)_{20}^{7\%} = \$9439$$

Using half at 4%, half at 10%,

$$a_2 = \text{repayment and interest} = \$50,000(a/p)_{20}^{4\%} + \$50,000(a/p)_{20}^{10\%}$$
$$= \$3679 + \$5873 = \$9552$$

The apparent conflict between a_1 and a_2 is caused by the repayment schedules; with Payments a_1 the debt-equity ratio remains 50% throughout the life of the property, while Payments a_2 result in such a repayment schedule that the proportion of debt capital decreases with each succeeding payment. The subsequent higher equity proportions with a_2 therefore lead to higher repayment and interest costs. Since an economy study is separable from the financial structure, our study should be based on continuing the structure given; under these conditions Solution a_1 is a correct one, Solution a_2 is incorrect. The changing financial structure with payments of size a_2 is apparent in Table F-1.

TABLE F-1. A Caution in Treating the Repayment of Composite Capital

	Debt				Equity				
Year	Outstanding balance at beginning of year	Interest accrued during year	Payment	Outstanding balance at end of year	Outstanding balance at beginning of year	Interest accrued during year	Payment	Outstanding balance at end of year	Debt ratio at beginning of year
0	$50,000	$2,000	$3,679	$48,321	$50,000	$5,000	$5,873	$49,127	0.500
1	48,321	1,933	3,679	46,575	49,127	4,913	5,873	48,167	0.497
2	46,575	1,863	3,679	44,762	48,167	4,817	5,873	47,111	0.493
3	44,762	·	·	·	47,111	·	·	·	0.487
·									
11	·	·	·	·	·	·	·	·	0.453
·									
16	·	·	·	·	·	·	·	·	0.424
·									
20	3,538	141	3,679	0	5,339	534	5,873	0	0.399

Given an opportunity to repay two outstanding obligations of unequal interest rates at our discretion, we could minimize payments by first repaying the higher cost capital (perhaps a second mortgage) and later repaying the lower cost capital (perhaps a first mortgage). If such a repayment plan had been permitted by the owners and creditors in the situation just treated, the equal annual payments, a_3, over each of the 20 years could be computed as follows:

$$a_3 (p/a)_n^{10\%} = \$50,000 = a_3 (p/a)_{20-n}^{4\%} (p/f)_n^{4\%}$$

$$= a_3 [(p/a)_{20}^{4\%} - (p/a)_n^{4\%}](p/a)_n^{10\%}$$

$$= (p/a)_{20}^{4\%} - (p/a)_n^{4\%} (p/a)_n^{10\%} + (p/a)_n^{4\%}$$

$$= (p/a)_{20}^{4\%} = 13.590$$

At $n = 9$, $5.759 + 7.435 = 13.194$

At $n = 10$, $6.145 + 8.111 = 14.256$

so

$$n = 9 + 396/1062 = 9.373$$

By interpolation

$$(a/p)_{9.373}^{10\%} = 0.16958$$

so

$$a_3 = \$50,000 \, (a/p)_{9.373}^{10\%} = \$8479$$

G

Effect of Anticipated Inflation on Investment Analysis

The text of Appendix G, dealing with the problem of inflation, is authored by George Terborgh and first appeared as Pages 5 through 14 of "Effect of Anticipated Inflation on Investment Analysis" (No. 2 in a series of studies in the *Analysis of Business Investment Projects*) copyrighted by Machinery & Allied Products Institute and Council for Technological Advancement, 1200 Eighteenth Street Northwest, Washington, D.C. It appears here by their permission.

The question is frequently asked how the expectation of inflation affects the analysis of investment projects. In an age when this expectation is prevalent it is an important question, the more so since there is wide disagreement on the answer. Indeed, opinions range from the view that prospective inflation has *no* effect on the investment merit of a project to the view that it vastly enhances its attractiveness. It is desirable therefore to take a closer look at the problem.[1]

Later in this study we propose to consider the effect of inflationary anticipations on the results of the MAPI system. As a preliminary, however, we are going to investigate the subject more broadly. How does inflation affect investment returns generally?

1. EFFECT OF INFLATION IN THE ABSENCE OF AN INCOME TAX

Since the answer to this question differs somewhat, depending on whether or not an income tax is applicable, we begin with the simplest case, where there is no

---

[1] The discussion is addressed to the effect of *inflation* simply because this is a far more likely prospect these days than *deflation*. Anyone concerned about the latter can apply the conclusions in reverse.

tax. This is unrealistic, of course, but it permits us to work up to the more complicated tax case by easy stages.

An investment consists essentially of an exchange of present dollars for what it is hoped will be a larger number of future dollars. If these future dollars are of the same size (in terms of purchasing power) as the dollars of investment, the prospective return on the commitment can readily be computed. If, however, they are expected to shrink progressively as time goes on, the calculation is more difficult.

There are two key questions in the analysis of this problem. (1) Do the future receipts from the investment (the amounts available for capital recovery and net return) respond to inflation? In other words, do they expand to offset the shrinkage of the dollar? (2) When the dollar is changing in size, how can the return on the investment be computed? Suppose we consider these questions in order.

Responsiveness of Future Receipts

Where the future investment receipts are predetermined by contract, as in the case of a bond or a fixed annuity, it is obvious that they do *not* respond to inflation. They remain completely unaffected. Here an unchanged flow of future dollars represents a reduced flow in terms of the dollars of investment. Where the future receipts are not predetermined, however, they may respond to inflation. This is, in fact, the normal expectation for business investments. The degree of response varies from case to case, of course, but for the purpose of forecasting the investment receipts from new projects it is usually assumed that the response will be complete. This implies that the increase in the flow of future dollars as a result of inflation will leave the flow unchanged in terms of the dollars of investment.

We can illustrate these two possibilities by a simple example. We have on the one hand a fixed (unresponsive) annuity of $1,000 a year for 10 years. On the other hand, we have a responsive annuity of the same duration yielding each year enough future dollars to equal $1,000 of the dollars of investment (present dollars). Assuming an inflation of 3 percent per annum, we have the following:[2]

TABLE 1

Year	A Fixed Annuity		B Responsive Annuity	
	In Future Dollars	In Present Dollars[3]	In Future Dollars[4]	In Present Dollars
	(1)	(2)	(3)	(4)
1	$1,000	$971	$1,030	$1,000
2	1,000	943	1,061	1,000
3	1,000	915	1,093	1,000
4	1,000	888	1,126	1,000
5	1,000	863	1,159	1,000
6	1,000	837	1,194	1,000
7	1,000	813	1,230	1,000
8	1,000	789	1,267	1,000
9	1,000	766	1,305	1,000
10	1,000	744	1,344	1,000

[2] This is merely an arbitrary assumption for illustrative purposes. Although roughly in line with the average inflation rate of recent years, it is not offered as a forecast of the future rate.

[3] Number of present dollars having the purchasing power of 1,000 future dollars received at the end of the year indicated.

[4] Number of future dollars required to equal the purchasing power of 1,000 present dollars received at the end of the year indicated.

When the receipts are constant in *future* dollars (unresponsive to inflation) their equivalent in present dollars declines over the 10-year interval to $744 in the final year. When they are constant in *present* dollars (responsive to inflation) their equivalent in future dollars rises to $1,344.

Measurement of Return

The second question propounded above is how to compute the rate of return on the investment under conditions of inflation. Here we wish to be quite emphatic. *No rational computation is possible without converting future receipts in varying dollars into their equivalent in the dollars of investment.*[5]

This is absolutely basic. It is no more possible to make reliable financial measurements in variable dollars than to make reliable physical measurements with a rubber yardstick. Yet this is precisely what some analysts have done in attempting to assess the effects of inflation on the return from investment projects. The results are illusory.

Consider the two annuities presented in the foregoing table. The rates of return must be computed from Columns 2 and 4, which translate the actual receipts into their equivalent in the dollars of investment. Suppose, for example, that both annuities cost $6,144, which would make the fixed annuity yield 10 percent *in the absence of inflation.* With the assumed 3 percent inflation its real return is only 6.8 percent. The responsive annuity, on the other hand, yields 10 percent notwithstanding inflation.[6]

We conclude that in the absence of an income tax, inflation reduces the real return on an investment with unresponsive receipts, while leaving this return *unchanged* on an investment with fully responsive receipts.[7]

Return on Equity

The conclusion just stated relates to the effect of inflation (in the absence of an income tax) on the return on the *total* investment. If the investment consists entirely of equity capital, the conclusion is of course applicable to the equity return as well. But the capital may be a mixture of debt and equity, in which case the equity may gain from the unresponsiveness to inflation of fixed charges for interest and amortization. Accordingly, the effect of inflation on the equity return may differ from its effect on the total return.

Given mixed capital, no income tax, and wholly *unresponsive* investment receipts, the real equity return is of course reduced by inflation in the same degree as the total return. But if the receipts are even slightly responsive, the equity suffers less than the total. Moreover, while it requires fully responsive receipts to offset the effect of inflation on the total return, it requires less than a full response to protect the equity. By the same token, a full response will yield a gain to the equity.

---

[5] The dollars both of investment and receipt can of course be converted into some other uniform measure if desired.

[6] These are the rates of discount at which the receipts in Columns 2 and 4 have a present worth equal to the investment of $6,144. Where the flow of receipts in the absence of inflation is level, as here, the rate of return with inflation is given by the expression:

$$\frac{i - k}{1 + k}$$

where i = the rate of return (in decimals) without inflation
   k = the rate of inflation (in decimals) per year

[7] It should be added that for inflation to be neutral in the latter case *all* future receipts must be responsive, including receipts from disposals, if any, at the termination of the investment. The example just given assumes that there are no such disposals. It can easily be demonstrated that with responsive disposal values the result is the same.

The impact of inflation on the real equity return depends on a variety of factors —the rate of inflation, the duration of the investment, the pattern of the receipts, the degree to which the receipts are responsive, the debt ratio, the debt interest rate, etc. It would be interesting to explore the permutations of these variables, but the situation of which we are speaking (no income tax) is so unreal for business investment that it would be an academic exercise. Since the effect of a given inflation rate on the equity return can be quite different with an income tax than without it, we shall develop the analysis beyond this point on a realistic basis.

2. EFFECT OF INFLATION WITH AN INCOME TAX

As stated at the outset, the purpose of discussing the effect of inflation in the no-tax case is simply to develop the first stage of the analysis on a less complex problem. The case itself is literally "out of this world" so far as business investment is concerned. We turn now to reality. As before, we shall consider first the effect of inflation on the return on *total* investment.

Return on Total Investment

We have seen that in the absence of an income tax a fully responsive flow of receipts will protect the real return on total investment—in other words, will leave it the same as it would be without inflation. This is not necessarily so, however, when an income tax is applicable. Here it is true only of a very limited and special case—a nondepreciable, all-equity investment held in perpetuity.[8] For all other cases, fully responsive receipts (including, of course, receipts from terminal disposals) will give *less than complete* protection of the real after-tax return.

The reason is not far to seek: tax deductions for depreciation, for "basis" in computing terminal gains, and for interest on borrowed capital are *unresponsive* to inflation. The only case in which none of these deductions is involved is the perpetual, all-equity, nondepreciable investment just referred to.

Under historical-cost depreciation accounting, the tax benefit from the deduction is *received* in future dollars, but is *measured* in the dollars of investment. There is therefore a shrinkage in the real benefit when the future dollars are smaller than the investment dollars. This is true also of the "basis" allowed in computing gain or loss on disposals. As for interest deductions, since the interest itself is fixed by contract, the deductions are fixed also. The use of such unresponsive deductions in computing tax liability necessarily results in a less-than-complete offset to inflation even when the pre-tax investment receipts are fully responsive.[9]

We can illustrate by modifying the example in Table 1 to include the tax effect. The investment is $6,144 and the expected receipts are $1,000 a year for 10 years. Let us assume that tax depreciation is straight-line, that 30 percent of the investment represents borrowed capital at 5 percent interest, and that the tax rate is 50 percent. The real after-tax return is derived below, assuming: A, that there is no inflation; B, that inflation is 3 percent a year, but with fully responsive receipts.

In this case, the correct after-tax return in the absence of inflation is 6.1 percent. With inflation and fully responsive receipts, it is 4.9 percent. The difference reflects the unresponsiveness of the tax deductions for depreciation and interest. *If these were also responsive, the two rates of return would of course be identical.*

---

[8] It is not even true in this case if part of the nondepreciable investment consists of inventory for which historical costing (FIFO) is used for tax purposes. Current costing (LIFO) must be employed. Moreover, there must be no net liquidation of the inventory at any time.

[9] In discussing the no-tax case, we defined "investment receipts" as the future amounts available for capital recovery and return on investment. In the tax case, the term "pre-tax investment receipts" denotes the amounts available for capital recovery, return, *and* income tax, the "after-tax receipts" being of course the balance for capital recovery and return alone.

TABLE 2

	A			B			
	No Inflation			Inflation with Fully Responsive Receipts			
Year	Pre-Tax Receipts in Present Dollars	Depreciation and Interest Deductions[10]	After-Tax Receipts in Present Dollars[11]	Pre-Tax Receipts in Future Dollars	Depreciation and Interest Deductions[10]	After-Tax Receipts in Future Dollars[12]	After-Tax Receipts in Present Dollars
	(1)	(2)	(3)	(4)	(5)	(6)	(7)
1	$1,000	$ 707	$ 853	$1,030	$ 707	$ 869	$ 844
2	1,000	697	849	1,061	697	879	829
3	1,000	688	844	1,093	688	891	815
4	1,000	679	839	1,126	679	903	802
5	1,000	670	835	1,159	670	915	790
6	1,000	661	830	1,194	661	928	777
7	1,000	651	826	1,230	651	941	765
8	1,000	642	821	1,267	642	955	754
9	1,000	633	816	1,305	633	969	742
10	1,000	624	812	1,344	624	984	732

Real after-tax return (percent) 6.1[13]   Real after-tax return (percent) 4.9[13]

The principal factor in the erosion of the real after-tax return is of course the unresponsiveness of the deduction for depreciation. This accents the importance of adjusting it for inflation, in line with the long-standing recommendation of the Institute.

Return on Equity

As indicated, the foregoing calculation yields the real return on the *total* investment outstanding, debt and equity combined. Suppose we now consider briefly the effect of inflation on the real return for equity capital alone.

Given fully responsive pre-tax receipts, and a *nondepreciable* investment of *unlimited* duration, financed in part by borrowing, inflation will normally improve the real after-tax return on the equity.[14] This is because the *gain* to the equity from fixed interest charges exceeds the *loss* from the unresponsiveness of interest deductions. At a 50-percent tax rate, the gain is obviously twice the loss. If, however, the investment, even though nondepreciable, is of *limited* duration, the effect of inflation on the real after-tax equity return can go either way. With a low debt ratio, for example, the loss from the unresponsiveness of the "basis" deductible in computing the gain from terminal disposals may override the advantage noted above, and the equity can wind up worse off from inflation.[15]

With mixed financing and *depreciable* assets, the chance of gain to the equity is greatly reduced. Even with fully responsive pre-tax receipts, the loss from unrespon-

---

[10] Assuming the debt is repaid in equal annual installments.
[11] Column 1, minus one half of the excess of Column 1 over Column 2.
[12] Column 4, minus one half of the excess of Column 4 over Column 5.
[13] The rate of discount at which the after-tax receipts have a present worth equal to the investment.
[14] There may be an exception to this result if all or a part of the investment is in inventory and if historical costing (FIFO) is used for tax purposes.
[15] The more so if historical costing is used for inventory.

sive tax deductions is very likely to outweigh the gain from fixed interest and debt amortization charges. This is true, for example, of the case presented a moment ago in Table 2, as we can see by breaking down Columns 3 and 6 of that table to disclose the balance available for equity.

TABLE 3

	A			B			
	No Inflation			Inflation with Fully Responsive Receipts			
Year	After-Tax Receipts in Present Dollars (Col. 3 of Table 2)	Debt Charges for Interest and Amorti-zation[16]	Balance for Equity in Present Dollars	After-Tax Receipts in Future Dollars (Col. 6 of Table 2)	Debt Charges for Interest and Amorti-zation[16]	Balance for Equity in Future Dollars	Balance for Equity in Present Dollars
	(1)	(2)	(3)	(4)	(5)	(6)	(7)
1	$ 853	$ 276	$ 577	$ 869	$ 276	$ 593	$ 576
2	849	267	582	879	267	612	577
3	844	258	586	891	258	633	579
4	839	249	590	903	249	654	581
5	835	240	595	915	240	675	582
6	830	230	600	928	230	698	585
7	826	221	605	941	221	720	585
8	821	212	609	955	212	743	587
9	816	203	613	969	203	766	587
10	812	194	618	984	194	790	587

Real return on equity (percent) 6.4[17]   Real return on equity (percent) 5.9[17]

The real equity return in the absence of inflation is 6.4 percent, but with the assumed inflation of 3 percent per annum, it is only 5.9 percent. Here the protection even of the equity return requires that the pre-tax receipts from the investment be *more than* fully responsive to inflation.

3. APPLICATION TO BUSINESS INVESTMENT

We have used these oversimplified examples to explore the general principles involved. These principles are applicable, however, to ordinary business investment projects, and it is unnecessary to develop elaborate realistic cases to confirm them.

A word is in order about the definition of investment receipts from business projects. They are the excess of the revenue generated by the project over the operating costs incurred by it, plus net realizations from asset disposals, if any.[18] In the absence of an income tax, these receipts represent the amounts available for recovery of capital and for return on investment. When a tax is applicable, they are the amounts available for capital recovery, return, *and* income tax, being equivalent to the "pre-tax receipts" of the previous discussion.

---

[16] It will be recalled that debt constitutes 30 percent of the initial capital and is amortized in equal annual installments over the life of the investment. The interest rate is 5 percent.

[17] The rate of discount at which the balance of after-tax receipts available for equity have a present worth equal to the initial equity investment of $4,301 ($6,144 x .70).

[18] Operating costs being figured *before* (exclusive of) depreciation, interest, and income tax. The excess of revenue over operating costs is the "absolute earnings" of a project as defined in *Business Investment Policy,* Chapter 4.

With this definition out of the way, we are now ready to summarize the findings developed from the examples and to suggest some guidelines for business investment analysis. We shall pass up entirely the purely academic case in which no income tax is applicable and deal with the real world. The following conclusions therefore assume an income tax. *They assume further that the pre-tax investment receipts are in all cases fully responsive to inflation.*

Nondepreciable Investment

1. For a nondepreciable, *all-equity* investment of *unlimited* duration, the real after-tax return will, with one exception, be completely protected.[19]

2. For a nondepreciable, *all-equity* investment of *limited* duration, the protection will invariably be less than complete.

3. For a nondepreciable *mixed* (debt and equity) investment of *unlimited* duration, the real after-tax return on the *total* capital will be less than fully protected, while the corresponding return on *equity* will probably (but not necessarily) be increased.[20]

4. For a nondepreciable *mixed* investment of *limited* duration, the protection of the return on *total* capital will also be less than complete, the protection of the equity return being either more or less than complete, depending on the debt-equity ratio and various other factors.

Depreciable Investment

When we turn to depreciable investment, the range of possibilities is narrower. In this case there can be no investment of unlimited duration, since by definition depreciable assets have a limited service life. We can therefore exclude the cases treated in conclusions 1 and 3 above.

With depreciables, we have another unresponsive tax deduction (depreciation) to add to the two affecting nondepreciables (debt interest and the "basis" for terminal gain calculations). This means that the effect of inflation is less favorable (or more unfavorable) than for a nondepreciable investment of like duration.[21]

1. For a depreciable *all-equity* investment, inflation will reduce the real after-tax return.

2. For a depreciable *mixed* (debt and equity) investment, inflation will also reduce the real after-tax return on *total* capital. Whether it will protect the return on *equity* alone depends on a number of factors—the inflation rate, the tax rate, the debt ratio, the debt interest rate, the service lives of the assets, the tax depreciation method, terminal salvage values, etc.—and no categorical answer can be given. In general, however, it requires rather extreme assumptions (principally as to debt ratios) to deduce a break-even for the equity at present tax rates.[22]

Take the standard MAPI earnings projection, for example (B.I.P., p. 72). If we assume a service life of 10 years, no terminal salvage, one-quarter debt capital at 3 percent interest, a 50-percent tax rate, and sum-of-digits tax depreciation, an inflation of 3 percent per annum with fully responsive receipts will reduce the real

---

[19] The exception being investment in inventory subject to historical costing for tax purposes.

[20] Again depending on the extent of investment in historically costed inventory.

[21] This statement is good for the stated assumption that the pre-tax investment receipts are fully responsive in both cases. We shall comment further on this assumption in a moment.

[22] These are cases in which the gain to the equity from the unresponsiveness of debt charges equals the loss from the unresponsiveness of depreciation, interest, and terminal "basis" deductions.

after-tax equity return from 10 percent to 9.3 percent. To break the equity even in this case, it is necessary to assume a debt ratio close to *50 percent*.[23] Below that ratio, it loses from inflation.

The Moral

Assuming (1) fully responsive pre-tax receipts, (2) an investment in depreciable assets, (3) a corporate income tax at something like the present level, and (4) tax depreciation and terminal "basis" computed on historical cost, it is quite unlikely that inflation will improve the real after-tax return *even on the equity*. The debt leverage required for a break-even is too high.

Unless it is assumed that the investment receipts are *more than* fully responsive to inflation—an assumption rarely made, and for which there seems no clear justification—the prospect of inflation makes most depreciable projects *less,* rather than more, attractive. There is thus no justification for the popular notion that inflation makes business investment in general more profitable.

As we pointed out, the real return is likely to be better protected against inflation when the investment is in nondepreciable assets. This conclusion, however, was on the stated assumption that the pre-tax receipts are fully responsive. In this connection we may add a word of caution about a class of nondepreciable assets for which the assumption is clearly excluded. We refer to dollar assets, such as cash, bonds, notes, receivables, etc. Not only does the value of such assets fail to respond to inflation at all (being fixed in future dollars); there is no reason to assume that their flow of receipts responds either.[24] Any equity investment therein must necessarily take a beating with or without an income tax. This is true even if part of the capital is borrowed.

There can be no doubt, of course, that if equity investment has enough debt leverage it can benefit from inflation under almost any circumstances. Inflation places a premium on borrowing. It places an even larger premium on the *leasing,* rather than the *ownership,* of assets (provided, of course, that the rentals are fixed in advance in dollars, and provided the lessors do not attempt to hedge anticipated inflation in these rentals).[25] But these escapes are limited in the one case by the disinclination of management to build up top-heavy debt ratios and in the other by various factors and considerations restricting the leasing of capital assets. In general inflation is hard to beat.

We may add that the entire picture would be transformed if tax deductions for depreciation and terminal "basis" were made to respond to inflation. This reform, which the Institute has advocated for years, would give a reasonable chance to hedge the effects of inflation over most of the range of business investment projects.[26]

---

[23] Whatever the flow of pre-tax receipts, the break-even debt ratio (in percent) is close to the income-tax rate (also in percent).

[24] This does not mean, of course, that the *amount* of these assets required to do a given physical volume of business is not increased by inflation. Obviously, it is. The required increase is, however, a reflection of the unresponsiveness itself.

[25] The analysis of lease-purchase alternatives will be discussed in the fifth issue of this series.

[26] See, for example, *Realistic Depreciation Policy* (MAPI, 1954), Chapters 12 and 14.

H

Derivation of and Capital Recovery Factors for Nonuniform (Dispersed) Service Lives

A GROUP PROPERTY consists of two or more similar units of property. The life of each unit may not be the same as the average life of all units within the property group. If not, there is said to be a *dispersion* of property lives. Knowledge of the average life of a property group may not be sufficient information for a precise economic analysis; the dispersion of lives may also be required. Figure H–1 shows two *survivor curves*. Here it can be observed that although the two groups have identical average lives their dispersions are different.

Group A has no dispersion of lives since each of the two units survives exactly four years. The annual equivalent of first cost on a per unit basis is:

$$(\text{First cost})(a/p)\,_n^{\,i}$$

and assuming $i = 8\%$ and a first cost of $1000 per unit,

$$(\$1000)(0.30192) = \$301.92 \text{ per year per unit}$$

That this annual charge will recover first cost can be shown by taking the present worth for four years for each of the two units:

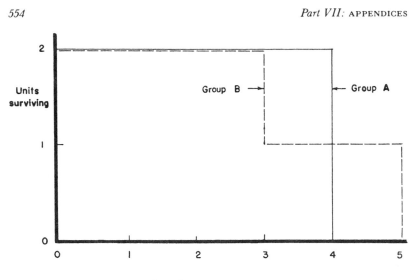

Fig. H-1. Survivor curves for two property groups with identical average lives but different survivor dispersions.

$$(2)(\$301.92)(p/a)_4^{8\%} = (\$603.84)(3.312) = \$2000$$

The annual charge just determined for Group A is not sufficient to recover costs for Group B however:

$$\$301.92[(p/a)_3^{8\%} + (p/a)_5^{8\%}]$$
$$= (\$301.92)(2.577 + 3.993)$$
$$= \$1984$$

To recover first cost of Group B the capital recovery factor, $_B(a/p)\frac{i}{n}$, should be such that

$$(\$1000)(2) = \$1000\ _B(a/p)_n^i[(p/a)_3^{8\%} + (p/a)_5^{8\%}]$$

or

$$_B(a/p)_n^i = \frac{2}{2.577 + 3.993} = 0.30441$$

or generally for dispersion XX

$$_{xx}(a/p)_n^i = \frac{\text{number of units, } U}{(p/a)_{n_1}^i + (p/a)_{n_2}^i + \cdots + (p/a)_{n_u}^i} \qquad \text{(H-1)}$$

where $n_j = $ life of the jth unit in ascending order of lives.

If all unit lives were integer,

$$_{xx}(a/p)_n^i = \frac{\text{number of units, } U}{(p/f)_1^i \, U_1 + (p/f)_2^i \, U_2 \, \cdots} \tag{H-2}$$

Where U_j = number of units surviving at end of year j.

If survivor information is given as a mathematical formula for a continuous function rather than as discrete data for individual units and year of retirement, either of the two equations may be used to determine the exact capital recovery factor:

$$_{xx}(a/p)_n^i = \frac{1}{\int_0^{z\,\text{max}} (p/a)_z^i \, y_z dz} \tag{H-3}$$

where

y_z = ordinate to the retirement frequency curve at age z (retirements during age interval dz)

z = attained age, in years

or

$$_{xx}(a/p)_n^i = \frac{1}{\int_0^{z\,\text{max}} (p/f)_z^i \int_{z\,\text{max}}^z y_z dz dz} \tag{H-4}$$

If discrete data are given and integer year intervals are used, either of two equations will yield an *approximate* capital recovery factor:

$$_{xx}(a/p)_n^i$$
$$\cong \frac{\text{number of units, } U}{(p/a)_1^i(\bar{U}_1 - \bar{U}_2) + (p/a)_2^i(\bar{U}_2 - \bar{U}_3) + (p/a)_m^i(\bar{U}_m)} \tag{H-5}$$

where

$\bar{U}_2$ = average number of units surviving during the second year

m = maximum life, in years

or

$$_{xx}(a/p)_n^i \cong \frac{\text{number of units, } U}{(p/f)_1^i(\bar{U}_1) + (p/f)_2^i(\bar{U}_2) + \cdots + (p/f)_m^i(\bar{U}_m)} \tag{H-6}$$

Because of their collective representation of a "family" of survivor curves, in the material which follows we have derived equations for determination of capital recovery factors for all of the Iowa type dispersions, L_z, O_z, R_z, and S_z. In addition we have derived an equation for determination of capital recovery factors for the negative exponential dispersion, NE, which is important because it is realistic, is mathematically convenient, and has some unusual characteristics. Computer solution of these equations for various interest rates and average lives has permitted the observations of Tables 14–1, 14–4, and 14–5. Table 14–5 emphasizes both the possible error and the strong tendency for the dispersions to be clustered in distinct groupings; it is apparent that selected "representative" dispersions employed as recommended in Chapter 14 will satisfy the usual requirements for accuracy. In the belief that unnecessary tabling of data only complicates the reader's task and diminishes his perspective of the impact of dispersion, we have tabled here only the capital recovery factors for the NE, O_2, O_1, S_0, S_1, and SQ dispersions. The others have been deferred to a publication planned for those who may sometime require still greater precision.

Survivor equations for the selected dispersions are:

1. Negative exponential (NE) where

$$\begin{pmatrix} \text{Percent surviving at age } z \\ \text{in a property group with} \\ \text{an average life of } n \text{ years} \end{pmatrix} = 100e^{-\frac{z}{n}}$$

where $e = $ the base of natural logarithms $= 2.718 \ldots$

2. Iowa Type O_2 (O2) where

$$\begin{pmatrix} \text{Percent surviving at age } z \\ \text{in a property group with} \\ \text{an average life of } n \text{ years} \end{pmatrix}$$

$$= 100 - \int_0^z \left\{ 5.1 \left[1 + \left(\frac{z}{n} - 0.4 \right)^6 \right]^{-0.6} + 0.52468485 \right\} dz$$

for $0 \leq z \leq 3.09n$

3. Iowa Type O_1 (O1) or straight-line where

$$\begin{pmatrix} \text{Percent surviving at age } z \\ \text{in a property group with} \\ \text{an average life of } n \text{ years} \end{pmatrix} = 100 - \frac{100z}{2n}$$

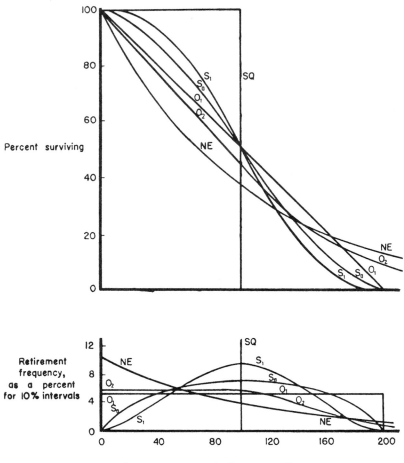

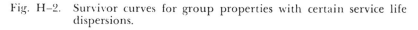

Fig. H–2.   Survivor curves for group properties with certain service life
dispersions.

4. Iowa Type S_0 (S0) where

$$\begin{pmatrix} \text{Percent surviving at age } z \\ \text{in a property group with} \\ \text{an average life of } n \text{ years} \end{pmatrix}$$

$$= 100 - \int_0^z 6.95219904 \left(\frac{2zn - z^2}{n^2} \right)^{0.74857140} dz$$

5. Iowa Type S_1 (S1) where

$$\begin{pmatrix} \text{Percent surviving at age } z \\ \text{in a property group with} \\ \text{an average life of } n \text{ years} \end{pmatrix}$$

$$= 100 - \int_0^z 9.08025966 \left(\frac{2zn - z^2}{n^2} \right)^{1.52839970} dz$$

6. Square (SQ) or rectangular where

$$\begin{pmatrix} \text{Percent surviving at age } z \\ \text{in a property group with} \\ \text{an average life of } n \text{ years} \end{pmatrix} \begin{aligned} &= 100 \text{ for } z \leq n \\ &= 0 \text{ for } z \geq n \end{aligned}$$

These six survivor curves are shown in Figure H–2.

Capital Recovery Factors for Group Properties With Negative Exponential (NE) Survivor Dispersion

The proportion of units retired during age interval dz from a property group with an average life of n years and survivor dispersion of the NE type

$$= y_z dz = \frac{e^{-\frac{z}{n}}}{n} dz$$

The present worth of a unit surviving exactly z years

$$= (p/a)_z^i = \frac{1 - (1 + i)^{-z}}{i}$$

Substitution into Equation H-3 yields

$$\text{NE}(a/p)_n^i = \frac{1}{\int_0^\infty \frac{e^{-\frac{z}{n}}}{n} \left[\frac{1 - (1 + i)^{-z}}{i} \right] dz}$$

$$= \frac{ni}{\int_0^\infty e^{-\frac{z}{n}} dz - \int_0^\infty e^{-\frac{z}{n}}(1 + i)^{-z} dz}$$

$$= \frac{ni}{-ne^{-\frac{z}{n}} \Big|_0 + \frac{[e^{\frac{1}{n}}(1 + i)]^{-z}}{\frac{1}{n} + ln(1 + i)} \Big|_0}$$

$$= \frac{ni}{n - \dfrac{1}{\dfrac{1}{n} + ln(1 + i)}}$$

$$= \frac{i}{1 - \dfrac{1}{1 + n\, ln(1 + i)}}$$

$$= \frac{i[1 + n\, ln(1 + i)]}{1 + n\, ln(1 + i) - 1}$$

Therefore

$$\text{NE}(a/p)_n^i = \frac{i}{n\, ln(1 + i)} + i \tag{H-7}$$

So, for example, if $i = 10\%$, $n = 10$

$$\text{NE}(a/p)_{10}^{10\%} = \frac{0.10}{10(0.09531)} + 0.10 = 0.2049$$

This compares with

$$\text{SQ}(a/p)_{10}^{10\%} = 0.16275$$

A rather interesting characteristic of the negative exponential dispersion is that the $\text{NE}(a'/f)_n^i$ factor is independent of i if the period is annual (that is, $m = 1$, $i_a = i$):

$$\text{NE}(a'/f)_n^i = \text{NE}(a/f)_n^i (a'/a)^i = [\text{NE}(a/p)_n^i - i](a'/a)^i$$

$$= \frac{i}{n\, ln(1 + i)} \cdot \frac{ln(1 + i)}{i}$$

$$= \frac{1}{n}$$

It follows that

$$\text{NE}(a'/p)_n^i = \frac{1}{n} + i$$

As a result of this characteristic the continuous annual equivalent of first cost and salvage for a negative exponential dispersion

$$= (\text{first cost})\, \text{NE}(a'/p)_n^i - \text{salvage}\, \text{NE}(a'/f)_n^i$$

$$= (\text{first cost})\left(\frac{1}{n} + i\right) - \text{salvage}\left(\frac{1}{n}\right)$$

$$= \frac{\text{first cost} - \text{salvage}}{n} + i\,(\text{first cost})$$

which is consistent with still another characteristic; the value of surviving units whose survivor dispersion is described by the negative exponential does not decrease with time.

Capital Recovery Factors for Group Properties With Iowa Type O_1 (O1) or Straight-Line Survivor Dispersion

The proportion of units retired during age interval, dz, from a property group with an average life of n years and survivor dispersion of the O_1 type

$$= y_z \, dz = \frac{dz}{2n}$$

The present worth of a unit surviving exactly z years

$$= (p/a)_z^i = \frac{1 - (1 + i)^{-z}}{i}$$

Substitution into Equation H-3 yields

$$_{01}(a/p)_n^i = \frac{1}{\int_0^{2n} \left[\frac{1 - (1 + i)^{-z}}{i} \right] \frac{dz}{2n}} = \frac{2ni}{\int_0^{2n} dz - \int_0^{2n} (1 + i)^{-z} \, dz}$$

$$= \frac{2ni}{2n - \left[\frac{(1 + i)^{-z}}{-ln(1 + i)} \Big|_0^{2n} \right]} = \frac{2ni}{2n + \frac{(1 + i)^{-2n}}{ln(1 + i)} - \frac{1}{ln(1 + i)}}$$

Therefore

$$_{01}(a/p)_n^i = \frac{2ni \, ln(1 + i)}{2n \, ln(1 + i) + (1 + i)^{-2n} - 1} \tag{H-8}$$

So, for example, if $i = 10\%$, $n = 10$

$$_{01}(a/p)_{10}^{10\%} = \frac{2(0.09531)}{20(0.09531) + 0.1486 - 1} = 0.1807$$

This compares with

$$_{SQ}(a/p)_{10}^{10\%} = 0.16275$$

Capital Recovery Factors for Group Properties With Iowa Type O_x, L_x, R_x, Survivor Dispersions

Equations for the retirement frequency of the Iowa Type O_x, L_x, and R_x survivor dispersions are not convenient to integrate; for this

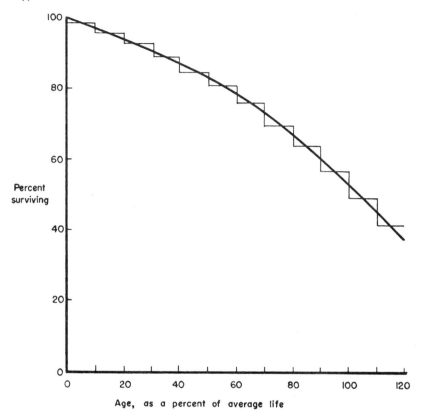

Fig. H–3. The R_1 survivor curve and an approximation of same.

reason an approximate method of determining capital recovery factors for such dispersions has been applied.

The continuous survivor curve of Figure H–3 may be approximated by the "stair-step" discrete curve shown there. Ordinate values could be taken as (1) the ordinate value at midpoint of the age interval or (2) the average of beginning and end of interval ordinate values; the ordinate values are readily available.[1]

Ordinate values at midpoint of the age interval have been used;

---

[1] See A. Marston, R. Winfrey, and J. C. Hempstead, *Engineering Valuation and Depreciation* (Ames, Iowa State Univ. Press, 1953), pp. 414–18.

R. Winfrey, *Statistical Analysis of Industrial Property Retirements.* Iowa State Univ. Eng. Exp. Sta. Bul. 125, 1935, pp. 102–6.

R. Winfrey, *Depreciation of Group Properties,* Iowa State Univ. Eng. Exp. Sta. Bul. 155, 1942, pp. 124–28.

F. Couch, *Classification of Type O Retirement Characteristics of Industrial Property.* Unpublished M.S. thesis, Iowa State Univ., 1957, pp. 51–90.

that is, the average survivors during the age interval 0% to 10% of average life are taken as 98.67319% for the R_1 curve, the percent surviving at age equals 5% of average life. Average survivors during the age interval 10% to 20% of average life are taken as 95.74739% (for R_1) the percent surviving for age equals 15% of average life.

The worth at the beginning of the interval of the average survivors during the interval

$$= (p/a)^i_{0.10n}(\text{average survivors})$$

and the present worth of survivors for an interval beginning z intervals from now

$$= (p/f)^i_{0.10nz}(p/a)^i_{0.10n}(\text{average survivors})$$

The survivors at 5%, 15%, 25%, . . . of average life have been used in a computer program to sum the present worth of survivors and from these, capital recovery factors are then computed.

Capital Recovery Factors for Group Properties With Iowa Type S_x Survivor Dispersions

Capital recovery factors for group properties following the Iowa Type S_x survivor dispersion can be expressed as Bessel functions. This interesting discovery and derivation are the work of Dr. Clair G. Maple, director of the Iowa State University Computation Center. This development greatly simplified the derivation of capital recovery factors for the S_x dispersions. In the material which follows, several "hand" calculations suffice to provide an accurate determination of the $_{sx}(a/p)^i_n$ factor. Without Dr. Maple's discovery, factor determination would have followed that employed for the O_x, L_x, and R_x dispersions which are extremely cumbersome if accomplished as "hand" calculations.

Because the derivation involves a large number of symbols, a number of those symbols take on a different meaning (for example, σ, B, C_1, C_2, I, a, m, k, t, and u) than that employed elsewhere in this book. For that reason the reader is cautioned to refer to definitions herein for the derivation and to refer to **Appendix A**, "Notation," for symbols used elsewhere in the book.

The retirement frequency of the Iowa Type S_x (symmetrical) curves follows the general equation:[2]

---

[2] A. Marston, R. Winfrey, and J. C. Hempstead, *Engineering Valuation and Depreciation* (Ames, Iowa State Univ. Press, 1953), pp. 410–13.

$$y_x = y_0 \left(1 - \frac{x^2}{a^2}\right)^m \tag{H-9}$$

in which

y_x = ordinate to the frequency curve at age x (origin at the mean age) in percent

y_0 = ordinate to the frequency curve at its mode

x = age (in units equal to 10% of average life) measured from the average life ordinate (If $x = -2.5$ the equivalent age is 75% of average life.)

a = a parameter of the equations of the Iowa type frequency curves

m = a parameter of the equations of the Iowa type frequency curves

The survivor curve is a reverse summation of the retirement frequency curve:

$$\text{Percent survivors at age } x = \int_a^x y_0 \left(1 - \frac{x^2}{a^2}\right)^m dx$$

Make the following substitutions into Equation H-9:

Let $v = \dfrac{x}{a}$

n = average life in years

y_v = retirements during interval dv in units, not percent

z = attained age in years

Then

$$z = \left(\frac{x}{10} + 1\right) n = \left(\frac{av}{10} + 1\right) n = \frac{nav}{10} + n = 0.1 \, nav + n$$

$$dx = adv$$

and

$$y_v = \frac{ay_0}{100} (1 - v^2)^m \, dv$$

The limits $-a \le x \le a$ become: $-1 \le v \le 1$, and the present worth of survivors is

$$\int_{-1}^1 (p/a)^i_{0.1nav+n} \, y_v$$

$$= \int_{-1}^1 \frac{1 - (1 + i)^{-0.1nav-n}}{i} \, y_v$$

Let $e^k = 1 + i$. Substituting $1 + i = e^k$ and substituting for y_v; the present worth of survivors is

$$= \int_{-1}^{1} \left(\frac{1 - e^{-0.1nakv - nk}}{i} \right) \left(\frac{ay_0}{100} \right) (1 - v^2)^m \, dv$$

$$= \frac{1}{i} \int_{-1}^{1} \frac{ay_0}{100} (1 - v^2)^m \, dv - \frac{1}{i} \int_{-1}^{1} \frac{ay_0}{100} (1 - v^2)^m \, e^{-0.1nakv - nk} \, dv$$

and for the S_r dispersions:

$$\int_{-1}^{1} \frac{ay_0}{100} (1 - v^2)^m \, dv = 1 = \int_{-a}^{a} \frac{y_0}{100} \left(1 - \frac{x^2}{a^2} \right)^m \, dx$$

which simply shows that the summation of all retirements equals 100% of those units originally placed. So

$$sx \, (a/p)_n^i = \cfrac{i}{1 - \cfrac{ay_0}{100e^{nk}} \int_1^1 (1 - v^2)^m \, e^{-0.1nakv} \, dv}$$

$$= \cfrac{i}{1 - \cfrac{ay_0}{100e^{nk}} \left[\int_{-1}^{0} e^{-0.1nakv} (1 - v^2)^m \, dv + \int_0^1 e^{-0.1nakv} (1 - v^2)^m \, dv \right]}$$

$$= \cfrac{i}{1 - \cfrac{ay_0}{100e^{nk}} \left[\int_0^1 e^{0.1nakv} (1 - v^2)^m \, dv + \int_0^1 e^{-0.1nakv} (1 - v^2)^m \, dv \right]}$$

$$= \cfrac{i}{1 - \cfrac{ay_0 2}{100e^{nk}} \int_0^1 \left(\frac{e^{0.1nakv} + e^{-0.1nakv}}{2} \right) (1 - v^2)^m \, dv}$$

Let

$$0.1 \, ank = \sigma$$

and

$$F(0.1ank) = \frac{i}{1 - K_2 \, F(\sigma)}$$

where

$$K_2 = \frac{ay_0 2}{100}$$

and

$$F(\sigma) = e^{\frac{-10\sigma}{a}} \int_0^1 \left(\frac{e^{\sigma v} + e^{-\sigma v}}{z} \right) (1 - v^2)^m \, dv$$

$$= e^{\frac{-10\sigma}{a}} \int_0^1 (\cosh \sigma v)(1 - v^2)^m \, dv$$

Differentiate with respect to σ:

$$F'(\sigma) = -\frac{10}{a} e^{\frac{-10\sigma}{a}} \int_0^1 (1 - v^2)^m (\cosh \sigma v) dv$$

$$+ e^{\frac{-10\sigma}{a}} \int_0^1 v(1 - v^2)^m \sinh \sigma v dv$$

$$= -\frac{10}{a} F(\sigma) + e^{\frac{-10\sigma}{a}} \int_0^1 v(1 - v^2)^m \sinh \sigma v dv$$

Differentiate again with respect to (σ) :

$$F''(\sigma) = -\frac{10}{a} F'(\sigma) - \frac{10}{a} e^{\frac{-10\sigma}{a}} \int_0^1 v(1 - v^2)^m (\sinh \sigma v) dv$$

$$+ e^{\frac{-10\sigma}{a}} \int_0^1 v^2(1 - v^2) \cosh \sigma v dv$$

$$= -\frac{10}{a} F'(\sigma) - \frac{10}{a} F'(\sigma) - \left(\frac{10}{a} \right)^2 F(\sigma)$$

$$+ e^{\frac{-10\sigma}{a}} \int_0^1 (1 - 1 + v^2)(1 - v^2)^m \cosh \sigma v dv$$

$$= -\frac{20}{a} F'(\sigma) - \left(\frac{10}{a} \right)^2 F(\sigma) + e^{\frac{-10\sigma}{a}} \int_0^1 (1 - v^2)^m \cosh \sigma v dv$$

$$- e^{\frac{-10\sigma}{a}} \int_0^1 (1 - v^2)^{m+1} \cosh \sigma v dv$$

$$= -\frac{20}{a} F'(\sigma) - \left[\left(\frac{10}{a} \right)^2 - 1 \right] F(\sigma)$$

$$- e^{\frac{-10\sigma}{a}} \int_0^1 (1 - v^2)^{m+1} \cosh \sigma v dv$$

Now consider the integral

$$I(\sigma) = \int_0^1 (1 - v^2)^{m+1} \cosh \sigma v \, dv$$

and integrate by parts taking

$$u = (1 - v^2)^{m+1} \qquad\qquad\qquad dt = \cosh \sigma v \, dv$$

$$du = (m + 1)(1 - v^2)^m(-2v \, dv) \qquad\qquad t = \frac{\sinh \sigma v}{\sigma}$$

Therefore

$$I(\sigma) = (1 - v^2)^{m+1} \frac{\sinh \sigma v}{\sigma} \Big|_0^1 - \int_0^1 \frac{\sinh \sigma v}{\sigma} (m + 1)(1 - v^2)^m(-2v \, dv)$$

$$= \frac{2(m + 1)}{\sigma} \int_0^1 v(1 - v^2)^m \sinh \sigma v \, dv$$

From the expression for $F'(\sigma)$

$$\int_0^1 v(1 - v^2)^m \sinh \sigma v \, dv = e^{\frac{10\sigma}{a}} \left[F'(\sigma) + \frac{10}{a} F(\sigma) \right]$$

So

$$I(\sigma) = \frac{2(m + 1)e^{\frac{10\sigma}{a}}}{\sigma} \left[F'(\sigma) + \frac{10}{a} F(\sigma) \right]$$

and substitution into the expression for $F''(\sigma)$ yields:

$$F''(\sigma) = -\frac{20}{a} F'(\sigma) - \left[\left(\frac{10}{a}\right)^2 - 1 \right] F(\sigma)$$

$$- \frac{2(m + 1)}{\sigma} \left[F'(\sigma) + \frac{10}{a} F(\sigma) \right]$$

$$= \left[-\frac{20}{a} - \frac{2(m + 1)}{\sigma} \right] F'(\sigma)$$

$$- \left[\left(\frac{10}{a}\right)^2 - 1 + \frac{2(m + 1)}{\sigma} \frac{10}{a} \right] F(\sigma)$$

So

$$F''(\sigma) + \left[\frac{20}{a} + \frac{2(m + 1)}{\sigma} \right] F'(\sigma)$$

$$+ \left[\left(\frac{10}{a}\right)^2 - 1 + \frac{2(m + 1)}{\sigma} \frac{10}{a} \right] F(\sigma) = 0$$

or

$$\sigma\, F''(\sigma) + \left(\frac{20\sigma}{a} + 2m + 2\right) F'(\sigma)$$

$$+ \left[\left(\frac{10}{a}\right)^2 \sigma - \sigma + (2m + 2)\left(\frac{10}{a}\right)\right] F(\sigma) = 0$$

Since this is a second order differential equation, two boundary conditions are needed to determine $F(\sigma)$ uniquely:

$$F(0) = \int_0^1 (1 - v^2)^m\, dv$$

$$F'(0) = -\frac{10}{a}\int_0^1 (1 - v^2)^m\, dv$$

Now simplify the differential equation by setting $2(m + 1) = B$. Then

$$\sigma\, F''(\sigma) + \left(\frac{20\sigma}{a} + B\right) F'(\sigma) + \left[\left(\frac{10}{a}\right)^2 \sigma - \sigma + B\frac{10}{a}\right] F(\sigma) = 0$$

This can be reduced to a Bessel's equation as follows. Set

$$F = ut$$

then

$$F' = ut' + u't$$

and

$$F'' = ut'' + 2u't' + u''t$$

Substitute

$$\sigma ut'' + 2\sigma u't' + \sigma u''t + \left(\frac{20\sigma}{a} + B\right)(ut' + u't)$$

$$+ \left[\left(\frac{10}{a}\right)^2 \sigma - \sigma + B\frac{10}{a}\right] ut = 0$$

Consider this as an equation for t:

$$(\sigma u)t'' + \left[2\sigma u' + \left(\frac{20\sigma}{a} + B\right)u\right] t$$

$$+ \left\{\sigma u'' + \left(\frac{20\sigma}{a} + B\right)u' + \left[\left(\frac{10}{a}\right)^2 \sigma - \sigma + B\frac{10}{a}\right]u\right\} t = 0$$

Choose u so that coefficient of t' vanishes

$$2\sigma u' + \left(\frac{20\sigma}{a} + B\right)u = 0$$

or

$$2\sigma\frac{du}{d(\sigma)} = -\left(\frac{20\sigma}{a} + B\right)u$$

$$\frac{du}{u} = -\left(\frac{10}{a} + \frac{B}{2\sigma}\right)d(\sigma)$$

Integration yields

$$ln\,u = -\frac{10}{a}\sigma - \frac{B}{2}ln\,\sigma$$

$$u = \sigma^{\frac{-B}{2}}e^{\frac{-10\sigma}{a}}$$

$$u' = \left(-\frac{10}{a}\sigma^{\frac{-B}{2}} - \frac{B}{2}\sigma^{-1-\frac{B}{2}}\right)e^{\frac{-10\sigma}{a}}$$

$$u'' = \left(\frac{10}{a}\right)^2\sigma^{\frac{-B}{2}}e^{\frac{-10\sigma}{a}} + \frac{10}{a}B\sigma^{-1-\frac{B}{2}}e^{\frac{-10\sigma}{a}} + \frac{B}{2}\left(1+\frac{B}{2}\right)\sigma^{-2-\frac{B}{2}}e^{\frac{-10\sigma}{a}}$$

$$= e^{\frac{-10\sigma}{a}}\sigma^{-1-\frac{B}{2}}\left[\left(\frac{10}{a}\right)^2\sigma + \frac{10}{a}B + \frac{B}{2}\left(1+\frac{B}{2}\right)\sigma^{-1}\right]$$

Coefficient of $t = \sigma u'' + \left(\frac{20\sigma}{a} + B\right)u' + \left[\left(\frac{10}{a}\right)^2\sigma - \sigma + B\frac{10}{a}\right]u$

$$= e^{\frac{-10\sigma}{a}}\sigma^{\frac{-B}{2}}\left[\left(\frac{10}{a}\right)^2\sigma + \left(\frac{10}{a}\right)B + \frac{B}{2}\left(1+\frac{B}{2}\right)\sigma^{-1}\right.$$

$$\left. + \left(\frac{20\sigma}{a} + B\right)\left(-\frac{10}{a} - \frac{B}{2\sigma}\right) + \left(\frac{10}{a}\right)^2\sigma - \sigma + B\frac{10}{a}\right]$$

$$= e^{\frac{-10\sigma}{a}}\sigma^{\frac{-B}{2}}\left[\left(\frac{10}{a}\right)^2\sigma + \frac{10}{a}B + \frac{B}{2\sigma} + \frac{B^2}{4\sigma} - 2\sigma\left(\frac{10}{a}\right)^2\right.$$

$$\left. - \frac{10}{a}B - \frac{10}{a}B - \frac{B^2}{2\sigma} + \left(\frac{10}{a}\right)^2\sigma - \sigma + \frac{10}{a}B\right]$$

$$= e^{\frac{-10\sigma}{a}}\sigma^{\frac{-B}{2}}\left\{\sigma\left[\left(\frac{10}{a}\right)^2 - 2\left(\frac{10}{a}\right)^2 + \left(\frac{10}{a}\right)^2 - 1\right]\right.$$

$$\left. + \left(\frac{10}{a}B - \frac{10}{a}B - \frac{10}{a}B + \frac{10}{a}B\right) + \frac{1}{\sigma}\left(\frac{B}{2} + \frac{B^2}{4} - \frac{B^2}{2}\right)\right\}$$

$$= -e^{\frac{-10\sigma}{a}}\sigma^{\frac{-B}{2}}\left(\sigma + \frac{B^2 - 2B}{4\sigma}\right)$$

The equation for t becomes

$$\sigma^{1-\frac{B}{2}} e^{\frac{-10\sigma}{a}} t'' - e^{\frac{-10}{a}} \sigma^{\frac{-B}{2}} \left(\sigma + \frac{B^2 - 2B}{4\sigma} \right) t = 0$$

or

$$\sigma t'' - \left(\sigma + \frac{B^2 - 2B}{4\sigma} \right) t = 0$$

set

$$t(\sigma) = \sqrt{\sigma}\, s$$

$$t' = \sqrt{\sigma}\, s' + \frac{s}{2\sqrt{\sigma}} = \sqrt{\sigma} \left(s' + \frac{s}{2\sigma} \right)$$

$$t'' = \sqrt{\sigma}\, s'' + \frac{s'}{\sqrt{\sigma}} - \frac{s}{4\sigma^{3/2}}$$

so

$$\sigma^{3/2} s'' + \sigma^{1/2} s' - \frac{s}{4\sigma^{1/2}} - \left(\sigma + \frac{B^2 - 2B}{4\sigma} \right) \sigma^{1/2} s = 0$$

or

$$\sigma^2 s'' + \sigma s' - s \left(\frac{1}{4} + \sigma^2 + \frac{B^2 - 2B}{4} \right) = 0$$

or

$$\sigma^2 s'' + \sigma s' - s \left[\sigma^2 + \left(\frac{B-1}{2} \right)^2 \right] = 0$$

This is Bessel's modified equation with

$$n = \frac{B-1}{2}$$

Now

$$B = 2(m+1)$$

so

$$n = \frac{2m + 2 - 1}{2} = m + \frac{1}{2}$$

So the general solution of the s equation is

$$s(\sigma) = C_1 I_n(\sigma) + C_2 I_{-n}(\sigma)$$

where

$$I_n(\sigma) = i^{-n} J_n(i\sigma)$$

so

$$t(\sigma) = \sqrt{\sigma}[C_1 I_n(\sigma) + C_2 I_{-n}(\sigma)]$$

and

$$F(\sigma) = u(\sigma)t(\sigma) = e^{\frac{-10\sigma}{a}}\sigma^{\frac{1-B}{2}}[C_1 I_n(\sigma) + C_2 I_{-n}(\sigma)]$$
$$= e^{\frac{-10\sigma}{a}}\sigma^{-(m+1/2)}[C_1 I_n(\sigma) + C_2 I_{-n}(\sigma)]$$

Find the constants of integration, C_1 and C_2, since

$$sx(a/p)_n^i$$

$$= F(0.1an,k) = \frac{i}{1 - \frac{ay_0 2}{100} F(\sigma)}$$

$$= \frac{i}{1 - \frac{ay_0 2}{100}\{e^{-nk}(0.1ank)^{-(m+1/2)}[C_1 I_j(0.1ank) + C_2 I_j(0.1ank)]\}}$$

$$= \frac{i}{1 - \frac{ay_0 2}{100}\{(0.1ank)^{-(m+1/2)}[C_1 I_{m+1/2}(0.1ank) + C_2 I_{-(m+1/2)}(0.1\,ank)]\}}$$

and for

$$I_{m+1/2}(0.1ank) = \sum_{j=0}^{\infty} \frac{(0.1ank)^{m+1/2+2j}}{2^{m+1/2+2j}j!\Gamma(m + \frac{1}{2} + j + 1)}$$

For the various dispersions:[3]

---

[3] See A. Marston, R. Winfrey, and J. C. Hempstead, *Engineering Valuation and Depreciation* (Ames, Iowa State Univ. Press, 1953), pp. 410–13.

Dispersion	a	y_0	m
S_0	10	6.952,199,04	0.748,571,40
S_1	10	9.080,259,66	1.828,399,70
S_2	10	11.911,038,82	3.700,093,74
S_3	10	15.610,487,97	6.901,591,8
S_4	9	22.329,360,82	11.935,379,40
S_5	8	33.220,515,75	21.437,821,70
S_6	7	52.472,591,69	41.634,142,20

Now

$$F(0.1\ an, 0) = \frac{i}{1 - \frac{ay_0 2}{100} \int_0^1 (1 - v^2)^m\, dv}$$

and

$$\frac{ay_0 2}{100} \int_0^1 (1 - v^2)^m\, dv \text{ is finite}$$

therefore $C_2 = 0$. So

$$F(0.1\ an, k) = \frac{i}{1 - \frac{ay_0 2}{100 e^{nk}} \left\{ 0.1\ ank^{-(m+1/2)}[C_1 I_{m+1/2}(0.1\ ank)] \right\}}$$

To evaluate C_1, observe that

$$I_{m+1/2} = \frac{(0.1\ ank)^{m+1/2}}{2^{m+1/2}\Gamma(m + 1\tfrac{1}{2})}$$

so

$$F(0.1\ an, 0) = \frac{i}{1 - \frac{ay_0 2}{100 e^{nk}} \left[\frac{C_1(0.1\ ank)^{m+1/2}}{(0.1\ ank)^{m+1/2} 2^{m+1/2}\Gamma(m + 1\tfrac{1}{2})} \right]}$$

$$= \frac{i}{1 - \frac{ay_0 2}{100} \left[\frac{C_1}{2^{m+1/2}\Gamma(m + 1\tfrac{1}{2})} \right]}$$

$$= \frac{i}{1 - \frac{ay_0 2}{100} \int_0^1 (1 - v^2)^m\, dv}$$

Since the summation of all retirements equals 100% of the units originally placed:

$$1 = \int_{-a}^{a} \frac{y_0}{100} \left(1 - \frac{x^2}{a^2}\right)^m dx$$

because the frequency curve is symmetrical (S_x) :

$$1 = 2 \int_{0}^{a} \frac{y_0}{100} \left(1 - \frac{x^2}{a^2}\right)^m dx$$

Since

$$\frac{x}{a} = v; \text{ and } dx = a\,dv$$

$$1 = \frac{ay_02}{100} \int_{0}^{1} (1 - v^2)\, dv$$

Therefore

$$C_1 = \frac{100 \; 2^{m+1/2}\Gamma(m + 1\frac{1}{2})}{av_02}$$

Since

$$F(0.1an, k) = {}_Sx(a/p)_n^i$$

Then

$${}_Sx(a/p)_n^i$$

$$= \frac{i}{1 - \dfrac{ay_02}{100e^{nk}} \left\{ 0.1 \; ank^{-\,(m+1/2)}\left[2^{m+1/2}\Gamma(m + 1\frac{1}{2}) \dfrac{100}{ay_02}\right] I_{m+1/2}(0.1 \; ank) \right\}}$$

$$= \frac{i}{1 - \left[\dfrac{2^{m+1/2}\Gamma(m + 1\frac{1}{2})}{e^{nk}(0.1 \; ank)^{m+1/2}}\right] \displaystyle\sum_{j=0}^{\infty} \dfrac{0.1 \; ank^{m+1/2+2j}}{2^{m+1/2+2j}\,{}_1!\Gamma(m + \frac{1}{2} + j + 1)}}$$

$$= \frac{i}{1 - \left(\dfrac{20}{ank}\right)^{m+1/2}\dfrac{(m + \frac{1}{2})!}{e^{nk}} \displaystyle\sum_{j=0}^{\infty} \left(\dfrac{ank}{20}\right)^{m+1/2+2j} \dfrac{1}{j!(m + \frac{1}{2} + j)!}}$$

$$= \frac{i}{1 - \left(\dfrac{20}{ank}\right)^{m+1/2}\dfrac{(m + \frac{1}{2})!}{e^{nk}}\left[\left(\dfrac{ank}{20}\right)^{m+1/2}\dfrac{1}{(m + \frac{1}{2})!} + \left(\dfrac{ank}{20}\right)^{m+5/2}\dfrac{1}{1!(m + 1\frac{1}{2})!}\cdots\right]}$$

$$= \frac{i}{1 - e^{-nk}\left[1 + \left(\dfrac{ank}{20}\right)^{2}\dfrac{1}{(m + 1\frac{1}{2})} + \left(\dfrac{ank}{20}\right)^{4}\dfrac{1}{2!(m + 1\frac{1}{2})(m + 2\frac{1}{2})}\cdots\right]} \qquad \text{(H-10)}$$

$$= \frac{i}{1 - e^{-nk}\left[1 + \left(\dfrac{ank}{20}\right)^{2}\dfrac{1}{(m + 1\frac{1}{2})}\left[1 + \left[\left(\dfrac{ank}{20}\right)^{2}\dfrac{1}{2(m + 2\frac{1}{2})}\left[1 + \cdots\right]\right]\right]\right]} \qquad \text{(H-11)}$$

So, for example, if $n = 10$, and $i = 10\%$, then

$$k = ln(1 + i) = 0.09531$$

If the S_1 dispersion is selected,

$$m = 1.828,399,70$$
$$a = 10$$

so

$$\frac{ank}{20} = 0.47655$$

Rounding data and using Equation H–10:

$$s_1(a/p)_{10}^{10\%} = \frac{0.10}{1 - e^{-0.9531}\left[1 + \dfrac{(0.47655)^2}{3.3284} + \dfrac{(0.47655)^4}{2(3.3284)(4.3284)}\cdots\right]}$$

$$= \frac{0.10}{1 - 0.38557(1 + 0.0682 + 0.0018 + 0.0000 \ldots)}$$

$$\cong 0.1702$$

Or if $n = 10$ and $i = 10\%$, then

$$k = ln(1 + i) = 0.09531$$

If the S_6 dispersion is selected,

$$m = 41.634,142,20$$
$$a = 7$$
$$\frac{ank}{20} = 0.333,585$$

R ;unding data and using Equation H–10:

$$s_6(a/p)_{10}^{10\%} = \cfrac{0.10}{1 - e^{-0.9531}\left[1 + \cfrac{(0.333,585)^2}{43.13} + \cfrac{(0.333,585)^4}{2(43.13)(44.13)} \cdots\right]}$$

$$= \frac{0.10}{1 - 0.38557\,(1 + 0.002,58 + 0.000,00 \ldots)}$$

$$\cong 0.1630$$

and for comparison:

$$s_Q(a/p)_{10}^{10\%} = 0.16275$$
$$s_6(a/p)_{10}^{10\%} = 0.1630$$
$$s_1(a/p)_{10}^{10\%} = 0.1702$$

$i = 6\%$

	Uniform series end-of-period amount a equivalent to present sum p (capital recovery factor) for a group property with survivors following the ____ dispersion. $$xx(a/p)_n^i = xx(a/f)_n^i + i$$						
n	Negative exponential **NE**	Iowa Type **O_2**	Straight-line or Iowa Type **O_1**	Iowa Type **S_0**	Iowa Type **S_1**	Square, rectangular, or no dispersion **SQ**	**n**
1	1.08971	1.07966	1.07009	1.06671	1.06452	1.06000	1
2	0.57485	0.56270	0.55562	0.55219	0.54998	0.54544	2
3	0.40324	0.39059	0.38437	0.38089	0.37866	0.37411	3
4	0.31743	0.30468	0.29893	0.29540	0.29316	0.28859	4
5	0.26594	0.25324	0.24781	0.24423	0.24197	0.23740	5
6	0.23162	0.21904	0.21383	0.21022	0.20794	0.20336	6
7	0.20710	0.19468	0.18966	0.18601	0.18372	0.17913	7
8	0.18871	0.17647	0.17162	0.16792	0.16562	0.16104	8
9	0.17441	0.16236	0.15765	0.15392	0.15160	0.14702	9
10	0.16297	0.15112	0.14653	0.14277	0.14044	0.13587	10
11	0.15361	0.14195	0.13749	0.13370	0.13136	0.12679	11
12	0.14581	0.13435	0.13000	0.12618	0.12383	0.11928	12
13	0.13921	0.12794	0.12371	0.11986	0.11750	0.11296	13
14	0.13355	0.12248	0.11835	0.11448	0.11212	0.10758	14
15	0.12865	0.11777	0.11374	0.10984	0.10748	0.10296	15
16	0.12436	0.11366	0.10973	0.10582	0.10345	0.09895	16
17	0.12057	0.11006	0.10622	0.10229	0.09992	0.09544	17
18	0.11721	0.10688	0.10313	0.09918	0.09681	0.09236	18
19	0.11420	0.10404	0.10038	0.09642	0.09405	0.08962	19
20	0.11149	0.10150	0.09793	0.09396	0.09159	0.08718	20
21	0.10903	0.09922	0.09573	0.09175	0.08938	0.08500	21
22	0.10680	0.09715	0.09375	0.08976	0.08739	0.08305	22
23	0.10477	0.09528	0.09196	0.08795	0.08559	0.08128	23
24	0.10290	0.09357	0.09032	0.08632	0.08396	0.07968	24
25	0.10119	0.09200	0.08884	0.08482	0.08247	0.07823	25
26	0.09960	0.09057	0.08747	0.08346	0.08111	0.07690	26
27	0.09814	0.08925	0.08622	0.08221	0.07987	0.07570	27
28	0.09678	0.08802	0.08507	0.08106	0.07873	0.07459	28
29	0.09551	0.08689	0.08401	0.08000	0.07768	0.07358	29
30	0.09432	0.08584	0.08303	0.07901	0.07670	0.07265	30
35	0.08942	0.08156	0.07905	0.07507	0.07281	0.06897	35
40	0.08574	0.07844	0.07619	0.07226	0.07007	0.06646	40
45	0.08288	0.07607	0.07404	0.07018	0.06806	0.06470	45
50	0.08059	0.07423	0.07239	0.06860	0.06656	0.06344	50
60	0.07716	0.07155	0.07000	0.06639	0.06452	0.06188	60

	NE	**O_2**	**O_1**	**S_0**	**S_1**	**SQ**	
n	These data are also suggested as an approximation for the Iowa type survivor dispersions shown below						**n**
	O_3		L_0	L_1, R_1	$L_2, L_3, R_2,$ R_3, S_2	$L_4, L_5, R_4, R_5,$ S_3, S_4, S_5, S_6	

$i = 7\%$

n	Uniform series end-of-period amount a equivalent to present sum p (capital recovery factor) for a group property with survivors following the ___ dispersion. $$xx(a/p)_n^i = xx(a/f)_n^i + i$$						n
	Negative exponential **NE**	Iowa Type O_2	Straight-line or Iowa Type O_1	Iowa Type S_0	Iowa Type S_1	Square, rectangular, or no dispersion **SQ**	
1	1.10460	1.09216	1.08179	1.07784	1.07528	1.07000	1
2	0.58730	0.57285	0.56500	0.56098	0.55839	0.55309	2
3	0.41487	0.40002	0.39307	0.38898	0.38637	0.38105	3
4	0.32865	0.31380	0.30734	0.30319	0.30056	0.29523	4
5	0.27692	0.26221	0.25609	0.25189	0.24923	0.24389	5
6	0.24243	0.22794	0.22208	0.21782	0.21514	0.20980	6
7	0.21780	0.20355	0.19790	0.19359	0.19090	0.18555	7
8	0.19933	0.18533	0.17988	0.17552	0.17281	0.16747	8
9	0.18496	0.17123	0.16595	0.16154	0.15882	0.15349	9
10	0.17346	0.16000	0.15488	0.15043	0.14770	0.14238	10
11	0.16405	0.15086	0.14589	0.14141	0.13866	0.13336	11
12	0.15622	0.14329	0.13846	0.13394	0.13119	0.12590	12
13	0.14959	0.13691	0.13222	0.12768	0.12492	0.11965	13
14	0.14390	0.13148	0.12692	0.12235	0.11958	0.11434	14
15	0.13897	0.12680	0.12237	0.11777	0.11501	0.10979	15
16	0.13466	0.12273	0.11842	0.11380	0.11104	0.10586	16
17	0.13086	0.11916	0.11497	0.11034	0.10757	0.10243	17
18	0.12748	0.11601	0.11193	0.10728	0.10452	0.09941	18
19	0.12445	0.11320	0.10924	0.10458	0.10182	0.09675	19
20	0.12173	0.11070	0.10684	0.10217	0.09942	0.09439	20
21	0.11927	0.10844	0.10469	0.10001	0.09727	0.09229	21
22	0.11703	0.10641	0.10276	0.09808	0.09534	0.09041	22
23	0.11498	0.10456	0.10101	0.09632	0.09360	0.08871	23
24	0.11311	0.10288	0.09942	0.09474	0.09202	0.08719	24
25	0.11138	0.10134	0.09798	0.09329	0.09059	0.08581	25
26	0.10979	0.09993	0.09666	0.09198	0.08928	0.08456	26
27	0.10832	0.09863	0.09545	0.09077	0.08809	0.08343	27
28	0.10695	0.09743	0.09433	0.08966	0.08700	0.08239	28
29	0.10568	0.09633	0.09331	0.08865	0.08600	0.08145	29
30	0.10449	0.09530	0.09236	0.08771	0.08507	0.08059	30
35	0.09956	0.09112	0.08853	0.08395	0.08140	0.07723	35
40	0.09587	0.08808	0.08578	0.08129	0.07885	0.07501	40
45	0.09299	0.08578	0.08372	0.07934	0.07701	0.07350	45
50	0.09069	0.08398	0.08212	0.07787	0.07565	0.07246	50
60	0.08724	0.08139	0.07983	0.07583	0.07383	0.07123	60

n	**NE**	O_2	O_1	S_0	S_1	**SQ**	n
	These data are also suggested as an approximation for the Iowa type survivor dispersions shown below						
	O_3		L_0	L_1, R_1	L_2, L_3, R_2, R_3, S_2	$L_4, L_5, R_4, R_5, S_3, S_4, S_5, S_6$	

$i = 8\%$

	Uniform series end-of-period amount a equivalent to present sum p (capital recovery factor) for a group property with survivors following the ___ dispersion. $$xx(a/p)_n^i = xx(a/f)_n^i + i$$						
n	Negative exponential NE	Iowa Type O_2	Straight-line or Iowa Type O_1	Iowa Type S_0	Iowa Type S_1	Square, rectangular, or no dispersion SQ	n
1	1.11949	1.10466	1.09350	1.08897	1.08604	1.08000	1
2	0.59974	0.58302	0.57442	0.56980	0.56684	0.56077	2
3	0.42650	0.40949	0.40182	0.39711	0.39412	0.38803	3
4	0.33987	0.32297	0.31582	0.31105	0.30802	0.30192	4
5	0.28790	0.27125	0.26447	0.25962	0.25656	0.25046	5
6	0.25325	0.23691	0.23042	0.22550	0.22242	0.21632	6
7	0.22850	0.21249	0.20625	0.20127	0.19818	0.19207	7
8	0.20994	0.19428	0.18826	0.18322	0.18011	0.17401	8
9	0.19550	0.18019	0.17438	0.16929	0.16616	0.16008	9
10	0.18395	0.16898	0.16336	0.15823	0.15509	0.14903	10
11	0.17450	0.15987	0.15443	0.14926	0.14611	0.14008	11
12	0.16662	0.15233	0.14706	0.14185	0.13869	0.13269	12
13	0.15996	0.14599	0.14089	0.13565	0.13248	0.12652	13
14	0.15425	0.14059	0.13565	0.13038	0.12722	0.12130	14
15	0.14930	0.13595	0.13116	0.12587	0.12271	0.11683	15
16	0.14497	0.13191	0.12728	0.12196	0.11880	0.11298	16
17	0.14115	0.12838	0.12389	0.11856	0.11540	0.10963	17
18	0.13775	0.12525	0.12091	0.11556	0.11242	0.10670	18
19	0.13471	0.12248	0.11827	0.11292	0.10978	0.10413	19
20	0.13197	0.12001	0.11592	0.11057	0.10745	0.10185	20
21	0.12950	0.11778	0.11382	0.10847	0.10536	0.09983	21
22	0.12725	0.11578	0.11194	0.10658	0.10349	0.09803	22
23	0.12520	0.11396	0.11023	0.10488	0.10181	0.09642	23
24	0.12331	0.11230	0.10869	0.10335	0.10029	0.09498	24
25	0.12158	0.11079	0.10729	0.10195	0.09891	0.09368	25
26	0.11998	0.10940	0.10600	0.10068	0.09766	0.09251	26
27	0.11850	0.10813	0.10483	0.09952	0.09653	0.09145	27
28	0.11712	0.10695	0.10375	0.09846	0.09549	0.09049	28
29	0.11584	0.10587	0.10275	0.09748	0.09453	0.08962	29
30	0.11465	0.10486	0.10184	0.09658	0.09366	0.08883	30
35	0.10970	0.10077	0.09813	0.09300	0.09021	0.08580	35
40	0.10599	0.09780	0.09547	0.09049	0.08784	0.08386	40
45	0.10310	0.09556	0.09348	0.08866	0.08615	0.08259	45
50	0.10079	0.09382	0.09194	0.08728	0.08492	0.08174	50
60	0.09732	0.09129	0.08971	0.08538	0.08329	0.08080	60

	NE	O_2	O_1	S_0	S_1	SQ	
n	These data are also suggested as an approximation for the Iowa type survivor dispersions shown below						n
	O_3		L_0	L_1, R_1	L_2, L_3, R_2, R_3, S_2	$L_4, L_5, R_4, R_5, S_3, S_4, S_5, S_6$	

$i = 10 \%$

n	Negative exponential NE	Iowa Type O_2	Straight-line or Iowa Type O_1	Iowa Type S_0	Iowa Type S_1	Square, rectangular, or no dispersion SQ	n
1	1.14921	1.12966	1.11692	1.11123	1.10756	1.10000	1
2	0.62460	0.60343	0.59333	0.58751	0.58379	0.57619	2
3	0.44974	0.42855	0.41945	0.41351	0.40973	0.40211	3
4	0.36230	0.34147	0.33298	0.32692	0.32310	0.31547	4
5	0.30984	0.28949	0.28145	0.27528	0.27143	0.26380	5
6	0.27487	0.25505	0.24737	0.24111	0.23723	0.22961	6
7	0.24989	0.23061	0.22326	0.21692	0.21301	0.20541	7
8	0.23115	0.21241	0.20536	0.19894	0.19502	0.18744	8
9	0.21658	0.19836	0.19159	0.18511	0.18117	0.17364	9
10	0.20492	0.18721	0.18071	0.17418	0.17023	0.16275	10
11	0.19538	0.17817	0.17192	0.16534	0.16139	0.15396	11
12	0.18743	0.17069	0.16469	0.15807	0.15412	0.14676	12
13	0.18071	0.16442	0.15865	0.15200	0.14806	0.14078	13
14	0.17494	0.15910	0.15355	0.14688	0.14294	0.13575	14
15	0.16995	0.15452	0.14919	0.14250	0.13858	0.13147	15
16	0.16558	0.15056	0.14542	0.13873	0.13482	0.12782	16
17	0.16172	0.14709	0.14215	0.13545	0.13157	0.12466	17
18	0.15829	0.14403	0.13928	0.13259	0.12873	0.12193	18
19	0.15522	0.14132	0.13675	0.13006	0.12623	0.11955	19
20	0.15246	0.13891	0.13450	0.12783	0.12403	0.11746	20
21	0.14996	0.13674	0.13249	0.12584	0.12207	0.11562	21
22	0.14769	0.13479	0.13069	0.12407	0.12033	0.11401	22
23	0.14562	0.13302	0.12907	0.12247	0.11877	0.11257	23
24	0.14372	0.13142	0.12760	0.12103	0.11737	0.11130	24
25	0.14197	0.12995	0.12627	0.11973	0.11611	0.11017	25
26	0.14035	0.12861	0.12505	0.11855	0.11497	0.10916	26
27	0.13886	0.12738	0.12394	0.11748	0.11393	0.10826	27
28	0.13747	0.12625	0.12292	0.11649	0.11299	0.10745	28
29	0.13618	0.12520	0.12198	0.11560	0.11214	0.10673	29
30	0.13497	0.12423	0.12111	0.11477	0.11135	0.10608	30
35	0.12998	0.12030	0.11761	0.11150	0.10831	0.10369	35
40	0.12623	0.11745	0.11509	0.10924	0.10627	0.10226	40
45	0.12332	0.11531	0.11319	0.10760	0.10485	0.10139	45
50	0.12098	0.11364	0.11172	0.10638	0.10383	0.10086	50
60	0.11749	0.11122	0.10958	0.10469	0.10251	0.10033	60

n	NE	O_2	O_1	S_0	S_1	SQ	n
	These data are also suggested as an approximation for the Iowa type survivor dispersions shown below						
	O_3		L_0	L_1, R_1	L_2, L_3, R_2, R_3, S_2	$L_4, L_5, R_4, R_5, S_3, S_4, S_5, S_6$	

$i = 12\%$

	Uniform series end-of-period amount a equivalent to present sum p (capital recovery factor) for a group property with survivors following the _____ dispersion. $$xx(a/p)_n^i = xx(a/f)_n^i + i$$						
n	Negative exponential NE	Iowa Type O_2	Straight-line or Iowa Type O_1	Iowa Type S_0	Iowa Type S_1	Square, rectangular, or no dispersion SQ	n
1	1.17887	1.15466	1.14035	1.13349	1.12908	1.12000	1
2	0.64943	0.62393	0.61236	0.60532	0.60082	0.59170	2
3	0.47296	0.44775	0.43727	0.43006	0.42550	0.41635	3
4	0.38472	0.36015	0.35037	0.34301	0.33840	0.32923	4
5	0.33177	0.30795	0.29872	0.29122	0.28656	0.27741	5
6	0.29648	0.27343	0.26466	0.25704	0.25235	0.24323	6
7	0.27127	0.24898	0.24063	0.23291	0.22819	0.21912	7
8	0.25236	0.23081	0.22285	0.21505	0.21031	0.20130	8
9	0.23765	0.21682	0.20923	0.20135	0.19661	0.18768	9
10	0.22589	0.20573	0.19850	0.19056	0.18582	0.17698	10
11	0.21626	0.19676	0.18986	0.18188	0.17714	0.16842	11
12	0.20824	0.18936	0.18277	0.17476	0.17004	0.16144	12
13	0.20145	0.18317	0.17688	0.16885	0.16415	0.15568	13
14	0.19563	0.17791	0.17191	0.16387	0.15920	0.15087	14
15	0.19059	0.17341	0.16767	0.15964	0.15500	0.14682	15
16	0.18618	0.16951	0.16403	0.15601	0.15141	0.14339	16
17	0.18229	0.16611	0.16086	0.15286	0.14831	0.14046	17
18	0.17883	0.16312	0.15810	0.15012	0.14561	0.13794	18
19	0.17573	0.16047	0.15566	0.14772	0.14326	0.13576	19
20	0.17294	0.15811	0.15350	0.14560	0.14119	0.13388	20
21	0.17042	0.15599	0.15157	0.14372	0.13937	0.13224	21
22	0.16813	0.15409	0.14985	0.14205	0.13775	0.13081	22
23	0.16604	0.15237	0.14829	0.14055	0.13631	0.12956	23
24	0.16412	0.15081	0.14688	0.13920	0.13503	0.12846	24
25	0.16235	0.14938	0.14561	0.13799	0.13387	0.12750	25
26	0.16073	0.14808	0.14444	0.13689	0.13284	0.12665	26
27	0.15922	0.14689	0.14338	0.13589	0.13190	0.12590	27
28	0.15782	0.14579	0.14240	0.13498	0.13105	0.12524	28
29	0.15651	0.14478	0.14150	0.13414	0.13029	0.12466	29
30	0.15530	0.14384	0.14066	0.13338	0.12959	0.12414	30
35	0.15025	0.14003	0.13730	0.13038	0.12691	0.12232	35
40	0.14647	0.13728	0.13487	0.12832	0.12514	0.12130	40
45	0.14353	0.13521	0.13304	0.12683	0.12393	0.12074	45
50	0.14118	0.13360	0.13161	0.12572	0.12307	0.12042	50
60	0.13765	0.13126	0.12952	0.12421	0.12199	0.12013	60

	NE	O_2	O_1	S_0	S_1	SQ	
n	These data are also suggested as an approximation for the Iowa type survivor dispersions shown below						n
	O_3		L_0	L_1, R_1	L_2, L_3, R_2, R_3, S_2	$L_4, L_5, R_4, R_5, S_3, S_4, S_5, S_6$	

$i = 15\%$

Uniform series end-of-period amount a equivalent to present sum p (capital recovery factor) for a group property with survivors following the ___ dispersion.

$$xx(a/p)_n^i = xx(a/f)_n^i + i$$

n	Negative exponential NE	Iowa Type O_2	Straight-line or Iowa Type O_1	Iowa Type S_0	Iowa Type S_1	Square, rectangular, or no dispersion SQ	n
1	1.22325	1.19218	1.17554	1.16691	1.16137	1.15000	1
2	0.68663	0.65483	0.64110	0.63220	0.62654	0.61512	2
3	0.50775	0.47678	0.46433	0.45517	0.44943	0.43798	3
4	0.41831	0.38847	0.37689	0.36752	0.36170	0.35027	4
5	0.36465	0.33600	0.32513	0.31558	0.30971	0.29832	5
6	0.32888	0.30139	0.29115	0.28146	0.27555	0.26424	6
7	0.30332	0.27696	0.26731	0.25749	0.25156	0.24036	7
8	0.28416	0.25885	0.24975	0.23984	0.23391	0.22285	8
9	0.26925	0.24495	0.23636	0.22638	0.22047	0.20957	9
10	0.25733	0.23398	0.22587	0.21585	0.20995	0.19925	10
11	0.24757	0.22512	0.21746	0.20741	0.20156	0.19107	11
12	0.23944	0.21783	0.21059	0.20055	0.19474	0.18448	12
13	0.23256	0.21175	0.20490	0.19487	0.18913	0.17911	13
14	0.22666	0.20660	0.20012	0.19013	0.18444	0.17469	14
15	0.22155	0.20219	0.19605	0.18611	0.18050	0.17102	15
16	0.21708	0.19839	0.19256	0.18268	0.17715	0.16795	16
17	0.21313	0.19508	0.18954	0.17973	0.17429	0.16537	17
18	0.20963	0.19217	0.18690	0.17717	0.17182	0.16319	18
19	0.20649	0.18960	0.18458	0.17494	0.16967	0.16134	19
20	0.20366	0.18731	0.18253	0.17298	0.16781	0.15976	20
21	0.20111	0.18526	0.18070	0.17124	0.16617	0.15842	21
22	0.19878	0.18342	0.17905	0.16970	0.16473	0.15727	22
23	0.19666	0.18176	0.17758	0.16833	0.16346	0.15628	23
24	0.19472	0.18026	0.17624	0.16710	0.16233	0.15543	24
25	0.19293	0.17888	0.17502	0.16600	0.16132	0.15470	25
26	0.19128	0.17763	0.17391	0.16500	0.16042	0.15407	26
27	0.18975	0.17648	0.17290	0.16409	0.15962	0.15353	27
28	0.18833	0.17542	0.17196	0.16327	0.15889	0.15306	28
29	0.18701	0.17445	0.17110	0.16252	0.15824	0.15265	29
30	0.18578	0.17354	0.17030	0.16184	0.15765	0.15230	30
35	0.18066	0.16989	0.16708	0.15915	0.15541	0.15113	35
40	0.17683	0.16724	0.16473	0.15732	0.15397	0.15056	40
45	0.17385	0.16525	0.16295	0.15600	0.15300	0.15028	45
50	0.17147	0.16370	0.16156	0.15502	0.15232	0.15014	50
60	0.16789	0.16145	0.15951	0.15368	0.15148	0.15003	60

n	NE	O_2	O_1	S_0	S_1	SQ	n
	These data are also suggested as an approximation for the Iowa type survivor dispersions shown below						
	O_3		L_0	L_1, R_1	L_2, L_3, R_2, R_3, S_2	$L_4, L_5, R_4, R_5, S_3, S_4, S_5, S_6$	

i = 20 %

n							n
	Uniform series end-of-period amount a equivalent to present sum p (capital recovery factor) for a group property with survivors following the ___ dispersion. $$_{xx}(a/p)_n^i = {_{xx}}(a/f)_n^i + i$$						
	Negative exponential NE	Iowa Type O_2	Straight-line or Iowa Type O_1	Iowa Type S_0	Iowa Type S_1	Square, rectangular, or no dispersion SQ	
1	1.29696	1.25471	1.23425	1.22262	1.21518	1.20000	1
2	0.74848	0.70668	0.68950	0.67742	0.66981	0.65455	2
3	0.56565	0.52573	0.51020	0.49773	0.48998	0.47473	3
4	0.47424	0.43638	0.42208	0.40930	0.40146	0.38629	4
5	0.41939	0.38353	0.37030	0.35728	0.34939	0.33438	5
6	0.38283	0.34885	0.33659	0.32339	0.31549	0.30071	6
7	0.35671	0.32449	0.31312	0.29980	0.29192	0.27742	7
8	0.33712	0.30652	0.29597	0.28260	0.27476	0.26061	8
9	0.32188	0.29279	0.28299	0.26960	0.26184	0.24808	9
10	0.30970	0.28199	0.27288	0.25952	0.25186	0.23852	10
11	0.29972	0.27331	0.26483	0.25153	0.24399	0.23110	11
12	0.29141	0.26619	0.25828	0.24508	0.23767	0.22526	12
13	0.28438	0.26027	0.25288	0.23979	0.23253	0.22062	13
14	0.27835	0.25528	0.24835	0.23541	0.22830	0.21689	14
15	0.27313	0.25102	0.24452	0.23172	0.22478	0.21388	15
16	0.26856	0.24735	0.24123	0.22860	0.22182	0.21144	16
17	0.26453	0.24416	0.23838	0.22593	0.21933	0.20944	17
18	0.26094	0.24136	0.23589	0.22363	0.21720	0.20781	18
19	0.25773	0.23889	0.23370	0.22163	0.21537	0.20646	19
20	0.25485	0.23670	0.23176	0.21988	0.21379	0.20536	20
21	0.25224	0.23474	0.23002	0.21835	0.21243	0.20444	21
22	0.24986	0.23298	0.22847	0.21699	0.21124	0.20369	22
23	0.24769	0.23139	0.22707	0.21578	0.21019	0.20307	23
24	0.24571	0.22995	0.22580	0.21470	0.20927	0.20255	24
25	0.24388	0.22864	0.22464	0.21373	0.20846	0.20212	25
26	0.24219	0.22744	0.22358	0.21286	0.20775	0.20176	26
27	0.24063	0.22634	0.22261	0.21208	0.20711	0.20147	27
28	0.23918	0.22533	0.22171	0.21136	0.20654	0.20122	28
29	0.23783	0.22439	0.22089	0.21071	0.20603	0.20102	29
30	0.23657	0.22353	0.22012	0.21012	0.20557	0.20085	30
35	0.23134	0.22003	0.21700	0.20780	0.20386	0.20034	35
40	0.22742	0.21750	0.21472	0.20623	0.20279	0.20014	40
45	0.22438	0.21559	0.21298	0.20510	0.20208	0.20005	45
50	0.22194	0.21410	0.21161	0.20426	0.20160	0.20002	50
60	0.21828	0.21195	0.20958	0.20312	0.20100	0.20000	60

n	NE	O_2	O_1	S_0	S_1	SQ	n
	These data are also suggested as an approximation for the Iowa type survivor dispersions shown below						
	O_3		L_0	L_1, R_1	L_2, L_3, R_2, R_3, S_2	$L_4, L_5, R_4, R_5, S_3, S_4, S_5, S_6$	

	Uniform series end-of-period amount a equivalent to present sum p (capital recovery factor) for a group property with survivors following the ____ dispersion. $$xx(a/p)_n^i = xx(a/f)_n^i + i$$						
n	Negative exponential NE	Iowa Type O_2	Straight-line or Iowa Type O_1	Iowa Type S_0	Iowa Type S_1	Square, rectangular, or no dispersion SQ	n
1	1.37035	1.31723	1.29303	1.27836	1.26901	1.25000	1
2	0.81018	0.75891	0.73846	0.72313	0.71353	0.69444	2
3	0.62345	0.57528	0.55693	0.54107	0.53130	0.51230	3
4	0.53009	0.48501	0.46834	0.45209	0.44223	0.42344	4
5	0.47407	0.43187	0.41668	0.40016	0.39028	0.37185	5
6	0.43673	0.39716	0.38331	0.36663	0.35680	0.33882	6
7	0.41005	0.37289	0.36024	0.34350	0.33376	0.31634	7
8	0.39004	0.35506	0.34350	0.32679	0.31718	0.30040	8
9	0.37448	0.34149	0.33090	0.31428	0.30485	0.28876	9
10	0.36204	0.33085	0.32113	0.30465	0.29543	0.28007	10
11	0.35185	0.32233	0.31336	0.29708	0.28809	0.27349	11
12	0.34336	0.31536	0.30707	0.29101	0.28227	0.26845	12
13	0.33618	0.30957	0.30187	0.28608	0.27759	0.26454	13
14	0.33003	0.30470	0.29753	0.28200	0.27377	0.26150	14
15	0.32469	0.30055	0.29384	0.27860	0.27064	0.25912	15
16	0.32002	0.29698	0.29068	0.27573	0.26803	0.25724	16
17	0.31590	0.29388	0.28793	0.27329	0.26585	0.25576	17
18	0.31224	0.29116	0.28553	0.27119	0.26400	0.25459	18
19	0.30897	0.28876	0.28342	0.26937	0.26243	0.25366	19
20	0.30602	0.28663	0.28154	0.26778	0.26109	0.25292	20
21	0.30335	0.28473	0.27986	0.26639	0.25994	0.25233	21
22	0.30093	0.28302	0.27835	0.26517	0.25894	0.25186	22
23	0.29871	0.28148	0.27698	0.26408	0.25807	0.25148	23
24	0.29668	0.28008	0.27574	0.26311	0.25731	0.25119	24
25	0.29481	0.27880	0.27461	0.26224	0.25665	0.25095	25
26	0.29309	0.27764	0.27358	0.26145	0.25606	0.25076	26
27	0.29149	0.27657	0.27262	0.26075	0.25554	0.25061	27
28	0.29001	0.27559	0.27175	0.26011	0.25508	0.25048	28
29	0.28863	0.27468	0.27093	0.25953	0.25467	0.25039	29
30	0.28735	0.27384	0.27018	0.25900	0.25430	0.25031	30
35	0.28201	0.27044	0.26710	0.25693	0.25294	0.25010	35
40	0.27801	0.26798	0.26484	0.25552	0.25211	0.25003	40
45	0.27490	0.26612	0.26310	0.25452	0.25156	0.25001	45
50	0.27241	0.26469	0.26173	0.25377	0.25119	0.25000	50
60	0.26867	0.26262	0.25970	0.25276	0.25074	0.25000	60

	NE	O_2	O_1	S_0	S_1	SQ	
n	These data are also suggested as an approximation for the Iowa type survivor dispersions shown below						n
	O_3		L_0	L_1, R_1	$L_2, L_3, R_2,$ R_3, S_2	$L_4, L_5, R_4, R_5,$ S_3, S_4, S_5, S_6	

$i = 30\%$

Uniform series end-of-period amount a equivalent to present sum p (capital recovery factor) for a group property with survivors following the ___ dispersion.

$$xx(a/p)_n^i = xx(a/f)_n^i + i$$

n	Negative exponential NE	Iowa Type O_2	Straight-line or Iowa Type O_1	Iowa Type S_0	Iowa Type S_1	Square, rectangular, or no dispersion SQ	n
1	1.44345	1.37975	1.35188	1.33412	1.32284	1.30000	1
2	0.87172	0.81147	0.78791	0.76928	0.75767	0.73478	2
3	0.68115	0.62533	0.60440	0.58511	0.57332	0.55063	3
4	0.58586	0.53423	0.51549	0.49575	0.48389	0.46163	4
5	0.52869	0.48085	0.46404	0.44404	0.43223	0.41058	5
6	0.49057	0.44613	0.43103	0.41094	0.39926	0.37839	6
7	0.46335	0.42194	0.40835	0.38831	0.37683	0.35687	7
8	0.44293	0.40424	0.39197	0.37209	0.36087	0.34192	8
9	0.42705	0.39081	0.37968	0.36004	0.34914	0.33124	9
10	0.41434	0.38031	0.37017	0.35084	0.34027	0.32346	10
11	0.40395	0.37191	0.36263	0.34365	0.33344	0.31773	11
12	0.39529	0.36506	0.35651	0.33792	0.32808	0.31345	12
13	0.38796	0.35937	0.35147	0.33328	0.32381	0.31024	13
14	0.38167	0.35460	0.34724	0.32947	0.32037	0.30782	14
15	0.37623	0.35053	0.34364	0.32630	0.31756	0.30598	15
16	0.37147	0.34703	0.34055	0.32363	0.31525	0.30458	16
17	0.36726	0.34399	0.33787	0.32136	0.31333	0.30351	17
18	0.36352	0.34132	0.33552	0.31942	0.31172	0.30269	18
19	0.36018	0.33897	0.33344	0.31774	0.31036	0.30207	19
20	0.35717	0.33689	0.33160	0.31628	0.30920	0.30159	20
21	0.35445	0.33502	0.32994	0.31499	0.30821	0.30122	21
22	0.35197	0.33335	0.32845	0.31386	0.30736	0.30094	22
23	0.34972	0.33183	0.32710	0.31286	0.30662	0.30072	23
24	0.34764	0.33046	0.32588	0.31197	0.30598	0.30055	24
25	0.34574	0.32921	0.32476	0.31117	0.30542	0.30043	25
26	0.34398	0.32807	0.32373	0.31046	0.30493	0.30033	26
27	0.34235	0.32702	0.32278	0.30981	0.30449	0.30025	27
28	0.34084	0.32606	0.32191	0.30922	0.30411	0.30019	28
29	0.33943	0.32517	0.32110	0.30869	0.30377	0.30015	29
30	0.33811	0.32434	0.32035	0.30820	0.30346	0.30011	30
35	0.33267	0.32101	0.31728	0.30631	0.30235	0.30003	35
40	0.32859	0.31860	0.31501	0.30503	0.30167	0.30001	40
45	0.32541	0.31679	0.31327	0.30411	0.30123	0.30000	45
50	0.32287	0.31539	0.31189	0.30343	0.30094	0.30000	50
60	0.31906	0.31341	0.30984	0.30251	0.30058	0.30000	60

n	NE	O_2	O_1	S_0	S_1	SQ	n
	These data are also suggested as an approximation for the Iowa type survivor dispersions shown below						
	O_3		L_0	L_1, R_1	$L_2, L_3, R_2,$ R_3, S_2	$L_4, L_5, R_4, R_5,$ S_3, S_4, S_5, S_6	

I

Interest Tables

$i = 0.25\%$

n	Uniform series end-of-period amount a equivalent to			Present sum p equivalent to			n
	future sum, f	present sum, p	gradient series, g	future sum, f	uniform series, a	gradient series, g	
	a/f	a/p	a/g	p/f	p/a	p/g	
1	1.00000	1.00250	0.000	0.9975	0.998	0.000	1
2	0.49938	0.50188	0.499	0.9950	1.993	0.995	2
3	0.33250	0.33500	0.998	0.9925	2.985	2.980	3
4	0.24906	0.25156	1.497	0.9901	3.975	5.950	4
5	0.19900	0.20150	1.995	0.9876	4.963	9.901	5
6	0.16563	0.16813	2.493	0.9851	5.948	14.826	6
7	0.14179	0.14429	2.990	0.9827	6.931	20.722	7
8	0.12391	0.12641	3.487	0.9802	7.911	27.584	8
9	0.11000	0.11250	3.983	0.9778	8.889	35.406	9
10	0.09888	0.10138	4.479	0.9753	9.864	44.184	10
11	0.08978	0.09228	4.975	0.9729	10.837	53.913	11
12	0.08219	0.08469	5.470	0.9705	11.807	64.589	12
13	0.07578	0.07828	5.965	0.9681	12.775	76.205	13
14	0.07028	0.07278	6.459	0.9656	13.741	88.759	14
15	0.06551	0.06801	6.953	0.9632	14.704	102.244	15
16	0.06134	0.06384	7.447	0.9608	15.665	116.657	16
17	0.05766	0.06016	7.940	0.9584	16.623	131.992	17
18	0.05438	0.05688	8.433	0.9561	17.580	148.245	18
19	0.05146	0.05396	8.925	0.9537	18.533	165.411	19
20	0.04882	0.05132	9.417	0.9513	19.484	183.485	20
21	0.04644	0.04894	9.908	0.9489	20.433	202.463	21
22	0.04427	0.04677	10.400	0.9466	21.380	222.341	22
23	0.04229	0.04479	10.890	0.9442	22.324	243.113	23
24	0.04048	0.04298	11.380	0.9418	23.266	264.775	24
25	0.03881	0.04131	11.870	0.9395	24.205	287.323	25
26	0.03727	0.03977	12.360	0.9371	25.143	310.752	26
27	0.03585	0.03835	12.849	0.9348	26.077	335.057	27
28	0.03452	0.03702	13.337	0.9325	27.010	360.233	28
29	0.03329	0.03579	13.825	0.9301	27.940	386.278	29
30	0.03214	0.03464	14.313	0.9278	28.868	413.185	30
31	0.03106	0.03356	14.800	0.9255	29.793	440.950	31
32	0.03006	0.03256	15.287	0.9232	30.717	469.570	32
33	0.02911	0.03161	15.774	0.9209	31.638	499.039	33
34	0.02822	0.03072	16.260	0.9186	32.556	529.353	34
35	0.02738	0.02988	16.745	0.9163	33.472	560.508	35
36	0.02658	0.02908	17.231	0.9140	34.386	592.499	36
48	0.01963	0.02213	23.021	0.8871	45.179	1040.055	48
60	0.01547	0.01797	28.751	0.8609	55.652	1600.085	60
120	0.00716	0.00966	56.508	0.7411	103.562	5852.112	120
180	0.00441	0.00691	82.781	0.6380	144.805	11987.173	180
240	0.00305	0.00555	107.586	0.5492	180.311	19398.985	240
300	0.00224	0.00474	130.946	0.4728	210.876	27613.517	300
360	0.00172	0.00422	152.890	0.4070	237.189	36263.930	360
∞	0.00000	0.00250	400.000	0.0000	400.000	160000.000	∞

n	Uniform series worth of a future sum or sinking fund factor	Uniform series worth of a present sum or capital recovery factor	Uniform series worth of a gradient series	Present worth of a future sum	Present worth of a uniform series	Present worth of a gradient series	n
	$\dfrac{i}{(1+i)^n-1}$	$\dfrac{i(1+i)^n}{(1+i)^n-1}$	$\dfrac{1}{i}-\dfrac{n}{(1+i)^n-1}$	$\dfrac{1}{(1+i)^n}$	$\dfrac{(1+i)^n-1}{i(1+i)^n}$	$\dfrac{1}{i}\left[\dfrac{(1+i)^n-1}{i(1+i)^n}-\dfrac{n}{(1+i)^n}\right]$	n
	Equation 4-4	Equation 4-6	Equation 4-8a	Equation 4-2	Equation 4-5	Equation 4-9a	

$i = 0.25\%$

n	Future sum f equivalent to			n
	present sum, p	uniform series, a	gradient series, g	
	f/p	f/a	f/g	
1	1.002	1.000	0.00	1
2	1.005	2.002	1.00	2
3	1.008	3.008	3.00	3
4	1.010	4.015	6.01	4
5	1.013	5.025	10.03	5
6	1.015	6.038	15.05	6
7	1.018	7.053	21.09	7
8	1.020	8.070	28.14	8
9	1.023	9.091	36.21	9
10	1.025	10.113	45.30	10
11	1.028	11.139	55.41	11
12	1.030	12.166	66.55	12
13	1.033	13.197	78.72	13
14	1.036	14.230	91.92	14
15	1.038	15.265	106.15	15
16	1.041	16.304	121.41	16
17	1.043	17.344	137.71	17
18	1.046	18.388	155.06	18
19	1.049	19.434	173.45	19
20	1.051	20.482	192.88	20
21	1.054	21.533	213.36	21
22	1.056	22.587	234.90	22
23	1.059	23.644	257.48	23
24	1.062	24.703	281.13	24
25	1.064	25.765	305.83	25
26	1.067	26.829	331.59	26
27	1.070	27.896	358.42	27
28	1.072	28.966	386.32	28
29	1.075	30.038	415.29	29
30	1.078	31.113	445.32	30
31	1.080	32.191	476.44	31
32	1.083	33.272	508.63	32
33	1.086	34.355	541.90	33
34	1.089	35.441	576.25	34
35	1.091	36.529	611.69	35
36	1.094	37.621	648.22	36
48	1.127	50.931	1172.48	48
60	1.162	64.647	1858.69	60
120	1.349	139.741	7896.57	120
180	1.567	226.973	18789.08	180
240	1.821	328.302	35320.80	240
300	2.115	446.008	58403.13	300
360	2.457	582.737	89094.75	360
∞	∞	∞	∞	∞

If the effective annual rate of interest is as shown above, and if interest is compounded continuously:

(1) An end-of-year amount of \$1 is equivalent to \$ _____ flowing uniformly and continuously throughout the year.

$$a'/a = c'/c = f'/f = g'/g = p'/p = \frac{\ln(1+i_a)}{i_a}$$

$$= 0.9988$$

(2) An end-of-year amount of \$ _____ is equivalent to \$1 flowing uniformly and continuously throughout the year.

$$a/a' = c/c' = f/f' = g/g' = p/p' = \frac{i_a}{\ln(1+i_a)}$$

$$= 1.001$$

n	Future worth of a present sum	Future worth of a uniform series	Future worth of a gradient series	n
	$(1+i)^n$	$\dfrac{(1+i)^n-1}{i}$	$\dfrac{1}{i}\left[\dfrac{(1+i)^n-1}{i} - n\right]$	
	Equation 4-1	Equation 4-3	Equation 4-7a	

$i = 0.5\%$

n	Uniform series end-of-period amount a equivalent to			Present sum p equivalent to			n
	future sum, f	present sum, p	gradient series, g	future sum, f	uniform series, a	gradient series, g	
	a/f	a/p	a/g	p/f	p/a	p/g	
1	1.00000	1.00500	0.000	0.9950	0.995	0.000	1
2	0.49875	0.50375	0.499	0.9901	1.985	0.990	2
3	0.33167	0.33667	0.997	0.9851	2.970	2.960	3
4	0.24813	0.25313	1.494	0.9802	3.950	5.901	4
5	0.19801	0.20301	1.990	0.9754	4.926	9.803	5
6	0.16460	0.16960	2.485	0.9705	5.896	14.655	6
7	0.14073	0.14573	2.980	0.9657	6.862	20.449	7
8	0.12283	0.12783	3.474	0.9609	7.823	27.176	8
9	0.10891	0.11391	3.967	0.9561	8.779	34.824	9
10	0.09777	0.10277	4.459	0.9513	9.730	43.386	10
11	0.08866	0.09366	4.950	0.9466	10.677	52.853	11
12	0.08107	0.08607	5.441	0.9419	11.619	63.214	12
13	0.07464	0.07964	5.930	0.9372	12.556	74.460	13
14	0.06914	0.07414	6.419	0.9326	13.489	86.583	14
15	0.06436	0.06936	6.907	0.9279	14.417	99.574	15
16	0.06019	0.06519	7.394	0.9233	15.340	113.424	16
17	0.05651	0.06151	7.880	0.9187	16.259	128.123	17
18	0.05323	0.05823	8.366	0.9141	17.173	143.663	18
19	0.05030	0.05530	8.850	0.9096	18.082	160.036	19
20	0.04767	0.05267	9.334	0.9051	18.987	177.232	20
21	0.04528	0.05028	9.817	0.9006	19.888	195.243	21
22	0.04311	0.04811	10.299	0.8961	20.784	214.061	22
23	0.04113	0.04613	10.781	0.8916	21.676	233.677	23
24	0.03932	0.04432	11.261	0.8872	22.563	254.082	24
25	0.03765	0.04265	11.741	0.8828	23.446	275.269	25
26	0.03611	0.04111	12.220	0.8784	24.324	297.228	26
27	0.03469	0.03969	12.698	0.8740	25.198	319.952	27
28	0.03336	0.03836	13.175	0.8697	26.068	343.433	28
29	0.03213	0.03713	13.651	0.8653	26.933	367.663	29
30	0.03098	0.03598	14.126	0.8610	27.794	392.632	30
31	0.02990	0.03490	14.601	0.8567	28.651	418.335	31
32	0.02889	0.03389	15.075	0.8525	29.503	444.762	32
33	0.02795	0.03295	15.548	0.8482	30.352	471.906	33
34	0.02706	0.03206	16.020	0.8440	31.196	499.758	34
35	0.02622	0.03122	16.492	0.8398	32.035	528.312	35
36	0.02542	0.03042	16.962	0.8356	32.871	557.560	36
48	0.01849	0.02349	22.544	0.7871	42.580	959.919	48
60	0.01433	0.01933	28.006	0.7414	51.726	1448.646	60
120	0.00610	0.01110	53.551	0.5496	90.073	4823.505	120
180	0.00344	0.00844	76.212	0.4075	118.504	9031.336	180
240	0.00216	0.00716	96.113	0.3021	139.581	13415.540	240
300	0.00144	0.00644	113.419	0.2240	155.207	17603.432	300
360	0.00100	0.00600	128.324	0.1660	166.792	21403.304	360
∞	0.00000	0.00500	200.000	0.0000	200.000	40000.000	∞

n	Uniform series worth of a future sum or sinking fund factor	Uniform series worth of a present sum or capital recovery factor	Uniform series worth of a gradient series	Present worth of a future sum	Present worth of a uniform series	Present worth of a gradient series	n
	$\dfrac{i}{(1+i)^n - 1}$	$\dfrac{i(1+i)^n}{(1+i)^n - 1}$	$\dfrac{1}{i} - \dfrac{n}{(1+i)^n - 1}$	$\dfrac{1}{(1+i)^n}$	$\dfrac{(1+i)^n - 1}{i(1+i)^n}$	$\dfrac{1}{i}\left[\dfrac{(1+i)^n - 1}{i(1+i)^n} - \dfrac{n}{(1+i)^n}\right]$	
	Equation 4-4	Equation 4-6	Equation 4-8a	Equation 4-2	Equation 4-5	Equation 4-9a	

$i = 0.5\%$

n	Future sum f equivalent to			n
	present sum, p	uniform series, a	gradient series, g	
	f/p	f/a	f/g	
1	1.005	1.000	0.00	1
2	1.010	2.005	1.00	2
3	1.015	3.015	3.00	3
4	1.020	4.030	6.02	4
5	1.025	5.050	10.05	5
6	1.030	6.076	15.10	6
7	1.036	7.106	21.18	7
8	1.041	8.141	28.28	8
9	1.046	9.182	36.42	9
10	1.051	10.228	45.61	10
11	1.056	11.279	55.83	11
12	1.062	12.336	67.11	12
13	1.067	13.397	79.45	13
14	1.072	14.464	92.85	14
15	1.078	15.537	107.31	15
16	1.083	16.614	122.85	16
17	1.088	17.697	139.46	17
18	1.094	18.786	157.16	18
19	1.099	19.880	175.94	19
20	1.105	20.979	195.82	20
21	1.110	22.084	216.80	21
22	1.116	23.194	238.89	22
23	1.122	24.310	262.08	23
24	1.127	25.432	286.39	24
25	1.133	26.559	311.82	25
26	1.138	27.692	338.38	26
27	1.144	28.830	366.07	27
28	1.150	29.975	394.90	28
29	1.156	31.124	424.88	29
30	1.161	32.280	456.00	30
31	1.167	33.441	488.28	31
32	1.173	34.609	521.72	32
33	1.179	35.782	556.33	33
34	1.185	36.961	592.12	34
35	1.191	38.145	629.08	35
36	1.197	39.336	667.22	36
48	1.270	54.098	1219.57	48
60	1.349	69.770	1954.01	60
120	1.819	163.879	8775.87	120
180	2.454	290.819	22163.74	180
240	3.310	462.041	44408.18	240
300	4.465	692.994	78598.79	300
360	6.023	1004.515	128903.01	360
∞	∞	∞	∞	∞

If the effective annual rate of interest is as shown above, and if interest is compounded continuously:

(1) An end-of-year amount of \$1 is equivalent to \$ _____ flowing uniformly and continuously throughout the year.

$$a'/a = c'/c = f'/f = g'/g = p'/p = \frac{\ln(1+i_a)}{i_a}$$

$$= 0.9975$$

(2) An end-of-year amount of \$ _____ is equivalent to \$1 flowing uniformly and continuously throughout the year.

$$a/a' = c/c' = f/f' = g/g' = p/p' = \frac{i_a}{\ln(1+i_a)}$$

$$= 1.002$$

n	Future worth of a present sum	Future worth of a uniform series	Future worth of a gradient series	n
	$(1+i)^n$	$\frac{(1+i)^n - 1}{i}$	$\frac{1}{i}\left[\frac{(1+i)^n - 1}{i} - n\right]$	
	Equation 4-1	Equation 4-3	Equation 4-7a	

$i = 1\%$

n	Uniform series end-of-period amount a equivalent to			Present sum p equivalent to			n
	future sum, f	present sum, p	gradient series, g	future sum, f	uniform series, a	gradient series, g	
	a/f	a/p	a/g	p/f	p/a	p/g	
1	1.00000	1.01000	0.000	0.9901	0.990	0.000	1
2	0.49751	0.50751	0.498	0.9803	1.970	0.980	2
3	0.33002	0.34002	0.993	0.9706	2.941	2.921	3
4	0.24628	0.25628	1.488	0.9610	3.902	5.804	4
5	0.19604	0.20604	1.980	0.9515	4.853	9.610	5
6	0.16255	0.17255	2.471	0.9420	5.795	14.321	6
7	0.13863	0.14863	2.960	0.9327	6.728	19.917	7
8	0.12069	0.13069	3.448	0.9235	7.652	26.381	8
9	0.10674	0.11674	3.934	0.9143	8.566	33.696	9
10	0.09558	0.10558	4.418	0.9053	9.471	41.843	10
11	0.08645	0.09645	4.901	0.8963	10.368	50.807	11
12	0.07885	0.08885	5.381	0.8874	11.255	60.569	12
13	0.07241	0.08241	5.861	0.8787	12.134	71.113	13
14	0.06690	0.07690	6.338	0.8700	13.004	82.422	14
15	0.06212	0.07212	6.814	0.8613	13.865	94.481	15
16	0.05794	0.06794	7.289	0.8528	14.718	107.273	16
17	0.05426	0.06426	7.761	0.8444	15.562	120.783	17
18	0.05098	0.06098	8.232	0.8360	16.398	134.996	18
19	0.04805	0.05805	8.702	0.8277	17.226	149.895	19
20	0.04542	0.05542	9.169	0.8195	18.046	165.466	20
21	0.04303	0.05303	9.635	0.8114	18.857	181.695	21
22	0.04086	0.05086	10.100	0.8034	19.660	198.566	22
23	0.03889	0.04889	10.563	0.7954	20.456	216.066	23
24	0.03707	0.04707	11.024	0.7876	21.243	234.180	24
25	0.03541	0.04541	11.483	0.7798	22.023	252.894	25
26	0.03387	0.04387	11.941	0.7720	22.795	272.196	26
27	0.03245	0.04245	12.397	0.7644	23.560	292.070	27
28	0.03112	0.04112	12.852	0.7568	24.316	312.505	28
29	0.02990	0.03990	13.304	0.7493	25.066	333.486	29
30	0.02875	0.03875	13.756	0.7419	25.808	355.002	30
35	0.02400	0.03400	15.987	0.7059	29.409	470.158	35
40	0.02046	0.03046	18.178	0.6717	32.835	596.856	40
45	0.01771	0.02771	20.327	0.6391	36.095	733.704	45
50	0.01551	0.02551	22.436	0.6080	39.196	879.418	50
60	0.01224	0.02224	26.533	0.5504	44.955	1192.806	60
∞	0.00000	0.01000	100.000	0.0000	100.000	10000.000	∞

n	Uniform series worth of a future sum or sinking fund factor	Uniform series worth of a present sum or capital recovery factor	Uniform series worth of a gradient series	Present worth of a future sum	Present worth of a uniform series	Present worth of a gradient series	n
	$\dfrac{i}{(1+i)^n - 1}$	$\dfrac{i(1+i)^n}{(1+i)^n - 1}$	$\dfrac{1}{i} - \dfrac{n}{(1+i)^n - 1}$	$\dfrac{1}{(1+i)^n}$	$\dfrac{(1+i)^n - 1}{i(1+i)^n}$	$\dfrac{1}{i}\left[\dfrac{(1+i)^n - 1}{i(1+i)^n} - \dfrac{n}{(1+i)^n}\right]$	
	Equation 4-4	Equation 4-6	Equation 4-8a	Equation 4-2	Equation 4-5	Equation 4-9a	

$$i = 1\%$$

| n | Future sum f equivalent to | | | n |
| | present sum, p | uniform series, a | gradient series, g | |
	f/p	f/a	f/g	
1	1.010	1.000	0.00	1
2	1.020	2.010	1.00	2
3	1.030	3.030	3.01	3
4	1.041	4.060	6.04	4
5	1.051	5.101	10.10	5
6	1.062	6.152	15.20	6
7	1.072	7.214	21.35	7
8	1.083	8.286	28.57	8
9	1.094	9.369	36.85	9
10	1.105	10.462	46.22	10
11	1.116	11.567	56.68	11
12	1.127	12.683	68.25	12
13	1.138	13.809	80.93	13
14	1.149	14.947	94.74	14
15	1.161	16.097	109.69	15
16	1.173	17.258	125.79	16
17	1.184	18.430	143.04	17
18	1.196	19.615	161.47	18
19	1.208	20.811	181.09	19
20	1.220	22.019	201.90	20
21	1.232	23.239	223.92	21
22	1.245	24.472	247.16	22
23	1.257	25.716	271.63	23
24	1.270	26.973	297.35	24
25	1.282	28.243	324.32	25
26	1.295	29.526	352.56	26
27	1.308	30.821	382.09	27
28	1.321	32.129	412.91	28
29	1.335	33.450	445.04	29
30	1.348	34.785	478.49	30
35	1.417	41.660	666.03	35
40	1.489	48.886	888.64	40
45	1.565	56.481	1148.11	45
50	1.645	64.463	1446.32	50
60	1.817	81.670	2166.97	60
∞	∞	∞	∞	∞

n	Future worth of a present sum	Future worth of a uniform series	Future worth of a gradient series	n
	$(1+i)^n$	$\dfrac{(1+i)^n - 1}{i}$	$\dfrac{1}{i}\left[\dfrac{(1+i)^n - 1}{i} - n\right]$	
	Equation 4-1	Equation 4-3	Equation 4-7a	

If the effective annual rate of interest is as shown above, and if interest is compounded continuously:

(1) An end-of-year amount of $1 is equivalent to $ _____ flowing uniformly and continuously throughout the year.

$$a'/a = c'/c = f'/f = g'/g = p'/p = \frac{\ln(1+i_a)}{i_a}$$

$$= 0.9950$$

(2) An end-of-year amount of $ _____ is equivalent to $1 flowing uniformly and continuously throughout the year.

$$a/a' = c/c' = f/f' = g/g' = p/p' = \frac{i_a}{\ln(1+i_a)}$$

$$= 1.005$$

i = 2%

n	Uniform series end-of-period amount *a* equivalent to			Present sum *p* equivalent to			n
	future sum, *f*	present sum, *p*	gradient series, *g*	future sum, *f*	uniform series, *a*	gradient series, *g*	
	a/f	*a/p*	*a/g*	*p/f*	*p/a*	*p/g*	
1	1.00000	1.02000	0.000	0.9804	0.980	0.000	1
2	0.49505	0.51505	0.495	0.9612	1.942	0.961	2
3	0.32675	0.34675	0.987	0.9423	2.884	2.846	3
4	0.24262	0.26262	1.475	0.9238	3.808	5.617	4
5	0.19216	0.21216	1.960	0.9057	4.713	9.240	5
6	0.15853	0.17853	2.442	0.8880	5.601	13.680	6
7	0.13451	0.15451	2.921	0.8706	6.472	18.903	7
8	0.11651	0.13651	3.396	0.8535	7.325	24.878	8
9	0.10252	0.12252	3.868	0.8368	8.162	31.572	9
10	0.09133	0.11133	4.337	0.8203	8.983	38.955	10
11	0.08218	0.10218	4.802	0.8043	9.787	46.998	11
12	0.07456	0.09456	5.264	0.7885	10.575	55.671	12
13	0.06812	0.08812	5.723	0.7730	11.348	64.948	13
14	0.06260	0.08260	6.179	0.7579	12.106	74.800	14
15	0.05783	0.07783	6.631	0.7430	12.849	85.202	15
16	0.05365	0.07365	7.080	0.7284	13.578	96.129	16
17	0.04997	0.06997	7.526	0.7142	14.292	107.555	17
18	0.04670	0.06670	7.968	0.7002	14.992	119.458	18
19	0.04378	0.06378	8.407	0.6864	15.678	131.814	19
20	0.04116	0.06116	8.843	0.6730	16.351	144.600	20
21	0.03878	0.05878	9.276	0.6598	17.011	157.796	21
22	0.03663	0.05663	9.705	0.6468	17.658	171.379	22
23	0.03467	0.05467	10.132	0.6342	18.292	185.331	23
24	0.03287	0.05287	10.555	0.6217	18.914	199.630	24
25	0.03122	0.05122	10.974	0.6095	19.523	214.259	25
26	0.02970	0.04970	11.391	0.5976	20.121	229.199	26
27	0.02829	0.04829	11.804	0.5859	20.707	244.431	27
28	0.02699	0.04699	12.214	0.5744	21.281	259.939	28
29	0.02578	0.04578	12.621	0.5631	21.844	275.706	29
30	0.02465	0.04465	13.025	0.5521	22.396	291.716	30
35	0.02000	0.04000	14.996	0.5000	24.999	374.883	35
40	0.01656	0.03656	16.889	0.4529	27.355	461.993	40
45	0.01391	0.03391	18.703	0.4102	29.490	551.565	45
50	0.01182	0.03182	20.442	0.3715	31.424	642.361	50
60	0.00877	0.02877	23.696	0.3048	34.761	823.698	60
∞	0.00000	0.02000	50.000	0.0000	50.000	2500.000	∞

n	Uniform series worth of a future sum or sinking fund factor	Uniform series worth of a present sum or capital recovery factor	Uniform series worth of a gradient series	Present worth of a future sum	Present worth of a uniform series	Present worth of a gradient series	n
	$\dfrac{i}{(1+i)^n-1}$	$\dfrac{i(1+i)^n}{(1+i)^n-1}$	$\dfrac{1}{i}-\dfrac{n}{(1+i)^n-1}$	$\dfrac{1}{(1+i)^n}$	$\dfrac{(1+i)^n-1}{i(1+i)^n}$	$\dfrac{1}{i}\left[\dfrac{(1+i)^n-1}{i(1+i)^n}-\dfrac{n}{(1+i)^n}\right]$	
	Equation 4-4	Equation 4-6	Equation 4-8a	Equation 4-2	Equation 4-5	Equation 4-9a	

$i = 2\%$

n	Future sum f equivalent to			n
	present sum, p	uniform series, a	gradient series, g	
	f/p	f/a	f/g	
1	1.020	1.000	0.00	1
2	1.040	2.020	1.00	2
3	1.061	3.060	3.02	3
4	1.082	4.122	6.08	4
5	1.104	5.204	10.20	5
6	1.126	6.308	15.41	6
7	1.149	7.434	21.71	7
8	1.172	8.583	29.15	8
9	1.195	9.755	37.73	9
10	1.219	10.950	47.49	10
11	1.243	12.169	58.44	11
12	1.268	13.412	70.60	12
13	1.294	14.680	84.02	13
14	1.319	15.974	98.70	14
15	1.346	17.293	114.67	15
16	1.373	18.639	131.96	16
17	1.400	20.012	150.60	17
18	1.428	21.412	170.62	18
19	1.457	22.841	192.03	19
20	1.486	24.297	214.87	20
21	1.516	25.783	239.17	21
22	1.546	27.299	264.95	22
23	1.577	28.845	292.25	23
24	1.608	30.422	321.09	24
25	1.641	32.030	351.51	25
26	1.673	33.671	383.55	26
27	1.707	35.344	417.22	27
28	1.741	37.051	452.56	28
29	1.776	38.792	489.61	29
30	1.811	40.568	528.40	30
35	2.000	49.994	749.72	35
40	2.208	60.402	1020.10	40
45	2.438	71.893	1344.64	45
50	2.692	84.579	1728.97	50
60	3.281	114.052	2702.58	60
∞	∞	∞	∞	∞

n	Future worth of a present sum	Future worth of a uniform series	Future worth of a gradient series	n
	$(1+i)^n$	$\dfrac{(1+i)^n - 1}{i}$	$\dfrac{1}{i}\left[\dfrac{(1+i)^n - 1}{i} - n\right]$	
	Equation 4-1	Equation 4-3	Equation 4-7a	

If the effective annual rate of interest is as shown above, and if interest is compounded continuously:

(1) An end-of-year amount of $1 is equivalent to $ _____ flowing uniformly and continuously throughout the year.

$$a'/a = c'/c = f'/f = g'/g = p'/p = \frac{\ln(1+i_a)}{i_a}$$

$$= 0.9901$$

(2) An end-of-year amount of $ _____ is equivalent to $1 flowing uniformly and continuously throughout the year.

$$a/a' = c/c' = f/f' = g/g' = p/p' = \frac{i_a}{\ln(1+i_a)}$$

$$= 1.010$$

$i = 3\%$

n	Uniform series end-of-period amount a equivalent to			Present sum p equivalent to			n
	future sum, f	present sum, p	gradient series, g	future sum, f	uniform series, a	gradient series, g	
	a/f	a/p	a/g	p/f	p/a	p/g	
1	1.00000	1.03000	0.000	0.9709	0.971	0.000	1
2	0.49261	0.52261	0.493	0.9426	1.913	0.943	2
3	0.32353	0.35353	0.980	0.9151	2.829	2.773	3
4	0.23903	0.26903	1.463	0.8885	3.717	5.438	4
5	0.18835	0.21835	1.941	0.8626	4.580	8.889	5
6	0.15460	0.18460	2.414	0.8375	5.417	13.076	6
7	0.13051	0.16051	2.882	0.8131	6.230	17.955	7
8	0.11246	0.14246	3.345	0.7894	7.020	23.481	8
9	0.09843	0.12843	3.803	0.7664	7.786	29.612	9
10	0.08723	0.11723	4.256	0.7441	8.530	36.309	10
11	0.07808	0.10808	4.705	0.7224	9.253	43.533	11
12	0.07046	0.10046	5.148	0.7014	9.954	51.248	12
13	0.06403	0.09403	5.587	0.6810	10.635	59.420	13
14	0.05853	0.08853	6.021	0.6611	11.296	68.014	14
15	0.05377	0.08377	6.450	0.6419	11.938	77.000	15
16	0.04961	0.07961	6.874	0.6232	12.561	86.348	16
17	0.04595	0.07595	7.294	0.6050	13.166	96.028	17
18	0.04271	0.07271	7.708	0.5874	13.754	106.014	18
19	0.03981	0.06981	8.118	0.5703	14.324	116.279	19
20	0.03722	0.06722	8.523	0.5537	14.877	126.799	20
21	0.03487	0.06487	8.923	0.5375	15.415	137.550	21
22	0.03275	0.06275	9.319	0.5219	15.937	148.509	22
23	0.03081	0.06081	9.709	0.5067	16.444	159.657	23
24	0.02905	0.05905	10.095	0.4919	16.936	170.971	24
25	0.02743	0.05743	10.477	0.4776	17.413	182.434	25
26	0.02594	0.05594	10.853	0.4637	17.877	194.026	26
27	0.02456	0.05456	11.226	0.4502	18.327	205.731	27
28	0.02329	0.05329	11.593	0.4371	18.764	217.532	28
29	0.02211	0.05211	11.956	0.4243	19.188	229.414	29
30	0.02102	0.05102	12.314	0.4120	19.600	241.361	30
35	0.01654	0.04654	14.037	0.3554	21.487	301.627	35
40	0.01326	0.04326	15.650	0.3066	23.115	361.750	40
45	0.01079	0.04079	17.156	0.2644	24.519	420.632	45
50	0.00887	0.03887	18.558	0.2281	25.730	477.480	50
60	0.00613	0.03613	21.067	0.1697	27.676	583.053	60
∞	0.00000	0.03000	33.333	0.0000	33.333	1111.111	∞

n	Uniform series worth of a future sum or sinking fund factor	Uniform series worth of a present sum or capital recovery factor	Uniform series worth of a gradient series	Present worth of a future sum	Present worth of a uniform series	Present worth of a gradient series	n
	$\dfrac{i}{(1+i)^n-1}$	$\dfrac{i(1+i)^n}{(1+i)^n-1}$	$\dfrac{1}{i}-\dfrac{n}{(1+i)^n-1}$	$\dfrac{1}{(1+i)^n}$	$\dfrac{(1+i)^n-1}{i(1+i)^n}$	$\dfrac{1}{i}\left[\dfrac{(1+i)^n-1}{i(1+i)^n}-\dfrac{n}{(1+i)^n}\right]$	
	Equation 4-4	Equation 4-6	Equation 4-8a	Equation 4-2	Equation 4-5	Equation 4-9a	

n	Future sum f equivalent to			n
	present sum, p	uniform series, a	gradient series, g	
	f/p	f/a	f/g	
1	1.030	1.000	0.00	1
2	1.061	2.030	1.00	2
3	1.093	3.091	3.03	3
4	1.126	4.184	6.12	4
5	1.159	5.309	10.30	5
6	1.194	6.468	15.61	6
7	1.230	7.662	22.08	7
8	1.267	8.892	29.74	8
9	1.305	10.159	38.64	9
10	1.344	11.464	48.80	10
11	1.384	12.808	60.26	11
12	1.426	14.192	73.07	12
13	1.469	15.618	87.26	13
14	1.513	17.086	102.88	14
15	1.558	18.599	119.96	15
16	1.605	20.157	138.56	16
17	1.653	21.762	158.72	17
18	1.702	23.414	180.48	18
19	1.754	25.117	203.90	19
20	1.806	26.870	229.01	20
21	1.860	28.676	255.88	21
22	1.916	30.537	284.56	22
23	1.974	32.453	315.10	23
24	2.033	34.426	347.55	24
25	2.094	36.459	381.98	25
26	2.157	38.553	418.43	26
27	2.221	40.710	456.99	27
28	2.288	42.931	497.70	28
29	2.357	45.219	540.63	29
30	2.427	47.575	585.85	30
35	2.814	60.462	848.74	35
40	3.262	75.401	1180.04	40
45	3.782	92.720	1590.66	45
50	4.384	112.797	2093.23	50
60	5.892	163.053	3435.11	60
∞	∞	∞	∞	∞

If the effective annual rate of interest is as shown above, and if interest is compounded continuously:

(1) An end-of-year amount of ¢1 is equivalent to ¢ _____ flowing uniformly and continuously throughout the year.

$$a'/a = c'/c = f'/f = g'/g = p'/p = \frac{\ln(1+i_a)}{i_a}$$

$$= 0.9853$$

(2) An end-of-year amount of ¢ _____ is equivalent to ¢1 flowing uniformly and continuously throughout the year.

$$a/a' = c/c' = f/f' = g/g' = p/p' = \frac{i_a}{\ln(1+i_a)}$$

$$= 1.015$$

n	Future worth of a present sum	Future worth of a uniform series	Future worth of a gradient series	n
	$(1+i)^n$	$\dfrac{(1+i)^n - 1}{i}$	$\dfrac{1}{i}\left[\dfrac{(1+i)^n - 1}{i} - n\right]$	
	Equation 4-1	Equation 4-3	Equation 4-7a	

$i = 4\%$

| n | Uniform series end-of-period amount a equivalent to | | | Present sum p equivalent to | | | n |
| | future sum, f | present sum, p | gradient series, g | future sum, f | uniform series, a | gradient series, g | |
	a/f	a/p	a/g	p/f	p/a	p/g	
1	1.00000	1.04000	0.000	0.9615	0.962	0.000	1
2	0.49020	0.53020	0.490	0.9246	1.886	0.925	2
3	0.32035	0.36035	0.974	0.8890	2.775	2.703	3
4	0.23549	0.27549	1.451	0.8548	3.630	5.267	4
5	0.18463	0.22463	1.922	0.8219	4.452	8.555	5
6	0.15076	0.19076	2.386	0.7903	5.242	12.506	6
7	0.12661	0.16661	2.843	0.7599	6.002	17.066	7
8	0.10853	0.14853	3.294	0.7307	6.733	22.181	8
9	0.09449	0.13449	3.739	0.7026	7.435	27.801	9
10	0.08329	0.12329	4.177	0.6756	8.111	33.881	10
11	0.07415	0.11415	4.609	0.6496	8.760	40.377	11
12	0.06655	0.10655	5.034	0.6246	9.385	47.248	12
13	0.06014	0.10014	5.453	0.6006	9.986	54.455	13
14	0.05467	0.09467	5.866	0.5775	10.563	61.962	14
15	0.04994	0.08994	6.272	0.5553	11.118	69.735	15
16	0.04582	0.08582	6.672	0.5339	11.652	77.744	16
17	0.04220	0.08220	7.066	0.5134	12.166	85.958	17
18	0.03899	0.07899	7.453	0.4936	12.659	94.350	18
19	0.03614	0.07614	7.834	0.4746	13.134	102.893	19
20	0.03358	0.07358	8.209	0.4564	13.590	111.565	20
21	0.03128	0.07128	8.578	0.4388	14.029	120.341	21
22	0.02920	0.06920	8.941	0.4220	14.451	129.202	22
23	0.02731	0.06731	9.297	0.4057	14.857	138.128	23
24	0.02559	0.06559	9.648	0.3901	15.247	147.101	24
25	0.02401	0.06401	9.993	0.3751	15.622	156.104	25
26	0.02257	0.06257	10.331	0.3607	15.983	165.121	26
27	0.02124	0.06124	10.664	0.3468	16.330	174.138	27
28	0.02001	0.06001	10.991	0.3335	16.663	183.142	28
29	0.01888	0.05888	11.312	0.3207	16.984	192.121	29
30	0.01783	0.05783	11.627	0.3083	17.292	201.062	30
35	0.01358	0.05358	13.120	0.2534	18.665	244.877	35
40	0.01052	0.05052	14.477	0.2083	19.793	286.530	40
45	0.00826	0.04826	15.705	0.1712	20.720	325.403	45
50	0.00655	0.04655	16.812	0.1407	21.482	361.164	50
60	0.00420	0.04420	18.697	0.0951	22.623	422.997	60
∞	0.00000	0.04000	25.000	0.0000	25.000	625.000	∞

n	Uniform series worth of a future sum or sinking fund factor	Uniform series worth of a present sum or capital recovery factor	Uniform series worth of a gradient series	Present worth of a future sum	Present worth of a uniform series	Present worth of a gradient series	n
	$\dfrac{i}{(1+i)^n-1}$	$\dfrac{i(1+i)^n}{(1+i)^n-1}$	$\dfrac{1}{i}-\dfrac{n}{(1+i)^n-1}$	$\dfrac{1}{(1+i)^n}$	$\dfrac{(1+i)^n-1}{i(1+i)^n}$	$\dfrac{1}{i}\left[\dfrac{(1+i)^n-1}{i(1+i)^n}-\dfrac{n}{(1+i)^n}\right]$	
	Equation 4-4	Equation 4-6	Equation 4-8a	Equation 4-2	Equation 4-5	Equation 4-9a	

n	Future sum f equivalent to			n
	present sum, p	uniform series, a	gradient series, g	
	f/p	f/a	f/g	
1	1.040	1.000	0.00	1
2	1.082	2.040	1.00	2
3	1.125	3.122	3.04	3
4	1.170	4.246	6.16	4
5	1.217	5.416	10.41	5
6	1.265	6.633	15.82	6
7	1.316	7.898	22.46	7
8	1.369	9.214	30.36	8
9	1.423	10.583	39.57	9
10	1.480	12.006	50.15	10
11	1.539	13.486	62.16	11
12	1.601	15.026	75.65	12
13	1.665	16.627	90.67	13
14	1.732	18.292	107.30	14
15	1.801	20.024	125.59	15
16	1.873	21.825	145.61	16
17	1.948	23.698	167.44	17
18	2.026	25.645	191.14	18
19	2.107	27.671	216.78	19
20	2.191	29.778	244.45	20
21	2.279	31.969	274.23	21
22	2.370	34.248	306.20	22
23	2.465	36.618	340.45	23
24	2.563	39.083	377.07	24
25	2.666	41.646	416.15	25
26	2.772	44.312	457.79	26
27	2.883	47.084	502.11	27
28	2.999	49.968	549.19	28
29	3.119	52.966	599.16	29
30	3.243	56.085	652.12	30
35	3.946	73.652	966.31	35
40	4.801	95.026	1375.64	40
45	5.841	121.029	1900.73	45
50	7.107	152.667	2566.68	50
60	10.520	237.991	4449.77	60
∞	∞	∞	∞	∞

n	Future worth of a present sum	Future worth of a uniform series	Future worth of a gradient series	n
	$(1+i)^n$	$\dfrac{(1+i)^n - 1}{i}$	$\dfrac{1}{i}\left[\dfrac{(1+i)^n - 1}{i} - n\right]$	
	Equation 4-1	Equation 4-3	Equation 4-7a	

If the effective annual rate of interest is as shown above, and if interest is compounded continuously:

(1) An end-of-year amount of $1 is equivalent to $ _____ flowing uniformly and continuously throughout the year.

$$a'/a = c'/c = f'/f = g'/g = p'/p = \frac{\ln(1+i_a)}{i_a}$$

$$= 0.9805$$

(2) An end-of-year amount of $ _____ is equivalent to $1 flowing uniformly and continuously throughout the year.

$$a/a' = c/c' = f/f' = g/g' = p/p' = \frac{i_a}{\ln(1+i_a)}$$

$$= 1.020$$

$i = 5\%$

n	Uniform series end-of-period amount a equivalent to			Present sum p equivalent to			n
	future sum, f	present sum, p	gradient series, g	future sum, f	uniform series, a	gradient series, g	
	a/f	a/p	a/g	p/f	p/a	p/g	
1	1.00000	1.05000	0.000	0.9524	0.952	0.000	1
2	0.48780	0.53780	0.488	0.9070	1.859	0.907	2
3	0.31721	0.36721	0.967	0.8638	2.723	2.635	3
4	0.23201	0.28201	1.439	0.8227	3.546	5.103	4
5	0.18097	0.23097	1.903	0.7835	4.329	8.237	5
6	0.14702	0.19702	2.358	0.7462	5.076	11.968	6
7	0.12282	0.17282	2.805	0.7107	5.786	16.232	7
8	0.10472	0.15472	3.245	0.6768	6.463	20.970	8
9	0.09069	0.14069	3.676	0.6446	7.108	26.127	9
10	0.07950	0.12950	4.099	0.6139	7.722	31.652	10
11	0.07039	0.12039	4.514	0.5847	8.306	37.499	11
12	0.06283	0.11283	4.922	0.5568	8.863	43.624	12
13	0.05646	0.10646	5.322	0.5303	9.394	49.988	13
14	0.05102	0.10102	5.713	0.5051	9.899	56.554	14
15	0.04634	0.09634	6.097	0.4810	10.380	63.288	15
16	0.04227	0.09227	6.474	0.4581	10.838	70.160	16
17	0.03870	0.08870	6.842	0.4363	11.274	77.140	17
18	0.03555	0.08555	7.203	0.4155	11.690	84.204	18
19	0.03275	0.08275	7.557	0.3957	12.085	91.328	19
20	0.03024	0.08024	7.903	0.3769	12.462	98.488	20
21	0.02800	0.07800	8.242	0.3589	12.821	105.667	21
22	0.02597	0.07597	8.573	0.3418	13.163	112.846	22
23	0.02414	0.07414	8.897	0.3256	13.489	120.009	23
24	0.02247	0.07247	9.214	0.3101	13.799	127.140	24
25	0.02095	0.07095	9.524	0.2953	14.094	134.228	25
26	0.01956	0.06956	9.827	0.2812	14.375	141.259	26
27	0.01829	0.06829	10.122	0.2678	14.643	148.223	27
28	0.01712	0.06712	10.411	0.2551	14.898	155.110	28
29	0.01605	0.06605	10.694	0.2429	15.141	161.913	29
30	0.01505	0.06505	10.969	0.2314	15.372	168.623	30
35	0.01107	0.06107	12.250	0.1813	16.374	200.581	35
40	0.00828	0.05828	13.377	0.1420	17.159	229.545	40
45	0.00626	0.05626	14.364	0.1113	17.774	255.315	45
50	0.00478	0.05478	15.223	0.0872	18.256	277.915	50
60	0.00283	0.05283	16.606	0.0535	18.929	314.343	60
∞	0.00000	0.05000	20.000	0.0000	20.000	400.000	∞

n	Uniform series worth of a future sum or sinking fund factor	Uniform series worth of a present sum or capital recovery factor	Uniform series worth of a gradient series	Present worth of a future sum	Present worth of a uniform series	Present worth of a gradient series	n
	$\dfrac{i}{(1+i)^n - 1}$	$\dfrac{i(1+i)^n}{(1+i)^n - 1}$	$\dfrac{1}{i} - \dfrac{n}{(1+i)^n - 1}$	$\dfrac{1}{(1+i)^n}$	$\dfrac{(1+i)^n - 1}{i(1+i)^n}$	$\dfrac{1}{i}\left[\dfrac{(1+i)^n - 1}{i(1+i)^n} - \dfrac{n}{(1+i)^n}\right]$	
	Equation 4-4	Equation 4-6	Equation 4-8a	Equation 4-2	Equation 4-5	Equation 4-9a	

$i = 5\%$

	Future sum f equivalent to			
n	present sum, p	uniform series, a	gradient series, g	n
	f/p	f/a	f/g	
1	1.050	1.000	0.00	1
2	1.102	2.050	1.00	2
3	1.158	3.152	3.05	3
4	1.216	4.310	6.20	4
5	1.276	5.526	10.51	5
6	1.340	6.802	16.04	6
7	1.407	8.142	22.84	7
8	1.477	9.549	30.98	8
9	1.551	11.027	40.53	9
10	1.629	12.578	51.56	10
11	1.710	14.207	64.14	11
12	1.796	15.917	78.34	12
13	1.886	17.713	94.26	13
14	1.980	19.599	111.97	14
15	2.079	21.579	131.57	15
16	2.183	23.657	153.15	16
17	2.292	25.840	176.81	17
18	2.407	28.132	202.65	18
19	2.527	30.539	230.78	19
20	2.653	33.066	261.32	20
21	2.786	35.719	294.39	21
22	2.925	38.505	330.10	22
23	3.072	41.430	368.61	23
24	3.225	44.502	410.04	24
25	3.386	47.727	454.54	25
26	3.556	51.113	502.27	26
27	3.733	54.669	553.38	27
28	3.920	58.403	608.05	28
29	4.116	62.323	666.45	29
30	4.322	66.439	728.78	30
35	5.516	90.320	1106.41	35
40	7.040	120.800	1616.00	40
45	8.985	159.700	2294.00	45
50	11.467	209.348	3186.96	50
60	18.679	353.584	5871.67	60
∞	∞	∞	∞	∞

If the effective annual rate of interest is as shown above, and if interest is compounded continuously:

(1) An end-of-year amount of $1 is equivalent to $ _____ flowing uniformly and continuously throughout the year.

$$a'/a = c'/c = f'/f = g'/g = p'/p = \frac{\ln(1+i_a)}{i_a}$$

$$= 0.9758$$

(2) An end-of-year amount of $ _____ is equivalent to $1 flowing uniformly and continuously throughout the year.

$$a/a' = c/c' = f/f' = g/g' = p/p' = \frac{i_a}{\ln(1+i_a)}$$

$$= 1.025$$

	Future worth of a present sum	Future worth of a uniform series	Future worth of a gradient series	
n	$(1+i)^n$	$\dfrac{(1+i)^n - 1}{i}$	$\dfrac{1}{i}\left[\dfrac{(1+i)^n - 1}{i} - n\right]$	n
	Equation 4-1	Equation 4-3	Equation 4-7a	

$i = 6\%$

| n | Uniform series end-of-period amount a equivalent to | | | Present sum p equivalent to | | | n |
| | future sum, f | present sum, p | gradient series, g | future sum, f | uniform series, a | gradient series, g | |
	a/f	a/p	a/g	p/f	p/a	p/g	
1	1.00000	1.06000	0.000	0.9434	0.943	0.000	1
2	0.48544	0.54544	0.485	0.8900	1.833	0.890	2
3	0.31411	0.37411	0.961	0.8396	2.673	2.569	3
4	0.22859	0.28859	1.427	0.7921	3.465	4.946	4
5	0.17740	0.23740	1.884	0.7473	4.212	7.935	5
6	0.14336	0.20336	2.330	0.7050	4.917	11.459	6
7	0.11914	0.17914	2.768	0.6651	5.582	15.450	7
8	0.10104	0.16104	3.195	0.6274	6.210	19.842	8
9	0.08702	0.14702	3.613	0.5919	6.802	24.577	9
10	0.07587	0.13587	4.022	0.5584	7.360	29.602	10
11	0.06679	0.12679	4.421	0.5268	7.887	34.870	11
12	0.05928	0.11928	4.811	0.4970	8.384	40.337	12
13	0.05296	0.11296	5.192	0.4688	8.853	45.963	13
14	0.04758	0.10758	5.564	0.4423	9.295	51.713	14
15	0.04296	0.10296	5.926	0.4173	9.712	57.555	15
16	0.03895	0.09895	6.279	0.3936	10.106	63.459	16
17	0.03544	0.09544	6.624	0.3714	10.477	69.401	17
18	0.03236	0.09236	6.960	0.3503	10.828	75.357	18
19	0.02962	0.08962	7.287	0.3305	11.158	81.306	19
20	0.02718	0.08718	7.605	0.3118	11.470	87.230	20
21	0.02500	0.08500	7.915	0.2942	11.764	93.114	21
22	0.02305	0.08305	8.217	0.2775	12.042	98.941	22
23	0.02128	0.08128	8.510	0.2618	12.303	104.701	23
24	0.01968	0.07968	8.795	0.2470	12.550	110.381	24
25	0.01823	0.07823	9.072	0.2330	12.783	115.973	25
26	0.01690	0.07690	9.341	0.2198	13.003	121.468	26
27	0.01570	0.07570	9.603	0.2074	13.211	126.860	27
28	0.01459	0.07459	9.857	0.1956	13.406	132.142	28
29	0.01358	0.07358	10.103	0.1846	13.591	137.310	29
30	0.01265	0.07265	10.342	0.1741	13.765	142.359	30
35	0.00897	0.06897	11.432	0.1301	14.498	165.743	35
40	0.00646	0.06646	12.359	0.0972	15.046	185.957	40
45	0.00470	0.06470	13.141	0.0727	15.456	203.110	45
50	0.00344	0.06344	13.796	0.0543	15.762	217.457	50
60	0.00188	0.06188	14.791	0.0303	16.161	239.043	60
∞	0.00000	0.06000	16.667	0.0000	16.667	277.778	∞

n	Uniform series worth of a future sum or sinking fund factor	Uniform series worth of a present sum or capital recovery factor	Uniform series worth of a gradient series	Present worth of a future sum	Present worth of a uniform series	Present worth of a gradient series	n
	$\dfrac{i}{(1+i)^n - 1}$	$\dfrac{i(1+i)^n}{(1+i)^n - 1}$	$\dfrac{1}{i} - \dfrac{n}{(1+i)^n - 1}$	$\dfrac{1}{(1+i)^n}$	$\dfrac{(1+i)^n - 1}{i(1+i)^n}$	$\dfrac{1}{i}\left[\dfrac{(1+i)^n - 1}{i(1+i)^n} - \dfrac{n}{(1+i)^n}\right]$	
	Equation 4-4	Equation 4-6	Equation 4-8a	Equation 4-2	Equation 4-5	Equation 4-9a	

$i = 6\%$

| n | Future sum f equivalent to | | | n |
| | present sum, p | uniform series, a | gradient series, g | |
	f/p	f/a	f/g	
1	1.060	1.000	0.00	1
2	1.124	2.060	1.00	2
3	1.191	3.184	3.06	3
4	1.262	4.375	6.24	4
5	1.338	5.637	10.62	5
6	1.419	6.975	16.26	6
7	1.504	8.394	23.23	7
8	1.594	9.897	31.62	8
9	1.689	11.491	41.52	9
10	1.791	13.181	53.01	10
11	1.898	14.972	66.19	11
12	2.012	16.870	81.17	12
13	2.133	18.882	98.04	13
14	2.261	21.015	116.92	14
15	2.397	23.276	137.93	15
16	2.540	25.673	161.21	16
17	2.693	28.213	186.88	17
18	2.854	30.906	215.09	18
19	3.026	33.760	246.00	19
20	3.207	36.786	279.76	20
21	3.400	39.993	316.55	21
22	3.604	43.392	356.54	22
23	3.820	46.996	399.93	23
24	4.049	50.816	446.93	24
25	4.292	54.865	497.74	25
26	4.549	59.156	552.61	26
27	4.822	63.706	611.76	27
28	5.112	68.528	675.47	28
29	5.418	73.640	744.00	29
30	5.743	79.058	817.64	30
35	7.686	111.435	1273.91	35
40	10.286	154.762	1912.70	40
45	13.765	212.744	2795.73	45
50	18.420	290.336	4005.60	50
60	32.988	533.128	7885.47	60
∞	∞	∞	∞	∞

If the effective annual rate of interest is as shown above, and if interest is compounded continuously:

(1) An end-of-year amount of ¢1 is equivalent to ¢ _____ flowing uniformly and continuously throughout the year.

$$a'/a = c'/c = f'/f = g'/g = p'/p = \frac{\ln(1+i_a)}{i_a}$$

$$= 0.9711$$

(2) An end-of-year amount of ¢ _____ is equivalent to ¢1 flowing uniformly and continuously throughout the year.

$$a/a' = c/c' = f/f' = g/g' = p/p' = \frac{i_a}{\ln(1+i_a)}$$

$$= 1.030$$

n	Future worth of a present sum	Future worth of a uniform series	Future worth of a gradient series	n
	$(1+i)^n$	$\frac{(1+i)^n - 1}{i}$	$\frac{1}{i}\left[\frac{(1+i)^n - 1}{i} - n\right]$	
	Equation 4-1	Equation 4-3	Equation 4-7a	

$i = 7\%$

n	Uniform series end-of-period amount a equivalent to			Present sum p equivalent to			n
	future sum, f	present sum, p	gradient series, g	future sum, f	uniform series, a	gradient series, g	
	a/f	a/p	a/g	p/f	p/a	p/g	
1	1.00000	1.07000	0.000	0.9346	0.935	0.000	1
2	0.48309	0.55309	0.483	0.8734	1.808	0.873	2
3	0.31105	0.38105	0.955	0.8163	2.624	2.506	3
4	0.22523	0.29523	1.416	0.7629	3.387	4.795	4
5	0.17389	0.24389	1.865	0.7130	4.100	7.647	5
6	0.13980	0.20980	2.303	0.6663	4.767	10.978	6
7	0.11555	0.18555	2.730	0.6227	5.389	14.715	7
8	0.09747	0.16747	3.147	0.5820	5.971	18.789	8
9	0.08349	0.15349	3.552	0.5439	6.515	23.140	9
10	0.07238	0.14238	3.946	0.5083	7.024	27.716	10
11	0.06336	0.13336	4.330	0.4751	7.499	32.466	11
12	0.05590	0.12590	4.703	0.4440	7.943	37.351	12
13	0.04965	0.11965	5.065	0.4150	8.358	42.330	13
14	0.04434	0.11434	5.417	0.3878	8.745	47.372	14
15	0.03979	0.10979	5.758	0.3624	9.108	52.446	15
16	0.03586	0.10586	6.090	0.3387	9.447	57.527	16
17	0.03243	0.10243	6.411	0.3166	9.763	62.592	17
18	0.02941	0.09941	6.722	0.2959	10.059	67.622	18
19	0.02675	0.09675	7.024	0.2765	10.336	72.599	19
20	0.02439	0.09439	7.316	0.2584	10.594	77.509	20
21	0.02229	0.09229	7.599	0.2415	10.836	82.339	21
22	0.02041	0.09041	7.872	0.2257	11.061	87.079	22
23	0.01871	0.08871	8.137	0.2109	11.272	91.720	23
24	0.01719	0.08719	8.392	0.1971	11.469	96.255	24
25	0.01581	0.08581	8.639	0.1842	11.654	100.676	25
26	0.01456	0.08456	8.877	0.1722	11.826	104.981	26
27	0.01343	0.08343	9.107	0.1609	11.987	109.166	27
28	0.01239	0.08239	9.329	0.1504	12.137	113.226	28
29	0.01145	0.08145	9.543	0.1406	12.278	117.162	29
30	0.01059	0.08059	9.749	0.1314	12.409	120.972	30
35	0.00723	0.07723	10.669	0.0937	12.948	138.135	35
40	0.00501	0.07501	11.423	0.0668	13.332	152.293	40
45	0.00350	0.07350	12.036	0.0476	13.606	163.756	45
50	0.00246	0.07246	12.529	0.0339	13.801	172.905	50
60	0.00123	0.07123	13.232	0.0173	14.039	185.768	60
∞	0.00000	0.07000	14.286	0.0000	14.286	204.082	∞

n	Uniform series worth of a future sum or sinking fund factor	Uniform series worth of a present sum or capital recovery factor	Uniform series worth of a gradient series	Present worth of a future sum	Present worth of a uniform series	Present worth of a gradient series	n
	$\dfrac{i}{(1+i)^n-1}$	$\dfrac{i(1+i)^n}{(1+i)^n-1}$	$\dfrac{1}{i}-\dfrac{n}{(1+i)^n-1}$	$\dfrac{1}{(1+i)^n}$	$\dfrac{(1+i)^n-1}{i(1+i)^n}$	$\dfrac{1}{i}\left[\dfrac{(1+i)^n-1}{i(1+i)^n}-\dfrac{n}{(1+i)^n}\right]$	
	Equation 4-4	Equation 4-6	Equation 4-8a	Equation 4-2	Equation 4-5	Equation 4-9a	

$i = 7 \%$

	Future sum f equivalent to			
n	present sum, p	uniform series, a	gradient series, g	n
	f/p	f/a	f/g	
1	1.070	1.000	0.00	1
2	1.145	2.070	1.00	2
3	1.225	3.215	3.07	3
4	1.311	4.440	6.28	4
5	1.403	5.751	10.72	5
6	1.501	7.153	16.48	6
7	1.606	8.654	23.63	7
8	1.718	10.260	32.28	8
9	1.838	11.978	42.54	9
10	1.967	13.816	54.52	10
11	2.105	15.784	68.34	11
12	2.252	17.888	84.12	12
13	2.410	20.141	102.01	13
14	2.579	22.550	122.15	14
15	2.759	25.129	144.70	15
16	2.952	27.888	169.83	16
17	3.159	30.840	197.72	17
18	3.380	33.999	228.56	18
19	3.617	37.379	262.56	19
20	3.870	40.995	299.94	20
21	4.141	44.865	340.93	21
22	4.430	49.006	385.80	22
23	4.741	53.436	434.80	23
24	5.072	58.177	488.24	24
25	5.427	63.249	546.41	25
26	5.807	68.676	609.66	26
27	6.214	74.484	678.34	27
28	6.649	80.698	752.82	28
29	7.114	87.347	833.52	29
30	7.612	94.461	920.87	30
35	10.677	138.237	1474.81	35
40	14.974	199.635	2280.50	40
45	21.002	285.749	3439.28	45
50	29.457	406.529	5093.27	50
60	57.946	813.520	10764.58	60
∞	∞	∞	∞	∞

	Future worth of a present sum	Future worth of a uniform series	Future worth of a gradient series	
n	$(1+i)^n$	$\dfrac{(1+i)^n-1}{i}$	$\dfrac{1}{i}\left[\dfrac{(1+i)^n-1}{i} - n\right]$	n
	Equation 4-1	Equation 4-3	Equation 4-7a	

If the effective annual rate of interest is as shown above, and if interest is compounded continuously:

(1) An end-of-year amount of $1 is equivalent to $ _____ flowing uniformly and continuously throughout the year.

$$a'/a = c'/c = f'/f = g'/g = p'/p = \frac{\ln(1+i_a)}{i_a}$$

$$= 0.9665$$

(2) An end-of-year amount of $ _____ is equivalent to $1 flowing uniformly and continuously throughout the year.

$$a/a' = c/c' = f/f' = g/g' = p/p' = \frac{i_a}{\ln(1+i_a)}$$

$$= 1.035$$

$i = 8\%$

n	Uniform series end-of-period amount a equivalent to			Present sum p equivalent to			n
	future sum, f	present sum, p	gradient series, g	future sum, f	uniform series, a	gradient series, g	
	a/f	a/p	a/g	p/f	p/a	p/g	
1	1.00000	1.08000	0.000	0.9259	0.926	0.000	1
2	0.48077	0.56077	0.481	0.8573	1.783	0.857	2
3	0.30803	0.38803	0.949	0.7938	2.577	2.445	3
4	0.22192	0.30192	1.404	0.7350	3.312	4.650	4
5	0.17046	0.25046	1.846	0.6806	3.993	7.372	5
6	0.13632	0.21632	2.276	0.6302	4.623	10.523	6
7	0.11207	0.19207	2.694	0.5835	5.206	14.024	7
8	0.09401	0.17401	3.099	0.5403	5.747	17.806	8
9	0.08008	0.16008	3.491	0.5002	6.247	21.808	9
10	0.06903	0.14903	3.871	0.4632	6.710	25.977	10
11	0.06008	0.14008	4.240	0.4289	7.139	30.266	11
12	0.05270	0.13270	4.596	0.3971	7.536	34.634	12
13	0.04652	0.12652	4.940	0.3677	7.904	39.046	13
14	0.04130	0.12130	5.273	0.3405	8.244	43.472	14
15	0.03683	0.11683	5.594	0.3152	8.559	47.886	15
16	0.03298	0.11298	5.905	0.2919	8.851	52.264	16
17	0.02963	0.10963	6.204	0.2703	9.122	56.588	17
18	0.02670	0.10670	6.492	0.2502	9.372	60.843	18
19	0.02413	0.10413	6.770	0.2317	9.604	65.013	19
20	0.02185	0.10185	7.037	0.2145	9.818	69.090	20
21	0.01983	0.09983	7.294	0.1987	10.017	73.063	21
22	0.01803	0.09803	7.541	0.1839	10.201	76.926	22
23	0.01642	0.09642	7.779	0.1703	10.371	80.673	23
24	0.01498	0.09498	8.007	0.1577	10.529	84.300	24
25	0.01368	0.09368	8.225	0.1460	10.675	87.804	25
26	0.01251	0.09251	8.435	0.1352	10.810	91.184	26
27	0.01145	0.09145	8.636	0.1252	10.935	94.439	27
28	0.01049	0.09049	8.829	0.1159	11.051	97.569	28
29	0.00962	0.08962	9.013	0.1073	11.158	100.574	29
30	0.00883	0.08883	9.190	0.0994	11.258	103.456	30
35	0.00580	0.08580	9.961	0.0676	11.655	116.092	35
40	0.00386	0.08386	10.570	0.0460	11.925	126.042	40
45	0.00259	0.08259	11.045	0.0313	12.108	133.733	45
50	0.00174	0.08174	11.411	0.0213	12.233	139.593	50
60	0.00080	0.08080	11.902	0.0099	12.377	147.300	60
∞	0.00000	0.08000	12.500	0.0000	12.500	156.250	∞

n	Uniform series worth of a future sum or sinking fund factor	Uniform series worth of a present sum or capital recovery factor	Uniform series worth of a gradient series	Present worth of a future sum	Present worth of a uniform series	Present worth of a gradient series	n
	$\dfrac{i}{(1+i)^n - 1}$	$\dfrac{i(1+i)^n}{(1+i)^n - 1}$	$\dfrac{1}{i} - \dfrac{n}{(1+i)^n - 1}$	$\dfrac{1}{(1+i)^n}$	$\dfrac{(1+i)^n - 1}{i(1+i)^n}$	$\dfrac{1}{i}\left[\dfrac{(1+i)^n - 1}{i(1+i)^n} - \dfrac{n}{(1+i)^n}\right]$	
	Equation 4-4	Equation 4-6	Equation 4-8a	Equation 4-2	Equation 4-5	Equation 4-9a	

n	Future sum f equivalent to			n
	present sum, p	uniform series, a	gradient series, g	
	f/p	f/a	f/g	
1	1.080	1.000	0.00	1
2	1.166	2.080	1.00	2
3	1.260	3.246	3.08	3
4	1.360	4.506	6.33	4
5	1.469	5.867	10.83	5
6	1.587	7.336	16.70	6
7	1.714	8.923	24.04	7
8	1.851	10.637	32.96	8
9	1.999	12.488	43.59	9
10	2.159	14.487	56.08	10
11	2.332	16.645	70.57	11
12	2.518	18.977	87.21	12
13	2.720	21.495	106.19	13
14	2.937	24.215	127.69	14
15	3.172	27.152	151.90	15
16	3.426	30.324	179.05	16
17	3.700	33.750	209.38	17
18	3.996	37.450	243.13	18
19	4.316	41.446	280.58	19
20	4.661	45.762	322.02	20
21	5.034	50.423	367.79	21
22	5.437	55.457	418.21	22
23	5.871	60.893	473.67	23
24	6.341	66.765	534.56	24
25	6.848	73.106	601.32	25
26	7.396	79.954	674.43	26
27	7.988	87.351	754.38	27
28	8.627	95.339	841.74	28
29	9.317	103.966	937.07	29
30	10.063	113.283	1041.04	30
35	14.785	172.317	1716.46	35
40	21.725	259.057	2738.21	40
45	31.920	386.506	4268.82	45
50	46.902	573.770	6547.13	50
60	101.257	1253.213	14915.17	60
∞	∞	∞	∞	∞

n	Future worth of a present sum	Future worth of a uniform series	Future worth of a gradient series	n
	$(1+i)^n$	$\dfrac{(1+i)^n - 1}{i}$	$\dfrac{1}{i}\left[\dfrac{(1+i)^n-1}{i} - n\right]$	
	Equation 4-1	Equation 4-3	Equation 4-7a	

If the effective annual rate of interest is as shown above, and if interest is compounded continuously:

(1) An end-of-year amount of $1 is equivalent to $ _____ flowing uniformly and continuously throughout the year.

$$a'/a = c'/c = f'/f = g'/g = p'/p = \frac{\ln(1+i_a)}{i_a}$$

$$= 0.9620$$

(2) An end-of-year amount of $ _____ is equivalent to $1 flowing uniformly and continuously throughout the year.

$$a/a' = c/c' = f/f' = g/g' = p/p' = \frac{i_a}{\ln(1+i_a)}$$

$$= 1.039$$

$i = 10\%$

n	Uniform series end-of-period amount a equivalent to			Present sum p equivalent to			n
	future sum, f	present sum, p	gradient series, g	future sum, f	uniform series, a	gradient series, g	
	a/f	a/p	a/g	p/f	p/a	p/g	
1	1.00000	1.10000	0.000	0.9091	0.909	0.000	1
2	0.47619	0.57619	0.476	0.8264	1.736	0.826	2
3	0.30211	0.40211	0.937	0.7513	2.487	2.329	3
4	0.21547	0.31547	1.381	0.6830	3.170	4.378	4
5	0.16380	0.26380	1.810	0.6209	3.791	6.862	5
6	0.12961	0.22961	2.224	0.5645	4.355	9.684	6
7	0.10541	0.20541	2.622	0.5132	4.868	12.763	7
8	0.08744	0.18744	3.004	0.4665	5.335	16.029	8
9	0.07364	0.17364	3.372	0.4241	5.759	19.421	9
10	0.06275	0.16275	3.725	0.3855	6.145	22.891	10
11	0.05396	0.15396	4.064	0.3505	6.495	26.396	11
12	0.04676	0.14676	4.388	0.3186	6.814	29.901	12
13	0.04078	0.14078	4.699	0.2897	7.103	33.377	13
14	0.03575	0.13575	4.996	0.2633	7.367	36.800	14
15	0.03147	0.13147	5.279	0.2394	7.606	40.152	15
16	0.02782	0.12782	5.549	0.2176	7.824	43.416	16
17	0.02466	0.12466	5.807	0.1978	8.022	46.582	17
18	0.02193	0.12193	6.053	0.1799	8.201	49.640	18
19	0.01955	0.11955	6.286	0.1635	8.365	52.583	19
20	0.01746	0.11746	6.508	0.1486	8.514	55.407	20
21	0.01562	0.11562	6.719	0.1351	8.649	58.110	21
22	0.01401	0.11401	6.919	0.1228	8.772	60.689	22
23	0.01257	0.11257	7.108	0.1117	8.883	63.146	23
24	0.01130	0.11130	7.288	0.1015	8.985	65.481	24
25	0.01017	0.11017	7.458	0.0923	9.077	67.696	25
26	0.00916	0.10916	7.619	0.0839	9.161	69.794	26
27	0.00826	0.10826	7.770	0.0763	9.237	71.777	27
28	0.00745	0.10745	7.914	0.0693	9.307	73.650	28
29	0.00673	0.10673	8.049	0.0630	9.370	75.415	29
30	0.00608	0.10608	8.176	0.0573	9.427	77.077	30
35	0.00369	0.10369	8.709	0.0356	9.644	83.987	35
40	0.00226	0.10226	9.096	0.0221	9.779	88.953	40
45	0.00139	0.10139	9.374	0.0137	9.863	92.454	45
50	0.00086	0.10086	9.570	0.0085	9.915	94.889	50
60	0.00033	0.10033	9.802	0.0033	9.967	97.701	60
∞	0.00000	0.10000	10.000	0.0000	10.000	100.000	∞

n	Uniform series worth of a future sum or sinking fund factor	Uniform series worth of a present sum or capital recovery factor	Uniform series worth of a gradient series	Present worth of a future sum	Present worth of a uniform series	Present worth of a gradient series	n
	$\dfrac{i}{(1+i)^n-1}$	$\dfrac{i(1+i)^n}{(1+i)^n-1}$	$\dfrac{1}{i}-\dfrac{n}{(1+i)^n-1}$	$\dfrac{1}{(1+i)^n}$	$\dfrac{(1+i)^n-1}{i(1+i)^n}$	$\dfrac{1}{i}\left[\dfrac{(1+i)^n-1}{i(1+i)^n}-\dfrac{n}{(1+i)^n}\right]$	
	Equation 4-4	Equation 4-6	Equation 4-8a	Equation 4-2	Equation 4-5	Equation 4-9a	

n	Future sum f equivalent to			n
	present sum, p	uniform series, a	gradient series, g	
	f/p	f/a	f/g	
1	1.100	1.000	0.00	1
2	1.210	2.100	1.00	2
3	1.331	3.310	3.10	3
4	1.464	4.641	6.41	4
5	1.611	6.105	11.05	5
6	1.772	7.716	17.16	6
7	1.949	9.487	24.87	7
8	2.144	11.436	34.36	8
9	2.358	13.579	45.79	9
10	2.594	15.937	59.37	10
11	2.853	18.531	75.31	11
12	3.138	21.384	93.84	12
13	3.452	24.523	115.23	13
14	3.797	27.975	139.75	14
15	4.177	31.772	167.72	15
16	4.595	35.950	199.50	16
17	5.054	40.545	235.45	17
18	5.560	45.599	275.99	18
19	6.116	51.159	321.59	19
20	6.727	57.275	372.75	20
21	7.400	64.002	430.02	21
22	8.140	71.403	494.03	22
23	8.954	79.543	565.43	23
24	9.850	88.497	644.97	24
25	10.835	98.347	733.47	25
26	11.918	109.182	831.82	26
27	13.110	121.100	941.00	27
28	14.421	134.210	1062.10	28
29	15.863	148.631	1196.31	29
30	17.449	164.494	1344.94	30
35	28.102	271.024	2360.24	35
40	45.259	442.593	4025.93	40
45	72.890	718.905	6739.05	45
50	117.391	1163.909	11139.09	50
60	304.482	3034.816	29748.16	60
∞	∞	∞	∞	∞

n	Future worth of a present sum	Future worth of a uniform series	Future worth of a gradient series	n
	$(1+i)^n$	$\dfrac{(1+i)^n-1}{i}$	$\dfrac{1}{i}\left[\dfrac{(1+i)^n-1}{i} - n\right]$	
	Equation 4-1	Equation 4-3	Equation 4-7a	

If the effective annual rate of interest is as shown above, and if interest is compounded continuously:

(1) An end-of-year amount of $1 is equivalent to $ _____ flowing uniformly and continuously throughout the year.

$$a'/a = c/c = f/f = g'/g = p/p = \frac{\ln(1+i_a)}{i_a}$$

$$= 0.9531$$

(2) An end-of-year amount of $ _____ is equivalent to $1 flowing uniformly and continuously throughout the year.

$$a/a' = c/c' = f/f' = g/g' = p/p' = \frac{i_a}{\ln(1+i_a)}$$

$$= 1.049$$

$i = 12\%$

n	Uniform series end-of-period amount a equivalent to			Present sum p equivalent to			n
	future sum, f	present sum, p	gradient series, g	future sum, f	uniform series, a	gradient series, g	
	a/f	a/p	a/g	p/f	p/a	p/g	
1	1.00000	1.12000	0.000	0.8929	0.893	0.000	1
2	0.47170	0.59170	0.472	0.7972	1.690	0.797	2
3	0.29635	0.41635	0.925	0.7118	2.402	2.221	3
4	0.20923	0.32923	1.359	0.6355	3.037	4.127	4
5	0.15741	0.27741	1.775	0.5674	3.605	6.397	5
6	0.12323	0.24323	2.172	0.5066	4.111	8.930	6
7	0.09912	0.21912	2.551	0.4523	4.564	11.644	7
8	0.08130	0.20130	2.913	0.4039	4.968	14.471	8
9	0.06768	0.18768	3.257	0.3606	5.328	17.356	9
10	0.05698	0.17698	3.585	0.3220	5.650	20.254	10
11	0.04842	0.16842	3.895	0.2875	5.938	23.129	11
12	0.04144	0.16144	4.190	0.2567	6.194	25.952	12
13	0.03568	0.15568	4.468	0.2292	6.424	28.702	13
14	0.03087	0.15087	4.732	0.2046	6.628	31.362	14
15	0.02682	0.14682	4.980	0.1827	6.811	33.920	15
16	0.02339	0.14339	5.215	0.1631	6.974	36.367	16
17	0.02046	0.14046	5.435	0.1456	7.120	38.697	17
18	0.01794	0.13794	5.643	0.1300	7.250	40.908	18
19	0.01576	0.13576	5.838	0.1161	7.366	42.998	19
20	0.01388	0.13388	6.020	0.1037	7.469	44.968	20
21	0.01224	0.13224	6.191	0.0926	7.562	46.819	21
22	0.01081	0.13081	6.351	0.0826	7.645	48.554	22
23	0.00956	0.12956	6.501	0.0738	7.718	50.178	23
24	0.00846	0.12846	6.641	0.0659	7.784	51.693	24
25	0.00750	0.12750	6.771	0.0588	7.843	53.105	25
26	0.00665	0.12665	6.892	0.0525	7.896	54.418	26
27	0.00590	0.12590	7.005	0.0469	7.943	55.637	27
28	0.00524	0.12524	7.110	0.0419	7.984	56.767	28
29	0.00466	0.12466	7.207	0.0374	8.022	57.814	29
30	0.00414	0.12414	7.297	0.0334	8.055	58.782	30
35	0.00232	0.12232	7.658	0.0189	8.176	62.605	35
40	0.00130	0.12130	7.899	0.0107	8.244	65.116	40
45	0.00074	0.12074	8.057	0.0061	8.283	66.734	45
50	0.00042	0.12042	8.160	0.0035	8.304	67.762	50
60	0.00013	0.12013	8.266	0.0011	8.324	68.810	60
∞	0.00000	0.12000	8.333	0.0000	8.333	69.444	∞

n	Uniform series worth of a future sum or sinking fund factor	Uniform series worth of a present sum or capital recovery factor	Uniform series worth of a gradient series	Present worth of a future sum	Present worth of a uniform series	Present worth of a gradient series	n
	$\dfrac{i}{(1+i)^n - 1}$	$\dfrac{i(1+i)^n}{(1+i)^n - 1}$	$\dfrac{1}{i} - \dfrac{n}{(1+i)^n - 1}$	$\dfrac{1}{(1+i)^n}$	$\dfrac{(1+i)^n - 1}{i(1+i)^n}$	$\dfrac{1}{i}\left[\dfrac{(1+i)^n - 1}{i(1+i)^n} - \dfrac{n}{(1+i)^n}\right]$	
	Equation 4-4	Equation 4-6	Equation 4-8a	Equation 4-2	Equation 4-5	Equation 4-9a	

n	Future sum f equivalent to			n
	present sum, p	uniform series, a	gradient series, g	
	f/p	f/a	f/g	
1	1.120	1.000	0.00	1
2	1.254	2.120	1.00	2
3	1.405	3.374	3.12	3
4	1.574	4.779	6.49	4
5	1.762	6.353	11.27	5
6	1.974	8.115	17.63	6
7	2.211	10.089	25.74	7
8	2.476	12.300	35.83	8
9	2.773	14.776	48.13	9
10	3.106	17.549	62.91	10
11	3.479	20.655	80.45	11
12	3.896	24.133	101.11	12
13	4.363	28.029	125.24	13
14	4.887	32.393	153.27	14
15	5.474	37.280	185.66	15
16	6.130	42.753	222.94	16
17	6.866	48.884	265.70	17
18	7.690	55.750	314.58	18
19	8.613	63.440	370.33	19
20	9.646	72.052	433.77	20
21	10.804	81.699	505.82	21
22	12.100	92.503	587.52	22
23	13.552	104.603	680.02	23
24	15.179	118.155	784.63	24
25	17.000	133.334	902.78	25
26	19.040	150.334	1036.12	26
27	21.325	169.374	1186.45	27
28	23.884	190.699	1355.82	28
29	26.750	214.583	1546.52	29
30	29.960	241.333	1761.11	30
35	52.800	431.663	3305.53	35
40	93.051	767.091	6059.10	40
45	163.988	1358.230	10943.58	45
50	289.002	2400.018	19583.49	50
60	897.597	7471.641	61763.68	60
∞	∞	∞	∞	∞

n	Future worth of a present sum	Future worth of a uniform series	Future worth of a gradient series	n
	$(1+i)^n$	$\dfrac{(1+i)^n - 1}{i}$	$\dfrac{1}{i}\left[\dfrac{(1+i)^n - 1}{i} - n\right]$	
	Equation 4-1	Equation 4-3	Equation 4-7a	

If the effective annual rate of interest is as shown above, and if interest is compounded continuously:

(1) An end-of-year amount of $1 is equivalent to $ _____ flowing uniformly and continuously throughout the year.

$$a'/a = c/c = f'/f = g'/g = p'/p = \frac{\ln(1+i_a)}{i_a}$$

$$= 0.9444$$

(2) An end-of-year amount of $ _____ is equivalent to $1 flowing uniformly and continuously throughout the year.

$$a/a' = c/c' = f/f' = g/g' = p/p' = \frac{i_a}{\ln(1+i_a)}$$

$$= 1.059$$

$i = 15\%$

n	Uniform series end-of-period amount a equivalent to			Present sum p equivalent to			n
	future sum, f	present sum, p	gradient series, g	future sum, f	uniform series, a	gradient series, g	
	a/f	a/p	a/g	p/f	p/a	p/g	
1	1.00000	1.15000	0.000	0.8696	0.870	0.000	1
2	0.46512	0.61512	0.465	0.7561	1.626	0.756	2
3	0.28798	0.43798	0.907	0.6575	2.283	2.071	3
4	0.20027	0.35027	1.326	0.5718	2.855	3.786	4
5	0.14832	0.29832	1.723	0.4972	3.352	5.775	5
6	0.11424	0.26424	2.097	0.4323	3.784	7.937	6
7	0.09036	0.24036	2.450	0.3759	4.160	10.192	7
8	0.07285	0.22285	2.781	0.3269	4.487	12.481	8
9	0.05957	0.20957	3.092	0.2843	4.772	14.755	9
10	0.04925	0.19925	3.383	0.2472	5.019	16.979	10
11	0.04107	0.19107	3.655	0.2149	5.234	19.129	11
12	0.03448	0.18448	3.908	0.1869	5.421	21.185	12
13	0.02911	0.17911	4.144	0.1625	5.583	23.135	13
14	0.02469	0.17469	4.362	0.1413	5.724	24.972	14
15	0.02102	0.17102	4.565	0.1229	5.847	26.693	15
16	0.01795	0.16795	4.752	0.1069	5.954	28.296	16
17	0.01537	0.16537	4.925	0.0929	6.047	29.783	17
18	0.01319	0.16319	5.084	0.0808	6.128	31.156	18
19	0.01134	0.16134	5.231	0.0703	6.198	32.421	19
20	0.00976	0.15976	5.365	0.0611	6.259	33.582	20
21	0.00842	0.15842	5.488	0.0531	6.312	34.645	21
22	0.00727	0.15727	5.601	0.0462	6.359	35.615	22
23	0.00628	0.15628	5.704	0.0402	6.399	36.499	23
24	0.00543	0.15543	5.798	0.0349	6.434	37.302	24
25	0.00470	0.15470	5.883	0.0304	6.464	38.031	25
26	0.00407	0.15407	5.961	0.0264	6.491	38.692	26
27	0.00353	0.15353	6.032	0.0230	6.514	39.289	27
28	0.00306	0.15306	6.096	0.0200	6.534	39.828	28
29	0.00265	0.15265	6.154	0.0174	6.551	40.315	29
30	0.00230	0.15230	6.207	0.0151	6.566	40.753	30
35	0.00113	0.15113	6.402	0.0075	6.617	42.359	35
40	0.00056	0.15056	6.517	0.0037	6.642	43.283	40
45	0.00028	0.15028	6.583	0.0019	6.654	43.805	45
50	0.00014	0.15014	6.620	0.0009	6.661	44.096	50
60	0.00003	0.15003	6.653	0.0002	6.665	44.343	60
∞	0.00000	0.15000	6.667	0.0000	6.667	44.444	∞

n	Uniform series worth of a future sum or sinking fund factor	Uniform series worth of a present sum or capital recovery factor	Uniform series worth of a gradient series	Present worth of a future sum	Present worth of a uniform series	Present worth of a gradient series	n
	$\dfrac{i}{(1+i)^n-1}$	$\dfrac{i(1+i)^n}{(1+i)^n-1}$	$\dfrac{1}{i}-\dfrac{n}{(1+i)^n-1}$	$\dfrac{1}{(1+i)^n}$	$\dfrac{(1+i)^n-1}{i(1+i)^n}$	$\dfrac{1}{i}\left[\dfrac{(1+i)^n-1}{i(1+i)^n}-\dfrac{n}{(1+i)^n}\right]$	
	Equation 4-4	Equation 4-6	Equation 4-8a	Equation 4-2	Equation 4-5	Equation 4-9a	

$i = 15\%$

n	Future sum f equivalent to			n
	present sum, p	uniform series, a	gradient series, g	
	f/p	f/a	f/g	
1	1.150	1.000	0.00	1
2	1.322	2.150	1.00	2
3	1.521	3.472	3.15	3
4	1.749	4.993	6.62	4
5	2.011	6.742	11.62	5
6	2.313	8.754	18.36	6
7	2.660	11.067	27.11	7
8	3.059	13.727	38.18	8
9	3.518	16.786	51.91	9
10	4.046	20.304	68.69	10
11	4.652	24.349	89.00	11
12	5.350	29.002	113.34	12
13	6.153	34.352	142.35	13
14	7.076	40.505	176.70	14
15	8.137	47.580	217.20	15
16	9.358	55.717	264.78	16
17	10.761	65.075	320.50	17
18	12.375	75.836	385.58	18
19	14.232	88.212	461.41	19
20	16.367	102.444	549.62	20
21	18.822	118.810	652.07	21
22	21.645	137.632	770.88	22
23	24.891	159.276	908.51	23
24	28.625	184.168	1067.79	24
25	32.919	212.793	1251.95	25
26	37.857	245.712	1464.75	26
27	43.535	283.569	1710.46	27
28	50.066	327.104	1994.03	28
29	57.575	377.170	2321.13	29
30	66.212	434.745	2698.30	30
35	133.176	881.170	5641.13	35
40	267.864	1779.090	11593.94	40
45	538.769	3585.128	23600.86	45
50	1083.657	7217.716	47784.78	50
60	4383.999	29219.992	194399.94	60
∞	∞	∞	∞	∞

n	Future worth of a present sum	Future worth of a uniform series	Future worth of a gradient series	n
	$(1+i)^n$	$\dfrac{(1+i)^n - 1}{i}$	$\dfrac{1}{i}\left[\dfrac{(1+i)^n - 1}{i} - n\right]$	
	Equation 4-1	Equation 4-3	Equation 4-7a	

If the effective annual rate of interest is as shown above, and if interest is compounded continuously:

(1) An end-of-year amount of $1 is equivalent to $ _____ flowing uniformly and continuously throughout the year.

$$a'/a = c'/c = f'/f = g'/g = p'/p = \frac{\ln(1+i_a)}{i_a}$$

$$= 0.9317$$

(2) An end-of-year amount of $ _____ is equivalent to $1 flowing uniformly and continuously throughout the year.

$$a/a' = c/c' = f/f' = g/g' = p/p' = \frac{i_a}{\ln(1+i_a)}$$

$$= 1.073$$

$i = 20\%$

n	Uniform series end-of-period amount a equivalent to			Present sum p equivalent to			n
	future sum, f	present sum, p	gradient series, g	future sum, f	uniform series, a	gradient series, g	
	a/f	a/p	a/g	p/f	p/a	p/g	
1	1.00000	1.20000	0.000	0.8333	0.833	0.000	1
2	0.45455	0.65455	0.455	0.6944	1.528	0.694	2
3	0.27473	0.47473	0.879	0.5787	2.106	1.852	3
4	0.18629	0.38629	1.274	0.4823	2.589	3.299	4
5	0.13438	0.33438	1.641	0.4019	2.991	4.906	5
6	0.10071	0.30071	1.979	0.3349	3.326	6.581	6
7	0.07742	0.27742	2.290	0.2791	3.605	8.255	7
8	0.06061	0.26061	2.576	0.2326	3.837	9.883	8
9	0.04808	0.24808	2.836	0.1938	4.031	11.434	9
10	0.03852	0.23852	3.074	0.1615	4.192	12.887	10
11	0.03110	0.23110	3.289	0.1346	4.327	14.233	11
12	0.02526	0.22526	3.484	0.1122	4.439	15.467	12
13	0.02062	0.22062	3.660	0.0935	4.533	16.588	13
14	0.01689	0.21689	3.817	0.0779	4.611	17.601	14
15	0.01388	0.21388	3.959	0.0649	4.675	18.509	15
16	0.01144	0.21144	4.085	0.0541	4.730	19.321	16
17	0.00944	0.20944	4.198	0.0451	4.775	20.042	17
18	0.00781	0.20781	4.298	0.0376	4.812	20.680	18
19	0.00646	0.20646	4.386	0.0313	4.843	21.244	19
20	0.00536	0.20536	4.464	0.0261	4.870	21.739	20
21	0.00444	0.20444	4.533	0.0217	4.891	22.174	21
22	0.00369	0.20369	4.594	0.0181	4.909	22.555	22
23	0.00307	0.20307	4.647	0.0151	4.925	22.887	23
24	0.00255	0.20255	4.694	0.0126	4.937	23.176	24
25	0.00212	0.20212	4.735	0.0105	4.948	23.428	25
26	0.00176	0.20176	4.771	0.0087	4.956	23.646	26
27	0.00147	0.20147	4.802	0.0073	4.964	23.835	27
28	0.00122	0.20122	4.829	0.0061	4.970	23.999	28
29	0.00102	0.20102	4.853	0.0051	4.975	24.141	29
30	0.00085	0.20085	4.873	0.0042	4.979	24.263	30
∞	0.00000	0.20000	5.000	0.0000	5.000	25.000	∞

n	Uniform series worth of a future sum or sinking fund factor	Uniform series worth of a present sum or capital recovery factor	Uniform series worth of a gradient series	Present worth of a future sum	Present worth of a uniform series	Present worth of a gradient series	n
	$\dfrac{i}{(1+i)^n - 1}$	$\dfrac{i(1+i)^n}{(1+i)^n - 1}$	$\dfrac{1}{i} - \dfrac{n}{(1+i)^n - 1}$	$\dfrac{1}{(1+i)^n}$	$\dfrac{(1+i)^n - 1}{i(1+i)^n}$	$\dfrac{1}{i}\left[\dfrac{(1+i)^n - 1}{i(1+i)^n} - \dfrac{n}{(1+i)^n}\right]$	
	Equation 4-4	Equation 4-6	Equation 4-8a	Equation 4-2	Equation 4-5	Equation 4-9a	

n	Future sum f equivalent to			n
	present sum, p	uniform series, a	gradient series, g	
	f/p	f/a	f/g	
1	1.200	1.000	0.00	1
2	1.440	2.200	1.00	2
3	1.728	3.640	3.20	3
4	2.074	5.368	6.84	4
5	2.488	7.442	12.21	5
6	2.986	9.930	19.65	6
7	3.583	12.916	29.58	7
8	4.300	16.499	42.50	8
9	5.160	20.799	58.99	9
10	6.192	25.959	79.79	10
11	7.430	32.150	105.75	11
12	8.916	39.581	137.90	12
13	10.699	48.497	177.48	13
14	12.839	59.196	225.98	14
15	15.407	72.035	285.18	15
16	18.488	87.442	357.21	16
17	22.186	105.931	444.65	17
18	26.623	128.117	550.58	18
19	31.948	154.740	678.70	19
20	38.338	186.688	833.44	20
21	46.005	225.026	1020.13	21
22	55.206	271.031	1245.15	22
23	66.247	326.237	1516.18	23
24	79.497	392.484	1842.42	24
25	95.396	471.981	2234.91	25
26	114.475	567.377	2706.89	26
27	137.371	681.853	3274.26	27
28	164.845	819.223	3956.12	28
29	197.814	984.068	4775.34	29
30	237.376	1181.882	5759.41	30
∞	∞	∞	∞	∞

n	Future worth of a present sum	Future worth of a uniform series	Future worth of a gradient series	n
	$(1+i)^n$	$\dfrac{(1+i)^n - 1}{i}$	$\dfrac{1}{i}\left[\dfrac{(1+i)^n - 1}{i} - n\right]$	
	Equation 4-1	Equation 4-3	Equation 4-7a	

If the effective annual rate of interest is as shown above, and if interest is compounded continuously:

(1) An end-of-year amount of $1 is equivalent to $ _____ flowing uniformly and continuously throughout the year.

$$a'/a = c'/c = f'/f = g'/g = p'/p = \frac{\ln(1+i_a)}{i_a}$$

$$= 0.9116$$

(2) An end-of-year amount of $ _____ is equivalent to $1 flowing uniformly and continuously throughout the year.

$$a/a' = c/c' = f/f' = g/g' = p/p' = \frac{i_a}{\ln(1+i_a)}$$

$$= 1.097$$

$i = 25\%$

n	Uniform series end-of-period amount a equivalent to			Present sum p equivalent to			n
	future sum, f	present sum, p	gradient series, g	future sum, f	uniform series, a	gradient series, g	
	a/f	a/p	a/g	p/f	p/a	p/g	
1	1.00000	1.25000	C.000	0.8000	0.800	0.000	1
2	0.44444	0.69444	0.444	0.6400	1.440	0.640	2
3	0.26230	0.51230	0.852	0.5120	1.952	1.064	3
4	0.17344	0.42344	1.225	0.4096	2.362	2.893	4
5	0.12185	0.37185	1.563	0.3277	2.689	4.204	5
6	0.08882	0.33882	1.868	0.2621	2.951	5.514	6
7	0.06634	0.31634	2.142	0.2097	3.161	6.773	7
8	0.05040	0.30040	2.387	0.1678	3.329	7.947	8
9	0.03876	0.28876	2.605	0.1342	3.463	9.021	9
10	0.03007	0.28007	2.797	0.1074	3.571	9.987	10
11	0.02349	0.27349	2.966	0.0859	3.656	10.846	11
12	0.01845	0.26845	3.115	0.0687	3.725	11.602	12
13	0.01454	0.26454	3.244	0.0550	3.780	12.262	13
14	0.01150	0.26150	3.356	0.0440	3.824	12.833	14
15	0.00912	0.25912	3.453	0.0352	3.859	13.326	15
16	0.00724	0.25724	3.537	0.0281	3.887	13.748	16
17	0.00576	0.25576	3.608	0.0225	3.910	14.108	17
18	0.00459	0.25459	3.670	0.0180	3.928	14.415	18
19	0.00366	0.25366	3.722	0.0144	3.942	14.674	19
20	0.00292	0.25292	3.767	0.0115	3.954	14.893	20
21	0.00233	0.25233	3.805	0.0092	3.963	15.078	21
22	0.00186	0.25186	3.836	0.0074	3.970	15.233	22
23	0.00148	0.25148	3.863	0.0059	3.976	15.362	23
24	0.00119	0.25119	3.886	0.0047	3.981	15.471	24
25	0.00095	0.25095	3.905	0.0038	3.985	15.562	25
26	0.00076	0.25076	3.921	0.0030	3.988	15.637	26
27	0.00061	0.25061	3.935	0.0024	3.990	15.700	27
28	0.00048	0.25048	3.946	0.0019	3.992	15.752	28
29	0.00039	0.25039	3.955	0.0015	3.994	15.796	29
30	0.00031	0.25031	3.963	0.0012	3.995	15.832	30
∞	0.00000	0.25000	4.000	0.0000	4.000	16.000	∞

n	Uniform series worth of a future sum or sinking fund factor	Uniform series worth of a present sum or capital recovery factor	Uniform series worth of a gradient series	Present worth of a future sum	Present worth of a uniform series	Present worth of a gradient series	n
	$\dfrac{i}{(1+i)^n - 1}$	$\dfrac{i(1+i)^n}{(1+i)^n - 1}$	$\dfrac{1}{i} - \dfrac{n}{(1+i)^n - 1}$	$\dfrac{1}{(1+i)^n}$	$\dfrac{(1+i)^n - 1}{i(1+i)^n}$	$\dfrac{1}{i}\left[\dfrac{(1+i)^n - 1}{i(1+i)^n} - \dfrac{n}{(1+i)^n}\right]$	
	Equation 4-4	Equation 4-6	Equation 4-8a	Equation 4-2	Equation 4-5	Equation 4-9a	

$i = 25\%$

n	Future sum f equivalent to			n
	present sum, p	uniform series, a	gradient series, g	
	f/p	f/a	f/g	
1	1.250	1.000	0.00	1
2	1.562	2.250	1.00	2
3	1.953	3.812	3.25	3
4	2.441	5.766	7.06	4
5	3.052	8.207	12.83	5
6	3.815	11.259	21.04	6
7	4.768	15.073	32.29	7
8	5.960	19.842	47.37	8
9	7.451	25.802	67.21	9
10	9.313	33.253	93.01	10
11	11.642	42.566	126.26	11
12	14.552	54.208	168.83	12
13	18.190	68.760	223.04	13
14	22.737	86.949	291.80	14
15	28.422	109.687	378.75	15
16	35.527	138.109	488.43	16
17	44.409	173.636	626.54	17
18	55.511	218.045	800.18	18
19	69.389	273.556	1018.22	19
20	86.736	342.945	1291.78	20
21	108.420	429.681	1634.72	21
22	135.525	538.101	2064.40	22
23	169.407	673.626	2602.51	23
24	211.758	843.033	3276.13	24
25	264.698	1054.791	4119.16	25
26	330.872	1319.489	5173.96	26
27	413.590	1650.361	6493.44	27
28	516.988	2063.952	8143.81	28
29	646.235	2580.939	10207.76	29
30	807.794	3227.174	12788.70	30
∞	∞	∞	∞	∞

n	Future worth of a present sum	Future worth of a uniform series	Future worth of a gradient series	n
	$(1+i)^n$	$\dfrac{(1+i)^n-1}{i}$	$\dfrac{1}{i}\left[\dfrac{(1+i)^n-1}{i}-n\right]$	
	Equation 4-1	Equation 4-3	Equation 4-7a	

If the effective annual rate of interest is as shown above, and if interest is compounded continuously:

(1) An end-of-year amount of $1 is equivalent to $ _____ flowing uniformly and continuously throughout the year.

$$a'/a = c'/c = f'/f = g'/g = p'/p = \frac{\ln(1+i_a)}{i_a}$$

$$= 0.8926$$

(2) An end-of-year amount of $ _____ is equivalent to $1 flowing uniformly and continuously throughout the year.

$$a/a' = c/c' = f/f' = g/g' = p/p' = \frac{i_a}{\ln(1+i_a)}$$

$$= 1.120$$

$i = 30\%$

n	Uniform series end-of-period amount a equivalent to			Present sum p equivalent to			n
	future sum, f	present sum, p	gradient series, g	future sum, f	uniform series, a	gradient series, g	
	a/f	a/p	a/g	p/f	p/a	p/g	
1	1.00000	1.30000	0.000	0.7692	0.769	0.000	1
2	0.43478	0.73478	0.435	0.5917	1.361	0.592	2
3	0.25063	0.55063	0.827	0.4552	1.816	1.502	3
4	0.16163	0.46163	1.178	0.3501	2.166	2.552	4
5	0.11058	0.41058	1.490	0.2693	2.436	3.630	5
6	0.07839	0.37839	1.765	0.2072	2.643	4.666	6
7	0.05687	0.35687	2.006	0.1594	2.802	5.622	7
8	0.04192	0.34192	2.216	0.1226	2.925	6.480	8
9	0.03124	0.33124	2.396	0.0943	3.019	7.234	9
10	0.02346	0.32346	2.551	0.0725	3.092	7.887	10
11	0.01773	0.31773	2.683	0.0558	3.147	8.445	11
12	0.01345	0.31345	2.795	0.0429	3.190	8.917	12
13	0.01024	0.31024	2.889	0.0330	3.223	9.314	13
14	0.00782	0.30782	2.969	0.0254	3.249	9.644	14
15	0.00598	0.30598	3.034	0.0195	3.268	9.917	15
16	0.00458	0.30458	3.089	0.0150	3.283	10.143	16
17	0.00351	0.30351	3.135	0.0116	3.295	10.328	17
18	0.00269	0.30269	3.172	0.0089	3.304	10.479	18
19	0.00207	0.30207	3.202	0.0068	3.311	10.602	19
20	0.00159	0.30159	3.228	0.0053	3.316	10.702	20
21	0.00122	0.30122	3.248	0.0040	3.320	10.783	21
22	0.00094	0.30094	3.265	0.0031	3.323	10.848	22
23	0.00072	0.30072	3.278	0.0024	3.325	10.901	23
24	0.00055	0.30055	3.289	0.0018	3.327	10.943	24
25	0.00043	0.30043	3.298	0.0014	3.329	10.977	25
26	0.00033	0.30033	3.305	0.0011	3.330	11.005	26
27	0.00025	0.30025	3.311	0.0008	3.331	11.026	27
28	0.00019	0.30019	3.315	0.0006	3.331	11.044	28
29	0.00015	0.30015	3.319	0.0005	3.332	11.058	29
30	0.00011	0.30011	3.322	0.0004	3.332	11.069	30
∞	0.00000	0.30000	3.333	0.0000	3.333	11.111	∞

n	Uniform series worth of a future sum or sinking fund factor	Uniform series worth of a present sum or capital recovery factor	Uniform series worth of a gradient series	Present worth of a future sum	Present worth of a uniform series	Present worth of a gradient series	n
	$\dfrac{i}{(1+i)^n - 1}$	$\dfrac{i(1+i)^n}{(1+i)^n - 1}$	$\dfrac{1}{i} - \dfrac{n}{(1+i)^n - 1}$	$\dfrac{1}{(1+i)^n}$	$\dfrac{(1+i)^n - 1}{i(1+i)^n}$	$\dfrac{1}{i}\left[\dfrac{(1+i)^n - 1}{i(1+i)^n} - \dfrac{n}{(1+i)^n}\right]$	
	Equation 4-4	Equation 4-6	Equation 4-8a	Equation 4-2	Equation 4-5	Equation 4-9a	

n	Future sum f equivalent to			n
	present sum, p	uniform series, a	gradient series, g	
	f/p	f/a	f/g	
1	1.300	1.000	0.00	1
2	1.690	2.300	1.00	2
3	2.197	3.990	3.30	3
4	2.856	6.187	7.29	4
5	3.713	9.043	13.48	5
6	4.827	12.756	22.52	6
7	6.275	17.583	35.28	7
8	8.157	23.858	52.86	8
9	10.604	32.015	76.72	9
10	13.786	42.619	108.73	10
11	17.922	56.405	151.35	11
12	23.298	74.327	207.76	12
13	30.288	97.625	282.08	13
14	39.374	127.913	379.71	14
15	51.186	167.286	507.62	15
16	66.542	218.472	674.91	16
17	86.504	285.014	893.38	17
18	112.455	371.518	1178.39	18
19	146.192	483.973	1549.91	19
20	190.050	630.165	2033.88	20
21	247.065	820.215	2664.05	21
22	321.184	1067.280	3484.27	22
23	417.539	1388.464	4551.55	23
24	542.801	1806.003	5940.01	24
25	705.641	2348.803	7746.01	25
26	917.333	3054.444	10094.81	26
27	1192.533	3971.778	13149.26	27
28	1550.293	5164.311	17121.04	28
29	2015.381	6714.604	22285.35	29
30	2619.996	8729.985	28999.95	30
∞	∞	∞	∞	∞

n	Future worth of a present sum	Future worth of a uniform series	Future worth of a gradient series	n
	$(1+i)^n$	$\dfrac{(1+i)^n - 1}{i}$	$\dfrac{1}{i}\left[\dfrac{(1+i)^n - 1}{i} - n\right]$	
	Equation 4-1	Equation 4-3	Equation 4-7a	

If the effective annual rate of interest is as shown above, and if interest is compounded continuously:

(1) An end-of-year amount of $1 is equivalent to $ _____ flowing uniformly and continuously throughout the year.

$$a'/a = c'/c = f/f = g'/g = p'/p = \frac{\ln(1+i_a)}{i_a}$$

$$= 0.8745$$

(2) An end-of-year amount of $ _____ is equivalent to $1 flowing uniformly and continuously throughout the year.

$$a/a' = c/c' = f/f' = g/g' = p/p' = \frac{i_a}{\ln(1+i_a)}$$

$$= 1.143$$

$i = 40\%$

n	Uniform series end-of-period amount a equivalent to			Present sum p equivalent to			n
	future sum, f	present sum, p	gradient series, g	future sum, f	uniform series, a	gradient series, g	
	a/f	a/p	a/g	p/f	p/a	p/g	
1	1.00000	1.40000	0.000	0.7143	0.714	0.000	1
2	0.41667	0.81667	0.417	0.5102	1.224	0.510	2
3	0.22936	0.62936	0.780	0.3644	1.589	1.239	3
4	0.14077	0.54077	1.092	0.2603	1.849	2.020	4
5	0.09136	0.49136	1.358	0.1859	2.035	2.764	5
6	0.06126	0.46126	1.581	0.1328	2.168	3.428	6
7	0.04192	0.44192	1.766	0.0949	2.263	3.997	7
8	0.02907	0.42907	1.919	0.0678	2.331	4.471	8
9	0.02034	0.42034	2.042	0.0484	2.379	4.858	9
10	0.01432	0.41432	2.142	0.0346	2.414	5.170	10
11	0.01013	0.41013	2.221	0.0247	2.438	5.417	11
12	0.00718	0.40718	2.285	0.0176	2.456	5.611	12
13	0.00510	0.40510	2.334	0.0126	2.469	5.762	13
14	0.00363	0.40363	2.373	0.0090	2.478	5.879	14
15	0.00259	0.40259	2.403	0.0064	2.484	5.969	15
16	0.00185	0.40185	2.426	0.0046	2.489	6.038	16
17	0.00132	0.40132	2.444	0.0033	2.492	6.090	17
18	0.00094	0.40094	2.458	0.0023	2.494	6.130	18
19	0.00067	0.40067	2.468	0.0017	2.496	6.160	19
20	0.00048	0.40048	2.476	0.0012	2.497	6.183	20
∞	0.00000	0.40000	2.500	0.0000	2.500	6.250	∞

n	Uniform series worth of a future sum or sinking fund factor	Uniform series worth of a present sum or capital recovery factor	Uniform series worth of a gradient series	Present worth of a future sum	Present worth of a uniform series	Present worth of a gradient series	n
	$\dfrac{i}{(1+i)^n - 1}$	$\dfrac{i(1+i)^n}{(1+i)^n - 1}$	$\dfrac{1}{i} - \dfrac{n}{(1+i)^n - 1}$	$\dfrac{1}{(1+i)^n}$	$\dfrac{(1+i)^n - 1}{i(1+i)^n}$	$\dfrac{1}{i}\left[\dfrac{(1+i)^n - 1}{i(1+i)^n} - \dfrac{n}{(1+i)^n}\right]$	
	Equation 4-4	Equation 4-6	Equation 4-8a	Equation 4-2	Equation 4-5	Equation 4-9a	

$i = 40\%$

n	Future sum f equivalent to			n
	present sum, p	uniform series, a	gradient series, g	
	f/p	f/a	f/g	
1	1.400	1.000	0.00	1
2	1.960	2.400	1.00	2
3	2.744	4.360	3.40	3
4	3.842	7.104	7.76	4
5	5.378	10.946	14.86	5
6	7.530	16.324	25.81	6
7	10.541	23.853	42.13	7
8	14.758	34.395	65.99	8
9	20.661	49.153	100.38	9
10	28.925	69.814	149.53	10
11	40.496	98.739	219.35	11
12	56.694	139.235	318.09	12
13	79.371	195.929	457.32	13
14	111.120	275.300	653.25	14
15	155.568	386.420	928.55	15
16	217.795	541.988	1314.97	16
17	304.913	759.784	1856.96	17
18	426.879	1064.697	2616.74	18
19	597.630	1491.576	3681.44	19
20	836.683	2089.206	5173.02	20
∞	∞	∞	∞	∞

n	Future worth of a present sum	Future worth of a uniform series	Future worth of a gradient series	n
	$(1+i)^n$	$\dfrac{(1+i)^n-1}{i}$	$\dfrac{1}{i}\left[\dfrac{(1+i)^n-1}{i}-n\right]$	
	Equation 4-1	Equation 4-3	Equation 4-7a	

If the effective annual rate of interest is as shown above, and if interest is compounded continuously:

(1) An end-of-year amount of $1 is equivalent to $ ___ flowing uniformly and continuously throughout the year.

$$a'/a = c'/c = f/f = g'/g = p'/p = \frac{\ln(1+i_a)}{i_a}$$

$$= 0.8412$$

(2) An end-of-year amount of $ ___ is equivalent to $1 flowing uniformly and continuously throughout the year.

$$a/a' = c/c' = f/f' = g/g' = p/p' = \frac{i_a}{\ln(1+i_a)}$$

$$= 1.189$$

$i = 50\%$

n	Uniform series end-of-period amount a equivalent to			Present sum p equivalent to			n
	future sum, f	present sum, p	gradient series, g	future sum, f	uniform series, a	gradient series, g	
	a/f	a/p	a/g	p/f	p/a	p/g	
1	1.00000	1.50000	0.000	0.6667	0.667	0.000	1
2	0.40000	0.90000	0.400	0.4444	1.111	0.444	2
3	0.21053	0.71053	0.737	0.2963	1.407	1.037	3
4	0.12308	0.62308	1.015	0.1975	1.605	1.630	4
5	0.07583	0.57583	1.242	0.1317	1.737	2.156	5
6	0.04812	0.54812	1.423	0.0878	1.824	2.595	6
7	0.03108	0.53108	1.565	0.0585	1.883	2.947	7
8	0.02030	0.52030	1.675	0.0390	1.922	3.220	8
9	0.01335	0.51335	1.760	0.0260	1.948	3.428	9
10	0.00882	0.50882	1.824	0.0173	1.965	3.584	10
11	0.00585	0.50585	1.871	0.0116	1.977	3.699	11
12	0.00388	0.50388	1.907	0.0077	1.985	3.784	12
13	0.00258	0.50258	1.933	0.0051	1.990	3.846	13
14	0.00172	0.50172	1.952	0.0034	1.993	3.890	14
15	0.00114	0.50114	1.966	0.0023	1.995	3.922	15
16	0.00076	0.50076	1.976	0.0015	1.997	3.945	16
17	0.00051	0.50051	1.983	0.0010	1.998	3.961	17
18	0.00034	0.50034	1.988	0.0007	1.999	3.973	18
19	0.00023	0.50023	1.991	0.0005	1.999	3.981	19
20	0.00015	0.50015	1.994	0.0003	1.999	3.987	20
∞	0.00000	0.50000	2.000	0.0000	2.000	4.000	∞

n	Uniform series worth of a future sum or sinking fund factor	Uniform series worth of a present sum or capital recovery factor	Uniform series worth of a gradient series	Present worth of a future sum	Present worth of a uniform series	Present worth of a gradient series	n
	$\dfrac{i}{(1+i)^n - 1}$	$\dfrac{i(1+i)^n}{(1+i)^n - 1}$	$\dfrac{1}{i} - \dfrac{n}{(1+i)^n - 1}$	$\dfrac{1}{(1+i)^n}$	$\dfrac{(1+i)^n - 1}{i(1+i)^n}$	$\dfrac{1}{i}\left[\dfrac{(1+i)^n - 1}{i(1+i)^n} - \dfrac{n}{(1+i)^n}\right]$	
	Equation 4-4	Equation 4-6	Equation 4-8a	Equation 4-2	Equation 4-5	Equation 4-9a	

$i = 50\%$

| n | Future sum f equivalent to | | | n |
| | present sum, p | uniform series, a | gradient series, g | |
	f/p	f/a	f/g	
1	1.500	1.000	0.00	1
2	2.250	2.500	1.00	2
3	3.375	4.750	3.50	3
4	5.063	8.125	8.25	4
5	7.594	13.187	16.37	5
6	11.391	20.781	29.56	6
7	17.086	32.172	50.34	7
8	25.629	49.258	82.52	8
9	38.443	74.887	131.77	9
10	57.665	113.330	206.66	10
11	86.498	170.995	319.99	11
12	129.746	257.493	490.99	12
13	194.620	387.239	748.48	13
14	291.929	581.859	1135.72	14
15	437.894	873.788	1717.58	15
16	656.841	1311.682	2591.36	16
17	985.261	1968.523	3903.05	17
18	1477.892	2953.784	5871.57	18
19	2216.838	4431.676	8825.35	19
20	3325.257	6648.513	13257.03	20
∞	∞	∞	∞	∞

n	Future worth of a present sum	Future worth of a uniform series	Future worth of a gradient series	n
	$(1+i)^n$	$\dfrac{(1+i)^n - 1}{i}$	$\dfrac{1}{i}\left[\dfrac{(1+i)^n - 1}{i} - n\right]$	
	Equation 4-1	Equation 4-3	Equation 4-7a	

If the effective annual rate of interest is as shown above, and if interest is compounded continuously:

(1) An end-of-year amount of $1 is equivalent to $ _____ flowing uniformly and continuously throughout the year.

$$a'/a = c'/c = f'/f = g'/g = p'/p = \frac{\ln(1+i_a)}{i_a}$$

= 0.8109

(2) An end-of-year amount of $ _____ is equivalent to $1 flowing uniformly and continuously throughout the year.

$$a/a' = c/c' = f/f' = g/g' = p/p' = \frac{i_a}{\ln(1+i_a)}$$

= 1.233

J

Selected References

Books

American Association of State Highway Officials, *Road-User Benefit Analyses for Highway Improvements.* Washington, D.C., AASHO, 1960.

American Telephone and Telegraph Company, *Engineering Economy*, 2d ed. New York, 1963.

BARGES, A., *The Effect of Capital Structure on the Cost of Capital.* Englewood Cliffs, N.J., Prentice-Hall, 1963.

BARISH, N. N., *Economic Analysis for Engineering and Managerial Decision Making.* New York, McGraw-Hill, 1962.

BIERMAN, H., AND S. SMIDT, *The Capital Budgeting Decision.* New York. MacMillan, 1971.

BOECKH, E. H., *Building Valuation Manual*, 3 vols. (1. Residential and Agricultural, 2. Commercial, 3. Industrial and Institutional). Milwaukee, Wis., American Appraisal Co., 1967.

BONBRIGHT, J. C., *Principles of Public Utility Rates.* New York, Columbia Univ. Press, 1961.

BOWER, J. L., *Managing the Resource Allocation Process.* Boston, Harvard Univ., 1970.

BULLINGER, C. E., *Engineering Economy*, 3d ed. New York, McGraw-Hill, 1958.

CANADA, J. R., *Intermediate Economic Analysis for Management and Engineering.* Englewood Cliffs, N. J., Prentice-Hall, 1971.

DEAN, J., *Capital Budgeting.* New York, Columbia Univ. Press, 1951.

_____ , *Managerial Economics.* Englewood Cliffs, N.J., Prentice-Hall, 1951.

DeGARMO, E. P., *Engineering Economy*, 4th ed. New York, Macmillan, 1967.

ENGLISH, J. M., ed., *Cost Effectiveness: The Economic Evaluation of Engineered Systems.* New York, Wiley, 1968.

EISNER, R., *Determinants of Capital Expenditures.* Urbana, Univ. Ill. Studies in Business Expectations and Planning No. 2, 1956.

Financial Compound Interest and Annuity Tables. Boston, Financial Publ. Co. See most recent edition.

FISH, J. C. L., *Engineering Economics.* New York, McGraw-Hill, 1923.

GARFIELD, P. J., AND W. F. LOVEJOY, *Public Utility Economics.* Englewood Cliffs, N.J., Prentice-Hall, 1964.

GORT, M., "The Planning of Investment." *J. Business*, Vol. 24, No. 1 (Apr. 1951).

GRANT, E. L., AND W. G. IRESON, *Principles of Engineering Economy*, 5th ed. New York, Ronald Press, 1970.

GRANT, E. L., AND P. T. NORTON, JR., *Depreciation.* New York, Ronald Press, 1955.

HAPPEL, J., *Chemical Process Economics.* New York, Wiley, 1958.

HEILBRONER, R. L., *The Making of Economic Society*, 3rd ed. Englewood Cliffs, N.J., Prentice-Hall, 1970.

HILLIER, F. S., *The Evaluation of Risky Interrelated Investments.* Amsterdam, North-Holland Publ. Co., 1969.

HIRSHLEIFER, J., J. C. DEHAVEN, AND J. W. MILLIMAN, *Water Supply: Economics, Technology, and Policy.* Chicago, Univ. Chicago Press, 1960.

ISTVAN, DONALD F., *Capital Expenditure Decisions—How They Are Made in Large Corporations.* Bloomington, Univ. Ind. Bus. Rept. No. 33, 1961.

JAMES, L. D., AND R. R. LEE, *Economics of Water Resource Planning.* New York, McGraw-Hill, 1971.

JELEN, F. C., ed., *Cost and Optimization Engineering.* New York, McGraw-Hill, 1970.

JEYNES, P. H., *Profitability and Economic Choice.* Ames, Iowa State Univ. Press, 1968.

LESSER, A., ed., "Planning and Justifying Capital Expenditures," *Engineering Economist.* Hoboken, N.J., Stevens Inst. Tech., 1959.

LUTZ, F., AND V. LUTZ. *The Theory of Investment of the Firm.* Princeton, N.J., Princeton Univ. Press, 1951.

MANNE, A. S., *Investments for Capacity Expansion.* Cambridge, M.I.T. Press, 1967.

McKEAN, R. N., *Efficiency in Government through Systems Analysis.* New York, Wiley, 1964.

MAO, J. C. T., *Quantitative Analysis of Financial Decisions.* New York, Macmillan, 1969.

MARSTON, A., R. WINFREY, AND J. C. HEMPSTEAD, *Engineering Valuation and Depreciation.* Ames, Iowa State Univ. Press, 1953.

MEREWITZ, L., AND S. H. SOSNICK, *The Budget's New Clothes: A Critique of Planning-Programming-Budgeting and Benefit-Cost Analysis.* Chicago, Markham Publ. Co., 1971.

MERRILL, H. F., ed., *Classics in Management.* New York, American Management Association, 1960.

MEYER, J. R., AND E. KUH, *The Investment Decision: An Empirical Study.* Cambridge, Harvard Univ. Press, 1957.

MOHR, L. E., *An Introduction to Compound Interest in Valuation Computations Using Programmed Learning.* Ames, Ind. Eng. Dept., Iowa State Univ., 1965.

MORRIS, W. T., *The Analysis of Managerial Decisions.* Homewood, Ill., Irwin, 1964.

National Bureau of Economic Research, *Regularization of Business Investments.* Princeton, N.J., Princeton Univ. Press, 1954.

OSBORN, J. O., AND K. KAMMERMEYER, *Money and the Chemical Engineer.* Englewood Cliffs, N.J., Prentice-Hall, 1958.

PERRY, J. H., ed., *Chemical Engineers' Handbook,* 4th ed. New York, McGraw-Hill, 1963.

PETERS, M. S., *Plant Design and Economics for Chemical Engineers,* 2d ed. New York, McGraw-Hill, 1968.

PFLOMM, N. E., *Managing Capital Expenditures,* no. 107. New York, National Industrial Conference Board, Inc., 1962.

PREIST, A. J. G., *Principles of Public Utility Regulation,* 2 vols. Charlottesville, Va., Michie Company, 1969.

REISMAN, A., *Managerial and Engineering Economics.* Boston, Allyn and Bacon, 1971.

ROSENTHAL, S. A., *Engineering Economics and Practice.* New York, Macmillan, 1964.

SAMUELSON, P. A., *Economics,* 8th ed. New York, McGraw-Hill, 1970.

SAVAGE, L. J., *The Foundations of Statistics.* New York, Wiley, 1954.

SCHLAIFER, R., *Probability and Statistics for Business Decisions.* New York, McGraw-Hill, 1959.

SHARPE, W. F., *Portfolio Theory and Capital Markets.* New York, McGraw-Hill, 1970.

SOLOMON, E., ed., *The Management of Corporate Capital.* New York, Macmillan, 1964.

Studies in Budgeting. New York, American-Elsevier Publ. Co., 1971.

TAYLOR, G. A., *Managerial and Engineering Economy.* Princeton, N.J., Van Nostrand, 1964.

TERBORGH, G., *Business Investment Management.* Washington, D.C., Machinery and Allied Products Inst., 1967.

———, *Dynamic Equipment Policy.* New York, McGraw-Hill, 1949.

THUESEN, H. G., W. J. FABRYCKY AND G. J. THUESEN, *Engineering Economy,* 4th ed. Englewood Cliffs, N.J., Prentice-Hall, 1971.

TYLER, C., AND C. H. WINTER, JR., *Chemical Engineering Economics,* 4th ed. New York, McGraw-Hill, 1959.

U.S. CONGRESS, Subcommittee on Economy in Government of the Joint Economic Committee, *The Analysis and Evaluation of Public Expenditures: The PPB System,* vol. 1. USGPO, 1969.

VANDELL, R. F., AND R. F. VANCIL, *Cases in Capital Budgeting.* Homewood, Ill., Irwin, 1962.

WEINGARTNER, H. M., *Mathematical Programming and the Analysis of Capital Budgeting Problems.* Englewood Cliffs, N.J., Prentice-Hall, 1963.

WELLINGTON, A. M., *The Economic Theory of Railway Location.* New York, Wiley, 1887.

WILLIAMS, J. D., *The Compleat Strategyst.* New York, McGraw-Hill, 1954.

WINFREY, R., *Economic Analysis for Highways.* Scranton, Pa., International Textbook Co., 1969.

WINFREY, R., *Statistical Analysis of Industrial Property Retirements.*
Iowa State Univ. Eng. Exp. Sta. Bul. 125, 1935.
WOODS, K. B., ed., *Highway Engineering Handbook.* New York, McGraw-
Hill, 1960.

Annual Publications

Capital Investments on the World Petroleum Industry. New York, The
Chase Manhattan Bank.
Economic Report of the President. Washington, D.C., USGPO.
Engineering Economy Abstracts presents abstracts from a search of more
than 60 periodicals. Ames, Ind. Eng. Dept., Iowa State Univ., annual
series beginning 1965.
Federal Tax Regulations, 2 vols. St. Paul, Minn., West Publ. Co.
Fortune 500 presents a tabulation of data for the largest industrial firms,
utilities, and others. By the publishers of *Fortune.*
Highway Research Board. Various reports and bulletins dealing with
highway economics are published after presentation at annual meetings.
Washington, D.C., HRB.
Lasser, J. L., *Your Income Tax.* New York, Simon and Schuster.
Moody's . . . Manual. New York, Moody's Investors Service, Inc. Annual
bound volumes for (1) Industrial, (2) Public Utility, (3) Bank and Fi-
nance, (4) Municipal and Government, and (5) Transportation. Also in-
cludes biweekly loose-leaf supplements.
*Proceedings of the Iowa State Conference on Public Utility Valuation and
the Rate Making Process.* Ames, Iowa State Univ., annually from 1962.
Standard Federal Tax Reporter 12 vols., Chicago, Commerce Clearing
House.
U.S. Master Tax Guide. Chicago, Commerce Clearing House.

Periodicals

Accounting Review, Evanston, Ill., American Accounting Assoc., quart.
AIIE Transactions, New York, American Institute of Industrial Engineers,
quart.
American Journal of Agricultural Economics, Lexington, Ky., American
Agricultural Economics Assoc., Univ. Kentucky, 5/yr.
Appraisal Journal, Chicago, American Institute of Real Estate Appraisers,
quart.
Bell Journal of Economics and Management Science, New York, American
Telephone and Telegraph Co., 2/yr.
Chemical Engineering, New York, McGraw-Hill, biweekly.
Engineering Economist, Hoboken, N.J., Stevens Institute of Technology,
Engineering Economy Division, American Society for Education, quart.
———. Summer symposium papers of EED-ASEE, "Decision-Making Cri-
teria for Capital Expenditures." Symposia are planned for three-year
intervals; copies of the symposium papers are available from *Engineering
Economist,* Stevens Institute of Technology, Hoboken, N.J.
Engineering News Record, New York, McGraw-Hill, weekly.
Harvard Business Review, Boston, Harvard Univ., 6/yr.

Industrial Engineering, (formerly *J. Ind. Eng.*), New York, American Institute of Industrial Engineers, monthly.

Journal of the American Waterworks Association, New York, American Waterworks Assoc., monthly.

Journal of Business, Chicago, Univ. Chicago Press, quart.

Journal of Finance, New York, American Finance Assoc. 5/yr.

Journal of Financial and Quantitative Analysis, Seattle, Univ. Wash., 5/yr.

Journal of the Sanitary Engineering Division, ASCE, New York, American Society of Civil Engineers, bimonthly.

Journal of the Water Pollution Control Federation, Lancaster, Pa., Water Pollution Control Federation, monthly.

Management Science, Baltimore, Institute of Management Sciences, quart.

Management Services, New York, American Institute of Certified Public Accountants, bimonthly.

Operations Research, Baltimore, Operations Research Society of America, bimonthly.

Public Utilities Fortnightly, Washington, D.C., Public Utilities Reports, fortnightly.

Western Economic Journal, Los Angeles, Univ. Calif., quart.

Index